Subcellular Biochemistry

Volume 110

The book series SUBCELLULAR BIOCHEMISTRY is a renowned and well recognized forum for disseminating advances of emerging topics in Cell Biology and related subjects. All volumes are edited by established scientists and the individual chapters are written by experts on the relevant topic. The individual chapters of each volume are fully citable and indexed in Medline/Pubmed to ensure maximum visibility of the work.

Akihiko Nakano • Jaakko Saraste
Editors

The Golgi Network, Volume I

Golgi Organization and Dynamics

Springer

Editors
Akihiko Nakano
RIKEN Center for Advanced Photonics
Wako, Saitama, Japan

Jaakko Saraste
Department of Biomedicine and Molecular Imaging Center
University of Bergen
Bergen, Norway

ISSN 0306-0225 ISSN 2542-8810 (electronic)
Subcellular Biochemistry
ISBN 978-3-032-06935-1 ISBN 978-3-032-06936-8 (eBook)
https://doi.org/10.1007/978-3-032-06936-8

This Springer imprint is published by the registered company Springer Nature Switzerland AG
The registered company address is: Gewerbestrasse 11, 6330 Cham, Switzerland

Preface

In 1898, Camillo Golgi discovered a novel structure in the cerebellar Purkinje cell, which had the appearance of an extensive reticulum around the nucleus. Now, after 127 years have passed, it is again timely to ask what we know about this still mysterious "Golgi apparatus." The title of this book, "The Golgi Network," is symbolic. It could refer to the structure of this organelle, which often looks reticular and network-like. The word network itself is indeed used in the case of the *cis*- and *trans*-Golgi networks. The two important main functions of the Golgi apparatus are: (1) glycosylation, which involves an intricate network of enzymes involved in glycan synthesis and modification, and (2) protein and lipid sorting, which critically regulates the complex cellular network of trafficking. Research in these fields is making tremendous progress thanks to the state-of-the-art technologies developed in the past few decades. Human networking, interaction among researchers, has increasingly helped in bringing together such novel techniques and collaboration to better tackle the many questions on this organelle that remain unanswered. The book is divided into two volumes: I. Golgi Organization and Dynamics and II. Golgi Functions. A variety of topics have been chosen to introduce the readers to the frontiers of advanced Golgi research, hoping that this collection will give an update on current ideas regarding many aspects of the Golgi apparatus and the organelles it closely interacts with, thereby providing a valuable resource for researchers and students.

Wako, Saitama, Japan — Akihiko Nakano
Bergen, Norway — Jaakko Saraste

Overview for Volume I

This book, titled "The Golgi Network," is divided into two volumes: Volume I focuses on Golgi organization and dynamics, while Volume II addresses Golgi functions. In Volume I, the structure and dynamics of the Golgi apparatus are explored from various perspectives. The beautiful electron microscopy images of the Golgi apparatus presented in Chap. 1 will captivate readers, showcasing the variability in the organization of this organelle across different mammalian systems. Furthermore, it is well established that Golgi organization varies significantly among eukaryotes. The classic textbook representation of the Golgi, featuring stacked layers of cisternae, is commonly observed in plant and invertebrate cells (Chaps. 5, 10, and 11). In higher animals, however, the multiple stacked units of the Golgi are often interconnected, forming continuous large structures known as Golgi ribbons (Chaps. 2 and 4). An intriguing exception is the budding yeast *Saccharomyces cerevisiae*, where the Golgi cisternae do not form stacks, but instead exist as dispersed separate entities in the cytoplasm (Chaps. 12–14). This configuration can facilitate live imaging analysis and raises questions about the advantages of cisternal stacking in higher eukaryotes. Despite these varied structural patterns, the molecular machinery of trafficking is highly conserved among eukaryotes (as discussed in Volume II), suggesting that there are common mechanisms at play to sustain the functions of the Golgi.

Structurally, the main body of the Golgi consists of distinct compartments known as *cis*, medial, and *trans* cisternae, each containing different sets of enzymes essential for the orderly modification of proteins and lipids (Chaps. 3 and 12). Additionally, neighboring compartments at both the entry and exit sites of the Golgi stacks play critical roles, particularly in the context of trafficking. A specific compartment located between the endoplasmic reticulum (ER) and the *cis*-Golgi has been referred to by various names, including the intermediate compartment (IC), ER-Golgi intermediate compartment (ERGIC), Golgi-entry core compartment (GECCO), and *cis*-Golgi network (CGN). All these terms likely refer to the same pre-Golgi organelle, which is crucial for sorting secretory cargo exported from the ER exit sites (ERES) and facilitating their subsequent delivery into the main body of the Golgi. On the opposite *trans* side of the Golgi, we find another important sorting platform, which is known as the *trans*-Golgi network (TGN). In addition to its sorting functions in cargo trafficking from the Golgi, the TGN has been shown to act as an early endosomal compartment in plant and yeast cells (Chaps. 11 and 14). Zonal differentiation within the TGN has been proposed (Chap. 12), and a close relationship between the TGN and recycling endosomes (RE) has been suggested (Chaps. 5 and 12). To prevent further confusion regarding the definitions of these compartments, a clarification of their nomenclature will be necessary in the near future. The concept of unconventional protein trafficking, which bypasses the Golgi stacks, is also an exciting development related to alternative transport routes across or around the Golgi (Chap. 4). Special features of Golgi organization are discussed in relation to the cell cycle and cancer development (Chap. 6), neuronal endomembrane networks (Chap. 7), and the emerging role of biomolecular condensates in trafficking (Chap. 9).

In the current Volume I, various research topics addressing the diverse structural features of the Golgi are reviewed, and several models are proposed to explain how these structures are generated and maintained (Chaps. 8, 12, and 13). Dynamics is undoubtedly a critical aspect to consider, given the lively characteristics of the Golgi and its neighboring compartments in living cells. However, capturing this dynamism in a print-based format poses challenges.

Akihiko Nakano
Jaakko Saraste

Contents

Chapter 1
Three-Dimensional Ultrastructure of the Golgi Apparatus In Vivo: Scanning Electron Microscopy of Osmium-Macerated Mammalian Cells

Daisuke Koga, Ryosuke Morinaga, and Satoshi Kusumi

Abstract The osmium maceration method remains the only technique that enables direct three-dimensional (3D) visualization of subcellular structures via scanning electron microscopy (SEM), and it is particularly effective for resolving spatially complex organelles such as the Golgi apparatus. Here, we describe the 3D ultrastructure of the Golgi apparatus in various cell types, including epididymal epithelial principal cells, intestinal goblet cells, pituitary gonadotropes, and spinal ganglion cells, processed using the osmium maceration method. The overall morphology of the Golgi apparatus varied between cell types, exhibiting a range of configurations, including cup-shaped and spherical forms. Within the Golgi stacks, the *cis*-most cisterna and the *trans*-Golgi network displayed especially distinct structural characteristics. The *cis*-most cisterna typically appeared as a flat sheet with numerous fenestrations, a feature consistently observed across all four cell types. In contrast, the 3D organization of the *trans*-Golgi network, comprising tubules and/or plate-like membranes frequently interconnected to form elaborate architectures, varied according to the cell type. The morphological diversity of the Golgi apparatus in three dimensions likely reflects functional and structural specializations among different cell types.

Keywords The Golgi apparatus · Osmium maceration method · SEM · 3D · Serial section SEM

D. Koga (✉) · R. Morinaga
Department of Microscopic Anatomy and Cell Biology, Asahikawa Medical University, Asahikawa City, Hokkaido, Japan
e-mail: daisukek@asahikawa-med.ac.jp; rmorinaga@asahikawa-med.ac.jp

S. Kusumi
Division of Morphological Sciences, Graduate School of Medical and Dental Sciences, Kagoshima University, Kagoshima, Japan
e-mail: sk666ks@morphology.jp

A. Nakano, J. Saraste (eds.), *The Golgi Network, Volume I*, Subcellular Biochemistry 110, https://doi.org/10.1007/978-3-032-06936-8_1

Introduction

Morphological analyses, particularly those employing electron microscopy (EM), have continually advanced our understanding of the ultrastructural complexity of the Golgi apparatus. Over a century ago, Camillo Golgi, an Italian pathologist, identified black-stained reticular structures in neurons, such as cerebellar Purkinje cells of owls [33] and spinal ganglion cells of cats, rabbits, dogs, and cows [34], using light microscopy (LM) of tissues treated with the silver chromate staining method he had developed [32]. He referred to these black-stained structures as the *apparato reticolare interno* (internal reticular apparatus). Golgi and his disciples extended these LM observations to various tissues, including pancreas, cartilage, striated muscle, and even plant cells, demonstrating the widespread presence of the internal reticular apparatus [17]. However, the silver impregnation technique was considered unstable, being highly sensitive to environmental factors such as water composition. Under these challenging conditions, Cajal [10, 11] was among the few researchers outside Golgi's laboratory to successfully reproduce the visualization of the internal reticular apparatus. His findings supported Golgi's view that the structure was a ubiquitous cellular component. In contrast, Kopsch introduced the osmium impregnation method as a more stable alternative to silver staining, facilitating consistent visualization of the same reticular structure [69]. Consequently, some researchers began referring to the organelle as the "Golgi–Kopsch apparatus" [129]. Osmium impregnation remains a favored technique in morphological research due to its reliability and simplicity [6, 24, 27, 61, 68]. In a separate line of investigation, Holmgren [47] proposed that the structure revealed by trichloroacetic acid fixation, which he termed the "Trophospongium," corresponded to the internal reticular apparatus observed in silver-stained tissues. Although this interpretation was later challenged, given that the Trophospongium was thought to derive from plasma membrane invaginations, Cajal [11] acknowledged part of Holmgren's perspective and referred to the structure as the "Golgi–Holmgren apparatus." The now-standard term "Golgi apparatus" was first proposed by Besta [7] and gained broader acceptance through its use in English-language literature [77, 88, 93]. Nevertheless, skepticism persisted regarding the organelle's authenticity, with some researchers considering it an artifact [3, 92], and debate over its existence continued for nearly five decades after its discovery.

The invention of the transmission electron microscope (TEM) in 1931 by M. Knoll and E. Ruska revolutionized subcellular imaging by enabling resolution beyond the limits of LM, and TEM has been employed for biological studies since the mid-1940s [94]. The development of ultrathin sectioning techniques [28] represented a pivotal advancement in the ultrastructural analysis of organelles such as the endoplasmic reticulum (ER; [94]), mitochondria ([90, 91, 111]), and the Golgi apparatus. A major milestone was reached in 1954 when several studies provided in vivo TEM images of the Golgi apparatus, offering definitive morphological evidence of its structure [14, 23, 112]. Among these, the high-resolution micrograph published by Dalton and Felix [14] is widely recognized as the first clear depiction of the Golgi apparatus at the ultrastructural level. They defined the Golgi apparatus as comprising three elements, large vacuoles, small granules, and lamellae, later collectively termed the "Golgi complex" [15, 22]. Over the ensuing decades, a large body of TEM studies across diverse cell types, from vertebrates and invertebrates to plants, progressively clarified the ultrastructural organization of the Golgi apparatus [15, 22, 56, 82]. These investigations established that its most defining feature is a stack of smooth-surfaced cisternae, typically displaying flattened, plate-like centers with dilated rims. The cisternae are often slightly curved and consistently aligned in a stacked arrangement with distinct polarity, described using terms such as proximal/distal, forming/mature, or convex/concave [2, 22, 36, 81]. These are now commonly referred to as the *cis* and *trans* faces of the Golgi apparatus [20].

In the early 1960s, enzyme cytochemistry was introduced as a morphological method for identifying the Golgi apparatus by LM [85]. This approach was later adapted for TEM, enabling the localization of specific cisternae within the Golgi stack based on distinct patterns of phosphatase activity. For instance, thiamine pyrophosphatase (TPPase) and nucleoside diphosphatase (NDPase) activities were predominantly detected in the *trans* cisternae [87], whereas nicotinamide adenine dinucleotide phosphatase (NADPase) activity was restricted to the medial cisternae [113]. In contrast, osmium impregnation, used since the era of LM studies, was shown to selectively label the *cis* cisternae [24]. Together, enzyme cytochemistry and EM provided key insights into the intrinsic polarity of the Golgi apparatus, which is now understood to comprise three distinct compartments: *cis*, medial, and *trans* cisternae.

Three-dimensional (3D) EM analysis of the Golgi apparatus began to receive increasing attention in the 1970s. Rambourg and collaborators were instrumental in advancing this field, employing two main strategies to elucidate the organelle's 3D architecture. One approach utilized high-voltage EM to visualize entire Golgi structures by capturing stereopairs from thick sections tilted at varying angles [104]. The other relied on thorough conventional TEM observations of ultrathin sections from tissues fixed with reduced osmium or processed via enzyme cytochemistry to reveal the 3D ultrastructure of individual cisternae [102, 105]. The 3D representations (i.e., intricate diagrams) of Golgi cisternae published in these studies, painstakingly derived from two-dimensional (2D) observations, are renowned for their accuracy and elegance, and have become foundational references in the morphological study of the Golgi apparatus. In parallel, another major advancement in 3D-EM was the development of electron tomography, which integrates high-voltage EM with computational 3D reconstruction. This technique enabled detailed visualization of individual cisternae and their spatial relationships with adjacent organelles such as ER and mitochondria [70, 73, 74, 80]. These innovations allowed researchers to examine the Golgi stack's intricate architecture at unprecedented resolution and identify variations in cisternal organization across different cell types. As a result, morphological analyses of the Golgi apparatus have increasingly emphasized its 3D organization, reflecting the spatial complexity of this organelle.

While LM and TEM have been widely used to study the Golgi apparatus, scanning electron microscopy (SEM) has been applied far less frequently. Due to its greater depth of field compared to TEM, SEM offers an effective means of 3D structural visualization and is especially useful for revealing the complex architecture of the Golgi apparatus. Among SEM-based approaches, the osmium maceration method remains the only technique capable of providing 3D ultrastructural information for cellular organelles such as the Golgi apparatus, ER, and mitochondria [117, 118]. Accordingly, this chapter introduces the osmium maceration method and highlights the 3D ultrastructure of the Golgi apparatus in mammalian cells in vivo, as visualized by SEM of osmium-macerated specimens.

The discussion focuses on several representative cell types with well-developed Golgi apparatuses that have been extensively investigated in prior EM studies: epithelial principal cells of the epididymis [14, 44], goblet cells in the jejunum [84], gonadotropes in the pituitary gland [50, 51], and dorsal root ganglion cells [34]. Particular attention is given to (1) the overall morphology of the Golgi apparatus, (2) the stacking pattern of Golgi cisternae, (3) the ultrastructure of the *cis* cisternae, (4) the ultrastructure of the *trans*-Golgi network (TGN), and (5) 3D morphological dynamics of the Golgi apparatus in response to secretory stimulation. Finally, the chapter will describe the 3D reconstruction of the entire Golgi apparatus using modern 3D-SEM techniques, including serial section SEM.

Osmium Maceration Method for Direct Observation of the 3D Ultrastructure of Organelles

SEM offers a long focal depth and a wide field of view, making it an invaluable tool for visualizing the 3D surface architecture of cells and tissues. However, subcellular structures are not typically discernible in SEM images of specimens prepared using conventional protocols. To overcome this limitation, electron microscopists have developed several tissue-fracturing techniques, such as freeze-cracking and resin-cracking, that allow precise cleavage of cellular structures [119, 121]. Although these approaches facilitated exposure of the intracellular milieu, they were insufficient for visualizing organelles embedded within the cytoplasmic matrix, particularly when using SEM to observe fractured cell surfaces. To address this challenge, Professor Keiichi Tanaka and colleagues at Tottori University developed an innovative SEM-based technique known as the osmium maceration method. This method remains the only approach that permits direct SEM visualization of subcellular structures without the need for 3D reconstruction [59, 65, 117, 118]. In this procedure, the cytoplasmic matrix is removed from freeze-fractured tissue surfaces using a diluted osmium tetroxide solution, selectively preserving organelles such as mitochondria, smooth and rough ER, secretory granules, the Golgi apparatus, ribosomes on the rough ER, lipid droplets, glycogen granules, and the nucleus. In contrast, centrioles and cytoskeletal components, including microfilaments, intermediate filaments, and microtubules, are dissolved during maceration and therefore cannot be observed. For detailed visualization of fine organellar architecture, an

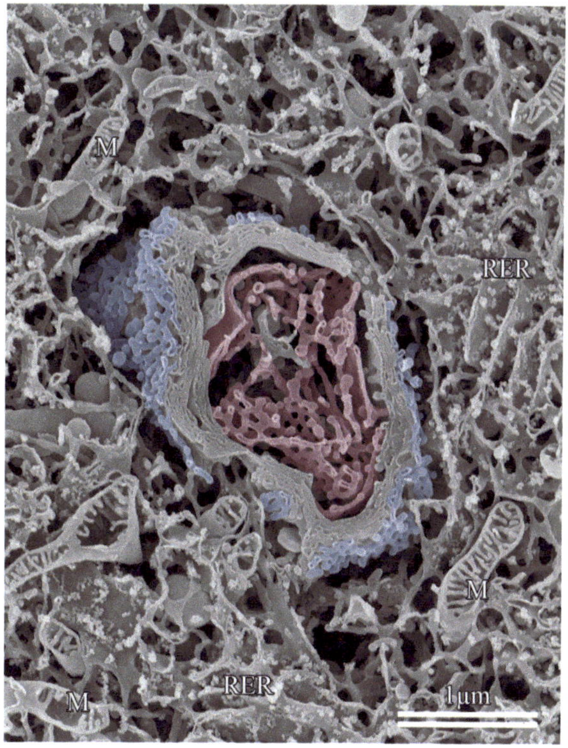

Fig. 1.1 Scanning electron microscopy (SEM) image of the Golgi apparatus in a rat spinal ganglion cell processed using the osmium maceration method. The three-dimensional ultrastructure, including the *cis*-most cisterna (blue) and the *trans*-Golgi network (red), is clearly visualized. The spatial relationship among the Golgi apparatus, rough endoplasmic reticulum (RER), and mitochondria (M) is distinctly revealed

ultra-high-resolution SEM equipped with a field-emission or Schottky-emission electron gun and a semi-in-lens or in-lens detection system is essential. Organelles visualized using this technique closely resemble schematic textbook illustrations (Fig. 1.1). The osmium maceration method has yielded significant insights into the 3D ultrastructure of mitochondria ([71, 72, 107]), ER [89, 132], and the Golgi apparatus [46, 116, 120, 130] in tissue specimens. Although primarily applied to tissue samples, the osmium maceration method has also been used to investigate organellar architecture in cultured cells, free-floating cells such as leukocytes, and microorganisms. In these contexts, embedding in chitosan or agarose facilitates direct SEM observation of intracellular structures [26, 52, 66]. More recently, a correlative light and scanning electron microscopy technique was developed by combining Tokuyasu's cryosectioning method [122] with the osmium maceration procedure. This integrated approach enables direct correlation between fluorescence signals and 3D ultrastructural features of organelles [62]. Thus, the osmium maceration method not only enhances morphological analysis but also holds promise for broader applications in cell biology.

For details regarding the procedure and its variations, readers are referred to the foundational and methodological studies ([117, 118]; [59, 62, 65, 66]).

The Spatial Location and 3D Structure of the Golgi Apparatus in Different Cells

EM of single ultrathin sections provides valuable insights into the ultrastructure of the Golgi apparatus; however, due to the organelle's spatial complexity and extent, this approach is insufficient for revealing its 3D architecture (Fig. 1.2). In contrast, SEM of osmium-macerated tissues is a powerful technique that enables direct visualization of both the 3D ultrastructure of Golgi stacks and the overall morphology of the Golgi apparatus (Fig. 1.1). Comparative analysis of 2D ultrathin sections and 3D three-dimensionally observed from osmium-macerated tissues underscores the advantages of the maceration method, although ultrathin sections remain valuable for analyzing subcellular detail.

This section presents the 3D morphology of the Golgi apparatus as observed by SEM in osmium-macerated tissues of various rat cell types, including epididymal epithelial principal cells, intestinal goblet cells, pituitary gonadotropes, and spinal ganglion cells (Figs. 1.3, 1.4, 1.5, 1.6, 1.7, 1.8, 1.9, 1.10, and 1.11). In epididymal epithelial principal cells, responsible for glycoprotein production and fluid absorption essential for sperm maturation and transport, the Golgi apparatus is centrally located in the cytoplasm between the apical membrane and the basally positioned

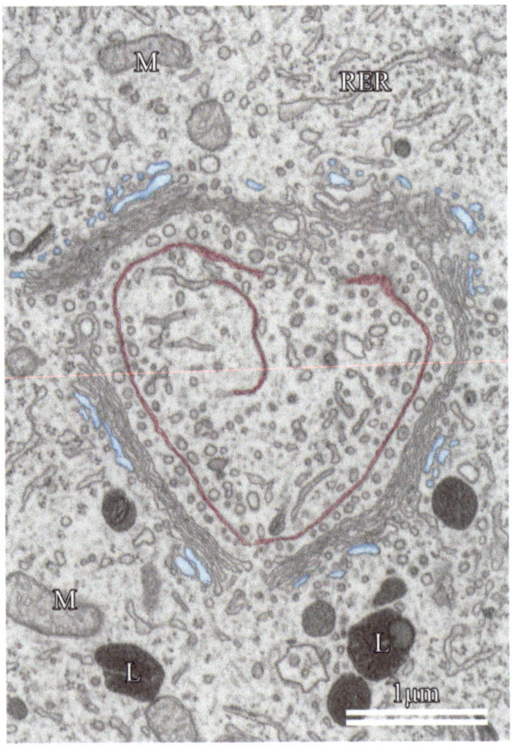

Fig. 1.2 Ultrathin section of a rat spinal ganglion cell showing the Golgi apparatus, imaged using backscattered electron-mode SEM of resin-embedded tissue. Blue: *cis*-most cisterna; red: *trans*-Golgi network; *M* mitochondria, *RER* rough endoplasmic reticulum, *L* lysosomes

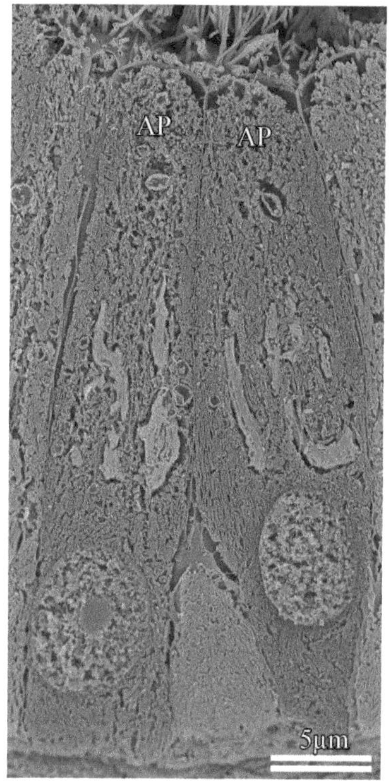

Fig. 1.3 Low-magnification SEM image of epithelial principal cells in the rat epididymal duct. The Golgi apparatus (green) is located centrally within the cytoplasm (magenta), positioned above the nucleus (blue). *AP* apical portion of the epididymal cells

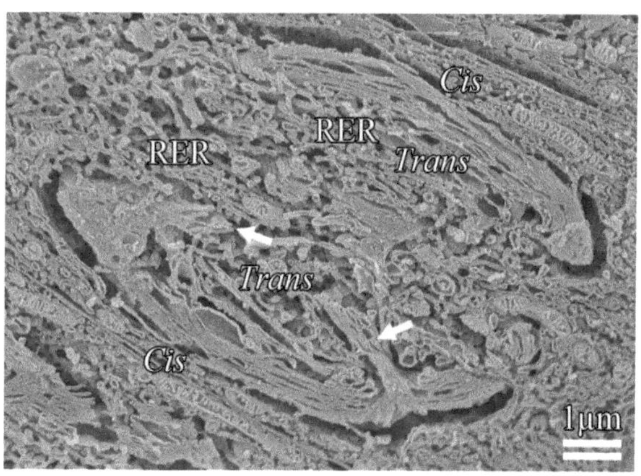

Fig. 1.4 High-resolution SEM image of the fractured surface of the Golgi apparatus in a rat epididymal epithelial principal cell. The Golgi stacks comprise several flat cisternae arranged in layers, forming a U-shaped configuration. Tubular elements of the rough endoplasmic reticulum (RER) and mitochondria (cyan) are observed within the Golgi region. The *trans*- and *cis*-sides are oriented toward the interior and exterior of the U, respectively. Arrows: Golgi cisternae extending toward the interior of the U-shaped stack. *Cis cis*-side, *Trans trans*-side of the Golgi apparatus

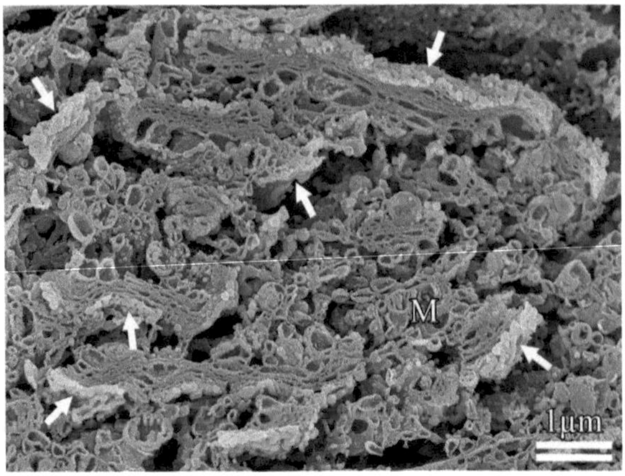

Fig. 1.5 Backscattered electron (BSE) image of the Golgi apparatus in a rat epididymal epithelial principal cell following osmium impregnation and maceration. The 3D ultrastructure is clearly delineated, with osmium deposits localized within the outer one or two cisternal layers (arrows). *M* mitochondrion

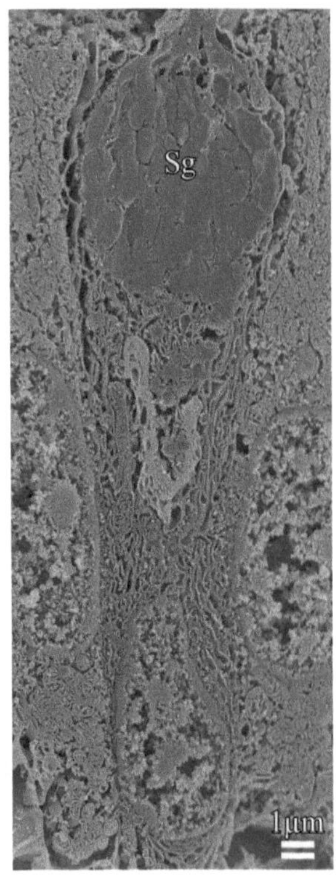

Fig. 1.6 Low-magnification SEM image of a goblet cell in the rat jejunum. The Golgi apparatus (green) is situated centrally and in the apical region of the nucleus (blue), which is positioned at the basal end of the cell. Secretory granules (Sg) are accumulated at the apical portion. Magenta: cytoplasm

nucleus (Fig. 1.3). Golgi stacks in these cells are composed of several flattened, sheet-like cisternae arranged in parallel layers, collectively forming a U-shaped, cup-shaped, or candle flame–like structure (Figs. 1.3 and 1.4). These cisternae fold or extend within the U- or cup-shaped configuration, generating a more complex 3D architecture. Mitochondria and tubular elements of the ER are also observed within

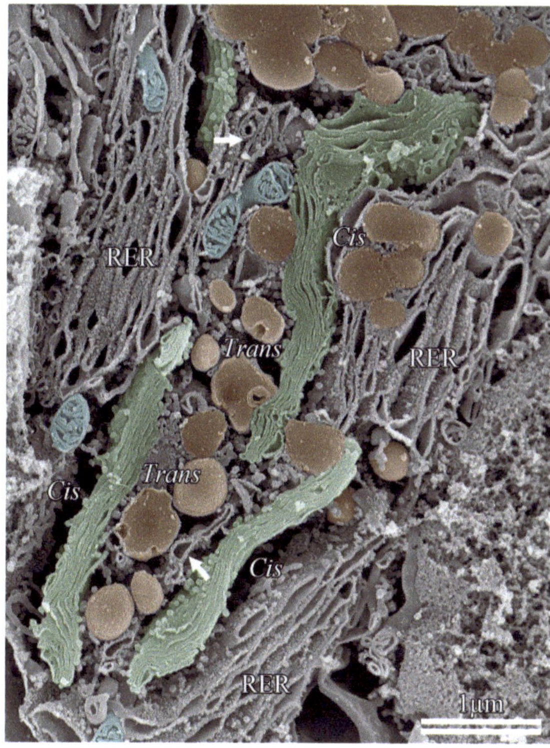

Fig. 1.7 High-resolution SEM image of the fractured surface of the Golgi apparatus in a goblet cell of the rat jejunum. The Golgi stacks consist of layered sheet-like cisternae forming a cup-shaped structure. The rough endoplasmic reticulum (arrows) and mitochondria (cyan) are visible within the Golgi region. Secretory granules (orange) are present both within and surrounding the Golgi apparatus. *RER* rough endoplasmic reticulum, *Cis cis*-side, *Trans trans*-side of the Golgi apparatus

this structure. The Golgi apparatus in these cells typically measures 8–10 μm along its long axis and approximately 5 μm along its short axis. The *cis*-side is positioned externally, while the TGN resides within the interior of the U- or cup-shaped structure. The *cis–trans* polarity is identifiable based on features characteristic of the *cis*-most cisterna and the TGN, further detailed in the sections "The 3D ultrastructure of the *cis*-side of the Golgi apparatus" and "The 3D ultrastructure of the *trans*-Golgi network." However, SEM analysis of osmium-macerated tissues alone may occasionally be insufficient for unambiguous identification of *cis–trans* polarity. In such cases, combining the classical osmium impregnation method, known for selectively staining *cis*-elements [24], with the osmium maceration technique provides a reliable approach for determining Golgi polarity [61]. As shown in Fig. 1.5, SEM imaging of epididymal epithelial principal cells subjected to this combined method (i.e., the osmium-impregnation/maceration method) reveals osmium deposits localized specifically within the outermost one or two cisternae, corresponding to the *cis*-Golgi. These osmiophilic *cis*-cisternae are also identifiable in the internal stacks of the U- or cup-shaped Golgi structure.

Goblet cells in the intestinal tract, specialized for mucus secretion, represent a prototypical exocrine cell type. In rat jejunal goblet cells, the Golgi apparatus is also centrally located between the apical membrane and the basally situated nucleus (Fig. 1.6). Similar to epididymal cells, the Golgi stacks comprise layered flat,

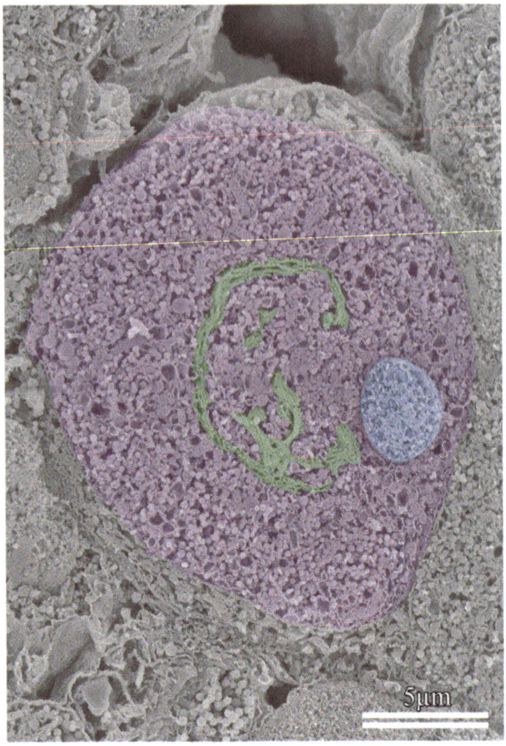

Fig. 1.8 Low-magnification SEM image of a gonadotrope in the rat pituitary gland. The Golgi apparatus (green) occupies a central position within the cell, while the nucleus is located peripherally. Magenta: cytoplasm

sheet-like cisternae forming a U- or cup-shaped structure (Figs. 1.6 and 1.7). The apical cytoplasm is densely packed with mucin-containing secretory granules, which are also distributed within the U-shaped Golgi region. Mitochondria and rough ER are likewise present within this configuration. As in epididymal cells, the *cis*-most cisterna is located externally, while the TGN lies internally. The dimensions of the Golgi apparatus in goblet cells typically range from 8–10 μm in length to 3–4 μm in width.

The gonadotropes of the anterior pituitary gland are endocrine cells responsible for producing the gonadotropins luteinizing hormone (LH) and follicle-stimulating hormone (FSH). In these cells, the Golgi stacks are composed of multiple sheet- or plate-like cisternae arranged in layered formations that collectively adopt a spherical configuration (Fig. 1.8). In cross-sectional images, these stacks appear as a continuous ring, enclosing a central, sealed space (Fig. 1.9). The nucleus, located peripherally, is situated in close proximity to the centrally positioned Golgi apparatus. Numerous secretory granules containing gonadotropins are distributed throughout the cytoplasm, including within the spherical structure of the Golgi apparatus. In addition to these granules, mitochondria and rough ER are also present within the spherical space. The flattened cisternae of the Golgi stacks extend and fold inward into the sphere. The *cis*-most cisterna is located on the outer surface of the spherical Golgi structure, whereas the TGN resides within it. The diameter of the sphere

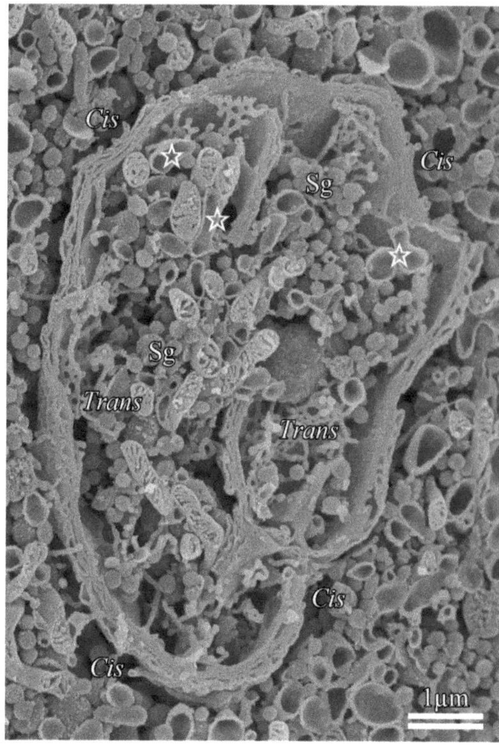

Fig. 1.9 High-magnification view of the fractured surface of the Golgi apparatus in a gonadotrope of the rat pituitary gland. The Golgi stack consists of several sheet-like cisternae concentrically arranged to form a spherical structure. The rough endoplasmic reticulum (☆), secretory granules, and mitochondria (cyan) are present within the Golgi region. *Cis* *cis*-side, *Trans* *trans*-side of the Golgi apparatus

ranges from 6 to 8 μm. A similar spherical morphology of the Golgi apparatus has been reported in other hormone-producing cells of the anterior pituitary gland and in tracheal epithelial cells, as demonstrated by SEM of osmium-macerated rat tissues [116, 120]. Stereoscopic thick-section images of prolactin-producing cells (i.e., mammotropes) subjected to enzyme cytochemistry for NADPase and TPPase, and examined using high-voltage EM, reveal that cisternae positive for NADPase and TPPase form ribbon- or thread-like structures that collectively assume a spheroidal configuration [96]. In both gonadotropes and mammotropes, endocrine cells of the pituitary gland, the Golgi apparatus adopts a spherical or spheroidal form. However, a critical morphological distinction exists: in mammotropes, the Golgi appears as a spheroidal basket with a coarse mesh, resulting from convoluted and branched Golgi ribbons, whereas in gonadotropes, it forms a compact, ball-like structure composed of plate-like cisternae interspersed with sparse, large openings. The characteristic spherical morphology of the Golgi apparatus has also been documented in other mammalian cells in vivo, including differentiating germ cells, leukocytes, and plasma cells [98].

In the rat dorsal root ganglia, at least three neuronal subtypes, large, smaller, and smallest, are classified based on cell size and organelle distribution [99]. In smaller-type neurons, numerous small Golgi stacks surround the centrally located nucleus, forming a ring-like arrangement reminiscent of the planet Saturn (Fig. 1.10). Rough

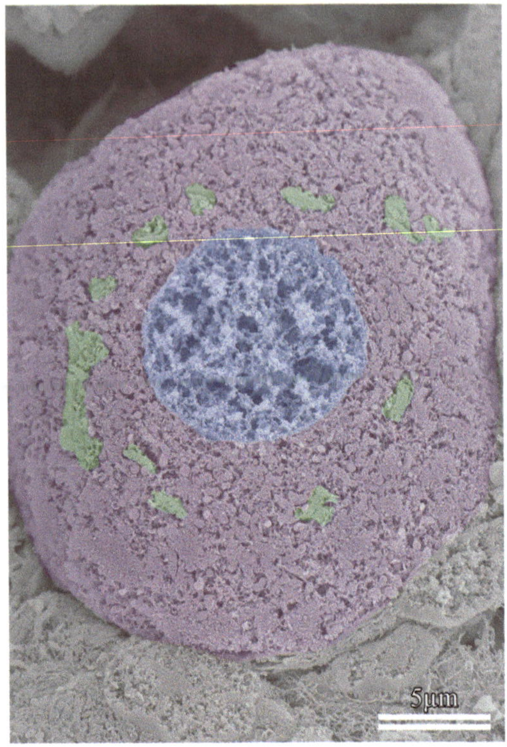

Fig. 1.10 Low-magnification SEM image of a smaller-type neuron of the rat spinal ganglion cell. The nucleus (blue) is centrally positioned within the soma (magenta) and is surrounded by numerous Golgi stacks (green)

ER (Nissl bodies) and mitochondria are localized around each Golgi stack (Fig. 1.11), with the *trans* side of each stack generally oriented toward the nucleus. By contrast, in large-type ganglion cells, the small Golgi stacks are dispersed throughout the cytoplasm [59]. The continuity between individual Golgi stacks in both neuron types remains unresolved in SEM studies of osmium-macerated tissues. Since the osmium maceration method relies on SEM imaging of freeze-fractured surfaces, visualizing deeper structures beyond the fractured planes, and thereby confirming the continuity of Golgi stacks, is inherently difficult in spinal ganglion cells. However, high-voltage EM of stereopairs from thick sections (3–7 μm) of osmium-impregnated spinal and trigeminal ganglion cells has revealed that *cis*-osmiophilic cisternae form a continuous perinuclear network [104]. High-voltage EM enables detailed 3D visualization of the Golgi apparatus by detecting dense deposits of osmium or lead within the stained cisternae, enhanced by osmium impregnation and enzyme histochemistry [96, 100, 104]. However, surrounding structures, such as unstained Golgi cisternae, mitochondria, and ER, often appear indistinct or are not visualized, and the physical limits of thick sections can hinder comprehensive imaging of large Golgi structures in neurons. To overcome this limitation, serial section TEM, a classical 3D EM technique, has been used to reconstruct the complete morphology of the Golgi apparatus [18, 115, 133]. Advanced 3D reconstruction methods, such as serial section TEM and modern 3D-SEM,

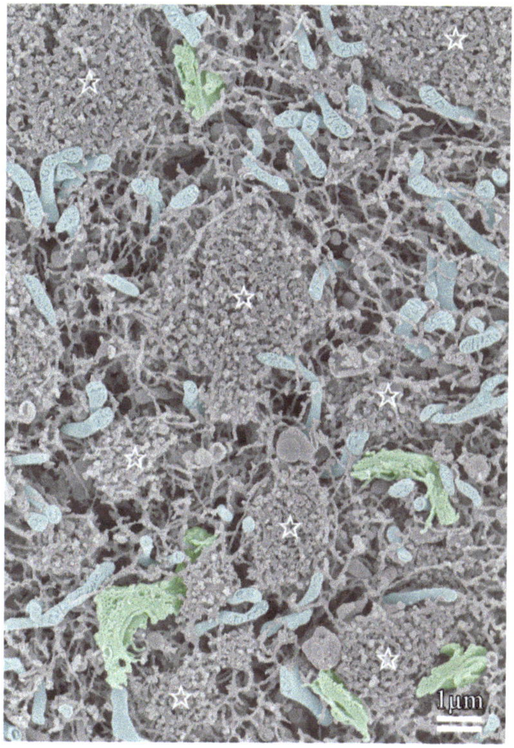

Fig. 1.11 High-magnification view of the Golgi apparatus in a large-type neuron of the rat dorsal root ganglion. The Golgi stacks (green) are composed of several flattened cisternae dispersed throughout the soma. The rough endoplasmic reticulum (Nissl bodies; ☆) and mitochondria (blue) are observed surrounding the Golgi stacks

focused ion beam SEM (FIB-SEM), serial block-face SEM (SBF-SEM), and array tomography (i.e., serial section SEM), are expected to elucidate the continuity of individual Golgi stacks in spinal ganglion cells. The complete 3D structure of the Golgi apparatus, as revealed by these techniques, will be discussed in the section "Elucidating the 3D morphology of the Golgi apparatus using serial section SEM and 3D reconstruction."

The Stack of the Golgi Cisternae

Numerous EM studies have demonstrated that Golgi stacks, comprising several flattened cisternae arranged in layers along the *cis*-to-*trans* axis (i.e., the compact or saccular region), are laterally connected by tubular membranes, referred to as the noncompact zone (or intersaccular region), forming a continuous ribbon-like structure [44, 58, 70, 75, 101, 105]. However, no tubular structures indicative of the noncompact zone have been observed in osmium-macerated tissues, at least in epididymal epithelial principal cells, intestinal goblet cells, pituitary gonadotropes, and spinal ganglion cells (Figs. 1.3, 1.4, 1.5, 1.6, 1.7, 1.8, 1.9, 1.10, and 1.11). In these cell types, the Golgi apparatus appears to consist solely of stacks of flattened cisternae arranged in layers, with compact regions forming a single continuous entity. This observation suggests that the noncompact zone may not be a universal feature of the Golgi apparatus in mammalian cells in vivo. Alternatively, this discrepancy may arise from differences in the interpretation of electron micrographs obtained from osmium-macerated specimens versus those from ultrathin sections of resin-embedded cells.

The Golgi stack is composed of discrete cisternal subcompartments: the *cis*-most cisterna, *cis* cisternae, medial cisternae, *trans* cisternae, and the TGN [83, 99]. High-magnification imaging of osmium-macerated tissues allows direct 3D visualization of the stack architecture formed by cisternae layered along the axis. For instance, in pituitary gonadotropes, the Golgi stack comprises five cisternae, including the *cis*-most cisterna and TGN (Fig. 1.12), whereas large-type spinal ganglion cells contain seven stacked cisternae (Fig. 1.13). In comparison, the Golgi stacks of rat epididymal epithelial principal cells and intestinal goblet cells contain 10 and 7 cisternae, respectively [59]. The number of cisternae forming the Golgi stack is cell-type dependent [98]: the cell with the fewest Golgi cisternae is the epithelial cells of the seminal vesicle (2–3 cisternae), while the cell with the most is the mucous cells in Brunner's glands (9–11 cisternae). Although this variability in cisternal number across cell types is well documented, the mechanisms regulating cisternal number in individual cells remain unclear and warrant further investigation.

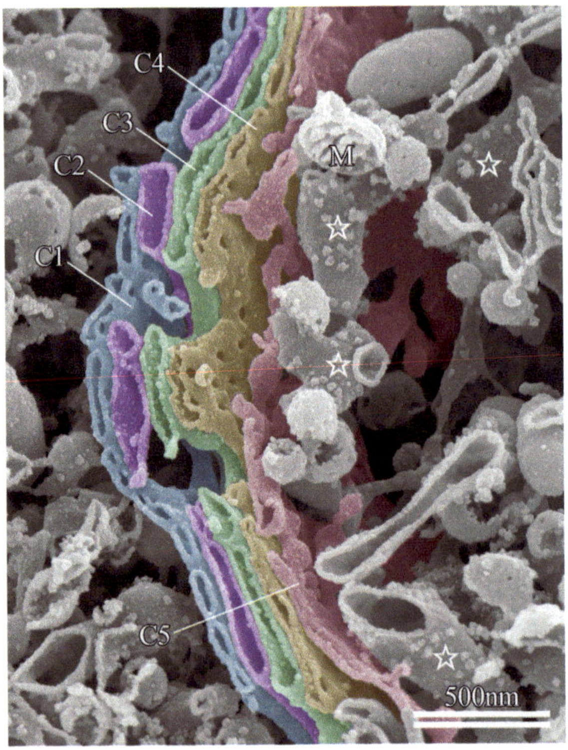

Fig. 1.12 High-magnification image of a Golgi stack in a gonadotrope of the rat pituitary gland. The stack consists of five cisternal layers (C1–C5). *C1 cis*-most cisterna, *C5 trans*-Golgi network, *M* mitochondrion; ☆: rough endoplasmic reticulum in the Golgi region. (Figure reproduced and modified with permission from *Archives of Histology and Cytology*; [59])

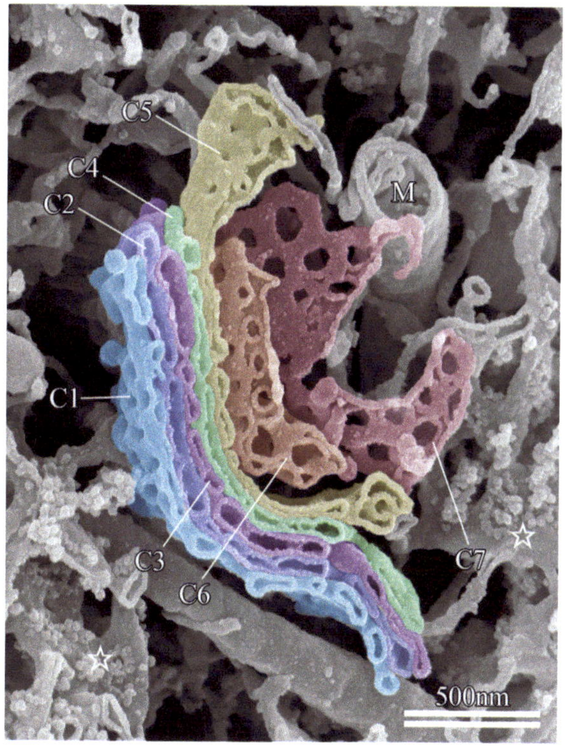

Fig. 1.13 High-magnification image of a Golgi stack in a large-type neuron of the rat dorsal root ganglion. The stack comprises seven cisternal layers (C1–C7). *C1 cis*-most cisterna, *C7 trans*-Golgi network, *M* mitochondrion; ☆: rough endoplasmic reticulum. (Figure reproduced and modified with permission from *Archives of Histology and Cytology*; [59])

Determining the presence or absence of direct continuity between heterotypic cisternae (e.g., *cis*-medial or medial-*trans* cisternae) is essential for understanding the mechanisms underlying intra-Golgi trafficking. Several trafficking models propose tubular connections between heterotypic cisternae [30, 31]. However, such intercisternal tubular connections are infrequently observed in mammalian cells using conventional TEM of ultrathin sections, SEM of osmium-macerated specimens, or EM tomography with 3D reconstruction [12, 74, 116]. It remains to be determined whether these connections are constitutive structures in normal cells or appear transiently under specific conditions, such as secretory stimulation. The osmium maceration method, which facilitates rapid acquisition of high-resolution 3D information without the need for time-consuming reconstruction techniques, offers a promising approach for visualizing rare structures such as intercisternal tubular continuity. Furthermore, this technique has uncovered a striking architectural feature of the Golgi apparatus in lacrimal gland cells: although appearing as a stack of multiple cisternae, the structure is, in fact, a single helically wound cisterna [120]. This remarkable observation provides novel insights into the diversity of Golgi architecture and may significantly advance our morphological understanding of this organelle.

The 3D Ultrastructure of the cis-*Side of the Golgi Apparatus*

The *cis*-most cisterna, also known as the *cis*-osmiophilic element or *cis*-osmiophilic tubular network due to its selective staining with osmium during prolonged osmication [101, 102], is located at the outermost region on the *cis*-side of the Golgi stack. It typically appears as an anastomosed network of interconnecting tubules, historically referred to as the *cis*-tubular network [101–103] or *cis*-Golgi network [78], based on observations from TEM of ultrathin sections and high-voltage EM of thick sections. In the present study, we use the term *cis*-most cisterna to describe the outermost Golgi cisterna, rather than *cis*-tubular network or *cis*-Golgi network.

When visualized using SEM of osmium-macerated tissues, the 3D ultrastructure of the *cis*-most cisternae is clearly discernible (Figs. 1.14 and 1.15). These cisternae typically exhibit a flat, sheet-like morphology with numerous fenestrations, each measuring 30–50 nm in diameter, a structural feature consistently observed in various mammalian cell types [59, 116, 120]. The osmium impregnation/maceration method reveals that osmium tetroxide preferentially accumulates in the outermost *cis*-Golgi cisterna during prolonged osmication (Fig. 1.5; [61]), suggesting that the sheet-like cisternae with fenestrations correspond to the *cis*-osmiophilic cisternae. The fenestrated sheet structure observed by SEM may have been previously

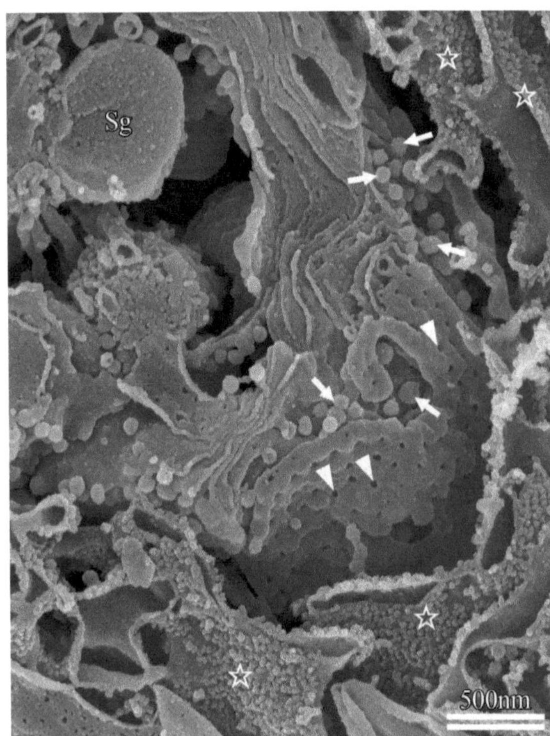

Fig. 1.14 High-magnification image of the *cis*-side of the Golgi apparatus in a goblet cell of the rat jejunum. The *cis*-most cisterna (blue) appears as a flat sheet perforated with numerous fenestrations (~30 nm; arrowheads). *Sg* secretory granules; arrows: small vesicles; magenta: rough endoplasmic reticulum

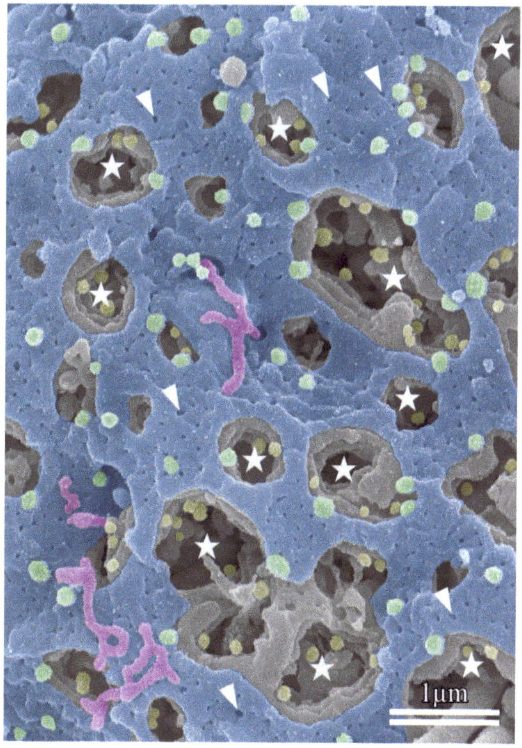

Fig. 1.15 High-magnification image of the *cis*-side of the Golgi apparatus in a gonadotrope of the rat pituitary gland. The *cis*-most cisterna (blue) is a flat sheet with numerous fenestrations (~30 nm; arrowheads). Large round openings (★; 200–500 nm) penetrate the underlying cisternae, forming passages between the *cis*-side (outer) and *trans*-side (inner) of the spherical Golgi apparatus. Small vesicles (yellow; 80–100 nm) are located at the periphery of these openings. A few tubular membranous structures (magenta) and budding profiles (green) are also observed on the *cis*-most cisterna. (Figure reproduced and modified with permission from *Biomedical Research (Tokyo)*; [60])

interpreted as an anastomosed tubular meshwork in TEM images of both thin and thick sections.

In addition to fenestrations, large circular openings, termed "wells," with diameters ranging from 200 to 500 nm are observed within the *cis*-most cisternae (Figs. 1.15 and 1.16). These structures, first described in the early spermatids by Hermo et al. [45], have since been reported by other EM studies [49, 70]. Some of these wells extend through the underlying medial cisternae or are aligned at equivalent positions across the Golgi stack, forming continuous tunnels through which small vesicles (80–100 nm) are frequently observed. These tunnels may facilitate direct communication between the *cis*- and *trans*-sides of the Golgi apparatus [59, 116]. Tubular elements of the rough ER, mitochondria, secretory granules, and small vesicles are often observed within these tunnels. In pituitary gonadotropes, microtubules extend isotropically from centrioles, the microtubule-organizing centers, located within the spherical core of the Golgi apparatus. These microtubules

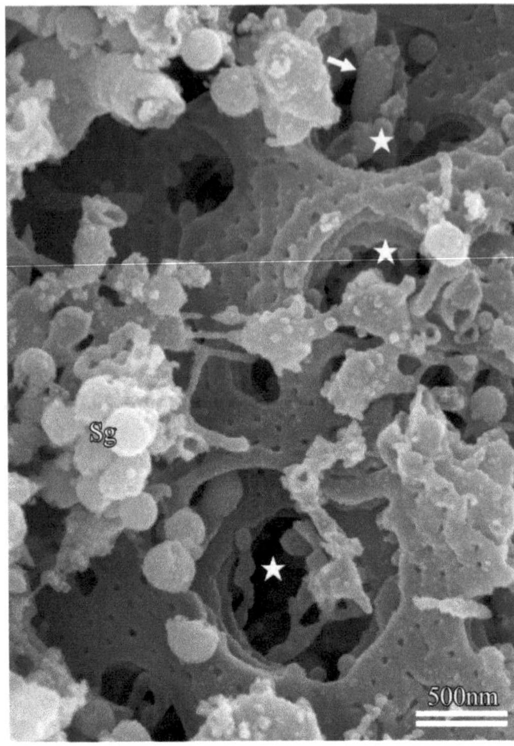

Fig. 1.16 High-magnification image of the *cis*-most cisterna in a gonadotrope of the rat pituitary gland. The rough endoplasmic reticulum (yellow) and mitochondrion (arrow) are connected through large openings (★), linking the outer and inner regions of the Golgi apparatus. *Sg* secretory granules. (Figure reproduced and modified with permission from *Archives of Histology and Cytology*; [59])

traverse the Golgi stack and project toward the cell periphery through tunnels that span from the *cis*- to the *trans*-side [125, 126]. These tunnels, which result from large wells aligned across successive cisternae, appear to function as conduits for microtubule passage. This structural arrangement supports the secretory pathway in gonadotropes and suggests that a spherical Golgi morphology may be especially suited to endocrine cells with limited polarity.

The 3D Ultrastructure of the trans-*Golgi Network*

The TGN, the terminal and exit compartment of the Golgi apparatus, functions as a critical sorting station in the protein trafficking pathway, directing cargo proteins to their appropriate destinations [21, 37]. The anastomosed tubular membrane system positioned a short distance from the *trans* cisternae was originally termed the *trans*-tubular network (TTN) by Rambourg et al. [101]. The term "*trans*-Golgi network," now widely adopted, originated from this earlier designation [37]. Part of the GERL (Golgi–Endoplasmic Reticulum–Lysosome) system proposed by [88] and Novikoff et al. ([87]) is generally considered functionally morphologically equivalent to the

TGN [8, 40, 86]. Unlike the relatively conserved structure of the *cis*-most cisterna, the 3D ultrastructure of the TGN varies markedly among different cell types [13, 59].

In rat epididymal epithelial principal cells, the TGN comprises a wide, intricately interconnected network of tubular membranes (Fig. 1.17). Nodule-like or spherical swellings are often observed along these tubules, suggesting the formation of secretory vesicles. Coated vesicles are not visible in SEM of osmium-macerated tissue as they are lost during the maceration process. Similar morphological features, including nodular and dilated regions, have been illustrated in schematic representations based on TEM images of ultrathin sections [44]. In goblet cells of the rat jejunum, the TGN comprises several small tubular membranes and prosecretory granules associated with short tubules (Fig. 1.18). These components are spatially distinct, and numerous small vesicles (~100 nm in diameter) are present in the surrounding area. Such features, tubular networks and prosecretory granules with associated tubules, are characteristic of exocrine cells exhibiting secretory polarity, such as mucous cells in Brunner's glands [103], pancreatic acinar cells [102], and intestinal goblet cells ([59]; Fig. 1.18). In contrast, the TGN of gonadotropes in the rat pituitary is highly developed, consisting of extensively interconnected sheets and tubular membranes forming a continuous 3D network (Fig. 1.19; [67]). Interestingly, in mammotropes, endocrine cells that secrete hormones, the TGN consists of small tubular membranes without associated prosecretory granules [13]. This

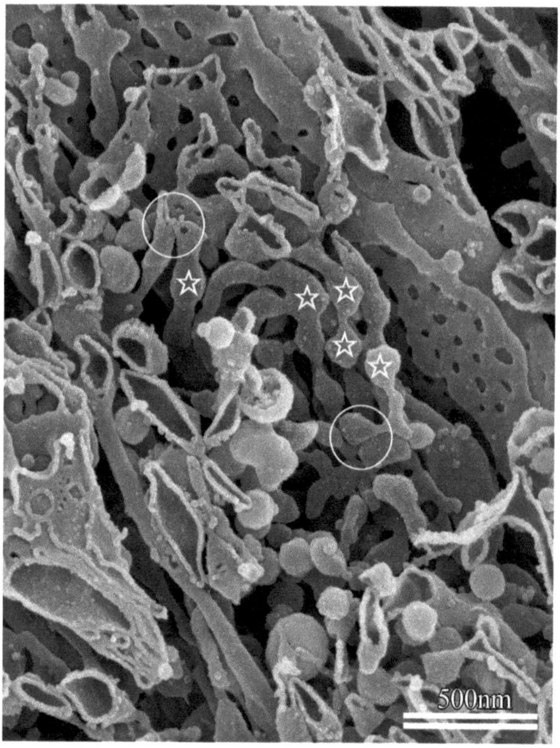

Fig. 1.17 High-magnification image of the *trans*-side of the Golgi apparatus in a rat epididymal epithelial principal cell. The *trans*-Golgi network (TGN; red) consists of interconnected tubular membranes with round swellings (☆). Tubular endoplasmic reticulum (ER; blue) is closely associated with the TGN at multiple contact sites (white-outlined circles), where ER membranes lack ribosomes. (Figure reproduced and modified with permission from *Archives of Histology and Cytology*; [59])

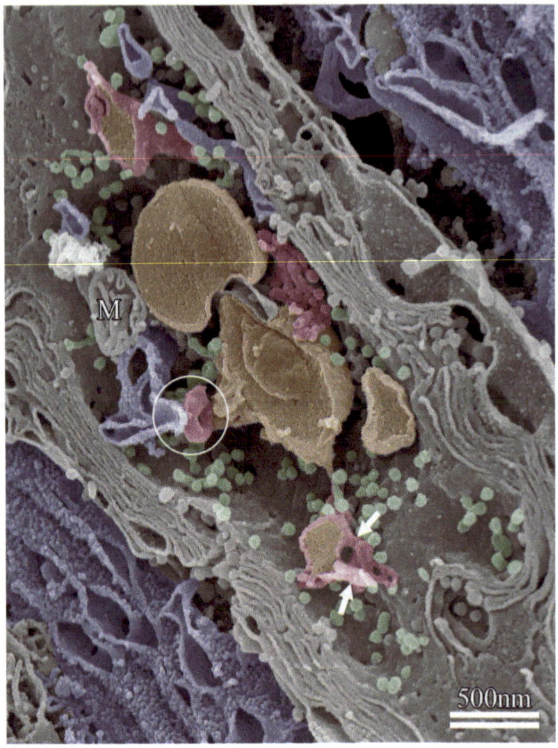

Fig. 1.18 High-resolution image of the *trans*-side of the Golgi apparatus in a goblet cell of the rat jejunum. The *trans*-Golgi network (TGN; red) is composed of tubular membranous structures. Tubules associated with prosecretory granules (arrows) and contact sites with the ER (blue; white-outlined circle) are observed. Numerous small vesicles (green) are present on the *trans* cisterna. *M* mitochondrion

configuration closely resembles the TGN morphology observed in exocrine cells such as goblet cells, mucous cells of Brunner's glands, and pancreatic acinar cells. Notably, despite both being hormone-producing endocrine cells of the pituitary, the TGN structure differs significantly between gonadotropes and mammotropes. In rat spinal ganglion cells, the TGN adopts a lace-like architecture composed of extensively anastomosed tubular membranes (Fig. 1.20), a pattern commonly observed in cells that do not produce secretory granules [13]. A similar 3D morphology is reported in Sertoli cells [101] and spinal ganglion cells [97]. Thus, the 3D organization of the TGN exhibits notable diversity among cell types, likely reflecting cell-specific functions. Supporting this, Clermont et al. [13] categorized 14 mammalian cell types into four morphological classes based on Golgi structure observed by TEM and concluded that the TGN is most extensive in cells not involved in secretory granule formation. In contrast, secretory cells that generate small or large granules possess a relatively small or absent TGN. The absence of a TGN in secretory cells of seminal vesicles and acinar cells of mammary glands is particularly noteworthy.

In epididymal epithelial principal cells, goblet cells, gonadotropes, and spinal ganglion cells, the TGN is closely associated with specialized domains of ER, termed the *trans*-ER. These consist of ribosome-free tubular membranes located at contact sites with the TGN (Figs. 1.17, 1.18, 1.19, and 1.20). Although these

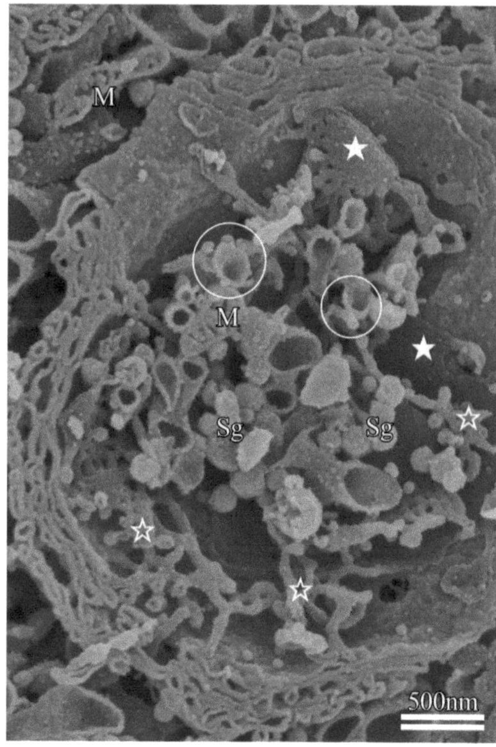

Fig. 1.19 High-magnification image of the *trans*-side of the Golgi apparatus in a gonadotrope of the rat pituitary gland. The *trans*-Golgi network (TGN; red) includes anastomosed tubules (☆) and membranous sheets (★). The ER (blue), mitochondrion (M), and secretory granules (Sg) are visible within the spherical Golgi apparatus. Although the ER and TGN are closely associated, no direct continuity is observed (white-outlined circles)

organelles are in close apposition, no direct continuity between the ER and the TGN has been observed. Despite considerable interest in their potential interconnection, ultrastructural continuity has not been demonstrated in TEM studies of various cell types [40, 41, 44, 50, 95]. Furthermore, 3D reconstructions using EM tomography have consistently shown intimate contacts between the *trans*-ER and *trans*-Golgi components, including the TGN [70, 73, 75, 80]. This spatial proximity is hypothesized to facilitate lipid biosynthesis, such as the conversion of ceramide into sphingomyelin by enzymes localized in *trans*-Golgi compartments [16, 39].

The 3D Morphological Dynamics of the Golgi Apparatus Following Secretory Stimulation: Special Attention to Gonadotropes

Gonadotropes are centrally positioned within the hypothalamic–pituitary–gonadal axis, and their functional activity is primarily regulated by gonadotropin-releasing hormone (GnRH), secreted by GnRH-producing neurons in the hypothalamus, and by gonadal steroids, including androgens in males and estrogens in females [60]. Following castration (i.e., orchiectomy or ovariectomy), the loss of gonadal

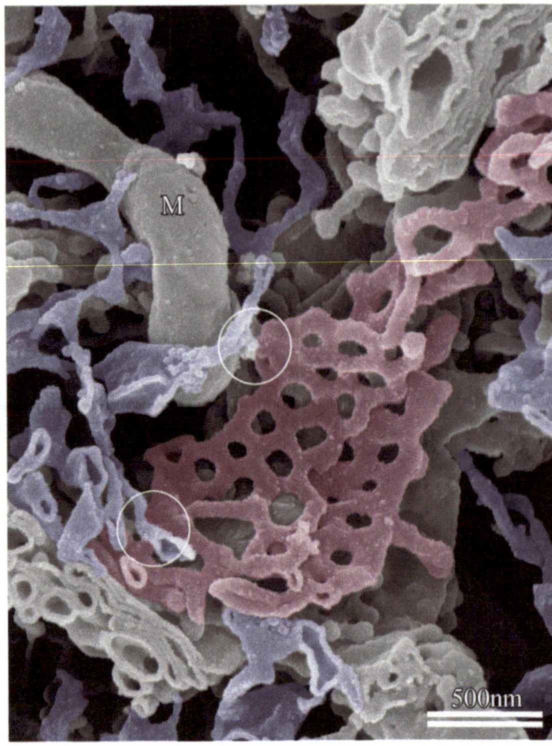

Fig. 1.20 High-magnification image of the *trans*-side of the Golgi apparatus in a spinal ganglion cell of the rat dorsal root ganglion. The *trans*-Golgi network (TGN; red) is composed of anastomosed membranous tubules forming a polygonal network. The endoplasmic reticulum (ER) and TGN are in close contact at several sites (white-outlined circles), although direct connections are not evident. *M* mitochondrion

negative feedback results in sustained stimulation of gonadotropes via elevated endogenous GnRH levels. In control rats, the Golgi apparatus in gonadotropes exhibits a spherical morphology; however, this structure disassembles into several smaller stacks 2 days post-castration (Fig. 1.21). A similar morphological change is observed during mitosis, where the Golgi ribbon undergoes unlinking in the late G_2 phase [127, 128]. Notably, in contrast to the unstacking and vesiculation of the Golgi apparatus that occurs during mitotic metaphase, the 3D ultrastructure of each Golgi stack in gonadotropes remains intact 2 days after castration, and cisternal unlinking is not observed. Disintegration of the Golgi apparatus has also been documented in pathological conditions, including neurodegenerative diseases such as Alzheimer's disease [54], amyotrophic lateral sclerosis [35, 38, 114], and Parkinson's disease [25, 123], as well as in cancer [9, 55, 108]. Thus, the morphology of the Golgi apparatus serves as a cellular indicator of physiological status in vivo. A precise understanding of both the baseline 3D structure of the Golgi apparatus in fully differentiated, unstimulated cells and the patterns of its morphological alterations under pathological conditions is essential for elucidating disease-related mechanisms. The characteristic 3D morphological changes in the Golgi apparatus of gonadotropes are detectable only during the acute phase, specifically 2 days after castration. In addition to Golgi disassembly, tubulovesicular clusters of ER membranes and spherical, enlarged mitochondria are observed. These alterations likely

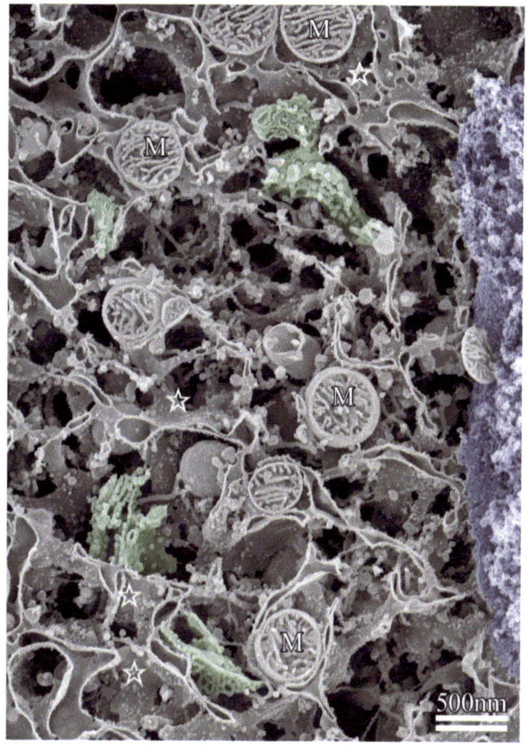

Fig. 1.21 High-magnification image of a pituitary gonadotrope in a rat 2 days post-castration. The Golgi apparatus (green) is dispersed into multiple small stacks. Mitochondria (M) appear swollen and assume a spherical shape. ☆: rough endoplasmic reticulum; Blue: nucleus

represent an acute response to increased secretory demand triggered by castration-induced stimulation. Interestingly, these changes are no longer detectable by the third day post-castration.

In the chronic phase (2 months post-castration), gonadotropes exhibit an extensive anastomosing tubular membrane network with vesicles that encircle the *cis*-most cisterna of the Golgi apparatus (Fig. 1.22). Under physiological conditions, only minimal tubular membrane structures are occasionally observed at this site (Fig. 1.15). The post-castration tubular network resembles schematic representations of *cis*-osmiophilic elements based on TEM observations [101–103]. This structure is identified as the ER–Golgi intermediate compartment (ERGIC; [42, 43]) due to its characteristic morphology and localization between the ER and Golgi apparatus. However, precise morphological identification of the ERGIC is challenging when using SEM alone on osmium-macerated tissues. A promising approach to address this limitation may involve combining the osmium maceration method with immunocytochemistry [62], targeting ERGIC53 (p58), a well-characterized ERGIC marker [109, 110]. The ERGIC, also referred to as vesicular tubular clusters (VTC) due to its tubular and vesicular morphology [4, 5], arises from the clustering of vesicles and tubules budding from ER exit sites (ERES). The distinct ERGIC morphology observed in gonadotropes 2 months post-castration likely reflects disrupted membrane trafficking dynamics, specifically an imbalance between anterograde

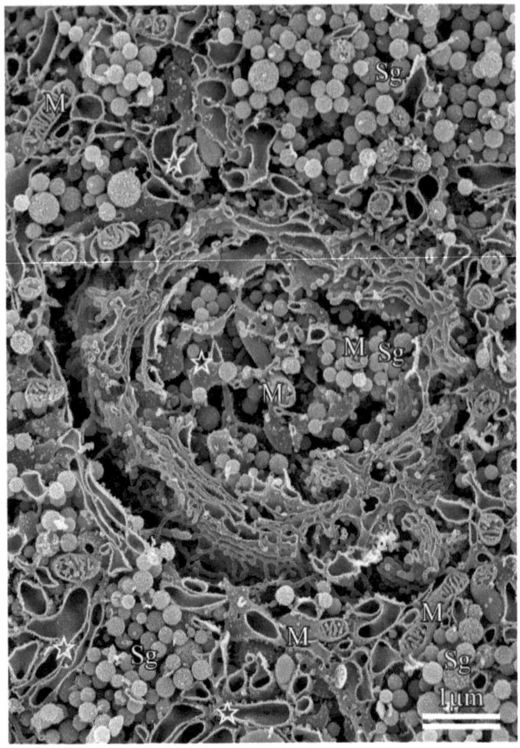

Fig. 1.22 High-magnification image of the Golgi apparatus in a gonadotrope of a rat 2 months post-castration. Tubular structures of the ER–Golgi intermediate compartment (magenta) are distributed along the outer surface of the spherical Golgi apparatus. *M* mitochondria, *Sg* secretory granules; ☆: rough endoplasmic reticulum

and retrograde transport and a sustained increase in secretory activity. Secretory proteins and membranes synthesized in the ER are exported via coat protein II (COPII)-coated vesicles that bud from specialized ER domains, variously referred to as transitional elements (TEs), transitional ER (tER), or ERES [29, 58]. These domains are morphologically defined by ribosome-free zones oriented toward the *cis*-face of the Golgi and exhibit budding (omega-shaped) profiles. Following coat loss, COPII vesicles fuse to form the ERGIC, which consists of tubulovesicular clusters [29, 57, 58]. These structures serve as intermediate stations for anterograde cargo trafficking to the Golgi apparatus. Moreover, the ERGIC also functions as a sorting hub for ER-resident proteins and lipids, which are returned to the ER via retrograde transport mediated by coat protein I (COPI)-coated vesicles [1, 57, 76]. In SEM images of gonadotropes 2 months after castration (Fig. 1.23), ~100 nm vesicles associated with the ERGIC and its tubular network likely represent COPII or COPI vesicles. Budding-like profiles observed on the ERGIC membranes appear consistent with the formation of COPI vesicles or the fusion of COPII vesicles. However, vesicles trafficking between the ER and Golgi is not visible in SEM images of osmium-macerated tissues, consistent with the general invisibility of coated vesicles in the TGN under similar conditions. No direct continuity between the ER and ERGIC is evident in SEM observations of gonadotropes post-castration, consistent with previous findings [5]. Together, these findings delineate distinct

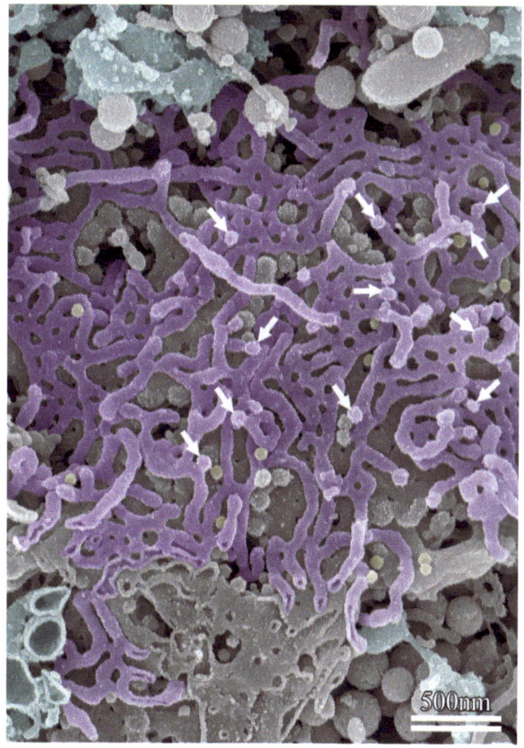

Fig. 1.23 High-magnification SEM image of the ER–Golgi intermediate compartment (ERGIC) in a gonadotrope of a rat 2 months post-castration. The ERGIC (magenta) forms a complex, anastomosing tubular network that extends across and envelops the outer surface of the Golgi apparatus. Numerous small vesicles (yellow) are observed on the ERGIC and the underlying *cis*-Golgi cisterna. Cyan: rough endoplasmic reticulum. Arrowheads indicate vesicle-like profiles evidently budding from the ERGIC

temporal phases of Golgi apparatus reorganization in gonadotropes following secretory stimulation and highlight the relationship between organelle morphology and cellular function.

Elucidating the 3D Morphology of the Golgi Apparatus Using Serial Section SEM and 3D Reconstruction

Serial section SEM combined with 3D reconstruction, commonly referred to as array tomography [48, 63, 79, 106, 124], is a modern 3D-SEM technique, alongside focused ion beam SEM (FIB-SEM) and serial block-face SEM (SBF-SEM). This approach involves acquiring backscattered electron (BSE) images from consecutive ultrathin sections of resin-embedded tissue mounted on solid substrates, such as glass slides or silicon wafers. By segmenting the contours of organelles in serial images and performing 3D reconstruction, surface-rendered models of structures such as the Golgi apparatus, ER, and mitochondria can be generated. Serial section SEM offers several advantages over FIB-SEM and SBF-SEM. (1) Sections mounted on solid substrates can be semi-permanently preserved and re-imaged as needed. (2)

The observation area in the x–y plane can be arbitrarily defined by adjusting the trimming size of the resin block to match the study objective. (3) The z-axis coverage depends on the number of serial sections obtained using an ultramicrotome. (4) Sections on solid substrates can be stained with heavy metals (e.g., uranyl acetate and lead citrate), allowing SEM observation under conditions comparable to those used in TEM. (5) Carbon coating of sections prevents charging artifacts during SEM imaging, thereby reducing image distortion. Despite these advantages, the primary limitation of serial section SEM is its lower z-axis resolution relative to FIB-SEM. Accordingly, selecting the appropriate 3D-SEM technique, serial section SEM, FIB-SEM, or SBF-SEM, requires careful consideration of the morphological features of interest and the specific objectives of the study.

Full 3D reconstructions of the Golgi apparatus in mammalian cells in vivo, specifically epididymal epithelial principal cells, cerebellar Purkinje cells, and pituitary gonadotropes, obtained using serial section SEM and 3D reconstruction, are presented in Fig. 1.24. In epididymal principal cells, the Golgi apparatus displays a spatially complex, convoluted architecture, with an overall form resembling tulip petals (Fig. 1.24a). In Purkinje cells, the Golgi apparatus comprises elongated, ribbon-like cisternae that form a continuous, basket-like structure encircling the nucleus (Fig. 1.24b). This distinctive 3D configuration closely resembles the *apparato reticolare interno* depicted in silver-stained Purkinje cells in Golgi's original study [33]. A similar morphology has been reported in neurons of the dorsal root and trigeminal ganglia using high-voltage EM of thick sections [104]. The ribbon-like configuration of the Golgi apparatus forming a perinuclear network appears to be a characteristic feature of neurons. A comparable net-like structure surrounding the nuclei is also observed in multinucleated osteoclasts using serial section SEM [131]. In contrast, the 3D reconstruction of the Golgi apparatus in pituitary gonadotropes reveals a compact, spherical morphology (Fig. 1.24c). These observations demonstrate marked variability in the overall 3D architecture of the Golgi apparatus across different cell types, reflecting not only cell type–specific functions but also the distinct overall morphology of each differentiated cell.

Although 3D-EM techniques, including serial section TEM and SEM, have consistently shown that the Golgi apparatus in mammalian cells in vivo is typically organized as a single continuous structure [18, 63, 115, 133], recent studies using FIB-SEM and SBF-SEM have reported the presence of multiple, unconnected Golgi stacks in certain cell types, such as megakaryocytes and hepatocytes [19, 53]. The mechanisms underlying the failure of these cells to maintain a continuous Golgi architecture, characteristic of most mammalian cells under physiological conditions, remain unclear. However, comparative analyses of continuous versus discontinuous Golgi morphologies, as revealed by modern 3D-SEM, may provide insight into the mechanisms of Golgi assembly under normal physiological conditions and the potential pathological implications of its disassembly in disease contexts, including neurodegenerative disorders and cancer.

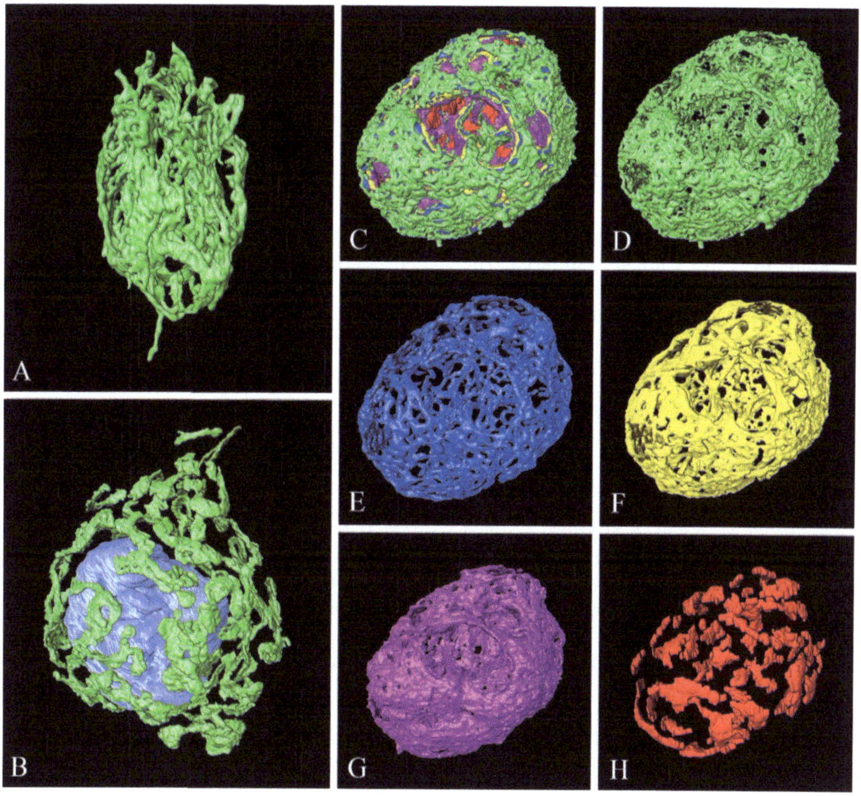

Fig. 1.24 (**a**) Surface-rendered image of the Golgi apparatus (green) in an epithelial principal cell of the rat epididymis. (**b**) Surface-rendered image of the Golgi apparatus (green) and nucleus (blue) in a Purkinje cell of the rat cerebellum. (**c**) Surface-rendered image of the Golgi apparatus in a gonadotrope of the rat pituitary gland. (**d–h**) Segmented components of the Golgi apparatus extracted from (**c**), depicting the *cis*-most cisterna (green; **d**), medial cisternae (blue, yellow; **e**, **f**), *trans* cisterna (magenta; **g**), and the *trans*-Golgi network (red; **h**). [(**a**) reproduced with permission from *Hitachi Scientific Instrument News* [63]; (**b**) reproduced with permission from *Biomedical Research (Tokyo)* [64]]

Conclusion and Future Perspectives

In this chapter, we described the 3D ultrastructure of the Golgi apparatus in various cell types in vivo, including epididymal epithelial principal cells, intestinal goblet cells, pituitary gonadotropes, and spinal ganglion cells, visualized by SEM following the osmium maceration method. This remains the only technique that enables direct 3D observation of organelles at high resolution. The 3D architecture of the Golgi apparatus exhibited marked variability among cell types, likely reflecting functional specializations in vivo. Future advancements in morphological analyses, particularly those integrating correlative light and electron microscopy (CLEM), the osmium maceration method, 3D-SEM with volumetric reconstruction, and

3D-CLEM, are expected to further elucidate the structural complexity and functional diversity of the Golgi apparatus in mammalian cells.

Acknowledgments The studies mentioned in this chapter were supported by a Japan Science and Technology Agency CREST program (JPMJCR21E3 to D.K.) and Grants-in-Aid for Scientific Research (JP19K07261 to D.K., JP24K09993 to D.K.). We thank Editage for editing a draft of this manuscript.

Compliance with Ethical Standards The authors declare no conflicts of interest. All animal experiments were approved by the Committee for Ethics on Animal Experiments of Asahikawa Medical University and conducted in accordance with the university's Animal Experimentation Regulations (approval no. R6-103-2, R6-104).

References

1. Aridor M, Bannykh SI, Rowe T, Balch WE (1995) Sequential coupling between COPII and COPI vesicle coats in endoplasmic reticulum to Golgi transport. J Cell Biol 131:875–893
2. Bainton DF, Farquhar MG (1966) Origin of granules in polymorphonuclear leukocytes: two types derived from opposite faces of the Golgi complex in developing granulocytes. J Cell Biol 28:277–301
3. Baker JR (1949) Further remarks on the Golgi element. Q J Microsc Sci 90:293–307
4. Balch WE, McCaffery JM, Plutner H, Farquhar MG (1994) Vesicular stomatitis virus glycoprotein is sorted and concentrated during export from the endoplasmic reticulum. Cell 76:841–852
5. Bannykh SI, Rowe T, Balch WE (1996) The organization of endoplasmic reticulum export complexes. J Cell Biol 135:19–35
6. Beams HW (1930) Studies on the vacuome and the Golgi apparatus in the acinar cells of the pancreas of the rat. Anat Rec 45:137–161
7. Besta C (1910) Sull'apparato reticolare interno (apparato di Golgi) della cellula nervosa. Anat Anz 36:476–486
8. Broadwell RD, Oliver C (1981) Golgi apparatus, GERL, and secretory granule formation within neurons of the hypothalamo-neurohypophysial system of control and hyperosmotically stressed mice. J Cell Biol 90:474–484
9. Bui S, Mejia I, Díaz B, Wang Y (2021) Adaptation of the Golgi apparatus in cancer cell invasion and metastasis. Front Cell Dev Biol 9:806482
10. Cajal SR (1903) Sobre la existencia de un aparato tubuliforme en el protoplasma de las células nerviosas y epiteliales de la lombriz de tierra. Bol Soc Esp Hist Nat 3:395–398
11. Cajal SR (1914) Algunas variaciones fisiológicas y patológicas del aparato reticular de Golgi. Trab Lab Invest Biol Univ Madrid 11:215–315
12. Clermont Y, Rambourg A, Hermo L (1994) Connections between the various elements of the *cis*- and mid-compartments of the Golgi apparatus of early rat spermatids. Anat Rec 240:469–480
13. Clermont Y, Rambourg A, Hermo L (1995) Trans-Golgi network (TGN) of different cell types: three-dimensional structural characteristics and variability. Anat Rec 242:289–301
14. Dalton AJ, Felix MD (1954) Cytologic and cytochemical characteristics of the Golgi substance of epithelial cells of the epididymis—in situ, in homogenates and after isolation. Am J Anat 94:171–207
15. Dalton AJ, Felix MD (1956) A comparative study of the Golgi complex. J Biophys Biochem Cytol 2(Suppl):79–84

16. De Matteis MA, Luini A (2008) Exiting the Golgi complex. Nat Rev Mol Cell Biol 9:273–284
17. Dröscher A (1998) Camillo Golgi and the discovery of the Golgi apparatus. Histochem Cell Biol 109:425–430
18. Dylewski DP, Haralick RM, Keenan TW (1984) Three-dimensional ultrastructure of the Golgi apparatus in bovine mammary epithelial cells during lactation. J Ultrastruct Res 87:75–85
19. Eckly A, Heijnen H, Pertuy F, Geerts W, Proamer F, Rinckel JY, Léon C, Lanza F, Gachet C (2014) Biogenesis of the demarcation membrane system (DMS) in megakaryocytes. Blood 123:921–930
20. Ehrenreich JH, Bergeron JJM, Siekevitz P, Palade GE (1973) Golgi fractions prepared from rat liver homogenates. I. Isolation procedure and morphological characterization. J Cell Biol 59:45–72
21. Farquhar MG, Hauri HP (1997) Protein sorting and vesicular traffic in the Golgi apparatus. In: Berger EG, Roth J (eds) The Golgi apparatus. Birkhäuser Verlag, Basel, pp 63–129
22. Farquhar MG, Palade GE (1981) The Golgi apparatus (complex)—1954–1981: from artifact to center stage. J Cell Biol 91:77s–103s
23. Farquhar MG, Rinehart JF (1954) Cytologic alterations in the anterior pituitary gland following thyroidectomy: an electron microscope study. Endocrinology 55:857–876
24. Friend DS, Murray MJ (1965) Osmium impregnation of the Golgi apparatus of several cell types of the rat. Am J Anat 117:135–170
25. Fujita Y, Ohama E, Takatama M, Al-Sarraj S, Okamoto K (2006) Fragmentation of Golgi apparatus of nigral neurons with α-synuclein-positive inclusions in patients with Parkinson's disease. Acta Neuropathol 112:261–265
26. Fukudome H, Tanaka K (1986) A method for observing intracellular structures of free cells by scanning electron microscopy. J Microsc 141(2):171–178
27. Gatenby JB (1931) The prozymogen granules ('vacuome') of R. R. Bensley in the Pseudotriton pancreas, and the modern neutral-red cytology. Am J Anat 48:421–477
28. Geren BB, McCulloch D (1951) Development and use of the Minot rotary microtome for thin sectioning. Exp Cell Res 2:97–102
29. Glick BS (2000) Organization of the Golgi apparatus. Curr Opin Cell Biol 12:450–456
30. Glick BS, Luini A (2011) Models for Golgi traffic: a critical assessment. Cold Spring Harb Perspect Biol 3:a005215
31. Glick BS, Nakano A (2009) Membrane traffic within the Golgi apparatus. Annu Rev Cell Dev Biol 25:113–132
32. Golgi C (1873) Sulla struttura della sostanza grigia del cervello. Gazz Med Ital Lombardia 33:244–246
33. Golgi C (1898a) Intorno alla struttura delle cellule nervose. Bollettino della Società Medico-Chirurgica di Pavia 13:1–14
34. Golgi C (1898b) Sulla struttura delle cellule nervose dei gangli spinali. Bollettino della Società Medico-Chirurgica di Pavia 13:53–63
35. Gonatas NK, Stieber A, Mourelatos Z, Chen Y, Gonatas JO, Appel SH, Hays AP, Hickey WF, Hauw JJ (1992) Fragmentation of the Golgi apparatus of motor neurons in amyotrophic lateral sclerosis. Am J Pathol 140:731–737
36. Grassé PP (1957) Ultrastructure, polarité et reproduction de l'appareil de Golgi. C R Acad Sci Hebd Seances Acad Sci D 245:1278–1281
37. Griffiths G, Simons K (1986) The *trans* Golgi network: sorting at the exit site of the Golgi complex. Science 234:438–443
38. Haase G, Rabouille C (2015) Golgi fragmentation in ALS motor neurons. New mechanisms targeting microtubules, tethers, and transport vesicles. Front Neurosci 9:448
39. Hanada K, Kumagai K, Yasuda S, Miura Y, Kawano M, Fukasawa M, Nishijima M (2003) Molecular machinery for non-vesicular trafficking of ceramide. Nature 426:803–809
40. Hand AR, Oliver C (1977a) Relationship between the Golgi apparatus, GERL, and secretory granules in acinar cells of the rat exorbital lacrimal gland. J Cell Biol 74:399–413

41. Hand AR, Oliver C (1977b) Cytochemical studies of GERL and its role in secretory granule formation in exocrine cells. Histochem J 9:375–392
42. Hauri HP, Schweizer A (1992) The endoplasmic reticulum–Golgi intermediate compartment. Curr Opin Cell Biol 4:600–608
43. Hauri HP, Appenzeller C, Andersson H, Kappeler F (2000) Lectins and traffic in the secretory pathway. FEBS Lett 476:32–37
44. Hermo L, Green H, Clermont Y (1991) Golgi apparatus of epithelial principal cells of the epididymal initial segment of the rat: structure, relationship with endoplasmic reticulum, and role in the formation of secretory vesicles. Anat Rec 229:159–176
45. Hermo L, Rambourg A, Clermont Y (1980) Three-dimensional architecture of the Golgi apparatus of rat spermatids. Am J Anat 157:357–373
46. Ho HC, Tang CY, Suarez SS (1999) Three-dimensional structure of the Golgi apparatus in mouse spermatids: a scanning electron microscopic study. Anat Rec 256:189–194
47. Holmgren E (1902) Einige Worte über das „Trophospongium" verschiedener Zellarten. Anat Anz 20:433–440
48. Horstmann H, Körber C, Sätzler K et al (2012) Serial section scanning electron microscopy (S^3EM) on silicon wafers for ultra-structural volume imaging of cells and tissues. PLoS One 7:e35172
49. Ichikawa M, Ichikawa A (1987) The fine structure of sublingual gland acinar cells of the Mongolian gerbil, Meriones unguiculatus, processed by rapid freezing followed by freeze-substitution fixation. Cell Tissue Res 250:305–314
50. Inoue K, Kurosumi K (1977) Cytochemical and three-dimensional studies on Golgi apparatus and GERL of rat anterior pituitary cells by transmission electron microscopy. Cell Struct Funct 2:171–186
51. Inoue K, Kurosumi K (1989) Ultrastructural observation of the trans-Golgi associated plate-like cisterna in the secretory cells of the rat anterior pituitary gland with special reference to the intracisternal skeleton. Anat Rec 225:272–278
52. Isola R, Riva A, Loy F, Loffredo F, Testa Riva F (2009) Fine structural analysis of Candida albicans by high-resolution scanning electron microscopy using a modified osmium maceration method. J Microbiol 47:260–264
53. Johkura K, Usuda N, Tanaka Y, Fukasawa M, Murata K, Noda T, Ohno N (2022) Whole-cell observation of ZIO-stained Golgi apparatus in rat hepatocytes with serial block-face scanning electron microscope, SBF-SEM. Microscopy 71:262–270
54. Joshi G, Bekier ME II, Wang Y (2015) Golgi fragmentation in Alzheimer's disease. Front Neurosci 9:340
55. Kellokumpu S, Sormunen R, Kellokumpu I (2002) Abnormal glycosylation and altered Golgi structure in colorectal cancer: dependence on intra-Golgi pH. FEBS Lett 516:217–224
56. Kessel RG, Beams HW (1965) An unusual configuration of the Golgi complex in pigment-producing "test" cells of the ovary of the tunicate, Styela. J Cell Biol 25:55–67
57. Klumperman J (2000) Transport between ER and Golgi. Curr Opin Cell Biol 12:445–449
58. Klumperman J (2011) Architecture of the mammalian Golgi. Cold Spring Harb Perspect Biol 3:a005181
59. Koga D, Ushiki T (2006) Three-dimensional ultrastructure of the Golgi apparatus in different cells: high-resolution scanning electron microscopy of osmium-macerated tissues. Arch Histol Cytol 69:357–374
60. Koga D, Bochimoto H, Kusumi S, Ushiki T, Watanabe T (2017a) Changes in the three-dimensional ultrastructure of membranous organelles in male rat pituitary gonadotropes after castration. Biomed Res (Tokyo) 38:1–18
61. Koga D, Bochimoto H, Watanabe T, Ushiki T (2016a) Backscattered electron image of osmium-impregnated/macerated tissues as a novel technique for identifying the *cis*-face of the Golgi apparatus by high-resolution scanning electron microscopy. J Microsc 263:87–96

62. Koga D, Kusumi S, Bochimoto H, Watanabe T, Ushiki T (2015) Correlative light and scanning electron microscopy for observing the three-dimensional ultrastructure of membranous cell organelles in relation to their molecular components. J Histochem Cytochem 63:968–979
63. Koga D, Kusumi S, Ushiki T (2016b) Three-dimensional shape of the Golgi apparatus in different cell types: serial section scanning electron microscopy of the osmium-impregnated Golgi apparatus. Microscopy 65:145–157
64. Koga D, Kusumi S, Ushiki T, Watanabe T (2017b) Integrative method for three-dimensional imaging of the entire Golgi apparatus by combining thiamine pyrophosphatase cytochemistry and array tomography using backscattered electron-mode scanning electron microscopy. Biomed Res (Tokyo) 38:285–296
65. Koga D, Kusumi S, Yagi H, Kato K (2024) Three-dimensional analysis of the intracellular architecture by scanning electron microscopy. Microscopy 73:215–225
66. Koga D, Nakajima M, Ushiki T (2012) A useful method for observing intracellular structures of free and cultured cells by scanning electron microscopy. J Electron Microsc 61:105–111
67. Koga D, Ushiki T, Watanabe T (2017c) Novel scanning electron microscopy methods for analyzing the 3D structure of the Golgi apparatus. Anat Sci Int 92:37–49
68. Kolatchev A (1916) Recherches cytologiques sur les cellules nerveuses des mollusques. Arch Russ Anat Hist Embryol 1:383–423
69. Kopsch F (1902) Die Darstellung des Binnennetzes in spinalen Ganglienzellen und anderen Körperzellen mittels Osmiumssäure. Sitzungsber KPreuss Akad Wiss 40:929–932
70. Ladinsky MS, Mastronarde DN, McIntosh JR, Howell KE, Staehelin LA (1999) Golgi structure in three dimensions: functional insights from the normal rat kidney cell. J Cell Biol 144:1135–1149
71. Lea PJ, Hollenberg MJ (1989) Mitochondrial structure revealed by high-resolution scanning electron microscopy. Am J Anat 184:245–257
72. Lea PJ, Temkin RJ, Freeman KB, Mitchell GA, Robinson BH (1994) Variations in mitochondrial ultrastructure and dynamics observed by high resolution scanning electron microscopy (HRSEM). Microsc Res Tech 27:269–277
73. Marsh BJ, Mastronarde DN, Buttle KF, Howell KE, McIntosh JR (2001) Organellar relationships in the Golgi region of the pancreatic beta cell line, HIT-T15, visualized by high resolution electron tomography. Proc Natl Acad Sci USA 98:2399–2406
74. Marsh BJ, Volkmann N, McIntosh JR, Howell KE (2004) Direct continuities between cisternae at different levels of the Golgi complex in glucose-stimulated mouse islet beta cells. Proc Natl Acad Sci USA 101:5565–5570
75. Martínez-Martínez N, Martínez-Alonso E, Tomás M, Neumüller J, Pavelka M, Martínez-Menárguez JA (2017) A new insight into the three-dimensional architecture of the Golgi complex: characterization of unusual structures in epididymal principal cells. PLoS One 12:e0185557
76. Martínez-Menárguez JA, Geuze HJ, Slot JW, Klumperman J (1999) Vesicular tubular clusters between the ER and Golgi mediate concentration of soluble secretory proteins by exclusion from COPI-coated vesicles. Cell 98:81–90
77. Mazzarello P, Garbarino C, Calligaro A (2009) How Camillo Golgi became "the Golgi". FEBS Lett 583:3732–3737
78. Mellman I, Simons K (1992) The Golgi complex: in vitro veritas? Cell 68:829–840
79. Micheva KD, Smith SJ (2007) Array tomography: a new tool for imaging the molecular architecture and ultrastructure of neural circuits. Neuron 55:25–36
80. Mogelsvang S, Marsh BJ, Ladinsky MS, Howell KE (2004) Predicting function from structure: 3D structure studies of the mammalian Golgi complex. Traffic 5:338–345
81. Mollenhauer HH, Whaley WG (1963) An observation on the functioning of the Golgi apparatus. J Cell Biol 18:222–225
82. Morre DJ, Mollenhauer HH (1964) Isolation of the Golgi apparatus from plant cells. J Cell Biol 23:295–305
83. Nakano A, Luini A (2010) Passage through the Golgi. Curr Opin Cell Biol 22:471–478

84. Neutra M, Leblond CP (1966) Synthesis of the carbohydrate of mucus in the Golgi complex as shown by electron microscope radioautography of goblet cells from rats injected with glucose-H^3. J Cell Biol 30:119–136
85. Novikoff AB, Goldfischer S (1961) Nucleosidediphosphatase activity in the Golgi apparatus and its usefulness for cytological studies. Proc Natl Acad Sci USA 47:802–810
86. Novikoff AB, Novikoff PM (1977) Cytochemical contributions to differentiating GERL from the Golgi apparatus. Histochem J 9:525–551
87. Novikoff PM, Novikoff AB, Quintana N, Hauw JJ (1971) Golgi apparatus, GERL, and lysosomes of neurons in rat dorsal root ganglia, studied by thick section and thin section cytochemistry. J Cell Biol 50:859–886
88. Novikoff AB (1964) GERL, its form and function in neurons of rat spinal ganglia. Biol Bull 127:358
89. Nusbaum J (1913) Über den sogenannten inneren Golgischen Netzapparat und sein Verhältnis zu den Mitochondrien, Chromidien und andern Zellstrukturen im Tierreich. Arch Zellforsch 10:359–367
90. Ogata T, Yamasaki Y (1997) Ultra-high-resolution scanning electron microscopy of mitochondria and sarcoplasmic reticulum arrangement in human red, white, and intermediate muscle fibers. Anat Rec 248:214–223
91. Palade GE (1952a) A study of fixation for electron microscopy. J Exp Med 95:285–298
92. Palade GE (1952b) The fine structure of mitochondria. Anat Rec 114:427–451
93. Palade GE, Claude A (1949) The nature of the Golgi apparatus. II. Identification of the Golgi apparatus with a complex of myelin figures. J Morphol 85:71–111
94. Penfield WG (1921) The Golgi apparatus and its relationship to Holmgren's trophospongium in nerve cells. Comparison during retispersion. J Comp Neurol 33:57–65
95. Porter KR, Claude A, Fullam EF (1945) A study of tissue culture cells by electron microscopy: methods and preliminary observations. J Exp Med 81:233–246
96. Rambourg A, Clermont Y (1986) Tridimensional structure of the Golgi apparatus in type A ganglion cells of the rat. Am J Anat 176:393–409
97. Rambourg A, Clermont Y (1990) Three-dimensional electron microscopy: structure of the Golgi apparatus. Eur J Cell Biol 51:189–200
98. Rambourg A, Clermont Y (1992) Three-dimensional structure of cytidine monophosphatase reactive trans-Golgi elements in spinal ganglion cells of the rat. Anat Rec 232:25–35
99. Rambourg A, Clermont Y (1997) Three-dimensional structure of the Golgi apparatus in mammalian cells. In: Berger EG, Roth J (eds) The Golgi apparatus. Birkhäuser Verlag, Basel, pp 37–61
100. Rambourg A, Clermont Y, Beaudet A (1983) Ultrastructural features of six types of neurons in rat dorsal root ganglia. J Neurocytol 12:47–66
101. Rambourg A, Clermont Y, Chretien M, Olivier L (1992) Formation of secretory granules in the Golgi apparatus of prolactin cells in the rat pituitary gland: a stereoscopic study. Anat Rec 232:169–179
102. Rambourg A, Clermont Y, Hermo L (1979) Three-dimensional architecture of the Golgi apparatus in Sertoli cells of the rat. Am J Anat 154:455–476
103. Rambourg A, Clermont Y, Hermo L (1988) Formation of secretion granules in the Golgi apparatus of pancreatic acinar cells of the rat. Am J Anat 183:187–199
104. Rambourg A, Clermont Y, Hermo L, Segretain D (1987) Tridimensional architecture of the Golgi apparatus and its components in mucous cells of Brunner's glands of the mouse. Am J Anat 179:95–107
105. Rambourg A, Clermont Y, Marraud A (1974) Three-dimensional structure of the osmium-impregnated Golgi apparatus as seen in the high voltage electron microscope. Am J Anat 140:27–45
106. Rambourg A, Segretain D, Clermont Y (1984) Tridimensional architecture of the Golgi apparatus in the atrial muscle cell of the rat. Am J Anat 170:163–179

107. Reichelt M, Joubert L, Perrino J, Koh AL, Phanwar I, Arvin AM (2012) 3D reconstruction of VZV infected cell nuclei and PML nuclear cages by serial section array scanning electron microscopy and electron tomography. PLoS Pathog 8:e1002740
108. Riva A, Tandler B, Ushiki T, Usai P, Isola R, Conti G, Loy F, Hoppel CL (2010) Mitochondria of human Leydig cells as seen by high resolution scanning electron microscopy. Arch Histol Cytol 73:37–44
109. Sano T, Kovacs K, Asa SL, Yamada S, Sanno N, Yokoyama S, Takami H (2002) Pituitary adenoma with "honeycomb Golgi" appearance showing a phenotypic change at recurrence from clinically nonfunctioning to typical Cushing disease. Endocr Pathol 13:125–130
110. Saraste J, Palade GE, Farquhar MG (1987) Antibodies to rat pancreas Golgi subfractions: identification of a 58-kD *cis*-Golgi protein. J Cell Biol 105:2021–2029
111. Schweizer A, Fransen JAM, Bächi T, Ginsel L, Hauri HP (1988) Identification, by a monoclonal antibody, of a 53-kD protein associated with a tubulo-vesicular compartment at the *cis*-side of the Golgi apparatus. J Cell Biol 107:1643–1653
112. Sjöstrand FS (1953) Electron microscopy of mitochondria and cytoplasmic double membranes. Nature 171:30–32
113. Sjöstrand FS, Hanzon V (1954) Ultrastructure of Golgi apparatus of exocrine cells of mouse pancreas. Exp Cell Res 7:415–429
114. Smith CE (1980) Ultrastructural localization of nicotinamide adenine dinucleotide phosphatase (NADPase) activity to the intermediate saccules of the Golgi apparatus in rat incisor ameloblasts. J Histochem Cytochem 28:16–26
115. Sundaramoorthy V, Sultana JM, Atkin JD (2015) Golgi fragmentation in amyotrophic lateral sclerosis, an overview of possible triggers and consequences. Front Neurosci 9:400
116. Tamaki H, Yamashina S (1997) Three-dimensional dynamics of the Golgi apparatus in mitotic parotid acinar cells: computer-aided reconstruction from cytochemically-marked ultrathin serial sections. Acta Histochem Cytochem 30:643–651
117. Tanaka K, Fukudome H (1991) Three-dimensional organization of the Golgi complex observed by scanning electron microscopy. J Electron Microsc Tech 17:15–23
118. Tanaka K, Mitsushima A (1984) A preparation method for observing intracellular structures by scanning electron microscopy. J Microsc 133:213–222
119. Tanaka K, Naguro T (1981) High resolution scanning electron microscopy of cell organelles by a new specimen preparation method. Biomed Res 2(Suppl):63–70
120. Tanaka K, Iino A, Naguro T (1974) Styrene resin cracking method for observing biological materials by scanning electron microscopy. J Electron Microsc 23:313–315
121. Tanaka K, Mitsushima A, Fukudome H, Kashima Y (1986) Three-dimensional architecture of the Golgi complex observed by high resolution scanning electron microscopy. J Submicrosc Cytol 18:1–9
122. Tokunaga J, Edanaga M, Fujita T, Adachi K (1974) Freeze cracking of scanning electron microscope specimens. A study of the kidney and spleen. Arch Histol Jpn 37:165–182
123. Tokuyasu KT (1989) Use of poly(vinylpyrrolidone) and poly(vinyl alcohol) for cryoultramicrotomy. Histochem J 21:163–171
124. Tomás M, Martínez-Alonso E, Martínez-Martínez N, Cara-Esteban M, Martínez-Menárguez JA (2021) Fragmentation of the Golgi complex of dopaminergic neurons in human substantia nigra: new cytopathological findings in Parkinson's disease. Histol Histopathol 36:47–60
125. Wacker I, Schroeder RR (2013) Array tomography. J Microsc 252:93–99
126. Watanabe T, Bochimoto H, Koga D, Hosaka M, Ushiki T (2014) Functional implications of the Golgi and microtubular network in gonadotropes. Mol Cell Endocrinol 385:88–96
127. Watanabe T, Sakai Y, Koga D, Bochimoto H, Hira Y, Hosaka M, Ushiki T (2012) A unique ball-shaped Golgi apparatus in the rat pituitary gonadotrope: its functional implications in relation to the arrangement of the microtubule network. J Histochem Cytochem 60:588–602
128. Wei JH, Seemann J (2010) Unraveling the Golgi ribbon. Traffic 11:1391–1400
129. Wei JH, Seemann J (2017) Golgi ribbon disassembly during mitosis, differentiation and disease progression. Curr Opin Cell Biol 47:43–51

130. Weigl R (1910) Zur Kenntnis des Golgi-Kopschschen Apparats in den Nervenzellen verschiedener Tiergruppen. Verh VIII Int Zool Kongr Graz 589–595
131. Yamamoto T, Hasegawa T, Hongo H, Amizuka N (2019) Three-dimensional morphology of the Golgi apparatus in osteoclasts: NADPase and arylsulfatase cytochemistry, and scanning electron microscopy using osmium maceration. Microscopy 68:243–253
132. Yamamoto T, Hasegawa T, Hongo H, Amizuka N (2021) Three-dimensional reconstruction of the Golgi apparatus in osteoclasts by a combination of NADPase cytochemistry and serial section scanning electron microscopy. Histochem Cell Biol 156:503–508
133. Yamasaki Y, Furuya Y, Araki K, Matsuura K, Kobayashi M, Ogata T (1997) Ultra-high-resolution scanning electron microscopy of the sarcoplasmic reticulum of the rat atrial myocardial cells. Anat Rec 248:70–75
134. Yamashina S (1995) Dynamic structure and function of Golgi apparatus in the salivary acinar cells. J Electron Microsc 44:124–134

Chapter 2
Golgi Units as Modules in the Ribbon

Akihiro Harada

Abstract Using super-resolution microscopies, SCLIM and STORM, and CRISPR/Cas9 knockin technology, we show that the Golgi complex is assembled by a number of small "Golgi units" that are 1–3 μm in diameter. Each Golgi unit can be considered a functional and morphological module because it contains all sets of glycosylation enzymes that occupy small domains named "zones." Golgi units change shape dynamically. They attach, detach, fuse, and separate from each other. The zones of glycosylation enzymes rapidly move near the rim of the unit. However, the behaviors of N- and O-linked glycosylation enzyme zones differ from those of glycosaminoglycan (GAG)-synthesizing enzyme zones in localization and dynamics. Since depletion of giantin dissociates the Golgi units, it is responsible for their attachment. Giantin depletion also prevents the movement of GAG-synthesizing enzymes between units, leading to incomplete GAG synthesis. Here, we present the model of the Golgi and its glycosylation enzymes that can explain a number of characteristics of the glycosylation processes.

Keywords Golgi apparatus · Glycosylation enzymes · CRISPR/Cas9 · Knockin

Introduction

In the Golgi, different proteins acquire distinct patterns of glycosylation. Namely, N-type, O-type, or mucin-type glycans, and glycosaminoglycans (GAGs) are covalently linked to proteins, and these sugar chains are important in determining the structure, destination, and function of the glycosylated proteins [1]. However, little is known about where the glycosylation enzymes localize within a single cisterna compared with their polarized *cis* to *trans* localization across the stacks. Because of

A. Harada (✉)
Department of Cell Biology, Graduate School of Medicine, The University of Osaka, Osaka, Japan
e-mail: aharada@acb.med.osaka-u.ac.jp

A. Nakano, J. Saraste (eds.), *The Golgi Network, Volume I*, Subcellular Biochemistry 110, https://doi.org/10.1007/978-3-032-06936-8_2

the lack of suitable antibodies for morphological analyses, localization and dynamics of glycosylation enzymes have been mainly investigated by transfection of the cDNAs coding for the enzymes of interest [2]. However, an overexpressed protein will occupy a larger area than its endogenous counterpart and show altered dynamics. For these reasons, the precise endogenous localization and dynamics of glycosylation enzymes within one Golgi cisternae have not been examined thus far. We can knock in exogenous tags into the desired genes of any species of organisms using CRISPR/Cas9 technology to monitor the localization and dynamics of the cognate endogenous proteins [3]. In parallel, high-speed super-resolution microscopy super-resolution confocal live imaging microscopy (SCLIM) developed by Nakano's group [4, 5] made it possible to observe the localization and dynamics of proteins and lipids at the nanoscale level in three dimensions. In these studies [3], summarized below, we demonstrated the basic structure and dynamics of the Golgi and its glycosylation enzymes using various super-resolution microscopies and immunoelectron microscopy.

Characterization of Golgi Units and Glycosylation Enzyme Zones

We mainly used three enzymes at the early stage of glycosylation in the Golgi, which are reported to localize in the *cis* to medial Golgi cisternae. These were the N-glycosylation enzyme MGAT1 (alpha-1,3-mannosyl-glycoprotein 2-beta-N-acetylglucosaminyltransferase 1,2), the O-glycosylation enzyme GALNT6 (polypeptide N-acetylgalactosaminyltransferase 6), and the GAG-synthesizing enzyme, XYLT2 (xylosyltransferase 2). In addition, we used two enzymes that are necessary at later stages of glycosylation. One is β4GalT1 (beta4-galactosyl transferase 1), an enzyme synthesizing polylactosamine on both N- and O-glycans. The other is NDST1 (N-deacetylase and N-sulfotransferase 1). To determine the endogenous localization of glycosylation enzymes, we knocked in various tags (9x Myc, 3x PA, 3x V5, Halo, and SNAP tags) at the C-terminus of the glycosylation enzymes (Fig. 2.1a) [3]. The signal of glycosylation enzymes obtained by knockin was concentrated and punctate within the Golgi, whereas the signal obtained by transfection localized more diffusely and throughout the entire Golgi, indicating that transfection is not a suitable means to assess the endogenous localizations of glycosylation enzymes. To assess the endogenous localizations of the enzymes in the Golgi in

Fig 2.1 (continued) (**b**) Comparison of GALNT6 and MGAT1 zones (top panels) and GALNT6 and XYLT2 zones (bottom panels) by SCLIM. Scale bar: 1 μm. (**c**) Comparison of GALNT6 zone and giantin (top) and XYLT2 zone and giantin (bottom) by STORM. Scale bar: 0.5 μm. (**d**) Visualization of the Golgi ribbon composed of the Golgi units stained by giantin using Airyscan microscopy. Scale bar: 2 μm. (**e**) Schematic representation of Golgi organization. Top: the Golgi ribbon composed of the Golgi units. Bottom: dynamics of the Golgi units

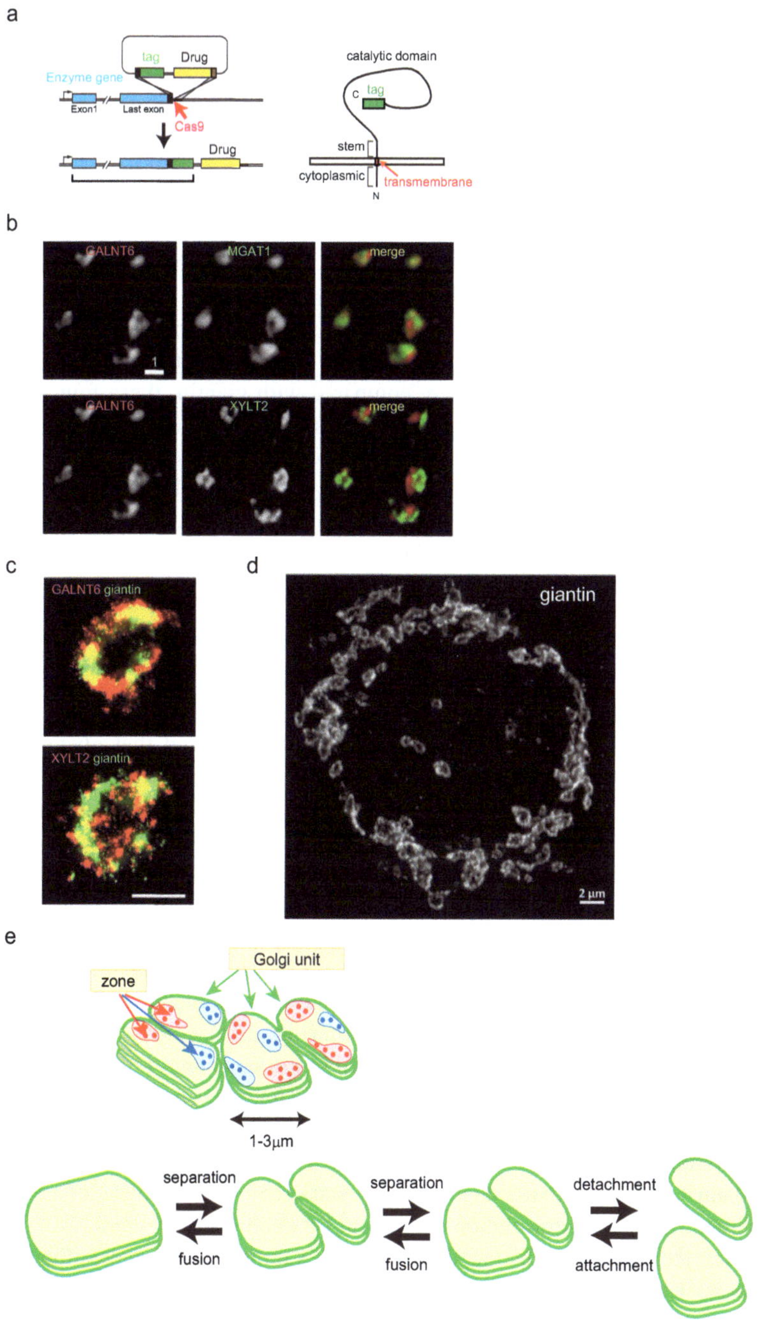

Fig. 2.1 (**a**) Strategy of tag knockin. Left: (top) wildtype allele of target gene and knockin construct that have a tag and a drug resistance gene (bottom) knocked-in allele. Right: model of knocked-in glycosylation enzyme, containing a tag inserted at the C-terminal end of the protein.

more detail, we dispersed the Golgi by nocodazole, which dissociates the large Golgi apparatus into smaller Golgi discs, or "mini-stacks," surrounded by giantin or other golgins (golgin-84 and TMF1). We referred to these Golgi discs as the Golgi units because they have a relatively uniform diameter and contain all sets of glycosylation enzymes examined, as described below. Then, we compared the localization of GALNT6, MGAT1, and XYLT2 in the same triple KI cell line (GALNT6-9 xMyc+MGAT1-3xPA+XYLT2-3xV5). All three enzymes were found to occupy discrete small areas within a Golgi unit. The correlation coefficient of GALNT6 and MGAT1 was significantly higher than that of GALNT6 and XYLT2 in nocodazole-treated and nocodazole-untreated cells. A high correlation coefficient between GALNT6 and MGAT1 indicates that they are almost completely colocalized. However, GALNT6 and XYLT2 overlap to a lesser extent (Fig. 2.1b). When GALNT6, MGAT1, and XYLT2 were compared with giantin, these four proteins localized almost to the same extent in side views, which indicates their close colocalization to the same cisternae. In *en face* views, although GALNT6 and MGAT1 colocalized extensively, GALNT6 and XYLT2 did not. As all these glycosylation enzymes occupy small areas, we designated these small areas as zones of glycosylation enzymes, as previously described [5].

Area and Localization of the Zones in Golgi Units

Previous results using Airyscan confocal microscopy showed that glycosylation enzymes are mainly localized at the center of the small Golgi units [2]. Thus, to investigate their localization at higher resolution, we utilized localization microscopy, STORM. We used giantin as a reference for the Golgi rims to align the Golgi units. Subsequently, we stained one of the glycosylation enzymes to measure the correlation coefficient between the enzyme and giantin to determine their localization relative to the rim of the Golgi unit. In nocodazole-treated cells, GALNT6 and MGAT1 appeared to localize diffusely and colocalize significantly with giantin. In contrast, XYLT2 localized as small puncta, some of which colocalized with giantin, while others were distributed away from giantin (Fig. 2.1c).

Additionally, all the enzymes were detected in the same unit, although their signals varied from one unit to another. Based on this finding, one Golgi unit appears to be a minimal functional module. When we calculated the areas based on the images obtained by STORM using image processing, the area of GALNT6 zones (12,460 nm^2) was larger than that of XYLT2 zones (6379 nm^2). To exclude the possibility that the reduced correlation coefficient between GALNT6 and XYLT2 was due to their localization to distinct cisternae, we examined their localization further by immunoelectron microscopy. Using this method, XYLT2 signals were identified only in the electron-dense GALNT6-positive cisternae, rather than GALNT6-negative cisternae, confirming the results obtained by SCLIM.

We also compared the localizations of the late-stage glycosylation enzymes by SCLIM. When we compared the localization of β4GalT1, a late-acting enzyme for both N- and O-glycosylation, and NDST1, a late-stage enzyme for GAG synthesis, β4GalT1 was found to localize closer to the center of the unit than NDST1 [3].

Dynamics of Units and Zones in the Intact Golgi Ribbon

When we examined the interphase Golgi, we observed that it consists of an assembly of units marked by giantin with diameters of 1–3 μm (Fig. 2.1d). The two glycosylation enzymes GALNT6 and XYLT2 were mainly localized near the rims of the units and both were contained in every unit. The diameter of these units was almost the same whether they were clustered or not. These observations indicate that the units of the cluster correspond to the Golgi units as well. They are usually connected side by side with each other or connected by tubules (Fig. 2.1e). We could verify that the Golgi is composed of units by examining other human cell lines, such as HeLa and T47D, and murine primary cells, such as mouse embryonic fibroblasts and neurons. Therefore, the Golgi is composed of Golgi units regardless of species or the type of tissue [3].

As we transfected cells with mNeonGreen-giantin and monitored the dynamics of the Golgi units by Airyscan confocal microscopy, we observed their active movements and deformation. A relatively large unit had a growing partition that ended up separating the unit into two smaller units. Subsequently, a newly generated small unit frequently detached and moved away from the other unit. Conversely, a unit attached to another unit and finally fused to become a larger unit (Fig. 2.1e). At higher temporal and spatial 3D resolutions by SCLIM, a partition appeared and disappeared repeatedly before the two units finally fused.

To determine the dynamics of glycosylation enzymes, we generated Halo-tag KI cell lines and observed enzyme dynamics by live imaging. GALNT6 was mainly localized and moved around the rims of the units very rapidly. GALNT6 did not appear to move out of the isolated units. We observed similar rapid movements for XYLT2 and NDST1 within the units. Finally, we generated and observed a double knockin clone in which the produced XYLT2 and GALNT6 were both tagged. The zones of XYLT2 and GALNT6 moved rapidly and in a mutually exclusive manner in the individual units confirming the result obtained with fixed cells.

To examine the movement of the enzymes between units, we performed fluorescence recovery after photobleaching (FRAP) analysis. The GALNT6 signal recovered very slowly after photobleaching either in the isolated or clustered units. This indicates that the movement of GALNT6 is restricted and occurs within individual units regardless of whether they form clusters. The signal of XYLT2 in an isolated unit recovered modestly, even less than that of GALNT6. However, the signal in clustered units recovered very rapidly. Therefore, the movement of XYLT2 is restricted within one unit only when the unit is isolated, whereas XYLT2 can move between units when they are in contact.

To determine their contribution to the behavior of the Golgi units, we knocked down various golgins in HeLa cells. Only giantin knockdown resulted in the dissociation of the Golgi units without causing considerable deformation of the units. The ensuing dissociation of units was rescued by introducing mNeonGreen-Giantin.

The Assembly of Golgi Units Optimizes GAG Synthesis

To determine the effect of giantin depletion on the localization of glycosylation enzymes, we measured the signals of GALNT6, MGAT1, and XYLT2 per Golgi unit in control and giantin knockdown cells. The intensity of the XYLT2 signal was very weak in some units, whereas in other units it was very high. This phenomenon was not observed in the case of signals for GALNT6. This implies that if the movement of XYLT2 was blocked by dissociation of the units, the difference in the amount of XYLT2 signal per Golgi unit was likely to increase. To determine the physiological consequence of giantin knockdown or knockout, we measured the formation of GAG chains between control and giantin-depleted Caco2 cells. In giantin knockdown cells, the amount of mature GAG was reduced compared to that in control cells, which is likely to reflect the insufficient amount of XYLT2 in part of the Golgi units. Based on these results, giantin is necessary for Golgi units to attach and form clusters. This allows XYLT2 to move between the units, averaging the amount of GAG-synthesizing enzymes in each unit to optimize normal GAG synthesis.

Summary and Conclusions

Our studies [3] provide structural basis to better understand the function of the Golgi complex. They showed that, in all types of cells examined, the Golgi complex is composed of units measuring 1–3 μm in diameter surrounded by giantin or golgin-84. Furthermore, by tag knockin, we succeeded in visualizing glycosylation enzymes that accurately reflect the localization and dynamics of their endogenous counterparts. From observations utilizing super-resolution microscopies, we discovered that GALNT6, MGAT1, and XYLT2 form small domains, which in our previous paper were named "zones" [5]. We also discovered that all these enzymes were included in one unit, confirming that the Golgi unit is a minimal functional module of the Golgi, and represents a distinct entity from a morphologically defined "mini stack." We showed similar localization of GALNT6 and MGAT1 zones in the Golgi units, whereas the localization of XYLT2 zone was distinct from those of GALNT6 and MGAT1. In addition, the XYLT2 zone is smaller than the GALNT6 zone based on quantitative analysis of the data obtained by STORM. It has been proposed that glycosylation enzymes are clustered by oligomerization or kin recognition [6]. However, as this concept is insufficient to explain distinct behaviors of the different zones, further investigations addressing the generation mechanisms of the zones will be required.

In addition to the differences in localization and area, the GALNT6 and XYLT2 zones showed different recovery rates after photobleaching. By FRAP experiments, the GALNT6 zone does not seem to move between units even when units are connected, whereas the XYLT2 zone can move rapidly. This may be explained by the fact that the XYLT2 zone is smaller than the GALNT6 zone. If a passage between two units is larger than the XYLT2 zone and smaller than the GALNT6 zone, only the XYLT2 zone can pass through. From electron microscopic observations, the Golgi was found to be composed of stacks connected by tubules side by side [7]. These tubules between Golgi cisternae may function as such passages that only allow the XYLT2 zone to go through. In addition, we demonstrated the dynamics of the units and zones by live imaging. We observed that the zones moved rapidly near the rim of the Golgi unit and that the GALNT6 zones and the XYLT2 zones moved largely in mutually exclusive manner.

Although the early glycosylation enzymes form zones around the rim of a single *cis* or medial cisterna, zones of late glycosylation enzymes, particularly β4GalT1, occupy a more central position within a *trans* cisterna. β4GalT1 is responsible for the synthesis of both N-linked and O-linked oligosaccharides in many glycoproteins, as well as the carbohydrate moieties of glycolipids. It might be convenient for this enzyme to localize at the center of the *trans* cisterna to modify a number of substrates that come from various positions in *cis*-medial cisternae. In contrast, as the GAG-synthesizing late-stage enzyme NDST1 is responsible only for the synthesis of GAGs, its localization near the rim could allow efficient recycling back to the previous cisterna in COPI vesicles.

There are several advantages in considering the tiny Golgi units as functional modules when we explain the function of the Golgi complex. The Golgi complex must perform different types of glycosylation reactions and other modifications very rapidly. If all sets of rapidly moving enzymes localize in one tiny unit, the chance of substrates entering the zone of their target enzymes is greatly increased even within a short period of time.

We observed greater divergence in the amount of XYLT2 among the isolated Golgi units and the reduced synthesis of GAG in the absence of giantin. This phenomenon is likely explained by the following logic. Giantin deficiency causes units to dissociate, which will prevent interunit movements of XYLT2. As the amounts of enzymes between units as a result will deviate, some units will contain insufficient amounts of enzymes for GAG synthesis. Interestingly, giantin knockout mice or rats were previously shown to develop defects in GAG or proteoglycan-rich tissues, such as bones or cartilages [8, 9]. Our observation is instructive to explain the origin of these phenotypes.

We found that giantin is essential for the attachment of the units. Previously, other proteins such as GM130 and GRASPs have also been known to link the Golgi units [10]. This indicates that the redundancy of these proteins in linking the Golgi units, particularly between GRASPs and giantin, ensures that the connection will not be broken easily.

In summary, we have discovered morphological characteristics of the Golgi and its glycosylation enzymes, which can help explain its various functions. Although we showed that giantin is important for the attachment of Golgi units, most of the molecular mechanisms underlying Golgi dynamics, such as unit detachment, fusion, and separation, remain unclear (Fig. 2.1e). We are currently trying to address these questions to obtain further insight into the molecular basis of the Golgi structure and function.

References

1. Varki A, Cummings RD, Esko JD, Stanley P, Hart GW, Aebi M, Mohnen D, Kinoshita T, Packer NH, Prestegard JH, Schnaar RL, Seeberger PH (2022) Essentials of glycobiology, 4th edition Ch. 4. Cold Spring Harbor Laboratory Press, New York
2. Tie HC, Ludwig A, Sandin S, Lu L (2018) The spatial separation of processing and transport functions to the interior and periphery of the Golgi stack. elife 7:e41301
3. Harada A, Kunii M, Kurokawa K, Sumi T, Kanda S, Zhang Y, Nadanaka S, Hirosawa KM, Tokunaga K, Tojima T, Taniguchi M, Moriwaki K, Yoshimura S, Yamamoto-Hino M, Goto S, Katagiri T, Kume S, Hayashi-Nishino M, Nakano M, Miyoshi E, Suzuki KGN, Kitagawa H, Nakano A (2024) Dynamic movement of the Golgi unit and its glycosylation enzyme zones. Nat Commun 15:4514
4. Kurokawa K, Ishii M, Suda Y, Ichihara A, Nakano A (2013) Live cell visualization of Golgi membrane dynamics by super-resolution confocal live imaging microscopy. Methods Cell Biol 118:235–242
5. Shimizu Y, Takagi J, Ito E, Ito Y, Ebine K, Komatsu Y, Goto Y, Sato M, Toyooka K, Ueda T, Kurokawa K, Uemura T, Nakano A (2021) Cargo sorting zones in the trans-Golgi network visualized by super-resolution confocal live imaging microscopy in plants. Nat Commun 12:1901
6. Nilsson T, Rabouille C, Hui N, Watson R, Warren G (1996) The role of the membrane-spanning domain and stalk region of N-acetylglucosaminyltransferase I in retention, kin recognition and structural maintenance of the Golgi apparatus in HeLa cells. J Cell Sci 109:1975–1989
7. Ladinsky MS, Mastronarde DN, McIntosh JR, Howell KE, Staehelin LA (1999) Golgi structure in three dimensions: functional insights from the normal rat kidney cell. J Cell Biol 144:1135–1149
8. Katayama K, Sasaki T, Goto S, Ogasawara K, Maru H, Suzuki K, Suzuki H (2011) Insertional mutation in the Golgb1 gene is associated with osteochondrodysplasia and systemic edema in the OCD rat. Bone 49:1027–1036
9. Lan Y, Zhang N, Liu H, Xu J, Jiang R (2016) Golgb1 regulates protein glycosylation and is crucial for mammalian palate development. Development 143:2344–2355
10. Puthenveedu MA, Bachert C, Puri S, Lanni F, Linstedt AD (2006) GM130 and GRASP65-dependent lateral cisternal fusion allows uniform Golgi-enzyme distribution. Nat Cell Biol 8:238–248

Chapter 3
Revealing Golgi Architecture: Novel Insights Through Light Microscopy

Divyanshu Mahajan and Lei Lu

Abstract The Golgi is a eukaryotic organelle that modifies, sorts, and transports proteins and lipids to their destinations. In mammals, the Golgi is a complex network of interconnected stacks, with each stack made up of closely packed flat membrane sacs called cisternae. The Golgi remains a poorly understood organelle, fraught with controversies despite decades of research. The challenge of visually resolving the cisternal organization of the Golgi under light or electron microscopy contributes to its enigmatic nature. We have developed new imaging methods, including GLIM, en face averaging, and side averaging, to resolve the nocodazole-induced Golgi ministacks. These tools provide new insights into the organization and functions of the Golgi. They enabled us to rapidly and quantitatively investigate cisternal Golgi organization and intra-Golgi transport of secretory cargos. New insights have been revealed about the spatial distribution of Golgi proteins involved in processing and trafficking, the differential partitioning of secretory cargos within cisternae based on their sizes, and the specific Golgi exit sites for constitutive versus signal-dependent secretory cargos. Additionally, the intra-Golgi transport kinetic data acquired from these tools provide a basis for evaluating different intra-Golgi transport models.

Keywords Golgi · Golgi cisternae · Imaging · Microscopy · Trafficking

Introduction

The Golgi complex (Golgi) is central to secretory and endocytic pathways, functioning as the post-translational processing and sorting hub of proteins and lipids (cargos). The fundamental structural unit of the Golgi is the Golgi stack, which consists of 4–7 tightly packed, flattened membrane sacs known as cisternae. In

D. Mahajan · L. Lu (✉)
School of Biological Sciences, Nanyang Technological University, Singapore, Singapore
e-mail: lulei@ntu.edu.sg

A. Nakano, J. Saraste (eds.), *The Golgi Network, Volume I*, Subcellular Biochemistry 110, https://doi.org/10.1007/978-3-032-06936-8_3

mammalian cells, the interconnected network of Golgi stacks establishes the distinctive Golgi ribbon organization, aggregated around the microtubule-organizing center. The stacked Golgi cisternae are subdivided into the *cis*-, medial-, and *trans*-Golgi. The *trans*-Golgi cisternae further develop into an anastomosing membrane network of tubules and vesicles referred to as the *trans*-Golgi network (TGN). Secretory cargos synthesized in the endoplasmic reticulum (ER) are received at the *cis*-Golgi, while material exchange with the endosomes and the plasma membrane (PM) occurs at the *trans*-Golgi and TGN.

Despite decades of research about the cisternal organization of the Golgi at the molecular and cellular levels, it remains one of the most enigmatic and controversial organelles [13]. The key to understanding the Golgi is elucidating the intra-Golgi transport mechanism, that is, how secretory cargos transport through the Golgi stack from the *cis* to the *trans* Golgi [19]. The intra-Golgi transport mechanism has been debated since the Golgi was imaged under electron microscopy (EM) in the 1950s. The cisternal maturation or progression model was first proposed, in which Golgi was thought to form via vesicles at the *cis*-cisterna and mature into the medial-cisterna, then *trans*-cisterna before eventually budding off as vesicles from the TGN [20]. This classical cisternal maturation model could not account for the new information obtained in the 1970s regarding the specific cisternal localization of Golgi enzymes [16]. A different view of the Golgi, the stable compartment model, then emerged, in which the cisternae were stable compartments with distinct compositions of Golgi enzymes and vesicular carriers transported the secretory cargos through these different cisternae [15, 55, 58].

However, the difficulty in explaining how large secretory cargos, such as procollagen fibrils and algal scales, are packaged into small transport vesicles led to a revival of the cisternal progression model [4, 34, 40]. The cisternal progression model could explain how large secretory cargos transverse the Golgi stacks as cisternae mature, without exiting the cisternal lumen [4, 42]. The model further gained acceptance when it introduced COPI-coated carriers to recycle Golgi resident proteins such as Golgi enzymes [36]. Finally, the direct visualization of cisternal maturation in the budding yeast *Saccharomyces cerevisiae* under light microscopy catapulted this model as the predominant one [35, 39]. However, similar findings have not been made in mammalian cells, which, unlike the budding yeast, do not have scattered and unstacked Golgi compartments.

Recently, a new intra-Golgi transport view, the rapid partitioning model [48], has emerged to account for the first-order exponential exit kinetics of secretory cargos observed in mammalian cells [26]. The kinetics contradicts the cisternal progression model [19], which predicts a linear Golgi exit kinetics. The rapid partitioning model suggests that upon entry secretory cargos rapidly partition throughout the Golgi stack into numerous processing and transport domains. There are other views of the Golgi that exist in the literature, and many of them are derived from the above cisternal progression and stable compartment models [33, 41, 43, 50, 76]. Further experimental data, particularly imaging data, is essential for the evaluation of these models, which demands new technological advancements in light and electron microscopy.

Microscopy in Studying the Golgi Organization

Since the Golgi was imaged under EM in the 1950s [9, 10, 17], the advancement in EM technologies has significantly contributed to our understanding of the Golgi. Immuno-EM technology, especially immunogold-EM, is widely used to investigate the specific cisternal localization of a Golgi protein with a resolution of ~30 nm [53]. Some of the drawbacks of conventional EM are that the instrument is expensive and requires technical expertise, special reagents, and tedious labeling protocols.

Fluorescence light microscopy (hereafter light microscopy) is the preferred tool for studying the organization and dynamics of the Golgi by cell biologists. Compared to immuno-EM, light microscopy is more user-friendly and can simultaneously image multiple proteins in 3D live cells, although conventional light microscopy has a much lower spatial resolution, ~250 nm in the xy-plane and 500–700 nm along the z-axis [49]. With the recent advent of new super-resolution microscopy, it is possible to visualize cellular structures much smaller than the diffraction limit of a conventional light microscope [2, 21, 24, 25, 28, 29, 56, 59–61]. The super-resolution microscopy holds great promise but is still challenging for studying the Golgi architecture and dynamics. For example, though stimulated emission depletion (STED) microscopy can achieve a spatial resolution of 50 nm in the xy-plane and ~100 nm along the z-axis [23, 30, 78], its spatial resolution is still insufficient to resolve Golgi cisternae, which are spaced at ~20 nm apart. Additionally, most super-resolution techniques have limitations when it comes to compatibility with multichannel, 3D, or live-cell imaging. Despite continuous technological advancements, most of the current imaging data tends to be primarily qualitative, rather than quantitative. Therefore, it is crucial to develop new analytical methods to address these challenges.

Nocodazole-Induced Golgi Ministacks as a Model of the Golgi

The Golgi proteome consists of ~1000 proteins [7], and knowing their cisternal localization is the first step to systematically understanding how the Golgi functions. Despite significant advances in understanding Golgi proteins' functions, precise mapping of their localizations within the Golgi cisternae remains challenging under light microscopy. This is primarily due to the complex configurations of native Golgi stacks and their heterogeneous morphology. In tissue-cultured mammalian cell lines, such as HeLa, the Golgi mostly appears like a perinuclear lump, and its cisternal features are poorly resolved even under advanced microscopy.

The mammalian Golgi sometimes exhibits a linear, side-view ribbon profile, which can aid in identifying sub-Golgi compartments using cisternal markers [1, 46]. Leveraging this Golgi configuration, an intensity line profile approach was developed to quantify the localization of Golgi proteins using *cis* and *trans*-Golgi markers as references [12, 63]. In this method, the position of a protein is indicated by the peak pixel position in the line intensity profile, and the relative localization of the protein is expressed as the ratio of the distance between the testing protein and the *cis*-Golgi marker to the distance between the *cis*- and *trans*-Golgi markers. However, this approach has some limitations. These include the rarity of single side-view ribbon profiles, 2D projection of unknown 3D cisternal orientation, and the subjective selection of the line for intensity profiles. Additionally, the method is limited by its inability to achieve sub-pixel resolution.

Nocodazole is a commonly used microtubule-depolymerizing drug [27]. Prolonged treatment with nocodazole in mammalian cells leads to the de novo assembly of ~100 Golgi ministacks (hereafter referred to as ministacks) at the ER exit sites (ERES) scattered throughout a cell [8, 64]. Therefore, nocodazole treatment disassembles the entangled Golgi lump into spatially separated individual ministacks. These ministacks have a cylindrical shape and a relatively uniform size. Similar small Golgi stacks that are dispersed but adjacent to the ERES naturally occur in plants (such as tobacco BY-2 cells and root columella cells of *Arabidopsis thaliana*), insects (such as *Drosophila melanogaster* S2 cells), and yeasts (such as *Pichia Pastoris*) [31, 32, 52, 62]. Like the native mammalian Golgi, the ministack has a similar polarized organization from the *cis* to *trans*-Golgi and further into the TGN [74]. Extensive studies have shown that the constitutive cargo trafficking and post-translational modification of proteins remain largely unaffected in nocodazole-induced ministacks [8, 51, 74, 75]. Therefore, ministacks offer a valid model for studying the Golgi. All our Golgi imaging tools have been developed using ministacks.

The Geometry of a Ministack

Within a ministack, the *cis*, medial, and *trans*-Golgi represent the stacked region while the pre-Golgi region and the TGN are non-stacked, consisting of tubular and vesicular membrane profiles. To describe the organization of a ministack, we have defined the following geometric and morphometric terms (Table 3.1).

We use the center of fluorescence mass (hereafter referred to as the center) to indicate the localization of a protein [71]. In our study, the centers of GM130 and C-terminally mCherry-tagged 1–81 amino acids of B4GALT1 (GalT-mCherry) serve as reference points for the *cis* and *trans*-Golgi, respectively [45, 54]. The vector extending from the center of GM130 to that of GalT-mCherry is defined as the 3D Golgi axis in the cellular space (Fig. 3.1a). The xy-plane represents the image plane, while the z-axis is the optical axis of the microscope. We define the Golgi axial angle as the angle between the 3D Golgi axis and the image plane. Although the Golgi axial angle is typically unknown, two angles are important for our analysis: a ministack with a Golgi axial angle of 90° is referred to as an en face view ministack, while a 0° angle corresponds to a side-view ministack. Lateral and axial localization refer to distribution orthogonal to or along the 3D Golgi axis, respectively. Thus, the en face view shows the lateral localization, while the side view

Table 3.1 List of morphometric terms defined in this article

Term	Definition
Axial localization	It refers to a protein's distribution along the Golgi axis
Distance between two proteins	It is the distance between their centers
En face view	A ministack with its Golgi axis (3D) orthogonal to the image plane or with a Golgi axial angle of 90°
Golgi axial angle	The angle formed between the Golgi axis (3D) and the image plane in 3D space
Golgi axis (3D or 2D)	The vector extending from the center of the GM130 to that of GalT-mCherry in 3D cellular space or 2D image plane
Lateral localization	It refers to a protein's distribution in a direction orthogonal to the Golgi axis
Lateral size (of side-averaged image)	In side averaging, it is the distance between the two outer half-maxima in the lateral line intensity profile. The lateral size is the diameter and should approach two times the radius measured in the radial mean intensity profile
Localization of a protein	It refers to the localization of the center of fluorescence mass
LQ	For a generic ministack with a random Golgi axial angle, *LQ* is defined as the distance from GM130 to the test protein (dx) divided by the distance from GM130 to GalT-mCherry (d_1) within the image plane
LQ^{side}	In side averaging, it is calculated as the distance from GM130 to the test protein divided by the distance from GM130 to GalT-mCherry. LQ^{side} should converge to the true *LQ* value with a sufficiently large number of analyzable side-view ministacks
Normalized lateral size (of side-averaged image)	In side averaging, it is the lateral size (diameter) normalized by that of giantin
Normalized radius (of en face-averaged image)	In the radial mean intensity profile of an en face average, the radius is the distance from the origin to the half-maximum position of the outer slope. The normalized radius is the one divided by the radius of giantin
Radial mean intensity profile	It plots the mean intensity of all pixels along the circumference of a circle against the circle's radius. The circle's center align with the center of the giantin ring or the image center
Side view	A ministack with its Golgi axis (3D) parallel to the image plane or with a Golgi axial angle of 0°

reveals both the lateral and axial localization. Note that most diagrams and EM images of the Golgi in the literature are side views as they can represent both axial and lateral localization and visually display the Golgi's stacked organization.

A generic ministack has a Golgi axial angle between 0 and 90°, and its fluorescence image represents a 2D projection onto the image plane (Fig. 3.1a). Within the image plane, we similarly use the center to refer to the localization of a protein and the Golgi axis as the vector from the center of GM130 to that of GalT-mCherry. The distance between two proteins is defined as the distance between their centers. These geometric and morphometric terms enable a detailed description of the spatial organization and localization of proteins within a Golgi ministack.

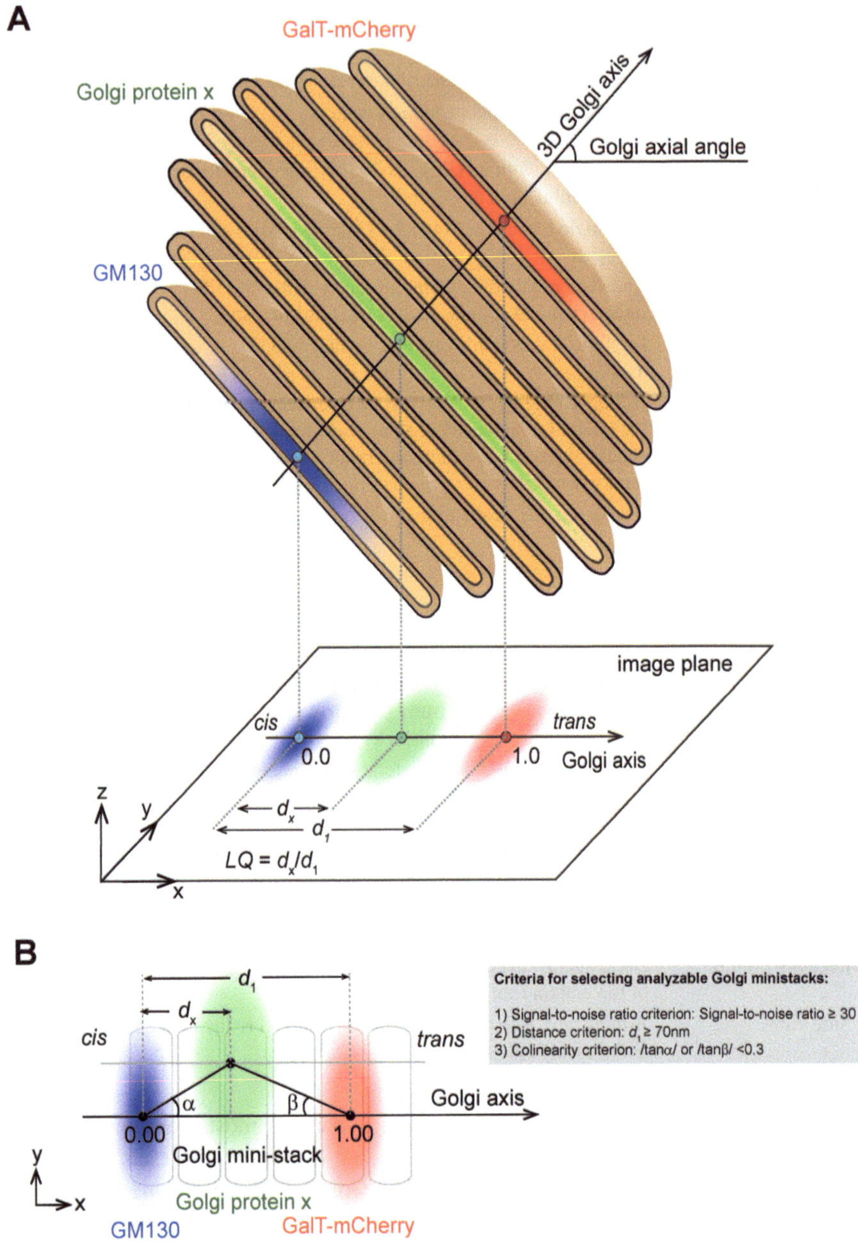

Fig. 3.1 The projection of a 3D ministack onto the 2D image plane. (**a**) A schematic diagram of a Golgi ministack in 3D space projected onto the 2D image plane. It illustrates the 3D Golgi axis, Golgi axial angle, distances (d_x and d_1), and the calculation of *LQ*. The ministack is triple-labeled for Golgi test protein x (green), GM130 (blue), and GalT-mCherry (red). (**b**) The Golgi ministack is projected into the xy-plane or image plane, and the center of Golgi protein x is off-axis. The schematic diagram illustrates the angle α, angle β, and the projected distance d_x. The three criteria for selecting the analyzable Golgi ministacks are shaded

Golgi Protein Localization by Imaging Centers of Mass (GLIM)

GLIM assumes that the ministack has a cylindrical structure with an infinite order of rotational symmetry around the 3D Golgi axis, and, hence, it can be modeled as a one-dimensional line along the Golgi axis [68, 71]. The axial localization of a test protein can be indicated using the metric, localization quotient or *LQ*. *LQ* is defined as the distance from GM130 to the test protein (d_x) divided by the distance from GM130 to GalT-mCherry (d_1) within the image plane. The *LQ* is independent of the Golgi axial angle. From our definition, *LQ*s of GM130 and GalT-mCherry are 0.00 and 1.00, respectively. To measure the *LQ* experimentally, we utilize ministacks triple-labeled with the test protein, GM130, and GalT-mCherry.

The proteins can be labeled endogenously or expressed exogenously. The center of fluorescence mass calculation, which can achieve sub-pixel super-resolution, and the chromatic shift aberration correction allow GLIM to achieve a high localization accuracy of nanometers. Since the projected axial distance is used for *LQ* calculation and due to the random orientation of ministacks in cytosolic 3D space, the following criteria were applied to ministacks to ensure the quality of the acquired data (Fig. 3.1b). The signal-to-noise ratio criterion ensures a high localization accuracy by the center of mass calculation. The distance criterion effectively eliminates those ministacks with a Golgi axial angle approaching 90°, such as en face view ministacks, which have a very small d_1. The collinearity criterion ensures that the centers of the three proteins in the ministack are sufficiently colinear to fit in our one-dimensional Golgi model. Ministacks meeting the above three criteria are classified as analyzable ministacks. The *LQ* is expressed as the *mean ± standard error of the mean* (*SEM*) with the number of analyzable ministacks (n) reported. The quality of the *LQ* critically depends on n—a larger n generally results in a smaller *SEM*.

In GLIM, the spatial resolution can be defined as the minimal distance between two proteins that can be resolved along the Golgi axis [68]. It is difficult to determine the spatial resolution of GLIM accurately, but a rough estimation is possible. Based on our experience, when n reaches above 100, a typical number we could achieve, the *SEM* can typically reach below 0.02. The two proteins can be considered resolved with $p < 0.05$ by t-test if the difference in their *LQ*s is ≥3 times the *SEM*, that is, $3 \times 0.02 = 0.06$. The axial size of the ministack is estimated to be ~300 nm [72, 74]; therefore, the spatial resolution of GLIM can be estimated as $0.06 \times 300 = 18$ nm. This estimation suggests that the resolution of GLIM is comparable to and even better than that of immunogold-EM, which is 30 nm.

As GLIM is a linear tool, any two Golgi proteins can be used as reference markers if their *LQ* differences are large enough to minimize calculation errors. Though *LQ* is dimensionless, it can be converted to a physical distance in nanometers from the center of GM130 by multiplying the unit *LQ* distance, 274 nm, which is the axial size of the Golgi ministack measured by side averaging [72]. The *LQ* of a given protein has been found to be highly reproducible from different labeling approaches, such as immunostaining and ectopic expression in either live or fixed conditions, and microscopic modalities, such as wide-field and spinning disk confocal. The *LQ* of a protein is also conserved across different cell lines [71]. According to *LQ*s of proteins with known sub-Golgi localization, regions of the Golgi be numerically defined as ERES/ERGIC, < − 0.25; *cis*, 0.00 ± 0.25; medial, 0.50 ± 0.25; *trans*-Golgi, 1.00 ± 0.25; and TGN, 1.25–2.00. The stacked region of the ministack has the *LQ* from −0.25 to 1.25, TGN represents *LQ* > 1.25, and *LQ* < −0.25 represents the pre-Golgi region. The *LQ* database is systematic and additive, enabling the addition of new proteins' *LQ*s across multiple imaging sessions to build a quantitative protein localization database of the Golgi. To date, >100 proteins (unpublished data) have been mapped. This is the first method to systematically, quantitatively, and accurately localize Golgi proteins.

However, GLIM assumes that proteins follow a unimodal distribution along the Golgi axis. It is unsuitable for Golgi proteins with multimodal or broad axial distribution. For example, a protein with an *LQ* value of 0.50 can have a compact medial-Golgi localization or an even distribution across the ministack from the *cis* to the *trans*-side. Additionally, although GLIM can be performed using various microscopic techniques, wide-field and spinning disk confocal microscopes are particularly suitable due to their fixed 2D image sensor arrays (CCD or CMOS), which simplify the chromatic shift aberration correction by using multiple-color beads. Though super-resolution techniques could enhance spatial resolution, challenges such as low signal-to-noise ratio, difficulties with three-color imaging, and issues in correcting chromatic shift aberration limit their application in GLIM.

Study of Ministack Organization by Averaging Super-Resolution Microscopic Images

Using super-resolution microscopy, we can gain a better insight into the structural details of the Golgi. Airyscan super-resolution microscopy (and possibly other structured illumination-based super-resolution techniques) is particularly suited for this purpose since it allows multicolor imaging and is compatible with conventional sample labeling protocols and fluorophores.

In the cytosol, the morphology of a Golgi ministack constantly changes, potentially influenced by the mechanical stress from the cytoskeleton or by the budding and fusion of transport carriers. These variations can introduce morphological noise into ministack images. To address this, we employed averaging to suppress the morphological noises from individual ministacks and enhance the common or native organization features. However, a limitation of averaging is the assumption of a single organization type (unimodal distribution) of ministacks, whereas their organization may have multimodal distributions due to factors that are not yet fully understood.

Under Airyscan microscopy, we observed Golgi proteins, such as *cis*-localized golgin84, medial-localized giantin, and *trans*-localized GPP130, appear as a ring or double-punctum. For a cisternal rim localized marker, the ring pattern corresponds to the en face view of ministacks while the double-punctum indicates the side view. By distinguishing these patterns, we identified en face and side-view ministacks and developed en face and side-averaging methods to reveal the lateral and axial distribution of Golgi proteins. Like GLIM, en face and side-averaging methods are systematic, and the normalization of resulting images enables the addition of averaged images to build a systematic en face or side-view composite image with multiple proteins. These two methods allow us to directly visualize the cisternal organization and intra-Golgi trafficking dynamics, as elaborated below.

En Face Averaging

En face averaging reveals a Golgi protein's lateral direction, that is, the distribution within disk-shaped cisternal membrane sheets of a ministack. Golgi proteins can localize to either the cisternal rim or the interior. In en face-averaged ministack images, the outer ring image patterns represent the cisternal rims, while the smaller and central disk-like patterns correspond to the cisternal interior. An additional inner ring pattern between the outer ring and interior disk can also be observed for proteins such as MGAT4B [70]. Therefore, en face averaging uncovers the concentric arrangement of proteins around the 3D Golgi axis within the ministack cisternae, consistent with the rotational symmetry of the ministack.

Airyscan imaging is required before performing en face averaging. The protein of interest can be either endogenous or exogenously transfected, and the cells must be immuno-stained for endogenous giantin, which serves to identify ministacks with en face views [69, 70]. The giantin ring pattern serves as a reference for processing individual ministack images, the procedure of which is shown in the following steps (Fig. 3.2a). (1) Cropping. A multicolor image of ministacks is cropped into small square images, each containing a single giantin ring corresponding to an en face view ministack. (2) Background subtraction. For an en face view ministack image, the background is removed from all channels. (3) Radius normalization. The ministack is radially expanded or shrunk to standardize the gyradius of giantin ring to a predetermined and fixed value. (4) Translation. The ministack is translated so that the center of the giantin ring aligns with the image center. (5) Intensity normalization. Total intensity values in each channel are normalized to ensure equal weightage during averaging. (6) Averaging. Multiple processed images are averaged to create the en face-averaged ministack image.

The en face-averaged ministack image can be characterized using the radial mean intensity profile (Fig. 3.2b), which plots the mean intensity of all pixels along the circumference of a circle against the circle's radius. The circle's center is aligned with the center of the giantin ring or the image center. The radius of the en face-averaged image is defined as the distance from the center to the half-maximum position of the outer slope in the radial mean intensity profile. The normalized

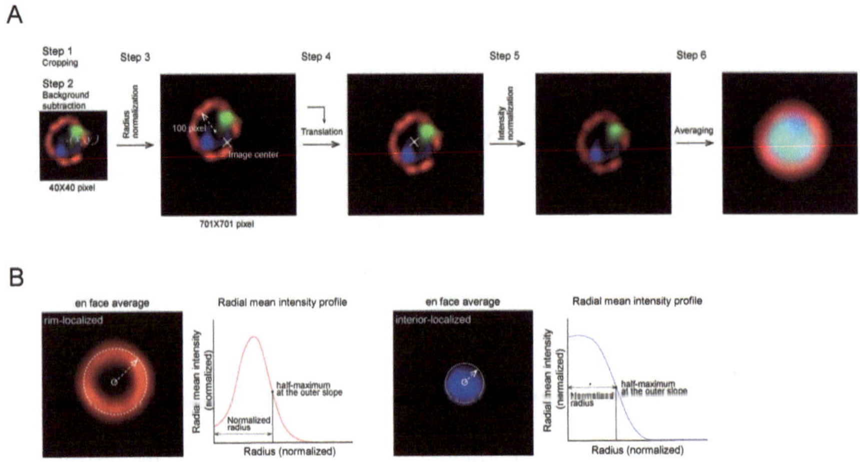

Fig. 3.2 En face averaging. (**a**) A representative ministack triple-labeled for giantin (red) and two Golgi proteins of interest (green and blue) showing the transformation procedure of en face averaging. The center of giantin fluorescence is represented as (x_c, y_c), and the center of the image is highlighted by a cross. (**b**) The radial mean intensity profile used to calculate the normalized radius of rim (left) or interior-localized (right) protein from en face-averaged image is illustrated. The dotted arrow and circle represent the radius and the corresponding circumference, along which pixels are averaged. The radius is defined as the distance from the center to the half-maximum of the outer slope

radius of a Golgi protein is the one divided by the radius of giantin. In the radial mean intensity profile, the circle's radius (x-axis) can be directly normalized by giantin radius (Fig. 3.2b).

Side Averaging

In side views, the cisternal rim markers appear as a double-punctum, while the cisternal interior markers appear as a short bar within this double-punctum (Fig. 3.3a). Ministacks with side views can be identified by the appearance of the giantin double-punctum. Our side-averaging method is based on the discovery that while the giantin lateral size of a ministack varies approximately from 400 to 1200 nm, its axial size, which is the distance from GM130 to GalT-mCherry, remains relatively constant at around 274 nm [72]. Therefore, to average ministacks of side views, we only need to normalize their varying lateral sizes.

Airyscan imaging is required before performing side averaging. Ministacks must be triple-labeled: the protein of interest, endogenous giantin to identify the side view, and a Golgi marker to differentiate between the *cis*- and *trans*-sides of the ministack [38, 72]. The giantin double-punctum pattern serves as a reference for processing ministacks with side views. The side-view-averaging procedure involves the following several steps (Fig. 3.3a). (1) Cropping. The multicolor ministack

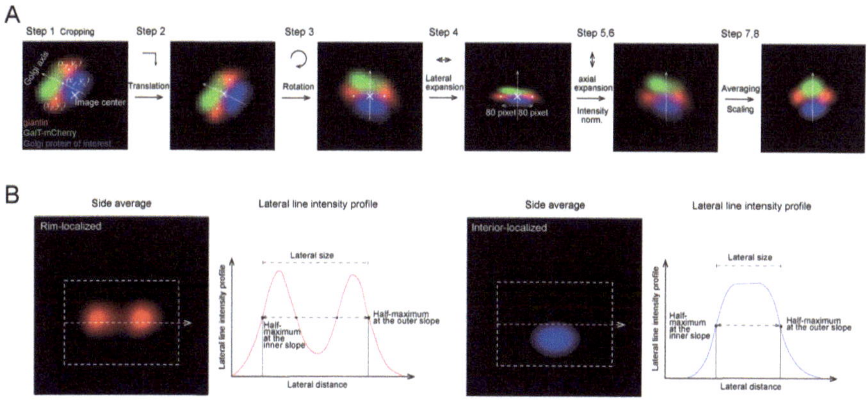

Fig. 3.3 Side averaging. (**a**) A representative ministack triple-labeled for giantin (red), GalT-mCherry (green), and a Golgi protein of interest (blue) showing the transformation procedure of side averaging. The center of giantin is represented as (x_c, y_c), and the centers of the left and right giantin punctum are represented as (x_1, y_1) and (x_2, y_2), respectively. The center of the image is marked by a cross. Intensity norm., intensity normalization. (**b**) The lateral line intensity profile used to calculate the lateral size of the rim (left) or interior-localized (right) protein from the side-averaged image is illustrated. The dotted box in the direction of the arrow represents the line for the intensity profile. The lateral size of the protein is defined as the distance between the two half-maxima at the inner and outer slopes

image is cropped into small square images, each containing a single giantin double-punctum corresponding to a side-view ministack. (2) Translation. The ministack is translated so that the center of the giantin double-punctum aligns with the image center. (3) Rotation. The ministack is rotated so that the giantin double-punctum is horizontal, with the *trans*-side of the ministack oriented upwards. (4) Lateral expansion. The image is laterally expanded along the x-axis to normalize the lateral size of the giantin double-punctum to a predetermined fixed value. (5) Axial expansion. The image is expanded by eightfold along the y-axis to amplify the axial dimension. (6) Intensity normalization. The total intensity of each channel is normalized to ensure equal weightage during averaging. (7) Averaging. Multiple processed images are averaged to create the side-averaged ministack image. (8) Scaling. The side-averaged image is laterally shrunk to ensure an isotropic scale.

Side averaging is highly informative as it enables visualization of both axial and lateral protein distributions. It enables the calculation of the *LQ* (LQ^{side}), lateral size, and the assessment of cisternal rim or interior localization. For instance, LQ^{side} can be calculated as the distance from GM130 to the test protein divided by the distance from GM130 to GalT-mCherry. A Golgi protein's lateral size can be calculated as the distances between the two half-maximal points of its lateral line intensity profile's outer slopes (Fig. 3.3b). However, side averaging cannot fully replace GLIM and en face averaging due to the difficulty in acquiring a sufficiently large number of high-quality side views. In HeLa cells, the number of ministacks with side views is approximately three times fewer than those with en face views. Additionally, oblique views are often misidentified as side views, which can compromise the quality and accuracy of the side averaging. Therefore, LQ^{side} and lateral size are generally less accurate than *LQ* and radius derived from GLIM and en face averaging, respectively.

Molecular Organization of the Ministack

The Golgi resident proteins can be functionally categorized into two groups: post-translational modification enzymes and trafficking machinery components, contributing to the processing and trafficking functions of the Golgi. En face and side-averaging methods have revealed that these functions are spatially organized within the Golgi [70] (Fig. 3.4a). Trafficking machinery components, such as small G proteins, coat proteins, golgins, and SNAREs, localize from the ERES and *cis*-Golgi to the TGN, based on their *LQ*s; at the medial and *trans*-Golgi, these proteins localize to the cisternal rim from en face and side averaging. In contrast, most Golgi enzymes localize to the medial and *trans*-Golgi, as shown by their *LQ*s; en face averaging further demonstrated that they are concentrated in the cisternal interior, with their lateral sizes smaller than rim-localized trafficking machinery components of similar *LQ*s. This spatial distribution suggests that Golgi enzymes and trafficking machinery components occupy different regions of the ministack.

These observations were supported by Airyscan and EM imaging of the native Golgi [70]. Our findings align with recent discoveries of protein arrays within the interior of the medial and *trans*-Golgi of green algae, which are thought to be Golgi enzymes [14]. We propose that the dense protein arrays form a Golgi enzyme matrix, which will be discussed later [70]. This arrangement requires that secretory cargos access the cisternal interior for post-translational modifications, such as glycosylation. Indeed, conventional secretory cargos like VSVG, E-cadherin, furin, and GPI-anchored proteins localize to the cisternal interior during their transition across Golgi ministacks (see below) [70, 72].

The en face and side averaging revealed that COPI coat subunits localize further outside the cisternal rim, with their radii being even larger than that of giantin (normalized radii >1) (Fig. 3.4b, c). The observation is consistent with the notion that COPI-coated vesicles bud from and then distribute around the cisternal rim. The GLIM and side averaging highlighted that COPI and COPII are axially adjacent to each other, with COPII at the pre-Golgi and COPI from *cis*- to *trans*-Golgi around the cisternal rim (Fig. 3.4a, c), consistent with previous EM studies [44, 47, 57, 77].

TGN-localized proteins ($LQ \geq 1.25$) exhibit two distinct distribution patterns. Vamp4, Rab6, Arl1, and golgin-97 display compact, punctate-like localization. In contrast, the clathrin coat and its associated adaptors and transmembrane cargos, such as GGAs, furin, and CI-M6PR, appear as extended puncta, forming a large crown capping the *trans*-side, where they decorate the buds, vesicles, and tubule tips (Fig. 3.4a, c) [70, 72], as expected from what we know of the TGN [11].

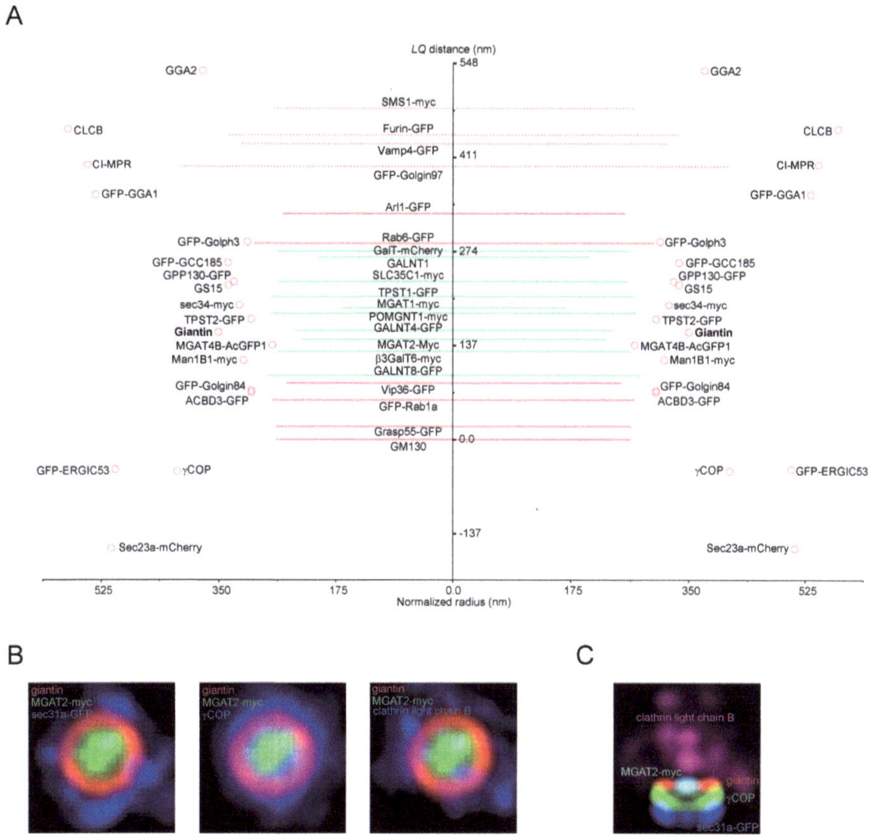

Fig. 3.4 Maps of an average Golgi ministack. (**a**) A quantitative molecular map of an average Golgi ministack showing the normalized radius against its corresponding *LQ* distance, both in nm. Line and red open circle denote the interior and ring localization pattern, respectively. The Golgi enzymes are represented by green lines, while other interior localized proteins are shown as red liens. The dotted red lines denote interior localized TGN proteins. (**b**) En face-averaged images showing the giantin ring (red) and MGAT2-Myc (green) along with coat proteins (blue), sec31a-GFP, γCOP, or clathrin light chain B. The coat proteins have a larger normalized radius and localize outside the giantin ring. (**c**) Side-averaged images showing the giantin double-punctum (red) and MGAT2-myc (cyan) along with coat proteins, sec31a-GFP (blue), γCOP (green), and clathrin light chain B (magenta)

The Morphometry of an Averaged Ministack

The morphometric data acquired from our three methods have allowed us to construct an averaged physical map of a Golgi ministack. The axial position of a protein is determined by multiplying its *LQ* by the unit *LQ* distance (274 nm). The lateral size of a protein is calculated by multiplying its normalized lateral size by the lateral size of the giantin ring, which is 700 nm. The lateral size of a rim-localized trafficking machinery component can be used to estimate the lateral size or diameter of the corresponding ministack cisternae (Fig. 3.3b). Our morphometric data indicates that from the *cis* to the medial region, the cisternal lateral size or diameter initially increases. It peaks at the medial Golgi, where giantin, a medial Golgi marker with an *LQ* of 0.57, localizes. Among all the proteins we have studied, giantin exhibits the largest lateral size, 700 nm. As the *LQ* increases from the medial to the *trans*-region, the lateral size or diameter of the corresponding cisternae decreases. Our morphometric data suggests that the ministack adopts a spindle shape, with the largest diameter at the medial region and smaller ones at the *cis* and *trans*-regions (Fig. 3.4a). This spindle-shaped morphology is consistent with previous EM-tomographic studies of native Golgi ministacks in green algae [6, 14] and plant cells [62], suggesting that it might be a universal feature of Golgi ministacks in eukaryotic cells.

The Observation of Synchronized Trafficking Waves

Understanding the mechanism of intra-Golgi transport at the molecular and cellular level is challenging, but GLIM and side averaging have enabled us to study synchronized intra-Golgi transport using RUSH (retention using synchronized hooks) secretory cargos [3]. Our three methods allow for quantitative measurement of RUSH secretory cargo movement along the Golgi axis, with side averaging providing a direct visualization.

We have tested various transmembrane secretory cargos in the RUSH system, including E-cadherin (type I transmembrane protein), tumor necrosis factor α (TNFα; type II transmembrane protein), and CD59 (GPI-anchored protein). According to the cisternal progression and stable compartment models [18], the secretory cargo synchronously released from the ER should remain as a traffic wave transiting from the *cis* to the *trans*-side of the Golgi. In contrast, the rapid partitioning model [48] predicts a homogeneous distribution of cargo throughout the Golgi upon entry. GLIM and side averaging directly demonstrated that, during intra-Golgi transport, RUSH secretory cargos remain as a synchronized traffic wave with a unimodal axial distribution. Therefore, GLIM and side averaging provide strong evidence against the rapid partitioning model.

Lateral Partitioning of Large Secretory Cargos to the Cisternal Rim

En face and side averaging uncovered that conventional RUSH secretory cargos, including transmembrane proteins, such as CD59, E-cadherin, VSVG, and CD8a-furin, as well as soluble proteins, such as GFP, localize to the cisternal interior during their intra-Golgi transport [70, 72]. Only upon reaching the *trans*-side of the ministack do they localize at the cisternal rim and appear as puncta and tubules, likely representing exocytic transport carriers targeting the PM. This interior localization during the intra-Golgi transport aligns with their post-translational modifications, such as glycosylation, which occur within the cisternal interior (Fig. 3.4a).

In contrast, large or bulky secretory cargos exhibit different lateral localization patterns during intra-Golgi transport. For instance, FM4-moxGFP and RUSH GFP-collagen X are soluble proteins that form aggregates of about 130 and 190 copies, respectively, within the Golgi lumen [70]. Similarly, GFP-FM4-CD8a, a transmembrane chimeric protein, forms an aggregate of ~180 copies. During their synchronized release, these large cargos remain at the cisternal rim while transiting through the Golgi stack from the *cis* to the *trans*-Golgi. This observation is consistent with previous EM studies of similar transmembrane and soluble FM4 aggregates [33].

Therefore, the differential lateral partitioning of large versus conventional or small secretory cargos during intra-Golgi transport is evident. We have recently proposed that the Golgi enzyme matrix (see above) in the cisternal interior might function as a gel filtration or size-exclusion chromatography column [73]. It is possible that large cargos are excluded from the cisternal interior due to the crowded molecular environment or the narrow spatial space of the enzyme matrix, while conventional secretory cargos, being much smaller, can enter and localize within the interior.

The Golgi Exit Site: The TGN or the *trans*-Golgi?

The TGN is widely regarded as the primary sorting site for the anterograde transport to various post-Golgi locations, such as the PM and endolysosomes [11]. Cargos with specific sorting signals are typically packaged into clathrin-coated vesicles, while those lacking these signals are generally directed into non-coated carriers for delivery to the PM via the constitutive pathway [11]. It has been commonly accepted that secretory cargos, regardless of sorting signals, exit the Golgi at the TGN. This view aligns with the cisternal progression model, which proposes that cargos exit the Golgi at the *trans*-most end of the Golgi, the TGN, for exocytic trafficking. However, the precise Golgi exit sites for secretory cargos have not been rigorously investigated at the cisternal levels.

Using GLIM to analyze the intra-Golgi transport kinetics of constitutive secretory membrane cargos, such as GFP-tagged E-cadherin, TNFα, and CD59 in RUSH format, we observed that their *LQ* vs. chase time plots plateau at ~1.0, corresponding to the *trans*-Golgi (Table 3.2). These cargos remain at their *LQ* plateau values for an extended period before exiting the Golgi as exocytic carriers targeting the PM. This *LQ* plateau indicates the Golgi exit site position and suggests that the *trans*-Golgi, rather than the TGN, serves as the Golgi exit site for constitutive secretory cargos. This finding challenges the conventional view of the TGN as the exit site for constitutive secretory cargos.

Table 3.2 List of the Golgi exit site (y_0) and intra-Golgi transport velocity d*LQ*/d*t* (at *LQ* = 0.40) values

RUSH reporter	y_0	Adjusted R^2	d*LQ*/d*t* (at *LQ* = 0.40)	d*LQ*/dt (at *LQ* = 0.40) (nm/min)
TNFα-SBP-GFP	1.09	0.94	0.075	20.5
SBP-GFP-CD59	1.03	0.97	0.034	9.2
SBP-GFP	0.74	0.98	0.022	6.1
SBP-GFP-Tac #1	0.86	0.98	0.053	14.6
SBP-GFP-Tac-TC #1	1.05	0.97	0.029	8.0
TfR-SBP-GFP	1.16	0.99	0.075	20.6
SBP-GFP-E-cadherin #1	0.98	0.99	0.053	14.5
SBP-GFP-E-cadherin #2	1.15	0.96	0.060	16.5
SBP-GFP-E-cadherin #3	1.11	0.98	0.075	20.6
SBP-GFP-CD8a-furin-WT #1	1.63	0.96	0.036	9.9
SBP-GFP-CD8a-furin-WT #2	1.58	0.97	0.044	12.0
SBP-GFP-CD8a-furin-WT #3	1.55	0.97	0.061	16.7
SBP-GFP-CD8a-furin-YA #1	1.49	0.98	0.128	35.0
SBP-GFP-CD8a-furin-YA #2	1.51	0.93	0.069	19.0
SBP-GFP-CD8a-furin-YA #3	1.48	0.95	0.046	12.6
SBP-GFP-CD8a-furin-AC #1	1.73	0.99	0.052	14.1
SBP-GFP-CD8a-furin-AC #2	1.50	1.00	0.028	7.8
SBP-GFP-CD8a-furin-AC #3	1.58	0.96	0.035	9.7
SBP-GFP-CD8a-furin-Y+AC #1	0.83	0.99	0.061	16.7
SBP-GFP-CD8a-furin-Y+AC #2	1.00	0.99	0.059	16.0
SBP-GFP-CD8a-furin-Y+AC #3	1.04	0.92	0.058	15.8

Data is from our previous work Ref. [73]

In contrast to constitutive secretory cargos, CD8a-furin in RUSH format, a chimeric fusion between CD8a and the C-terminal cytosolic tail of furin, is a TGN resident transmembrane protein. The furin cytosolic tail has two endocytic TGN targeting motifs, tyrosine (YKGL) and acidic cluster (SDSEEDE). These motifs interact with AP1 clathrin adaptors or Mint3 to target CD8a-furin to the TGN [22, 67]. The newly synthesized RUSH CD8a-furin demonstrates different intra-Golgi transport behavior from constitutive secretory cargos. During the intra-Golgi transport of RUSH CD8a-furin, its *LQ* plateaus at ~1.6, indicating localization at the TGN (Table 3.2). Singly mutating either motif had no effect on the exit site, but doubly mutating both motifs shifted the exit site of the resulting mutant to the *trans*-Golgi. Our data suggests that cargos might have divergent pathways at the *trans*-Golgi: those without TGN targeting signals exit the Golgi at the *trans*-Golgi by the constitutive pathway, while those with the TGN targeting signals can enter the TGN and are packaged into clathrin-coated carriers.

It is known that proteins can also enter the TGN via the endocytic TGN targeting pathway from the PM and endolysosomes [5, 37]. We propose using the term "secretory TGN targeting" to describe the signal-dependent TGN targeting via the secretory pathway, distinguishing it from "endocytic TGN targeting," which operates in the retrograde direction. It is noteworthy that both secretory and endocytic TGN targeting pathways utilize similar signal motifs, such as the tyrosine and acidic cluster motifs in furin, suggesting that they might involve overlapping machinery components. We propose that secretory TGN targeting uses the same clathrin adaptors as those involved in endocytic TGN targeting. This hypothesis warrants further investigation.

The Intra-Golgi Transport Kinetics Follow the Exponential Decay Equation

The cisternal progression model predicts a constant intra-Golgi transport velocity and Golgi exit kinetics for all cargos, while the stable compartment model allows for distinct intra-Golgi transport velocities and Golgi exit kinetics for different cargos [19, 48]. GLIM enabled us to quantitatively measure the intra-Golgi transport kinetics of synchronized RUSH secretory cargos. Our extensive GLIM kinetics data on various membrane and soluble secretory proteins showed that their *LQ* vs. chase time plots fit well with the first-order exponential decay equation (Fig. 3.5a). The instantaneous intra-Golgi transport velocity, defined as the derivative of the *LQ* with respect to the chase time, also follows a first-order exponential decay equation (Fig. 3.5b). This velocity can be converted to nm/min by multiplying the unit *LQ* distance (274 nm). According to the exponential decay equation, a secretory cargo's transport velocity decreases as it progresses from the *cis* to the *trans*-Golgi, with the velocity approaching zero at the Golgi exit site. For RUSH secretory cargos at the medial-Golgi position with an *LQ* value of 0.40, the instantaneous velocities were calculated to range from 6.1 to 20.5 nm/min [73]. Therefore, our intra-Golgi transport kinetic data challenges the cisternal progression model, which predicts an equal and constant intra-Golgi transport velocity for all cargos.

Additionally, previous findings that the Golgi exit kinetics of secretory cargos follow a first-order exponential decay equation rather than a constant rate [26, 66] further challenge the cisternal progression model. The Golgi residence time, defined as the time a secretory cargo spends at the Golgi, is a key metric for exit kinetics and

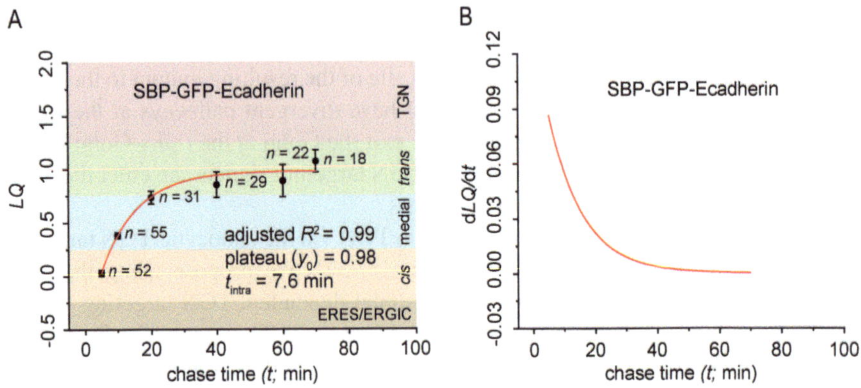

Fig. 3.5 Kinetic data from GLIM indicates that secretory cargos might not move at a constant velocity during intra-Golgi transport. The figure was from our previous published work [73]. (**a**) The intra-Golgi transport kinetics for RUSH reporter E-cadherin is plotted as *LQ* vs. chase time with *n* (the number of analyzable Golgi ministacks), adjusted R^2, y_0 (plateau), and t_{intra} indicated. The kinetic data fits well with the first-order exponential decay equation, which is indicated. Different Golgi regions are color-shaded. (**b**) The d*LQ*/d*t* vs. chase time plot. d*LQ*/d*t* represents the instantaneous intra-Golgi transport velocity. It is derived from the fitted first-order exponential decay equation of the left plot

is calculated as the half-time of the fitted exponential decay equation [66]. Distinct Golgi residence times have been observed for different RUSH cargos (Table 3.3) [65, 66, 73]. Particularly, we found that luminal domain truncation mutants of Tac and CD8a, Tac-TC and CD8a-TC, have prolonged Golgi residence times of 3.4 and 1.1 h, respectively, compared to 16 and 7.8 min for Tac and CD8a [73]. The Golgi residence times of Tac-TC and CD8a-TC are comparable to those of Golgi enzymes, such as MGAT2 (4.9 h) and ST6GAL1 (5.3 h). En face averaging coupled with GLIM further demonstrated the cisternal interior localization of Tac-TC at the *trans*-Golgi region [66]. Since Tac and CD8a should not possess known intra-Golgi recycling mechanisms, the unusually long Golgi residence times of Tac-TC and CD8a-TC suggest the stable nature of the *trans*-cisternal interior.

Our data supports a modified version of the stable compartment model, termed the rim progression model [33, 76]. This model proposes a stable interior and a dynamic rim domain of Golgi cisternae, where the rim membrane mediates intra-Golgi transport through constant en bloc fission and fusion events. It can account for the different intra-Golgi transport velocities, diverse Golgi residence times, and retention within the cisternal interior of secretory cargos. It also supports the enzyme matrix hypothesis (see above), which suggests that conventional cargos enter the cisternal interior while large ones are excluded from the cisternal interior and localized to the cisternal rim.

Table 3.3 Golgi residence times of secretory cargos and Golgi resident proteins

Transmembrane protein	Golgi residence time ($t_{1/2}$)	n	SEM
TNFα-SBP-GFP	6.0 min	73	0.4 min
GFP-CD8a	7.8 min	26	0.7 min
TfR-GFP	10 min	12	1 min
GFP-Tac	16 min	45	2 min
SBP-GFP-CD59	16 min	29	2 min
GFP-Tac (5A)	47 min	57	3 min
GFP-CD8a-TC	66 min	19	7 min
GFP-Tac-TC	3.4 h	22	0.3 h
MGAT2-GFP	4.9 h	26	0.5 h
ST6GAL1-GFP	5.3 h	21	0.6 h

Data is from our previous work Ref. [73]

Conclusion

New microscopic imaging tools often offer fresh perspectives in cell biology. The three imaging techniques we developed provide new insights into the Golgi organization and intra-Golgi transport dynamics. These methods enabled us to visualize the spatial separation of processing and trafficking functions, the waves of secretory transport, and the cisternal segregation of large and small cargos. Additionally, they allow for the quantitative measurement of intra-Golgi transport kinetics and Golgi exit sites. Our data supports a modified version of the stable compartment model. However, we believe that the current data is not yet sufficient to resolve the debate between the two intra-Golgi transport models or to settle the long-standing questions about how the Golgi works. We hope that our tools will be further refined and utilized by the Golgi research community to generate more systematic and insightful data about Golgi organization and functions.

Compliance with Ethical Standards The authors declare no conflicts of interest or competing interests.

References

1. Antony C, Cibert C, Géraud G, Maria AS, Maro B, Mayau V, Goud B (1992) The small GTP-binding protein rab6p is distributed from medial Golgi to the trans-Golgi network as determined by a confocal microscopic approach. J Cell Sci 103:785–796
2. Betzig E, Patterson GH, Sougrat R, Lindwasser OW, Olenych S, Bonifacino JS, Davidson MW, Lippincott-Schwartz J, Hess HF (2006) Imaging intracellular fluorescent proteins at nanometer resolution. Science 313:1642–1645
3. Boncompain G, Divoux S, Gareil N, de Forges H, Lescure A, Latreche L, Mercanti V, Jollivet F, Raposo G, Perez F (2012) Synchronization of secretory protein traffic in populations of cells. Nat Methods 9:493–498

4. Bonfanti L, Mironov AA Jr, Martínez-Menárguez JA, Martella O, Fusella A, Baldassarre M, Buccione R, Geuze HJ, Mironov AA, Luini A (1998) Procollagen traverses the Golgi stack without leaving the lumen of cisternae: evidence for cisternal maturation. Cell 95:993–1003
5. Bonifacino JS, Rojas R (2006) Retrograde transport from endosomes to the trans-Golgi network. Nat Rev Mol Cell Biol 7:568–579
6. Bykov YS, Schaffer M, Dodonova SO, Albert S, Plitzko JM, Baumeister W, Engel BD, Briggs JAG (2017) The structure of the COPI coat determined within the cell. elife 6:e32493
7. Chen X, Simon ES, Xiang Y, Kachman M, Andrews PC, Wang Y (2010) Quantitative proteomics analysis of cell cycle-regulated Golgi disassembly and reassembly. J Biol Chem 285:7197–7207
8. Cole NB, Sciaky N, Marotta A, Song J, Lippincott-Schwartz J (1996) Golgi dispersal during microtubule disruption: regeneration of Golgi stacks at peripheral endoplasmic reticulum exit sites. Mol Biol Cell 7:631–650
9. Dalton A, Felix MD (1956) A comparative study of the Golgi complex. J Biophys Biochem Cytol 2:79
10. Dalton AJ, Felix MD (1954) Cytologic and cytochemical characteristics of the Golgi substance of epithelial cells of the epididymis–in situ, in homogenates and after isolation. Am J Anat 94:171–207
11. De Matteis MA, Luini A (2008) Exiting the Golgi complex. Nat Rev Mol Cell Biol 9:273–284
12. Dejgaard SY, Murshid A, Dee KM, Presley JF (2007) Confocal microscopy-based linescan methodologies for intra-Golgi localization of proteins. J Histochem Cytochem 55:709–719
13. Emr S, Glick BS, Linstedt AD, Lippincott-Schwartz J, Luini A, Malhotra V, Marsh BJ, Nakano A, Pfeffer SR, Rabouille C, Rothman JE, Warren G, Wieland FT (2009) Journeys through the Golgi—taking stock in a new era. J Cell Biol 187:449–453
14. Engel BD, Schaffer M, Kuhn Cuellar L, Villa E, Plitzko JM, Baumeister W (2015) Native architecture of the Chlamydomonas chloroplast revealed by in situ cryo-electron tomography. elife 4:e04889
15. Farquhar MG (1985) Progress in unraveling pathways of Golgi traffic. Annu Rev Cell Biol 1:447–488
16. Farquhar MG, Palade GE (1981) The Golgi apparatus (complex)-(1954-1981)-from artifact to center stage. J Cell Biol 91:77s–103s
17. Farquhar MG, Rinehart JF (1954) Electron microscopic studies of the anterior pituitary gland of castrate rats. Endocrinology 54:516–541
18. Glick BS, Luini A (2011) Models for Golgi traffic: a critical assessment. Cold Spring Harb Perspect Biol 3:a005215
19. Glick BS, Nakano A (2009) Membrane traffic within the Golgi apparatus. Annu Rev Cell Dev Biol 25:113–132
20. Grasse PP (1957) Ultrastructure, polarity and reproduction of Golgi apparatus. C R Hebd Seances Acad Sci 245:1278–1281
21. Gustafsson MG (2005) Nonlinear structured-illumination microscopy: wide-field fluorescence imaging with theoretically unlimited resolution. Proc Natl Acad Sci USA 102:13081–13086
22. Han J, Wang Y, Wang S, Chi C (2008) Interaction of Mint3 with Furin regulates the localization of Furin in the trans-Golgi network. J Cell Sci 121:2217–2223
23. Hein B, Willig KI, Hell SW (2008) Stimulated emission depletion (STED) nanoscopy of a fluorescent protein-labeled organelle inside a living cell. Proc Natl Acad Sci 105:14271–14276
24. Hell SW (2003) Toward fluorescence nanoscopy. Nat Biotechnol 21:1347–1355
25. Hess ST, Girirajan TP, Mason MD (2006) Ultra-high resolution imaging by fluorescence photoactivation localization microscopy. Biophys J 91:4258–4272
26. Hirschberg K, Miller CM, Ellenberg J, Presley JF, Siggia ED, Phair RD, Lippincott-Schwartz J (1998) Kinetic analysis of secretory protein traffic and characterization of Golgi to plasma membrane transport intermediates in living cells. J Cell Biol 143:1485–1503
27. Hoebeke J, Van Nijen G, De Brabander M (1976) Interaction of oncodazole (R 17934), a new anti-tumoral drug, with rat brain tubulin. Biochem Biophys Res Commun 69:319–324

28. Huang B, Bates M, Zhuang X (2009) Super-resolution fluorescence microscopy. Annu Rev Biochem 78:993–1016
29. Huang B, Wang W, Bates M, Zhuang X (2008) Three-dimensional super-resolution imaging by stochastic optical reconstruction microscopy. Science 319:810–813
30. Klar TA, Jakobs S, Dyba M, Egner A, Hell SW (2000) Fluorescence microscopy with diffraction resolution barrier broken by stimulated emission. Proc Natl Acad Sci 97:8206–8210
31. Klumperman J (2011) Architecture of the mammalian Golgi. Cold Spring Harb Perspect Biol 3:a005181
32. Kondylis V, Rabouille C (2009) The Golgi apparatus: lessons from Drosophila. FEBS Lett 583:3827–3838
33. Lavieu G, Zheng H, Rothman JE (2013) Stapled Golgi cisternae remain in place as cargo passes through the stack. elife 2:e00558
34. Leblond CP (1989) Synthesis and secretion of collagen by cells of connective tissue, bone, and dentin. Anat Rec 224:123–138
35. Losev E, Reinke CA, Jellen J, Strongin DE, Bevis BJ, Glick BS (2006) Golgi maturation visualized in living yeast. Nature 441:1002–1006
36. Love HD, Lin CC, Short CS, Ostermann J (1998) Isolation of functional Golgi-derived vesicles with a possible role in retrograde transport. J Cell Biol 140:541–551
37. Lu L, Hong W (2014) From endosomes to the trans-Golgi network. Semin Cell Dev Biol 31:30–39
38. Mahajan D, Tie HC, Lu L (2023) Visualizing the cisternal organization of Golgi ministacks in HeLa cells by side-averaging. Bio Protocol 13:e4658
39. Matsuura-Tokita K, Takeuchi M, Ichihara A, Mikuriya K, Nakano A (2006) Live imaging of yeast Golgi cisternal maturation. Nature 441:1007–1010
40. Melkonian M, Becker B, Becker D (1991) Scale formation in algae. J Electron Microsc Tech 17:165–178
41. Mironov AA, Beznoussenko GV (2012) The kiss-and-run model of intra-Golgi transport. Int J Mol Sci 13:6800–6819
42. Mironov AA, Beznoussenko GV, Nicoziani P, Martella O, Trucco A, Kweon HS, Di Giandomenico D, Polishchuk RS, Fusella A, Lupetti P, Berger EG, Geerts WJ, Koster AJ, Burger KN, Luini A (2001) Small cargo proteins and large aggregates can traverse the Golgi by a common mechanism without leaving the lumen of cisternae. J Cell Biol 155:1225–1238
43. Mironov AA, Beznoussenko GV, Polishchuk RS, Trucco A (2005) Intra-Golgi transport: a way to a new paradigm? Biochim Biophys Acta (BBA)-Mol Cell Res 1744:340–350
44. Moelleken J, Malsam J, Betts MJ, Movafeghi A, Reckmann I, Meissner I, Hellwig A, Russell RB, Söllner T, Brügger B, Wieland FT (2007) Differential localization of coatomer complex isoforms within the Golgi apparatus. Proc Natl Acad Sci USA 104:4425–4430
45. Nakamura N, Rabouille C, Watson R, Nilsson T, Hui N, Slusarewicz P, Kreis TE, Warren G (1995) Characterization of a cis-Golgi matrix protein, GM130. J Cell Biol 131:1715–1726
46. Nilsson T, Pypaert M, Hoe MH, Slusarewicz P, Berger EG, Warren G (1993) Overlapping distribution of two glycosyltransferases in the Golgi apparatus of HeLa cells. J Cell Biol 120:5–13
47. Orci L, Stamnes M, Ravazzola M, Amherdt M, Perrelet A, Söllner TH, Rothman JE (1997) Bidirectional transport by distinct populations of COPI-coated vesicles. Cell 90:335–349
48. Patterson GH, Hirschberg K, Polishchuk RS, Gerlich D, Phair RD, Lippincott-Schwartz J (2008) Transport through the Golgi apparatus by rapid partitioning within a two-phase membrane system. Cell 133:1055–1067
49. Pawley J (2006) Handbook of biological confocal microscopy. Springer, New York
50. Pfeffer SR (2010) How the Golgi works: a cisternal progenitor model. Proc Natl Acad Sci 107:19614–19618
51. Rogalski AA, Bergmann JE, Singer S (1984) Effect of microtubule assembly status on the intracellular processing and surface expression of an integral protein of the plasma membrane. J Cell Biol 99:1101–1109

52. Rossanese OW, Soderholm J, Bevis BJ, Sears IB, O'Connor J, Williamson EK, Glick BS (1999) Golgi structure correlates with transitional endoplasmic reticulum organization in Pichia pastoris and Saccharomyces cerevisiae. J Cell Biol 145:69–81
53. Roth J, Bendayan M, Orci L (1978) Ultrastructural localization of intracellular antigens by the use of protein A-gold complex. J Histochem Cytochem 26:1074–1081
54. Roth J, Berger EG (1982) Immunocytochemical localization of galactosyltransferase in HeLa cells: codistribution with thiamine pyrophosphatase in trans-Golgi cisternae. J Cell Biol 93:223–229
55. Rothman JE (1994) Mechanisms of intracellular protein transport. Nature 372:55–63
56. Rust MJ, Bates M, Zhuang X (2006) Sub-diffraction-limit imaging by stochastic optical reconstruction microscopy (STORM). Nat Methods 3:793–795
57. Scales SJ, Pepperkok R, Kreis TE (1997) Visualization of ER-to-Golgi transport in living cells reveals a sequential mode of action for COPII and COPI. Cell 90:1137–1148
58. Schekman R, Orci L (1996) Coat proteins and vesicle budding. Science 271:1526–1533
59. Schermelleh L, Heintzmann R, Leonhardt H (2010) A guide to super-resolution fluorescence microscopy. J Cell Biol 190:165–175
60. Sengupta P, Jovanovic-Talisman T, Skoko D, Renz M, Veatch SL, Lippincott-Schwartz J (2011) Probing protein heterogeneity in the plasma membrane using PALM and pair correlation analysis. Nat Methods 8:969–975
61. Sharonov A, Hochstrasser RM (2006) Wide-field subdiffraction imaging by accumulated binding of diffusing probes. Proc Natl Acad Sci 103:18911–18916
62. Staehelin LA, Kang BH (2008) Nanoscale architecture of endoplasmic reticulum export sites and of Golgi membranes as determined by electron tomography. Plant Physiol 147:1454–1468
63. Starr T, Forsten-Williams K, Storrie B (2007) Both post-Golgi and intra-Golgi cycling affect the distribution of the Golgi phosphoprotein GPP130. Traffic 8:1265–1279
64. Storrie B, White J, Röttger S, Stelzer EH, Suganuma T, Nilsson T (1998) Recycling of Golgi-resident glycosyltransferases through the ER reveals a novel pathway and provides an explanation for nocodazole-induced Golgi scattering. J Cell Biol 143:1505–1521
65. Sun X, Mahajan D, Chen B, Song Z, Lu L (2021) A quantitative study of the Golgi retention of glycosyltransferases. J Cell Sci 134:jcs258564
66. Sun X, Tie HC, Chen B, Lu L (2020) Glycans function as a Golgi export signal to promote the constitutive exocytic trafficking. J Biol Chem 295:14750–14762
67. Teuchert M, Schäfer W, Berghöfer S, Hoflack B, Klenk H-D, Garten W (1999) Sorting of Furin at the Trans-Golgi network: interaction of the cytoplasmic tail sorting signals with ap-1 golgi-specific assembly proteins *. J Biol Chem 274:8199–8207
68. Tie HC, Chen B, Sun X, Cheng L, Lu L (2017) Quantitative localization of a Golgi protein by imaging its center of fluorescence mass. J Vis Exp 126:55996
69. Tie HC, Lu L (2023) Studying the organization of the Golgi by super-resolution microscopy. Methods Mol Biol 2557:113–125
70. Tie HC, Ludwig A, Sandin S, Lu L (2018) The spatial separation of processing and transport functions to the interior and periphery of the Golgi stack. elife 7:e41301
71. Tie HC, Mahajan D, Chen B, Cheng L, VanDongen AM, Lu L (2016) A novel imaging method for quantitative Golgi localization reveals differential intra-Golgi trafficking of secretory cargoes. Mol Biol Cell 27:848–861
72. Tie HC, Mahajan D, Lu L (2022) Visualizing intra-Golgi localization and transport by side-averaging Golgi ministacks. J Cell Biol 221:e202109114
73. Tie HC, Wang H, Mahajan D, Sun X, Chen B, Lu L (2024) Quantitative intra-Golgi transport and organization data suggest the stable compartment nature of the Golgi. *bioR xiv*:2024.2004.2011.589010
74. Trucco A, Polishchuk RS, Martella O, Pentima AD, Fusella A, Giandomenico DD, Pietro ES, Beznoussenko GV, Polishchuk EV, Baldassarre M (2004) Secretory traffic triggers the formation of tubular continuities across Golgi sub-compartments. Nat Cell Biol 6:1071–1081

75. Van De Moortele S, Picart R, Tixier-Vidal A, Tougard C (1993) Nocodazole and taxol affect subcellular compartments but not secretory activity of GH3B6 prolactin cells. Eur J Cell Biol 60:217–227
76. Volchuk A, Amherdt M, Ravazzola M, Brügger B, Rivera VM, Clackson T, Perrelet A, Söllner TH, Rothman JE, Orci L (2000) Megavesicles implicated in the rapid transport of intracisternal aggregates across the Golgi stack. Cell 102:335–348
77. Weigel AV, Chang CL, Shtengel G, Xu CS, Hoffman DP, Freeman M, Iyer N, Aaron J, Khuon S, Bogovic J, Qiu W, Hess HF, Lippincott-Schwartz J (2021) ER-to-Golgi protein delivery through an interwoven, tubular network extending from ER. Cell 184:2412–2429.e2416
78. Westphal V, Hell SW (2005) Nanoscale resolution in the focal plane of an optical microscope. Phys Rev Lett 94:143903

Chapter 4
Unconventional Protein Transport Across the Golgi Ribbon

Jaakko Saraste and Kristian Prydz

Abstract In addition to the classical secretory pathway, which involves the transport of endoplasmic reticulum (ER)-derived cargo through the cisternal stacks of the Golgi apparatus, several Golgi-independent transport routes contribute to the delivery of membrane-bound and soluble proteins to the plasma membrane (PM) or the extracellular space. Some of these transport events bypassing the Golgi stacks—collectively referred to as unconventional protein secretion (UcPS)—employ vesicular intermediates. However, the nature of these intermediates and the intracellular pathways utilized by various unconventional cargo are only partly understood. Notably, beyond its role in conventional secretory trafficking at the ER–Golgi interface, the intermediate compartment (IC) has been identified at the crossroads of two distinct vesicular UcPS pathways, participating in the sorting and Golgi-independent trafficking of both ER-synthesized (ER translocation signal-containing) and leaderless cytoplasmic cargo. Additionally, findings indicating that the IC directly interacts with the endocytic recycling system may help clarify how various unconventional cargo proteins reach the cell surface. Given that the central IC elements and recycling endosomes (REs)—defined by the GTPases Rab1 and Rab11, respectively—are closely associated with, and may even link the Golgi stacks, this chapter explores the connection between the classical and unconventional secretory routes and their potential divergence at the level of the Golgi ribbon.

Keywords Unconventional protein secretion (UcPS) · Golgi ribbon · Golgi bypass · ER–Golgi intermediate compartment (IC or ERGIC) · Recycling endosome (RE) · Brefeldin A (BFA) · Golgi reassembly and stacking proteins (GRASPs or GORASPs) · Rab1 · Rab11

J. Saraste (✉)
Department of Biomedicine and Molecular Imaging Center, University of Bergen, Bergen, Norway
e-mail: jaakko.saraste@uib.no

K. Prydz (✉)
Department of Biosciences, University of Oslo, Oslo, Norway
e-mail: kristian.prydz@ibv.uio.no

A. Nakano, J. Saraste (eds.), *The Golgi Network, Volume I*, Subcellular Biochemistry 110, https://doi.org/10.1007/978-3-032-06936-8_4

Introduction

Studies of professional secretory cells conducted mostly in the 1960s led to the original description of the secretory pathway [104]. For several decades, this classical ER–Golgi–PM route was regarded as the general mechanism that different eukaryotic cells employ not only for constitutive or regulated secretion of soluble proteins via exocytosis of post-Golgi carriers, but also for the delivery of integral membrane proteins to the cell surface. Both types of proteins were shown to be equipped with hydrophobic signal sequences that guide their insertion in the ER membrane or translocation into the ER lumen. The export of cargo proteins occurs at ER exit sites (ERES) through a process involving membrane-bound COPII coats [72, 86]. The proteins are subsequently transported via the ER–Golgi intermediate compartment (IC or ERGIC; [121]) to the Golgi apparatus where they undergo critical post-translational modifications, such as terminal glycosylation [132], before being delivered to the cell surface or the extracellular environment.

The basic unit of the Golgi apparatus is the cisterna, a membrane-enclosed flattened sac that occurs in varying numbers, forming the compact cisternal Golgi stacks. In cells of most vertebrates and certain invertebrates, multiple Golgi stacks are laterally connected to form a continuous ribbon of alternating compact and non-compact zones [8, 69, 81]. In contrast, other eukaryotes—including yeasts, plants, and most invertebrates—lack a Golgi ribbon structure [98]. For example, in cells of higher plants and the fruit fly *Drosophila melanogaster*, the Golgi stacks are distributed throughout the cytoplasm and closely associate with ERES. The yeast *Saccharomyces cerevisiae* features tubular Golgi networks, while protozoan parasites, such as *Giardia lamblia*, lack identifiable Golgi structures [143]. These observations underscore significant variations among eukaryotes in both Golgi organization and the overall secretory process.

The first papers questioning the monopoly of the classical secretory pathway in biosynthetic trafficking and protein secretion were published about 30–40 years ago. Strikingly, these pioneering studies showed that both yeast and mammalian cells are capable of releasing to their exterior soluble cytoplasmic proteins, such as α-factor, interleukin-1β (IL-1β), and galectins, that are devoid of signals for ER translocation [3, 23, 67, 114]. Currently, an ever-increasing number of this type of leaderless secretory proteins has been recognized [30, 60, 130]. Studies carried out a decade later provided evidence that even integral membrane glycoproteins synthesized at the ER can reach the PM without passing through the Golgi apparatus, as indicated by their lack of Golgi-specific sugar modifications. In addition, it was demonstrated that the delivery of certain proteins to the cell surface is unaffected by brefeldin A (BFA), which, by releasing membrane-bound COPI coats and disassembling the cisternal Golgi stacks, blocks transport along the classical secretory pathway ([4, 13, 92, 127]; see [46, 90, 109] for reviews). Additional evidence for Golgi bypass of secretory cargo proteins came from studies based on knockdown of machinery components acting in conventional secretory trafficking [41, 127, 161].

Today, several conserved pathways involved in unconventional protein secretion (UcPS) are known to significantly contribute to the secretory repertoire of eukaryotic cells. More detailed presentations of these Golgi-independent routes can be found in previous reviews [5, 25, 32, 40, 99, 109, 110, 166]. Following a brief introduction to the different types of UcPS, we focus here on recent studies that have revealed the universality of the pre-Golgi IC in eukaryotic cells and demonstrated its involvement in two unconventional secretory events based on membrane traffic, referred to as type III and IV UcPS. Given the functional connections of the IC with recycling endosomes (REs) and the Golgi stacks [89, 119, 122], and its emerging dual role as a sorting organelle in both classical and unconventional protein secretion, we propose that these pathways converge at the pre-Golgi level and diverge at the noncompact zones of the Golgi ribbon.

Brief Overview of UcPS

Four distinct categories of UcPS are generally recognized [112]. The type I and II UcPS pathways have been designated as *nonvesicular*, since they negotiate pore-mediated direct translocation of leaderless cytoplasmic proteins across the PM. The type I UcPS pathway has been most extensively studied in the case of fibroblast growth factor 2 (FGF2). It is based on the interaction of FGF2 with Na,K-ATPase and Tec kinase and its ability to assemble into self-made pores at specific domains of the PM. These domains are enriched in cholesterol and phosphatidylinositol 4,5-bisphosphate [PI(4,5)P_2], which binds to FGF2 triggering its homo-oligomerization. The FGF2 oligomers are released from PI(4,5)P_2 and captured by the heparan sulfate chains of glypican-1 in the extracellular leaflet of the PM [79, 131]. The type I UcPS pathway is triggered by inflammation, resulting in the release from monocytes of multiple leaderless proteins, including the prototypical UcPS cargo protein interleukin-1β (IL-1β), which can also employ type III UcPS for its release ([130], see below). In type II UcPS, the translocation of cytoplasmic substrates, such as acylated peptides, takes place via ABC transporters in a process that remains poorly characterized.

The two remaining categories of unconventional secretion—type III and IV UcPS—are also referred to as *vesicular* pathways. Namely, in contrast to direct PM passage, cargo proteins are translocated into the lumen of intracellular membrane compartments and rely on membrane traffic for their cellular release [163]. Notably, in type III UcPS the IC has been identified as the site of membrane translocation ([164]; see below), while in the type IV pathway cargo proteins pass in the conventional manner across the ER membrane. In both cases, subsequent transport to the cell surface bypasses the Golgi stacks and involves "vesicular" intermediates, whose nature remains enigmatic. Both secretory autophagosomes, various endosomal compartments, and secretory lysosomes have been implicated as way stations during cell surface delivery of cargo along these pathways [32, 110, 166]. In plant cells, unconventional Golgi-independent trafficking has been observed not only for proteins destined to the cell surface, but also for those transported from the ER to intracellular organelles [7].

Most extensively characterized cargo proteins of the type III UcPS pathway are cytokines of the interleukin family, particularly IL-1β. As mentioned above, monocytes can utilize alternative pathways for IL-1β secretion depending on their level of stimulation [130]. In pioneering studies of type IV UcPS, two glycoforms of CD45, a protein tyrosine phosphatase located at the PM, were shown to follow distinct Golgi-independent or -dependent pathways, which are differentially affected by BFA. The BFA-resistant, high-mannose form of CD45 reached the PM more rapidly than the BFA-sensitive, terminally glycosylated form [4]. Research on the cystic fibrosis transmembrane conductance regulator (CFTR) and its transport defective mutant (ΔF508 CFTR) provided evidence for the involvement of endocytic compartments in UcPS [161] and revealed the role GRASPs (also referred to as GORASPs)—IC/*cis*-Golgi-localized membrane tethering proteins (see below)—in this process [41, 111]. Previous studies of the slime mold *Dictyostelium discoideum* and the fruit fly *Drosophila melanogaster* had identified the orthologs of the mammalian GRASPs—GrpA and dGRASP—as key players in Golgi-independent trafficking of the leaderless cytoplasmic protein AcbA (type III UcPS), and membrane-bound α-integrins (type IV UcPS), respectively [63, 127].

Studies on the export of Acb1—the yeast counterpart of *Dictyostelium* AcbA—have led to the identification of a novel membrane compartment known as CUPS (Compartment for Unconventional Protein Secretion) as an important sorting station in type III UcPS [16, 24]. In response to nutrient starvation, Grh1—the yeast ortholog of mammalian GRASPs—relocates to CUPS, playing a key role in its formation. CUPS has been proposed to represent a hybrid compartment comprised of membranes derived from the Golgi apparatus and endosomes [25]. However, its close association with ERES and saccular and tubulovesicular morphology [26]—resembling that of the IC in mammalian cells—as well as recent identification of Grh1 as a component of the IC/ERGIC in yeast [141] suggests an alternative scenario. Thus, the formation of CUPS could be based on a functional link between the IC and endosomes, a connection that has also been suggested to facilitate Golgi bypass in higher eukaryotes ([89, 121; 122] see below). Recent studies showing that unconventional protein secretion in yeast involves direct contacts between the Grh1-positive CUPS and a modified TGN—which also functions as an endocytic recycling compartment—support this possibility [27].

Current State of the IC Network

The pleiomorphic IC comprises saccular and tubular subdomains that facilitate bidirectional transport in the early secretory pathway between the widespread ERES and the central *cis*-Golgi apparatus. Generally, the IC has been viewed as a transient structure, despite accumulating evidence supporting a more permanent nature of this compartment [121]. For instance, its stability has been demonstrated by studies on the effects of BFA, demonstrating that IC/*cis*-Golgi markers—such as p58/ERGIC53, Rab1/Ypt1, KDEL-receptor, GM130, and GRASP65/Grh1—persist in the tubulovesicular IC clusters in the drug-treated cells [36, 59, 76, 89, 97, 117, 124, 128, 139, 141, 150, 153]. This is in sharp contrast to the Golgi stacks that in most cell types are completely disassembled by BFA, resulting in the redistribution of Golgi enzymes to the ER. The introduction of GFP-Rab1 as a fluorescent reporter in live cell imaging revealed that the IC retains its dynamic properties in BFA-treated cells. The ongoing communication between peripheral and central IC elements via Rab1-positive tubules [89] is inconsistent with the prevailing notion that the drug completely blocks ER exit. Recent live imaging studies have provided important additional information on the distinct saccular and tubular IC subdomains, defined by the cargo receptors p58/ERGIC-53 and SURF4, as well as on fast ER–Golgi trafficking, mediated by Rab1-positive tubules involved in the transport of specialized cargo [156, 159].

The use of GFP-Rab1 in live-cell imaging further demonstrated that the dynamic IC network is associated with the centrosome, revealing the direct functional connection between the IC and the endocytic recycling system [89]. Namely, in dividing and motile cells—coinciding with centrosome repositioning—the Rab1-positive IC tubules and Rab11-positive REs coalesce in the pericentrosomal region as a result of their separation from the Golgi ribbon ([122]; Fig. 4.1). Additionally, these pericentrosomal "mirror compartments" persist and maintain their connection in BFA-treated cells, which allowed the demonstration of their role as way stations in Golgi-independent trafficking of CFTR to the cell surface ([89]; see below).

The enigmatic nature of the IC partly stems from studies of various model organisms demonstrating considerable variation in the organization of the ER–Golgi interface among different eukaryotes. Vertebrate cells typically feature a widespread ER network spanning the cytoplasm and numerous Golgi stacks that form a continuous ribbon in the pericentrosomal region adjacent to the nucleus. In invertebrates (e.g., *D. melanogaster* and *C. elegans*)—with some exceptions [8]—as well as in plants (e.g., *Arabidopsis*) and fission yeasts (e.g., *Pichia pastoris*), the widely distributed ERES and Golgi stacks associate to create compact secretory units. Furthermore, the budding yeast *Saccharomyces cerevisiae* contains tubular networks but lacks cisternal Golgi stacks [98]. The prevailing view has been that the IC evolved to fulfill the specific requirements of larger animal cells, where the pleiomorphic IC elements serve as transient carriers to facilitate microtubule (MT)-based long-distance transport between peripheral ERES and the central Golgi ribbon. In contrast, the IC has been thought to be unnecessary in invertebrates, plants, and yeasts, where short-range ER–Golgi communication is expected to rely on vesicular trafficking.

However, recent high-resolution imaging studies of the various model organisms are changing the picture. Research on ERES-Golgi units in *Drosophila*, utilizing focused ion beam scanning electron microscopy (FIB-SEM), has identified a compact pre-Golgi compartment located between the ERES and the *cis*-face of the Golgi stacks. This compartment contains the IC-localized dGRASP, the Rab1 effector GM130, as well as COPI and COPII protein coats that facilitate two-way

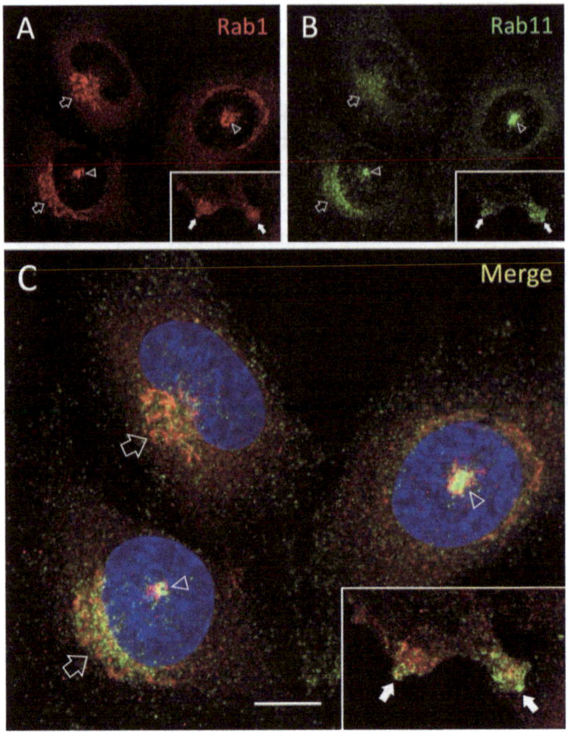

Fig. 4.1 Spatial connections between the IC elements and REs, defined by the GTPases Rab1 and Rab11, respectively. (**a**, **b**) Normal rat kidney (NRK) cells were double-stained for immunofluorescence using antibodies against Rab1 and Rab11and imaged with a confocal microscope. In two of the cells, the centrosomes have repositioned from the Golgi region to the cell center beneath the nucleus due to cell motility or onset of mitosis. (**c**) Note the extensive overlap of the IC elements and REs in the pericentrosomal area (open arrowheads) and the Golgi ribbon (open arrows) adjacent to the nucleus. The insets show the co-distribution of these elements (white arrows) at the cell periphery. See the text for further discussion. Bar 5 μm

ER–Golgi trafficking, thus sharing characteristics with the IC of vertebrate cells [160]. Furthermore, similar to BFA treatment of mammalian cells [59, 97, 150], GM130 persists in discrete foci in *Drosophila* S2 cells in response to COPI disassembly, while Golgi enzymes redistribute to the ER [35]. Additionally, the localization of p58/ERGIC53 (also known as ILE-1) and RER1 has identified the IC as a distinct compartment in the epithelial cells of *C. elegans,* suggesting its involvement in unconventional trafficking of membrane proteins (type IV UcPS) to the apical cell surface ([152]; see below).

Regarding plants, Ito and colleagues examined BFA-treated tobacco BY-2 cells and discovered that RER1 and SYP31—the plant counterpart of the mammalian IC/*cis*-Golgi SNARE syntaxin 5—persist in punctate structures, which also serve as a platform for Golgi reassembly after the removal of the drug. Similar to the BFA-resistant IC in mammalian cells, this compartment functions as the site of cargo arrival from the ER and has been designated Golgi Entry Core Compartment (GECCO) [55–57]. Notably, recent live-cell imaging studies of *Arabidopsis* suggest that the plant GECCO comprises two distinct subdomains, a widespread tubular network and a more stable domain that closely associates with the *cis*-Golgi [33]. Finally, super-resolution confocal live imaging microscopy (SCLIM) of the yeast counterparts of mammalian IC components Grh1 (mammalian GRASP65), Emp47 (p58/ERGIC53), and Ypt1 (Rab1) has led to the identification of the IC as a distinct compartment in the early secretory pathway of the budding yeast *S. cerevisiae*. Similar to the GECCO in plant cells, the yeast ERGIC persists as distinct structures in BFA-treated cells [141].

The studies discussed above indicate the presence in yeasts, plants, and invertebrates of a pre-Golgi compartment that is compositionally and functionally analogous to the mammalian IC. The universality of the IC is consistent with the high conservation of transport machineries involved in ER–Golgi trafficking across the eukaryotic kingdom [12, 72, 121]. It also aligns with the demonstration of the function of the IC as a membrane source in autophagy, an evolutionarily conserved degradation pathway in eukaryotic cells [37, 50]. More specifically, upon stimulation of autophagy, the IC forms specialized contacts with ERES, facilitating the relocation of COPII coats and the formation of IC-derived COPII vesicles that serve as precursors for autophagosome formation [38, 39, 75]. Furthermore, as will be discussed in the following, the ubiquitous nature of the IC is also relevant in terms of understanding the unconventional secretory pathways—particularly vesicular type III and IV UcPS—which are known to operate in different eukaryotic cells [5, 32, 166].

Role of IC in Translocation of Leaderless Proteins (Type III UcPS)

Although the role of membrane traffic in type III UcPS has been known for some time, where and how cytoplasmic leaderless proteins are translocated across membranes has remained unknown. Recently, in an elegant study, Zhang et al. [164] developed a mass spectrometry assay to search for proteins involved in the unconventional secretion of IL-1β from macrophages. Importantly, this study identified TMED10 (transmembrane Emp24 domain-containing protein 10) as a key component of the translocation machinery. TMED10 (also known as p23) is a single-spanning membrane protein that belongs to the p24 family of proteins with well-established functions in bi-directional ER–Golgi trafficking and organization of the early secretory pathway [11, 106, 134]. Interestingly, comparison of the localizations of TMED10 and IL-1β with that of ERGIC53, along with immuno-EM studies, identified the IC as the site of translocation for leaderless proteins [164]. TMED10 forms homo-oligomers to create transmembrane channels in the IC membrane, facilitating the passage of unfolded cargo proteins into the IC lumen in a process that is assisted by molecular chaperones.

In a follow-up study, the same research group set out to investigate the molecular machineries of type III UcPS, with special interest in possible regulatory roles of the IC-associated Rab proteins—Rab1A, Rab1B, and Rab2A—in unconventional secretion of IL-1β [135]. Strikingly, in addition to the previously established traditional roles of these small GTPases in ER–Golgi trafficking and Golgi organization [45, 121], both Rabs turned out to play specific roles in UcPS, participating in different transport steps. The two Rab1 isoforms were found to influence the entry of IL-1β into the IC lumen by promoting homo-oligomerization of TMED10. On the other hand, Rab2A was found to interact with both TMED10 and its effector, the MT-based motor protein kinesin KIF5B, thereby affecting MT-dependent reorganization of the IC to support its role in unconventional secretion. Obviously, such compartmentalization relates to the question of how the IC is able to differentiate between its dual functions in conventional and unconventional secretory trafficking. Notably, correlative light and electron microscopy (CLEM) provided evidence suggesting that these two different secretory events take place at spatially distinct regions of the IC network [135].

Following its exit from a specialized IC subdomain, IL-1β is delivered to the cell surface along a pathway that bypasses the Golgi apparatus [114], but remains otherwise poorly understood. Secretory autophagosomes have been proposed as candidates to mediate this transport step [31], in accordance with the function of the IC as a membrane source in the formation of autophagosomes [37, 50]. As IL-1β secretion is also controlled by the RE-associated Rab8A [31], the possibility exists that a specific subdomain of the IC interacts with the endosomal system, resulting in the formation of a compartment analogous to the CUPS operating in type III UcPS in yeast [25]. Similar to the yeast Grh1, the IC/*cis*-Golgi-localized GRASPs, which are also required for type III UcPS in mammalian cells [31, 163], could function in membrane tethering events required for the biogenesis of such a hybrid compartment (see below).

Golgi Bypass Based on Direct IC-Endosome Connection (Type IV UcPS)

Based on their communication with the *trans*-Golgi network (TGN), REs have been shown to participate in post-Golgi sorting and biosynthetic transport of newly synthesized proteins to the cell surface [2, 78, 136]. For example, in macrophages REs represent the principal sorting station for the secretion of cytokines [96]. In addition, the exchange of soluble and membrane-bound components between the secretory and endocytic pathways can take place at the pre-Golgi level. As summarized in the following, there is a growing body of evidence supporting direct communication between the IC and the endocytic recycling apparatus, opening up for unconventional secretory pathway(s) that allow membrane proteins to bypass the Golgi apparatus (type IV UcPS).

The best characterized integral membrane protein that is synthesized at the ER and follows a Golgi bypass route to the PM is CFTR, a chloride channel at the apical surface of epithelial cells [40]. Dissection of the molecular machineries required for cell surface delivery of CFTR first suggested the involvement of an endosomal intermediate in type IV UcPS [161]. Subsequent microscopic studies on CFTR transport under synchronized conditions demonstrated its localization to a tubular subcompartment of the IC, defined by Rab1 [89]. As mentioned above, this IC domain closely associates with the Rab11-positive REs in the pericentrosomal region (Fig. 4.1) and maintains this connection in BFA-treated cells. Accordingly, subsequent to its ER exit, CFTR was found to transit through this pericentrosomal membrane system and follow a Golgi-independent route to the cell surface [89]. Supporting this conclusion, the RE-associated GTPases Rab8 and Rab11 have been shown to regulate CFTR trafficking [100, 148], which also involves the GRASP proteins [41]. Moreover, similar to type III UcPS (see above), the ER stress-induced unconventional secretion of transmembrane proteins, including CFTR, has been shown to depend on selected TMED proteins [105].

The employment of CFTR as a prototype UcPS cargo suggested a pathway that certain ER-derived membrane proteins could follow during their BFA-resistant delivery to the PM [89]. Subsequent studies of membrane protein trafficking in neuronal dendrites have provided strong additional support for the concept of Golgi bypass, based on a direct connection between the early biosynthetic pathway and the endosomal system [14, 51]. Namely, the secretory apparatus at the periphery of dendrites consists of an ER network, ERES, and pleiomorphic IC elements and REs, which are distributed throughout the dendrites and closely associate with synaptic terminals. However, it lacks the so-called Golgi outposts, which, when present, are typically restricted to the proximal dendritic branch points close to the cell body [48, 61]. As a result, the neighboring IC elements and REs, residing close to ERES, can establish a Golgi-independent secretory route for local delivery of proteins to the synaptic PM. Indeed, a multitude of newly synthesized membrane glycoproteins—including ion channels, adhesion proteins, and neurotransmitter receptors—were found to reach the synaptic surface in high-mannose form; that is, lacking Golgi-specific modifications. Moreover, cell surface delivery of these proteins, particularly in mature neurons, was unaffected by BFA [14, 51]. Importantly, while Golgi bypass of transmembrane proteins has often been linked to special conditions, such as cellular stress [40, 44], the studies on neuronal secretion revealed an important role of type IV UcPS as a constitutive cellular process, mediating protein traffic under normal physiological conditions.

Type IV UcPS is also involved in constitutive biosynthetic membrane traffic in polarized epithelia, as demonstrated by studies introducing the intestine of the nematode *C. elegans* as an in vivo multicellular model [152]. The P-glycoprotein PGP-1—an ATP-binding cassette transporter used as a model cargo protein—was found to arrive at the apical PM of epithelial cells carrying high-mannose glycans, following a Golgi bypass pathway that is resistant to BFA. Its unconventional cell surface delivery was shown to be regulated by SMGL-1, a guanidine nucleotide exchange factor (GEF) that activates the RE-associated Rab8. Strikingly, SMGL-1 turned out to localize to the IC, revealing the key role of this compartment in Golgi bypass of PGP-1. In addition, SMGL-1 is present in Rab8-positive REs, which are spatially and functionally connected with the IC elements [152]. Further work by the same team elucidated a complex regulatory cascade between Rab8 and Rab11, controlling the activity of SMGL-1 and the function of successive subpopulations of REs in the late stages of unconventional PGP-1 trafficking [74].

The IC has also been introduced as a sorting site in polarized mammalian cells. In BFA-treated epithelial MDCK cells, where the Golgi does not undergo typical disassembly [117], secretory cargo molecules are still forced to take Golgi-independent transport routes to the apical and basolateral cell surface domains. For the model protein serglycin, this transport occurred with the same polarity as its Golgi-dependent transport in nontreated cells, indicating that sorting had taken place before Golgi entry [108, 145].

The function of a specialized glucose transporter GLUT4 storage compartment (GSC) in adipocytes and skeletal muscle cells provides another example regarding direct communication between the IC and the endocytic recycling apparatus [73]. The newly synthesized GLUT4 employs a Golgi-bypass pathway during its transport from the IC to the tubulovesicular GSC [18, 103]. This compartment also receives GLUT4 from the REs, following internalization of the transporter from the PM. The formation of the GSC and insulin-stimulated delivery of GLUT4 from this storage compartment to the PM are both Golgi-independent processes that are unaffected by BFA [93]. A similar situation has been recently described in the highly conserved Notch-signaling pathway. The Notch receptor is intracellularly stored in the IC, but upon cell activation becomes mobilized and rapidly delivered via REs to the cell surface [10]. Furthermore, newly synthesized membrane proteins of the primary cilium, including polycystin-1 and -2, are transported to this specialized domain of the PM following a BFA-resistant pathway, which is regulated by the RE-associated Rab8, Rab10, and Rab11 [43, 53, 64].

Divergence of Secretory Pathways at the Golgi Ribbon

The employment of Rab1 and Rab11 as markers has revealed the overlapping distributions of IC elements and REs in the pericentrosomal region and the cell periphery [88, 89, 118, 137, 158]. Furthermore, we proposed that these compartments meet at the level of the Golgi ribbon [122], as also evidenced by the localization of endogenous Rab1 and Rab11 in cultured cells (Fig. 4.1). The close connection of central IC elements and *cis*-Golgi cisternae has been demonstrated by earlier EM studies [69, 121, 125], and is supported by the examination of IC dynamics in living cells [157]. On the other hand, imaging studies of sea urchin, *Drosophila,* and mammalian cells have revealed the attachment of Rab11-positive REs with *trans*-Golgi, suggesting their close relationship with the TGN [34, 35, 167]. Indeed, the recently described meeting of secretory, endosomal, and autophagic pathways at the *trans*-Golgi/TGN [167] calls for reevaluation of the GERL (Golgi–ER–Lysosome) hypothesis, which suggested a direct functional connection between the early secretory pathway, *trans*-Golgi, and lysosomes—bypassing the Golgi stacks [101].

Challenging the prevailing notion that Golgi stacks are interconnected by thin tubules, we proposed that both the IC elements and REs localize to the noncompact zones of the Golgi ribbon, creating neighboring "linker compartments" that dynamically connect the cisternal stacks ([122]; Fig. 4.2a). This model is supported by EM tomographic analysis on Golgi organization in cultured and tissue cells, showing that the noncompact regions of the ribbon are occupied by pleiomorphic structures resembling IC elements or endosomes [69, 94]. In addition, it incorporates the idea that at the onset of mitosis, coinciding with Golgi fragmentation (unlinking of the stacks), the linker compartments separate from the Golgi ribbon and accumulate around the centrosome, establishing the pericentrosomal membrane system that has been shown to persist throughout mitosis [52, 88, 138]. As mentioned above, similar events take place during Golgi disassembly by BFA, resulting in the expansion of these pericentrosomal compartments defined by Rab1 and Rab11 [89]. Related to the present discussion, our model suggested that the interconnected IC elements and REs at the noncompact zones create a pathway that traverses the ribbon structure, effectively bypassing the Golgi stacks ([122]; Fig. 4.2a).

The positioning of the linker compartments at the intersection of various secretory pathways is supported by new information on the localization of GRASP proteins in IC/*cis*-Golgi, as well as their dual roles in unconventional secretion and organization of the Golgi ribbon. As previously noted, GRASPs are known to function across the eukaryotic kingdom in both type III and IV UcPS pathways that involve the IC to facilitate the secretion of cargoes such as Acb1, IL-1β, and CFTR [27, 31, 41, 63, 87, 163]. Importantly, studies employing acute GRASP depletion appear to have resolved a long-standing controversy regarding the function of these proteins in Golgi organization. Rather than facilitating the stacking of Golgi cisternae, as previously thought, the fundamental role of GRASPs in mammalian cells is to connect the individual stacks into a ribbon structure [49, 58, 165]. This newly clarified role places GRASPs at the rims of the cisternal stacks, consistent with results obtained using super-resolution light microscopy [140]. Furthermore, studies in yeast, *Drosophila* and mammalian cells have established that the IC is the primary localization for the various GRASP homologs [65, 88, 91, 141, 160]. Collectively, these findings open the possibility that GRASPs—rather than directly interconnecting the Golgi stacks via tubular protrusions—mediate the tethering of IC elements at the noncompact zones, thereby facilitating both ribbon formation and unconventional transport that bypasses the Golgi stacks (Fig. 4.2a).

The branching of the secretory process at the noncompact zones of the Golgi ribbon may clarify the origins of different glycoforms of certain PM proteins delivered to the cell surface via conventional secretion or type IV UcPS. As previously discussed, the protein tyrosine phosphatase CD45 reaches the PM with either high mannose or complex glycans, following pathways that demonstrate differential sensitivity to BFA. Moreover, the high-mannose form of the protein arrives at the cell

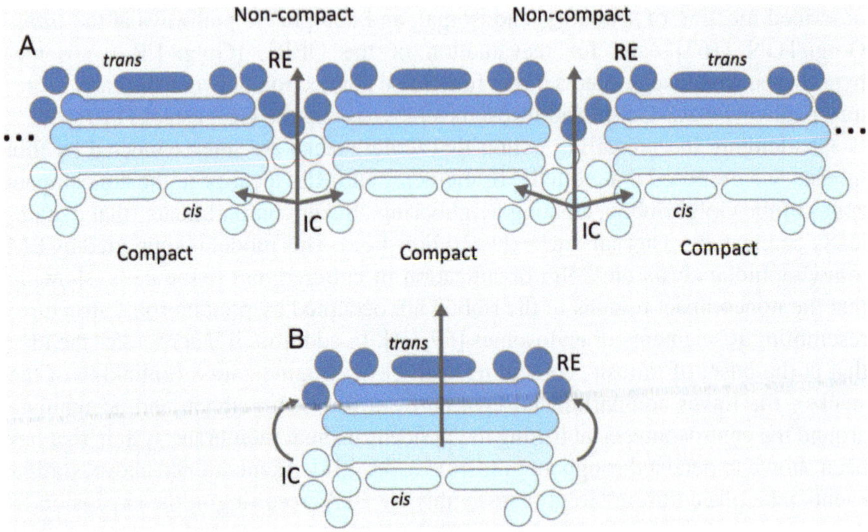

Fig. 4.2 Schematic models for Golgi organization and trafficking in vertebrate (**a**) and invertebrate (**b**) cells. (**a**) In vertebrates, multiple cisternal Golgi stacks (three are shown) are typically organized into a ribbon structure with alternating compact and noncompact zones. At the noncompact zones, communication between the IC elements and REs creates an unconventional pathway across the Golgi ribbon that bypasses the Golgi stacks. The Golgi-dependent and -independent secretory routes bifurcate at the level of the IC elements that associate with the *cis*-side of the Golgi ribbon. (**b**) In invertebrates, such as *Drosophila*, the Golgi stacks are actively kept as separate structures. Nonetheless, direct communication between the IC elements and REs, associated with the *cis*- and *trans*-sides of the Golgi, could generate a pathway that bypasses the Golgi stacks

surface approximately three times faster than the complex form, which receives Golgi-specific modifications [4]. Another similar example is provided by the double-stranded RNA receptor SID-2 in *C. elegans*, which utilizes both conventional and unconventional trafficking to reach the apical surface of intestinal epithelial cells [74]. The models depicted in Fig. 4.2 suggest that the central IC elements serve as sorting sites, where the decision between Golgi stack-dependent and -independent PM delivery is made, based on the segregation of cargo into different subdomains of these structures. Generally, by responding to intra- or extracellular signals, the conditions in the IC lumen could regulate the sorting process and determine the mode of secretory trafficking under various physiological conditions and cellular stress.

The vesicular pathways of UcPS are often triggered by cellular stress, potentially compensating for its inhibitory effects on conventional secretion [40, 102, 110]. Exposure of mammalian cells to various stress conditions typically leads to fragmentation of the Golgi ribbon [84] and may involve the relocation of GRASPs [62]. In addition to Golgi fragmentation, the stress conditions prevalent in many cancer cells result in the missorting of proteins, such as Golgi enzymes, to the cell surface and abnormal glycosylation of PM components [85, 113, 154, 162]. Notably, the delivery of proteins to the surface of cancer cells appears to involve unconventional pathways [54, 147]. For instance, the KDEL receptor-mediated retention of luminal ER chaperones—including GRP78/BiP, GRP94, and calreticulin—within the early secretory pathway is impaired in certain cancer cells, leading to their escape to the cell surface [144]. Importantly, these proteins are transported to the PM in a BFA-resistant manner, mediated by the RE-associated Rab4 and Rab11, thereby likely utilizing the type IV UcPS pathway [146]. Similarly, although their precise intracellular itineraries remain to be clarified, key proteins linked to cancer progression—including metalloproteinases such as MMP2 and MMP14 (also called MT1-MMP) and the cell adhesion protein E-cadherin—can reach the cell surface in a Golgi-independent manner via the endocytic recycling system defined by Rab8, Rab10, and Rab11 [29, 78, 80, 82, 102]. Moreover, MMP14 is recruited to the PM from an endosome-related storage compartment [15], resembling the mobilization of the insulin-responsive glucose transporter GLUT4.

It is commonly stated that cellular stress leads to the induction of additional pathways of protein secretion. However, the above observations suggest that stress conditions in certain cancer cells could—among other effects—alter the acidic pH prevailing in the lumen of IC elements and REs. These changes may result in the disconnection of linker compartments from the Golgi stacks, thereby disrupting Golgi ribbon integrity and conventional protein secretion. Instead, they may facilitate Golgi stack-independent delivery of proteins via UcPS to the cell surface and the tumor environment. Similar considerations could apply to virus egress, as discussed below.

Unconventional Cellular Exit of Coronaviruses and Other Large Cargo

Coronaviruses (CoVs) constitute a large family of enveloped RNA viruses that assemble by budding into the lumen of the IC. Due to their intracellular mode of maturation, these viruses depend on the host cell's secretory machinery for their cellular release. Early EM observations identified progeny CoV particles within the dilated rims of Golgi cisternae, suggesting that they are transported from the IC to the cell surface following the classical Golgi-dependent pathway ([142]; Fig. 4.3). However, neutralization of the acidic IC and Golgi lumen by the viral envelope (E) protein—an ion channel—results in major changes in the Golgi organization, including fragmentation of the Golgi ribbon and dilation of the cisternal stacks [129, 151, 155]. This contradiction could be solved if CoVs exploit alternative Golgi-independent route(s) for their cellular exit [107, 115, 123].

Light microscopy studies indeed provided evidence that following their synchronized exit from the Rab1-positive IC elements, newly assembled CoV particles are transported to the cell periphery in Rab11-containing endosomal carriers ([126]; Fig. 4.3). These findings support the notion that they exploit the type IV UcPS

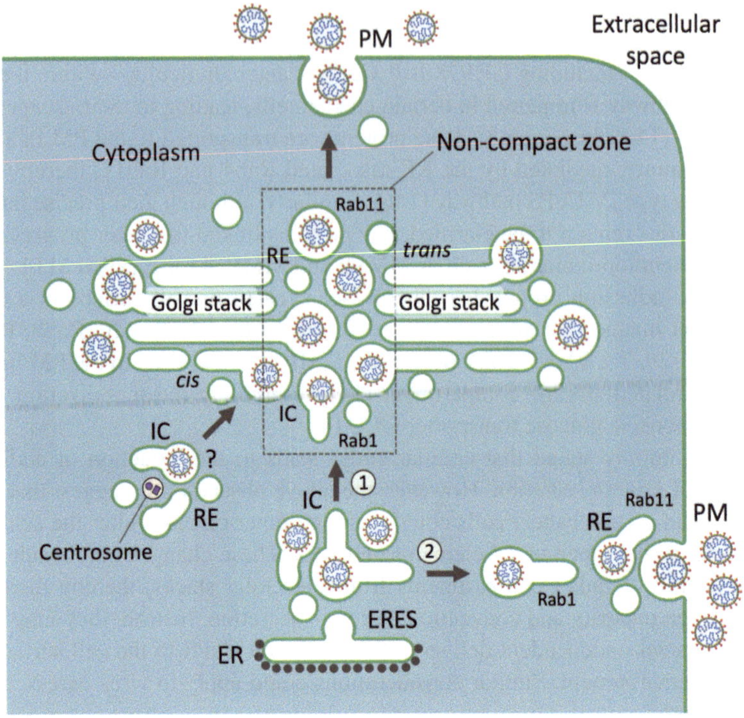

Fig. 4.3 Potential unconventional secretory routes exploited by CoVs during their cellular exit. Only two stacks of the Golgi ribbon are shown. After budding into the lumen of the IC near ERES, the progeny CoVs are packaged into saccular IC carriers that move along MT tracks, preferentially arriving at the noncompact zones of the ribbon (**1**). Subsequently, the virus particles are transferred from the pre-Golgi carriers to REs, entering the endocytic recycling circuit, which mediates their cellular exit. The mechanism of IC-RE communication may involve a GTPase cascade that functionally links Rab1 and Rab11. At the noncompact zones, the IC elements and REs also interact with the Golgi stacks, explaining the occasional presence of CoVs at the dilated rims of the Golgi cisternae, particularly at early times of infection before Golgi breakdown. The newly assembled viruses may also hijack a direct pathway (**2**) to the cell periphery [118], where the connection between IC carriers with REs leads to virus release. It remains unknown whether CoV assembly also takes place at the pericentrosomal IC elements. (Adapted from Ref. [123])

pathway, which involves a direct connection between the IC and the endocytic recycling system [123]. Similar to cancer cells, CoV infection induces cellular stress, which potentially separates the linker compartments from the Golgi stacks and boosts an "unconventional" secretory pathway for efficient virus egress (Fig. 4.3). However, as the state of endomembranes changes in the course of CoV infection, progeny viruses that have entered the endosomal system may later exit the cells via alternative Golgi-independent pathways, such as lysosomal secretion ([42]; reviewed by [129]).

Intriguingly, the E proteins produced by severe CoVs, such as the pandemic SARS-CoV-2, have evolved the ability to interact with TMED10, the protein translocation channel located in the IC membranes that functions in type III UcPS [164].

This molecular interaction activates the oligomerization of TMED10 and enhances the translocation of leaderless cargo into the IC lumen, thereby stimulating the unconventional secretion of various proinflammatory factors, including interleukins such as IL-1β [77]. The outcome is a state of hyperinflammation characterized by the development of a "cytokine cloud," which leads to dysfunction in lungs and other organs, contributing to the increased lethality of the infection.

Studying the egress of intracellularly budding viruses may shed light on the secretory mechanisms for large cargo in general [83, 123]. However, unlike virus release, the efficient secretion of exogenous and endogenous large cargo—such as engineered protein aggregates and extracellular matrix proteins like collagen I—has been reported to depend on an intact Golgi ribbon structure [71]. It should be noted, however, that the methods used in this study to experimentally fragment the Golgi ribbon—the depolymerization of MTs using nocodazole or the knockdown of GRASPs—can exert their cellular effects by disrupting the MT-dependent positioning of the linker compartments at the noncompact zones of the Golgi ribbon ([122]; Fig. 4.2a), or by impairing their role in unconventional transport of large cargo (Fig. 4.3). Nonetheless, the preferential localization of bulky cargoes, such as procollagens, to the periphery of the Golgi stacks [140] is consistent with the rim progression hypothesis, which proposes that the passage of large cargoes across the Golgi relies on dynamic lateral connections between neighboring stacks [71].

Furthermore, EM studies have shown that engineered protein aggregates [149], as well as endogenous very low-density lipoprotein (VLDL) particles [28], are excluded from the central compact regions of the Golgi stacks, where cargo-modifying enzymes are enriched [140]. Instead, they localize to the dilated rims of the cisternae, or to free membrane-enclosed structures between adjacent Golgi stacks. A striking observation was that engineered protein aggregates accumulating intracellularly at 15 °C could traverse the Golgi apparatus in just 10 min after the temperature was raised to 20 °C. Incubation at 15 °C causes secretory cargo to become arrested in the IC [120], while maintaining cells at 20 °C prevents cargo from exiting the TGN [47]. At both temperatures, the large protein aggregates were observed at the periphery of the Golgi stacks, where the linker compartments—IC and RE—are also expected to reside (Fig. 4.3). In hepatocytes, the VLDL-associated protein apoE was localized to both endosomes and the dilated rims of the Golgi stacks, indicating a functional relationship between these structures, but was absent from the compact interior of the Golgi stacks [28]. Additionally, an EM study of intestinal absorptive cells from fasting rats revealed the localization of endogenous lipoprotein particles in large membrane-bounded structures near the Golgi stacks [116].

To determine whether large secretory cargo is transported vectorially across the Golgi ribbon simply by partitioning into the dilated rims of the moving Golgi cisternae or utilizes a direct IC-to-RE route that bypasses the Golgi stacks, it would be informative to compare by EM the localizations of relevant cargoes with markers of the linker compartments and the TGN. However, the fast transit of protein aggregates from the *cis*- to the *trans*-side of the Golgi apparatus [149] suggests that large cargo follows the same Golgi-bypass route that has been proposed for CoVs ([123]; Fig. 4.3).

Conclusions and Perspectives

Building on the newly discovered role of the IC as a common sorting organelle in both classical and unconventional secretion [135, 164], this review explores the possibility that these processes are functionally connected in the early secretory pathway before diverging at the Golgi level. At the noncompact zones of the Golgi ribbon, the central IC elements are proposed to functionally interact not only with *cis*-Golgi membranes but also with *trans*-Golgi-associated REs, establishing a pathway for rapid PM delivery of specific cargo that bypasses the cisternal Golgi stacks (Fig. 4.2a). The convergence of these alternative pathways at the IC could explain how cells are able to switch between Golgi-dependent and -independent modes of secretion in response to various physiological conditions or cellular stress. Moreover, given the predominant association of GRASPs with the IC and *cis*-Golgi, our current model may help clarify the seemingly contradictory roles of these proteins in unconventional secretion and linking of the Golgi ribbon [17].

This discussion primarily focuses on eukaryotic cells in which the Golgi stacks are organized into a ribbon structure. But what about cells containing separate Golgi stacks? Recent studies on *Drosophila* have shown that the IC elements and REs closely associate with the *cis*- and *trans*-sides of the Golgi stacks, respectively [34, 160, 167]. Furthermore, live-cell imaging demonstrated the dynamic nature of the dispersed Golgi stacks, revealing that they are actively kept apart, but can frequently merge before separating again. Notably, the REs in the cells of the fruit fly alternate between Golgi-associated or free states [34]. If their detachment from *trans*-Golgi is inhibited by BFA, the Golgi stacks form ribbon-like structures while retaining their *cis-trans* polarity [35]. Besides suggesting a role for REs in the formation of the Golgi ribbon, these findings suggest that direct communication between the IC elements and REs takes place also in invertebrates such as *Drosophila*, creating an unconventional pathway that bypasses the Golgi stacks (Fig. 4.2b).

The vesicular UcPS pathways are highly conserved across eukaryotes, despite significant variations in the organization of their endomembrane systems [25, 32, 99, 109, 166]. These pathways are functional not only in cells with a continuous Golgi ribbon or isolated Golgi stacks, but also in those exhibiting dispersed Golgi cisternae, such as the budding yeast *Saccharomyces cerevisiae* [27]. These pathways are also expressed in primitive eukaryotes that lack identifiable Golgi structures, as exemplified by the unicellular parasite *Giardia lamblia* [5], indicating that passage through Golgi cisternae is not a prerequisite for successful protein secretion. Notably, orthologs of Rab1 and Rab2 are among the eight conserved Rabs identified in *Giardia* [70, 133], suggesting that these cells—much like those of other eukaryotes—contain a post-ER secretory compartment with IC characteristics. The core Rabs identified in *Giardia* also include Rab11, which defines the REs [20]. Moreover, the simple endomembrane system of *Giardia* has been proposed to include a tubulovesicular compartment performing both secretory and endocytic functions [1].

In conclusion, there is compelling evidence to suggest that the UcPS pathways predated the emergence of Golgi-dependent protein secretion in eukaryotic precursor cells around 2 billion years ago [6]. The observation that RNA viruses, such as CoVs, utilize unconventional transport mechanisms for cellular egress further underscores the ancient origins of these pathways [123, 126]. From an evolutionary standpoint, although these pathways were identified later than the classical Golgi-dependent secretory route, they may not be so unconventional after all [5]. In fact, instead of merely serving in the disposal of overexpressed proteins or providing additional means for protein secretion during cellular stress, these pathways are likely to represent fundamental transport mechanisms with significant physiological roles.

Differences in the transport kinetics of secretory and PM proteins have been observed for decades, both for proteins following the classical transport route and for those utilizing Golgi-independent mechanisms [9, 19, 68, 159]. Unicellular eukaryotes and cells of complex multicellular organisms often face the need for rapid transport of membrane components to specialized subdomains of the PM. Local membrane expansion is essential for processes like phagocytosis and cell migration, as well as the formation of autophagosomes. The ability to perform phagocytosis emerged early in eukaryotic evolution [66, 95], while autophagy has been observed in simple unicellular eukaryotes, including *Giardia lamblia* [22]. Consequently, the development of pathways that facilitate membrane delivery for these processes likely predates the appearance of the Golgi apparatus. In dendritic cells, the IC-associated SNARE Sec22b is essential for MHC class I-mediated antigen cross-presentation, regulating the delivery of membrane components such as p58/ERGIC-53 from the IC to the forming phagosomes [21]. The IC also serves as a membrane source for the formation of the phagophore, which marks the initial step in autophagy [37, 50]. As a result, the IC is emerging as a sorting hub that operates at the crossroads of classical and unconventional secretory pathways, distributing membrane components and soluble cargo proteins to various intracellular and extracellular destinations.

Acknowledgments We are grateful to Dr. Carolyn Machamer for critical reading and constructive comments on the manuscript.

Conflict of Interest The authors declare no conflicts of interest.

References

1. Abodeely M, DuBois KN, Hehl A, Stefanic S, Sajid M, deSouza W, Attias M, Engel JC, Hsieh I, Fetter RD, McKerrow JH (2009) A contiguous compartment functions as endoplasmic reticulum and endosome/lysosome in *Giardia lamblia*. Eukaryot Cell 8:1665–1676. https://doi.org/10.1128/EC.00123-09
2. Ang AL, Taguchi T, Francis S, Fölsch H, Murrells LJ, Pypaert M, Warren G, Mellman I (2004) Recycling endosomes can serve as intermediates during transport from the Golgi to

the plasma membrane of MDCK cells. J Cell Biol 167:531–543. https://doi.org/10.1083/jcb.200408165
3. Auron PE, Webb AC, Rosenwasser LJ, Mucci SF, Rich A, Wolff SM, Dinarello CA (1984) Nucleotide sequence of human monocyte interleukin 1 precursor cDNA. Proc Natl Acad Sci USA 81:7907–7911. https://doi.org/10.1073/pnas.81.24.7907
4. Baldwin TA, Ostergaard HL (2002) The protein-tyrosine phosphatase CD45 reaches the cell surface via Golgi-dependent and -independent pathways. J Biol Chem 277:50333–50340. https://doi.org/10.1074/jbc.M209075200
5. Balmer EA, Faso C (2021) The road less traveled? Unconventional protein secretion at parasite-host interfaces. Front Cell Dev Biol 9:662711. https://doi.org/10.3389/fcell.2021.662711
6. Barlow LD, Nyvltova E, Aguilar M, Tachezy J, Dacks JB (2018) A sophisticated, differentiated Golgi in the ancestor of eukaryotes. BMC Biol 16:27. https://doi.org/10.1186/s12915-018-0492-9. Correction: BMC Biol 16:35. doi: 10.1186/s12915-018-0510-y
7. Bellucci M, De Marchis F, Pompa A (2017) The endoplasmic reticulum is a hub to sort proteins toward unconventional traffic pathways and endosymbiotic organelles. J Exp Bot 69:7–20. https://doi.org/10.1093/jxb/erx286
8. Benvenuto G, Leone S, Astoricchio E, Bormke S, Jasek S, D'Aniello E, Kittelmann M, McDonald K, Hartenstein V, Baena V, Escrivà H, Bertrand S, Schierwater B, Burkhardt P, Ruiz-Trillo I, Jékely G, Ullrich-Lüter J, Lüter C, D'Aniello S, Arnone MI, Ferraro F (2024) Evolution of the ribbon-like organization of the Golgi apparatus in animal cells. Cell Rep 43:113791. https://doi.org/10.1016/j.celrep.2024.113791
9. Beznoussenko GV, Parashuraman S, Rizzo R, Polishchuk R, Martella O, Di Giandomenico D, Fusella A, Spaar A, Sallese M, Capestrano MG, Pavelka M, Vos MR, Rikers YG, Helms V, Mironov AA, Luini A (2014) Transport of soluble proteins through the Golgi occurs by diffusion via continuities across cisternae. elife 3:e02009. https://doi.org/10.7554/eLife.02009
10. Bian W, Jiang H, Yao L, Hao W, Wu L, Li X (2023) A spatially defined human Notch receptor interaction network reveals Notch intracellular storage and Ataxin-2-mediated fast recycling. Cell Rep 42:112819. https://doi.org/10.1016/j.celrep.2023.112819
11. Blum R, Pfeiffer F, Feick P, Nastainczyk W, Kohler B, Schäfer KH, Schulz I (1999) Intracellular localization and *in vivo* trafficking of p24A and p23. J Cell Sci 112:537–548. https://doi.org/10.1242/jcs.112.4.537
12. Bonifacino JS, Glick BS (2004) The mechanisms of vesicle budding and fusion. Cell 116:153–166. https://doi.org/10.1016/s0092-8674(03)01079-1
13. Bonnon C, Goutebroze L, Denisenko-Nehrbass N, Girault JA, Faivre-Sarrailh C (2003) The paranodal complex of F3/contactin and caspr/paranodin traffics to the cell surface via a non-conventional pathway. J Biol Chem 278:48339–48347. https://doi.org/10.1074/jbc.M309120200
14. Bowen AB, Bourke AM, Hiester BG, Hanus C, Kennedy MJ (2017) Golgi-independent secretory trafficking through recycling endosomes in neuronal dendrites and spines. elife 6:e27362. https://doi.org/10.7554/eLife.27362
15. Bravo-Cordero JJ, Marrero-Diaz R, Megías D, Genís L, García-Grande A, García MA, Arroyo AG, Montoya MC (2007) MT1-MMP proinvasive activity is regulated by a novel Rab8-dependent exocytic pathway. EMBO J 26:1499–1510. https://doi.org/10.1038/sj.emboj.7601606
16. Bruns C, McCaffery JM, Curwin AJ, Duran JM, Malhotra V (2011) Biogenesis of a novel compartment for autophagosome-mediated unconventional protein secretion. J Cell Biol 195:979–992. https://doi.org/10.1083/jcb.201106098
17. Burd CG (2021) GRASPing for consensus about the Golgi apparatus. J Cell Biol 220:e202103117. https://doi.org/10.1083/jcb.202103117
18. Camus SM, Camus MD, Figueras-Novoa C, Boncompain G, Sadacca LA, Esk C, Bigot A, Gould GW, Kioumourtzoglou D, Perez F, Bryant NJ, Mukherjee S, Brodsky FM (2020) CHC22 clathrin mediates traffic from early secretory compartments for human GLUT4 pathway biogenesis. J Cell Biol 219:e201812135. https://doi.org/10.1083/jcb.201812135

19. Castello-Serrano I, Heberle FA, Diaz-Rohrer B, Ippolito R, Shurer CR, Lujan P, Campelo F, Levental KR, Levental I (2024) Partitioning to ordered membrane domains regulates the kinetics of secretory traffic. elife 12:RP89306. https://doi.org/10.7554/eLife.89306
20. Castillo-Romero A, Leon-Avila G, Wang CC, Perez Rangel A, Camacho Nuez M, Garcia Tovar C, Ayala-Sumuano JT, Luna-Arias JP, Hernandez JM (2010) Rab11 and actin cytoskeleton participate in *Giardia lamblia* encystation, guiding the specific vesicles to the cyst wall. PLoS Negl Trop Dis 4:e697. https://doi.org/10.1371/journal.pntd.0000697
21. Cebrian I, Visentin G, Blanchard N, Jouve M, Bobard A, Moita C, Enninga J, Moita LF, Amigorena S, Savina A (2011) Sec22b regulates phagosomal maturation and antigen cross-presentation by dendritic cells. Cell 147:1355–1368. https://doi.org/10.1016/j.cell.2011.11.021
22. Cernikova L, Faso C, Hehl AB (2019) Roles of phosphoinositides and their binding proteins in parasitic protozoa. Trends Parasitol 35:996–1008. https://doi.org/10.1016/j.pt.2019.08.008
23. Cooper DN, Barondes SH (1990) Evidence for export of a muscle lectin from cytosol to extracellular matrix and for a novel secretory mechanism. J Cell Biol 110:1681–1691. https://doi.org/10.1083/jcb.110.5.1681
24. Cruz-Garcia D, Curwin AJ, Popoff JF, Bruns C, Duran JM, Malhotra V (2014) Remodeling of secretory compartments creates CUPS during nutrient starvation. J Cell Biol 207:695–703. https://doi.org/10.1083/jcb.201407119
25. Cruz-Garcia D, Malhotra V, Curwin AJ (2018) Unconventional protein secretion triggered by nutrient starvation. Semin Cell Dev Biol 83:22–28. https://doi.org/10.1016/j.semcdb.2018.02.021
26. Curwin AJ, Brouwers N, Alonso Y, Adell M, Teis D, Turacchio G, Parashuraman S, Ronchi P, Malhotra V (2016) ESCRT-III drives the final stages of CUPS maturation for unconventional protein secretion. elife 5:e16299. https://doi.org/10.7554/eLife.16299
27. Curwin AJ, Kurokawa K, Bigliani G, Brouwers N, Nakano A, Malhotra V (2025) The pathway of unconventional protein secretion involves CUPS and a modified *trans*-Golgi network. J Cell Biol 224:e202312120. https://doi.org/10.1083/jcb.202312120
28. Dahan S, Ahluwalia JP, Wong L, Posner BI, Bergeron JJ (1994) Concentration of intracellular hepatic apolipoprotein E in Golgi apparatus saccular distensions and endosomes. J Cell Biol 127:1859–1869. https://doi.org/10.1083/jcb.127.6.1859
29. Deryugina EI, Ratnikov BI, Yu Q, Baciu C, Rozanov DV, Strongin AY (2004) Prointegrin maturation follows rapid trafficking and processing of MT1-MMP in furin-negative colon carcinoma LoVo cells. Traffic 5:627–641. https://doi.org/10.1111/j.1600-0854.2004.00206.x
30. Dimou E, Nickel W (2018) Unconventional mechanisms of eukaryotic protein secretion. Curr Biol 28:R406–R410. https://doi.org/10.1016/j.cub.2017.11.074
31. Dupont N, Jiang S, Pilli M, Ornatowski W, Bhattacharya D, Deretic V (2011) Autophagy-based unconventional secretory pathway for extracellular delivery of IL-1β. EMBO J 30:4701–4711. https://doi.org/10.1038/emboj.2011.398
32. Filaquier A, Marin P, Parmentier ML, Villeneuve J (2022) Roads and hubs of unconventional protein secretion. Curr Opin Cell Biol 75:102072. https://doi.org/10.1016/j.ceb.2022.02.006
33. Fougère L, Grison M, Laquel P, Montrazi M, Cordelières F, Fernández-Monreal M, Poujol C, Uemura T, Nakano A, Ito Y, Boutté Y (2023) ER-to-Golgi trafficking via a dynamic intermediate *cis*-Golgi tubular network in *Arabidopsis*. Nat Cell Biol 27:424–437. https://doi.org/10.1038/s41556-025-01624-x
34. Fujii S, Kurokawa K, Inaba R, Hiramatsu N, Tago T, Nakamura Y, Nakano A, Satoh T, Satoh AK (2020a) Recycling endosomes attach to the *trans*-side of Golgi stacks in *Drosophila* and mammalian cells. J Cell Sci 133:jcs236935. https://doi.org/10.1242/jcs.236935
35. Fujii S, Kurokawa K, Tago T, Inaba R, Takiguchi A, Nakano A, Satoh T, Satoh AK (2020b) Sec71 separates Golgi stacks in *Drosophila* S2 cells. J Cell Sci 133:jcs245571. https://doi.org/10.1242/jcs.245571
36. Füllekrug J, Sönnichsen B, Schäfer U, Nguyen Van P, Söling HD, Mieskes G (1997) Characterization of brefeldin A induced vesicular structures containing cycling proteins

of the intermediate compartment/*cis*-Golgi network. FEBS Lett 404:75–81. https://doi.org/10.1016/s0014-5793(97)00097-5
37. Ge L, Melville D, Zhang M, Schekman R (2013) The ER-Golgi intermediate compartment is a key membrane source for the LC3 lipidation step of autophagosome biogenesis. elife 2:e00947. https://doi.org/10.7554/eLife.00947
38. Ge L, Zhang M, Kenny SJ, Liu D, Maeda M, Saito K, Mathur A, Xu K, Schekman R (2017) Remodeling of ER-exit sites initiates a membrane supply pathway for autophagosome biogenesis. EMBO Rep 18:1586–1603. https://doi.org/10.15252/embr.201744559
39. Ge L, Zhang M, Schekman R (2014) Phosphatidylinositol 3-kinase and COPII generate LC3 lipidation vesicles from the ER-Golgi intermediate compartment. elife 3:e04135. https://doi.org/10.7554/eLife.04135
40. Gee HY, Kim J, Lee MG (2018) Unconventional secretion of transmembrane proteins. Semin Cell Dev Biol 83:59–66. https://doi.org/10.1016/j.semcdb.2018.03.016
41. Gee HY, Noh SH, Tang BL, Kim KH, Lee MG (2011) Rescue of ΔF508-CFTR trafficking via a GRASP-dependent unconventional secretion pathway. Cell 146:746–760. https://doi.org/10.1016/j.cell.2011.07.021
42. Ghosh S, Dellibovi-Ragheb TA, Kerviel A, Pak E, Qiu Q, Fisher M, Takvorian PM, Bleck C, Hsu VW, Fehr AR, Perlman S, Achar SR, Straus MR, Whittaker GR, de Haan CAM, Kehrl J, Altan-Bonnet G, Altan-Bonnet N (2020) β-Coronaviruses use lysosomes for egress instead of the biosynthetic secretory pathway. Cell 183:1520–1535.e14. https://doi.org/10.1016/j.cell.2020.10.039
43. Gilder AL, Chapin HC, Padovano V, Hueschen CL, Rajendran V, Caplan MJ (2018) Newly synthesized polycystin-1 takes different trafficking pathways to the apical and ciliary membranes. Traffic 19:933–945. https://doi.org/10.1111/tra.12612
44. Giuliani F, Grieve A, Rabouille C (2011) Unconventional secretion: a stress on GRASP. Curr Opin Cell Biol 23:498–504. https://doi.org/10.1016/j.ceb.2011.04.005
45. Goud B, Liu S, Storrie B (2018) Rab proteins as major determinants of the Golgi complex structure. Small GTPases 9:66–75. https://doi.org/10.1080/21541248.2017.1384087
46. Grieve A, Rabouille C (2011) Golgi bypass: skirting around the heart of classical secretion. Cold Spring Harb Perspect Biol 3:a005298. https://doi.org/10.1101/cshperspect.a005298
47. Griffiths G, Fuller SD, Back R, Hollinshead M, Pfeiffer S, Simons K (1989) The dynamic nature of the Golgi complex. J Cell Biol 108:277–297. https://doi.org/10.1083/jcb.108.2.277
48. Grochowska KM, Andres-Alonso M, Karpova A, Kreutz MR (2022) The needs of a synapse—how local organelles serve synaptic proteostasis. EMBO J 41:e110057. https://doi.org/10.15252/embj.2021110057
49. Grond R, Veenendaal T, Duran JM, Raote I, van Es JH, Corstjens S, Delfgou L, El Haddouti B, Malhotra V, Rabouille C (2020) The function of GORASPs in Golgi apparatus organization *in vivo*. J Cell Biol 219:e202004191. https://doi.org/10.1083/jcb.202004191
50. Gui X, Yang H, Li T, Tan X, Shi P, Li M, Du F, Chen ZJ (2019) Autophagy induction via STING trafficking is a primordial function of the cGAS pathway. Nature 567:262–266. https://doi.org/10.1038/s41586-019-1006-9
51. Hanus C, Geptin H, Tushev G, Garg S, Alvarez-Castelao B, Sambandan S, Kochen L, Hafner AS, Langer JD, Schuman EM (2016) Unconventional secretory processing diversifies neuronal ion channel properties. elife 5:e20609. https://doi.org/10.7554/eLife.20609
52. Hehnly H, Doxsey S (2014) Rab11 endosomes contribute to mitotic spindle organization and orientation. Dev Cell 28:497–507
53. Hoffmeister H, Babinger K, Gürster S, Cedzich A, Meese C, Schadendorf K, Osten L, de Vries U, Rascle A, Witzgall R (2011) Polycystin-2 takes different routes to the somatic and ciliary plasma membrane. J Cell Biol 192:631–645. https://doi.org/10.1083/jcb.201007050
54. Iglesia RP, Prado MB, Alves RN, Escobar MIM, Fernandes CFL, Fortes ACDS, Souza MCDS, Boccacino JM, Cangiano G, Soares SR, de Araújo JPA, Tiek DM, Goenka A, Song X, Keady JR, Hu B, Cheng SY, Lopes MH (2022) Unconventional protein secretion in brain

tumor biology: enlightening the mechanisms for tumor survival and progression. Front Cell Dev Biol 10:907423. https://doi.org/10.3389/fcell.2022.907423
55. Ito Y, Boutté Y (2020) Differentiation of trafficking pathways at Golgi entry core compartments and post-Golgi subdomains. Front Plant Sci 11:609516. https://doi.org/10.3389/fpls.2020.609516
56. Ito Y, Uemura T, Nakano A (2018) The Golgi entry core compartment functions as a COPII-independent scaffold for ER-to-Golgi transport in plant cells. J Cell Sci 131:jcs203893. https://doi.org/10.1242/jcs.203893
57. Ito Y, Uemura T, Shoda K, Fujimoto M, Ueda T, Nakano A (2012) Cis-Golgi proteins accumulate near the ER exit sites and act as the scaffold for Golgi regeneration after brefeldin A treatment in tobacco BY-2 cells. Mol Biol Cell 23:3203–3214. https://doi.org/10.1091/mbc.E12-01-0034
58. Jarvela T, Linstedt AD (2014) Isoform-specific tethering links the Golgi ribbon to maintain compartmentalization. Mol Biol Cell 25:133–144. https://doi.org/10.1091/mbc.E13-07-0395
59. Jiang S, Rhee SW, Gleeson PA, Storrie B (2006) Capacity of the Golgi apparatus for cargo transport prior to complete assembly. Mol Biol Cell 17:4105–4117. https://doi.org/10.1091/mbc.e05-12-1112
60. Kandaswamy KK, Pugalenthi G, Hartmann E, Kalies KU, Möller S, Suganthan PN, Martinetz T (2010) SPRED: a machine learning approach for the identification of classical and non-classical secretory proteins in mammalian genomes. Biochem Biophys Res Commun 391:1306–1311. https://doi.org/10.1016/j.bbrc.2009.12.019
61. Kennedy MJ, Hanus C (2019) Architecture and dynamics of the neuronal secretory network. Annu Rev Cell Dev Biol 35:543–566. https://doi.org/10.1146/annurev-cellbio-100818-125418
62. Kim J, Noh SH, Piao H, Kim DH, Kim K, Cha JS, Chung WY, Cho HS, Kim JY, Lee MG (2016) Monomerization and ER relocalization of GRASP is a requisite for unconventional secretion of CFTR. Traffic 17:733–753. https://doi.org/10.1111/tra.12403
63. Kinseth MA, Anjard C, Fuller D, Guizzunti G, Loomis WF, Malhotra V (2007) The Golgi-associated protein GRASP is required for unconventional protein secretion during development. Cell 130:524–534. https://doi.org/10.1016/j.cell.2007.06.029
64. Knödler A, Feng S, Zhang J, Zhang X, Das A, Peränen J, Guo W (2010) Coordination of Rab8 and Rab11 in primary ciliogenesis. Proc Natl Acad Sci USA 107:6346–6351. https://doi.org/10.1073/pnas.1002401107
65. Kondylis V, Spoorendonk KM, Rabouille C (2005) dGRASP localization and function in the early exocytic pathway in *Drosophila* S2 cells. Mol Biol Cell 16:4061–4072. https://doi.org/10.1091/mbc.e04-10-0938
66. Koonin EV (2015) Origin of eukaryotes from within archea, archeal eukaryome and bursts of gene gain: eukaryogenesis just made easier? Philos Trans R Soc Lond Ser B Biol Sci 370:20140333. https://doi.org/10.1098/rstb.2014.0333
67. Kuchler K, Sterne RE, Thorner J (1989) *Saccharomyces cerevisiae* STE6 gene product: a novel pathway for protein export in eukaryotic cells. EMBO J 8:3973–3984. https://doi.org/10.1002/j.1460-2075.1989.tb08580.x
68. Kvalvaag AH, Tollersrud OK, Helgeland L (1988) A study on the intracellular transport of prothrombin, albumin and transferrin in rat. Biochim Biophys Acta 937:319–327. https://doi.org/10.1016/0005-2736(88)90254-4
69. Ladinsky MS, Mastronarde DN, McIntosh JR, Howell KE, Staehelin LA (1999) Golgi structure in three dimensions: functional insights from the normal rat kidney cell. J Cell Biol 144:1135–1149. https://doi.org/10.1083/jcb.144.6.1135
70. Langford TD, Silberman JD, Weiland MEL, Svärd SG, McCaffery JM, Sogin ML, Gillin FD (2002) *Giardia lamblia*: identification and characterization of Rab and GDI proteins in a genome survey of the ER to Golgi endomembrane system. Exp Parasitol 101:13–24. https://doi.org/10.1016/s0014-4894(02)00037-1

71. Lavieu G, Dunlop MH, Lerich A, Zheng H, Bottanelli F, Rothman JE (2014) The Golgi ribbon structure facilitates anterograde transport of large cargoes. Mol Biol Cell 25:3028–3036. https://doi.org/10.1091/mbc.E14-04-0931
72. Lee MC, Miller EA, Goldberg J, Orci L, Schekman R (2004) Bi-directional protein transport between the ER and Golgi. Annu Rev Cell Dev Biol 20:87–123. https://doi.org/10.1146/annurev.cellbio.20.010403.105307
73. Li DT, Habtemichael EN, Julca O, Sales CI, Westergaard XO, DeVries SG, Ruiz D, Sayal B, Bogan JS (2019) GLUT4 storage vesicles: specialized organelles for regulated trafficking. Yale J Biol Med 92:453–470. PMID: 31543708
74. Li X, Liu B, Wen Y, Wang J, Guo YR, Shi A, Lin L (2024) Coordination of RAB8 and RAB11 during unconventional protein secretion. J Cell Biol 223:e202306107. https://doi.org/10.1083/jcb.202306107
75. Li S, Yan R, Xu J, Zhao S, Ma X, Sun Q, Zhang M, Li Y, Liu JG, Chen L, Li S, Xu K, Ge L (2022) A new type of ERGIC-ERES membrane contact mediated by TMED9 and SEC12 is required for autophagosome biogenesis. Cell Res 32:119–138. https://doi.org/10.1038/s41422-021-00563-0
76. Lippincott-Schwartz J, Donaldson JG, Schweizer A, Berger EG, Hauri H-P, Yuan LC, Klausner RD (1990) Microtubule-dependent retrograde transport of proteins into the ER in the presence of brefeldin A suggests an ER recycling pathway. Cell 60:821–836. https://doi.org/10.1016/0092-8674(90)90096-w
77. Liu L, Zhang L, Hao X, Wang Y, Zhang X, Ge L, Wang P, Tian B, Zhang M (2024) Coronavirus envelope protein activates TMED10-mediated unconventional secretion of inflammatory factors. Nat Commun 15:8708. https://doi.org/10.1038/s41467-024-52818-0
78. Lock JG, Stow JL (2005) Rab11 in recycling endosomes regulates the sorting and basolateral transport of E-cadherin. Mol Biol Cell 16:1744–1755. https://doi.org/10.1091/mbc.e04-10-0867
79. Lolicato F, Saleppico R, Griffo A, Meyer A, Scollo F, Pokrandt B, Müller HM, Ewers H, Hähl H, Fleury JB, Seemann R, Hof M, Brügger B, Jacobs K, Vattulainen I, Nickel W (2022) Cholesterol promotes clustering of PI(4,5)P2 driving unconventional secretion of FGF2. J Cell Biol 221:e202106123. https://doi.org/10.1083/jcb.202106123
80. Low SH, Tang BL, Wong SH, Hong W (1992) Selective inhibition of protein targeting to the apical domain of MDCK cells by brefeldin A. J Cell Biol 118:51–62. https://doi.org/10.1083/jcb.118.1.51
81. Lowe M (2011) Structural organization of the Golgi apparatus. Curr Opin Cell Biol 23:85–93. https://doi.org/10.1016/j.ceb.2010.10.004
82. Lucken-Ardjomande Häsler S, Vallis Y, Pasche M, McMahon HT (2020) GRAF2, WDR44, and MICAL1 mediate Rab8/10/11-dependent export of E-cadherin, MMP14, and CFTR ΔF508. J Cell Biol 219:e201811014. https://doi.org/10.1083/jcb.201811014
83. Machamer CE (2013) Accommodation of large cargo within Golgi cisternae. Histochem Cell Biol 140:261–269. https://doi.org/10.1007/s00418-013-1120-y
84. Machamer CE (2015) The Golgi complex in stress and death. Front Neurosci 9:421. https://doi.org/10.3389/fnins.2015.00421
85. Makhoul C, Gosavi P, Gleeson PA (2019) Golgi dynamics: the morphology of the mammalian Golgi apparatus in health and disease. Front Cell Dev Biol 7:112. https://doi.org/10.3389/fcell.2019.00112
86. Malhotra V (2024) Tailored assemblies of COPII proteins in secretion. J Cell Biol 223:e202404013. https://doi.org/10.1083/jcb.202404013
87. Manjithaya R, Anjard C, Loomis WF, Subramani S (2010) Unconventional secretion of Pichia pastoris Acb1 is dependent on GRASP protein, peroxisomal functions, and autophagosome formation. J Cell Biol 188:537–546. https://doi.org/10.1083/jcb.200911149
88. Marie M, Dale HA, Kouprina N, Saraste J (2012) Division of the intermediate compartment at the onset of mitosis provides a mechanism for Golgi inheritance. J Cell Sci 125:5403–5416. https://doi.org/10.1242/jcs.108100

89. Marie M, Dale HA, Sannerud R, Saraste J (2009) The function of the intermediate compartment in pre-Golgi trafficking involves its stable connection with the centrosome. Mol Biol Cell 20:4458–4470. https://doi.org/10.1091/mbc.e08-12-1229
90. Marie M, Sannerud R, Dale HA, Saraste J (2008) Take the 'A' train: on fast tracks to the cell surface. Cell Mol Life Sci 65:2859–2874. https://doi.org/10.1007/s00018-008-8355-0
91. Marra P, Maffucci T, Daniele T, Tullio GD, Ikehara Y, Chan EK, Luini A, Beznoussenko G, Mironov A, De Matteis MA (2001) The GM130 and GRASP65 Golgi proteins cycle through and define a subdomain of the intermediate compartment. Nat Cell Biol 3:1101–1113. https://doi.org/10.1038/ncb1201-1101
92. Martin PE, Blundell G, Ahmad S, Errington RJ, Evans WH (2001) Multiple pathways in the trafficking and assembly of connexin 26, 32 and 43 into gap junction intercellular communication channels. J Cell Sci 114:3845–3855. https://doi.org/10.1242/jcs.114.21.3845
93. Martin S, Ramm G, Lyttle CT, Meerloo T, Stoorvogel W, James DE (2000) Biogenesis of insulin-responsive GLUT4 vesicles is independent of brefeldin A-sensitive trafficking. Traffic 1:652–660. https://doi.org/10.1034/j.1600-0854.2000.010809.x
94. Martínez-Martínez N, Martínez-Alonso E, Tomás M, Neumüller J, Pavelka M, Martínez-Menárguez JA (2017) A new insight into the three-dimensional architecture of the Golgi complex: characterization of unusual structures in epididymal principal cells. PLoS One 12:e0185557. https://doi.org/10.1371/journal.pone.0185557
95. Mills DB (2020) The origin of phagocytosis in earth history. Interface Focus 10:20200019. https://doi.org/10.1098/rsfs.2020.0019
96. Murray RZ, Stow JL (2014) Cytokine secretion in macrophages: SNAREs, Rabs, and membrane trafficking. Front Immunol 5:538. https://doi.org/10.3389/fimmu.2014.00538
97. Nakamura N, Rabouille C, Watson R, Nilsson T, Hui N, Slusarewicz P, Kreis TE, Warren G (1995) Characterization of a *cis*-Golgi matrix protein GM130. J Cell Biol 131:1715–1726. https://doi.org/10.1083/jcb.131.6.1715
98. Nakano A (2022) The Golgi apparatus and its next-door neighbors. Front Cell Dev Biol 10:884360. https://doi.org/10.3389/fcell.2022.884360
99. Nickel W, Seedorf M (2008) Unconventional mechanisms of protein transport to the cell surface of eukaryotic cells. Annu Rev Cell Dev Biol 24:287–308. https://doi.org/10.1146/annurev.cellbio.24.110707.175320
100. Noh SH, Gee HY, Kim Y, Piao H, Kim J, Kang CM, Lee G, Mook-Jung I, Lee Y, Cho JW, Lee MG (2018) Specific autophagy and ESCRT components participate in the unconventional secretion of CFTR. Autophagy 14:1761–1778. https://doi.org/10.1080/15548627.2018.1489479
101. Novikoff AB, Novikoff PM (1977) Cytochemical contributions to differentiating GERL from the Golgi apparatus. Histochem J 9:525–551. https://doi.org/10.1007/BF01002901
102. Nüchel J, Tauber M, Nolte JL, Mörgelin M, Türk C, Eckes B, Demetriades C, Plomann M (2021) An mTORC1-GRASP55 signaling axis controls unconventional secretion to reshape the extracellular proteome upon stress. Mol Cell 81:3275–3293.e12. https://doi.org/10.1016/j.molcel.2021.06.017
103. Orme CM, Bogan JS (2012) The ubiquitin regulatory X (UBX) domain-containing protein TUG regulates the p97 ATPase and resides at the endoplasmic reticulum-Golgi intermediate compartment. J Biol Chem 287:6679–6692. https://doi.org/10.1074/jbc.M111.284232
104. Palade GE (1975) Intracellular aspects of the process of protein synthesis. Science 189:347–358. https://doi.org/10.1126/science.1096303
105. Park H, Seo SK, Sim JR, Hwang SJ, Kim YJ, Shin DH, Jang DG, Noh SH, Park PG, Ko SH, Shin MH, Choi JY, Ito Y, Kang CM, Lee JM, Lee MG (2022) TMED3 complex mediates ER stress-associated secretion of CFTR, pendrin, and SARS-CoV-2 spike. Adv Sci (Weinh) 9:e2105320. https://doi.org/10.1002/advs.202105320
106. Pastor-Cantizano N, Montesinos JC, Bernat-Silvestre C, Marcote MJ, Aniento F (2016) p24 family proteins: key players in the regulation of trafficking along the secretory pathway. Protoplasma 253:967–985. https://doi.org/10.1007/s00709-015-0858-6

107. Prydz K, Saraste J (2022) The life cycle and enigmatic egress of coronaviruses. Mol Microbiol 117:1308–1316. https://doi.org/10.1111/mmi.14907
108. Prydz K, Dick G, Tveit H (2008) How many ways through the Golgi maze? Traffic 9:299–304. https://doi.org/10.1111/j.1600-0854.2007.00690.x
109. Prydz K, Tveit H, Vedeler A, Saraste J (2013) Arrivals and departures at the plasma membrane: direct and indirect transport routes. Cell Tissue Res 352:5–20. https://doi.org/10.1007/s00441-012-1409-5
110. Rabouille C (2017) Pathways of unconventional protein secretion. Trends Cell Biol 27:230–240. https://doi.org/10.1016/j.tcb.2016.11.007
111. Rabouille C, Linstedt AD (2016) GRASP: a multitasking tether. Front Cell Dev Biol 4:1. https://doi.org/10.3389/fcell.2016.00001
112. Rabouille C, Malhotra V, Nickel WJ (2012) Diversity in unconventional protein secretion. Cell Sci 125:5251–5255. https://doi.org/10.1242/jcs.103630
113. Rivinoja A, Pujol FM, Hassinen A, Kellokumpu S (2012) Golgi pH, its regulation and roles in human disease. Ann Med 44:542–554. https://doi.org/10.3109/07853890.2011.579150
114. Rubartelli A, Cozzolino F, Talio M, Sitia R (1990) A novel secretory pathway for interleukin-1 beta, a protein lacking a signal sequence. EMBO J 9:1503–1510. https://doi.org/10.1002/j.1460-2075.1990.tb08268.x
115. Ruch TR, Machamer CE (2012) The coronavirus E protein: assembly and beyond. Viruses 4:363–382. https://doi.org/10.3390/v4030363
116. Sabesin SM, Frase S (1977) Electron microscopic studies of the assembly, intracellular transport, and secretion of chylomicrons by rat intestine. J Lipid Res 18:496–511. PMID: 894141
117. Sandvig K, Prydz K, Hansen SH, van Deurs B (1991) Ricin transport in brefeldin A-treated cells: correlation between Golgi structure and toxic effect. J Cell Biol 115:971–981. https://doi.org/10.1083/jcb.115.4.971
118. Sannerud R, Marie M, Nizak C, Dale HA, Pernet-Gallay K, Perez F, Goud B, Saraste J (2006) Rab1 defines a novel pathway connecting the pre-Golgi intermediate compartment with the cell periphery. Mol Biol Cell 17:1514–1526. https://doi.org/10.1091/mbc.e05-08-0792
119. Saraste J, Goud B (2007) Functional symmetry of endomembranes. Mol Biol Cell 18:1430–1436. https://doi.org/10.1091/mbc.e06-10-0933
120. Saraste J, Kuismanen E (1984) Pre- and post-Golgi vacuoles operate in the transport of Semliki Forest virus membrane glycoproteins to the cell surface. Cell 38:535–549. https://doi.org/10.1016/0092-8674(84)90508-7
121. Saraste J, Marie M (2018) Intermediate compartment (IC): from pre-Golgi vacuoles to a semi-autonomous membrane system. Histochem Cell Biol 150:407–430. https://doi.org/10.1007/s00418-018-1717-2
122. Saraste J, Prydz K (2019) A new look at the functional organization of the Golgi ribbon. Front Cell Dev Biol 7:171. https://doi.org/10.3389/fcell.2019.00171
123. Saraste J, Prydz K (2021) Assembly and cellular exit of coronaviruses: hijacking an unconventional secretory pathway from the pre-Golgi intermediate compartment via the Golgi ribbon to the extracellular space. Cells 10:503. https://doi.org/10.3390/cells10030503
124. Saraste J, Svensson K (1991) Distribution of the intermediate elements operating in ER to Golgi transport. J Cell Sci 100:415–430. https://doi.org/10.1242/jcs.100.3.415
125. Saraste J, Palade GE, Farquhar MG (1987) Antibodies to rat pancreas Golgi subfractions: identification of a 58-kD *cis*-Golgi protein. J Cell Biol 105:2021–2029. https://doi.org/10.1083/jcb.105.5.2021
126. Saraste J, Enyioko M, Dale H, Prydz K, Machamer C (2022) Evidence for the role of Rab11-positive recycling endosomes as intermediates in coronavirus egress from epithelial cells. Histochem Cell Biol 158:241–251. https://doi.org/10.1007/s00418-022-02115-y
127. Schotman H, Karhinen L, Rabouille C (2008) dGRASP-mediated noncanonical integrin secretion is required for *Drosophila* epithelial remodeling. Dev Cell 14:171–182. https://doi.org/10.1016/j.devcel.2007.12.006

128. Seemann J, Jokitalo E, Pypaert M, Warren G (2000) Matrix proteins can generate the higher order architecture of the Golgi apparatus. Nature 407:1022–1026. https://doi.org/10.1038/35039538
129. Sergio MC, Ricciardi S, Guarino AM, Giaquinto L, De Matteis MA (2024) Membrane remodeling and trafficking piloted by SARS-CoV-2. Trends Cell Biol 34:785–800. https://doi.org/10.1016/j.tcb.2023.12.006
130. Sitia R, Rubartelli A (2018) The unconventional secretion of IL-1β: handling a dangerous weapon to optimize inflammatory responses. Semin Cell Dev Biol 83:12–21. https://doi.org/10.1016/j.semcdb.2018.03.011
131. Sparn C, Meyer A, Saleppico R, Nickel W (2022) Unconventional secretion mediated by direct protein self-translocation across the plasma membranes of mammalian cells. Trends Biochem Sci 47:699–709. https://doi.org/10.1016/j.tibs.2022.04.001
132. Stanley P (2011) Golgi glycosylation. Cold Spring Harb Perspect Biol 3:a005199. https://doi.org/10.1101/cshperspect.a005199
133. Stefanic S, Morf L, Kulangara C, Regös A, Sonda S, Schraner E, Spycher C, Wild P, Hehl AB (2009) Neogenesis and maturation of transient Golgi-like *cisternae* in a simple eukaryote. J Cell Sci 122:2846–2856. https://doi.org/10.1242/jcs.049411
134. Strating JR, Martens GJ (2009) The p24 family and selective transport processes at the ER-Golgi interface. Biol Cell 101:495–509. https://doi.org/10.1042/BC20080233
135. Sun Y, Tao X, Han Y, Lin X, Tian R, Wang H, Chang P, Sun Q, Ge L, Zhang M (2024) A dual role of ERGIC-localized Rabs in TMED10-mediated unconventional protein secretion. Nat Cell Biol 26:1077–1092. https://doi.org/10.1038/s41556-024-01445-4
136. Taguchi T (2013) Emerging roles of recycling endosomes. J Biochem 153:505–510. https://doi.org/10.1093/jb/mvt034
137. Takahashi S, Kubo K, Waguri S, Yabashi A, Shin HW, Katoh Y, Nakayama K (2012) Rab11 regulates exocytosis of recycling vesicles at the plasma membrane. J Cell Sci 125:4049–4057. https://doi.org/10.1242/jcs.102913
138. Takatsu H, Katoh Y, Ueda T, Waguri S, Murayama T, Takahashi S, Shin HW, Nakayama K (2013) Rab11 regulates exocytosis of recycling vesicles at the plasma membrane. Mitosis-coupled, microtubule-dependent clustering of endosomal vesicles around centrosomes. Cell Struct Funct 38:31–41. https://doi.org/10.1242/jcs.102913
139. Tang BL, Low SH, Hong W (1995) Differential response of resident proteins and cycling proteins of the Golgi to brefeldin A. Eur J Cell Biol 68:199–205
140. Tie HC, Ludwig A, Sandin S, Lu L (2018) The spatial separation of processing and transport functions to the interior and periphery of the Golgi stack. elife 7:e41301. https://doi.org/10.7554/eLife.41301
141. Tojima T, Suda Y, Jin N, Kurokawa K, Nakano A (2024) Spatiotemporal dissection of the Golgi apparatus and the ER-Golgi intermediate compartment in budding yeast. elife 13:e92900. https://doi.org/10.7554/eLife.92900
142. Tooze J, Tooze SA, Fuller SD (1987) Sorting of progeny coronavirus from condensed secretory proteins at the exit from the *trans*-Golgi network of AtT20 cells. J Cell Biol 105:1215–1226. https://doi.org/10.1083/jcb.105.3.1215
143. Touz MC, Zamponi N (2017) Sorting without a Golgi complex. Traffic 18:637–645. https://doi.org/10.1111/tra.12500
144. Tsai YL, Ha DP, Zhao H, Carlos AJ, Wei S, Pun TK, Wu K, Zandi E, Kelly K, Lee AS (2018) Endoplasmic reticulum stress activates SRC, relocating chaperones to the cell surface where GRP78/CD109 blocks TGF-β signaling. Proc Natl Acad Sci USA 115:E4245–E4254. https://doi.org/10.1073/pnas.1714866115
145. Tveit H, Akslen LK, Fagereng GL, Tranulis MA, Prydz K (2009) A secretory Golgi bypass route to the apical surface domain of epithelial MDCK cells. Traffic 10:1685–1695. https://doi.org/10.1111/j.1600-0854.2009.00984.x

146. Van Krieken R, Tsai YL, Carlos AJ, Ha DP, Lee AS (2021) ER residential chaperone GRP78 unconventionally relocalizes to the cell surface via endosomal transport. Cell Mol Life Sci 78:5179–5195. https://doi.org/10.1007/s00018-021-03849-z
147. Vats S, Galli T (2022) Role of SNAREs in unconventional secretion-focus on the VAMP7-dependent secretion. Front Cell Dev Biol 10:884020. https://doi.org/10.3389/fcell.2022.884020
148. Vogel GF, Klee KM, Janecke AR, Müller T, Hess M, Huber LA (2015) Cargo-selective apical exocytosis in epithelial cells is conducted by Myo5B, Slp4a, Vamp7, and Syntaxin 3. J Cell Biol 211:587–604. https://doi.org/10.1083/jcb.201506112
149. Volchuk A, Amherdt M, Ravazzola M, Brügger B, Rivera VM, Clackson T, Perrelet A, Söllner TH, Rothman JE, Orci L (2000) Megavesicles implicated in the rapid transport of intracisternal aggregates across the Golgi stack. Cell 102:335–348. https://doi.org/10.1016/s0092-8674(00)00039-8
150. Walenta JH, Didier AJ, Liu X, Krämer H (2001) The Golgi-associated hook3 protein is a member of a novel family of microtubule-binding proteins. J Cell Biol 152:923–934. https://doi.org/10.1083/jcb.152.5.923
151. Wang WA, Carreras-Sureda A, Demaurex N (2023) SARS-CoV-2 infection alkalinizes the ERGIC and lysosomes through the viroporin activity of the viral envelope protein. J Cell Sci 136:jcs260685. https://doi.org/10.1242/jcs.260685
152. Wang X, Li X, Wang J, Wang J, Hu C, Zeng J, Shi A, Lin L (2022) SMGL-1/NBAS acts as a RAB-8 GEF to regulate unconventional protein secretion. J Cell Biol 221:e202111125. https://doi.org/10.1083/jcb.202111125
153. Ward TH, Polishchuk RS, Caplan S, Hirschberg K, Lippincott-Schwartz J (2001) Maintenance of Golgi structure and function depends on the integrity of ER export. J Cell Biol 155:557–570. https://doi.org/10.1083/jcb.200107045
154. Wei JH, Seemann J (2017) Golgi ribbon disassembly during mitosis, differentiation and disease progression. Curr Opin Cell Biol 47:43–51. https://doi.org/10.1016/j.ceb.2017.03.008
155. Westerbeck JW, Machamer CE (2019) The infectious bronchitis coronavirus envelope protein alters Golgi pH to protect the spike protein and promote the release of infectious virus. J Virol 93:e00015–e00019. https://doi.org/10.1128/JVI.00015-19
156. Westrate LM, Hoyer MJ, Nash MJ, Voeltz GK (2020) Vesicular and uncoated Rab1-dependent cargo carriers facilitate ER to Golgi transport. J Cell Sci 133:jcs239814. https://doi.org/10.1242/jcs.239814
157. Wong-Dilworth L, Rodilla-Ramirez C, Fox E, Restel SD, Stockhammer A, Adarska P, Bottanelli FJ (2023) STED imaging of endogenously tagged ARF GTPases reveals their distinct nanoscale localizations. J Cell Biol 222:e202205107. https://doi.org/10.1083/jcb.202205107
158. Xu J, Liang Y, Li N, Dang S, Jiang A, Liu Y, Guo Y, Yang X, Yuan Y, Zhang X, Yang Y, Du Y, Shi A, Liu X, Li D, He K (2024) Clathrin-associated carriers enable recycling through a kiss-and-run mechanism. Nat Cell Biol 26:1652–1668. https://doi.org/10.1038/s41556-024-01499-4
159. Yan R, Chen K, Wang B, Xu K (2022) SURF4-induced tubular ERGIC selectively expedites ER-to-Golgi transport. Dev Cell 57:512–525.e8. https://doi.org/10.1016/j.devcel.2021.12.018
160. Yang K, Liu M, Feng Z, Rojas M, Zhou L, Ke H, Pastor-Pareja JC (2021) ER exit sites in *Drosophila* display abundant ER-Golgi vesicles and pearled tubes but no megacarriers. Cell Rep 36:109707. https://doi.org/10.1016/j.celrep.2021.10970
161. Yoo JS, Moyer BD, Bannykh S, Yoo HM, Riordan JR, Balch WE (2002) Non-conventional trafficking of the cystic fibrosis transmembrane conductance regulator through the early secretory pathway. J Biol Chem 277:11401–11409. https://doi.org/10.1074/jbc.M110263200
162. Zhang X (2021) Alterations of Golgi structural proteins and glycosylation defects in cancer. Front Cell Dev Biol 9:665289. https://doi.org/10.3389/fcell.2021.665289
163. Zhang M, Kenny SJ, Ge L, Xu K, Schekman R (2015) Translocation of IL-1β into a vesicle intermediate in autophagy-mediated secretion. elife 4:e11205. https://doi.org/10.7554/eLife.11205

164. Zhang M, Liu L, Lin X, Wang Y, Li Y, Guo Q, Li S, Sun Y, Tao X, Zhang D, Xiachen L, Zheng L, Ge L (2020) A translocation pathway for vesicle-mediated unconventional protein secretion. Cell 181:637–652.e15. https://doi.org/10.1016/j.cell.2020.03.031
165. Zhang Y, Seemann J (2021) Rapid degradation of GRASP55 and GRASP65 reveals their immediate impact on the Golgi structure. J Cell Biol 220:e202007052. https://doi.org/10.1083/jcb.202007052
166. Zheng J, Ge L (2022) Diverse cellular strategies for the export of leaderless proteins. Natl Sci Open 1:20220018. https://doi.org/10.1360/nso/20220018
167. Zhou L, Xue X, Yang K, Feng Z, Liu M, Pastor-Pareja JC (2023) Convergence of secretory, endosomal, and autophagic routes in *trans*-Golgi-associated lysosomes. J Cell Biol 222:e202203045. https://doi.org/10.1083/jcb.202203045

Chapter 5
Association of Golgi Stacks and Recycling Endosomes Ensures Post-Golgi Trafficking

Akiko K. Satoh

Abstract The *trans*-Golgi network (TGN) has traditionally been considered the primary sorting hub for newly synthesized proteins, whereas recycling endosomes (REs) are recognized as intermediate stations for endocytosed materials before their return to the plasma membrane. However, over the past two decades, functional overlaps between the TGN and RE have become evident, with both playing roles in the exocytic and endocytic pathways. Recent studies have shown that REs can transition between two distinct states: Golgi-associated RE (GA-RE) and free-RE. Detachment and reattachment of REs to Golgi stacks are commonly observed. In plants, it has been established that there are two types of TGNs: Golgi-associated TGN and Golgi-independent TGN. The dynamics of REs in animal cells strongly resemble those in plant TGNs. Together with their molecular similarities, these findings suggest that REs may be equivalent to at least some parts of the TGN. Given that the TGN is now understood to be composed of distinct zones, REs could represent the *trans*-most zone of the TGN.

GA-REs receive newly synthesized glycosylphosphatidylinositol-anchored proteins (GPI-APs) but not vesicular stomatitis virus G protein (VSV-G) from the *trans*-Golgi cisternae. Upon detachment from the Golgi stacks, free-REs serve as post-Golgi carriers for GPI-APs. Besides cargo sorting, the TGN/RE system may play a key role in regulating the spatial distribution of Golgi stacks within cells. We propose that the dynamic balance between the association and separation of Golgi stacks and TGNs/REs is crucial for determining the cell-wide arrangement of Golgi stacks, along with the organization of microtubules and the lateral-linking activity of Golgi stacks.

Keywords *trans*-Golgi network · Recycling endosome · Golgi-ribbon · *Drosophila* · Brefeldin A · Sec71

A. K. Satoh (✉)
Program of Life and Environmental Sciences, Graduate School of Integrated Sciences for Life, Hiroshima University, Hiroshima, Japan
e-mail: aksatoh@hiroshima-u.ac.jp

A. Nakano, J. Saraste (eds.), *The Golgi Network, Volume I*, Subcellular Biochemistry 110, https://doi.org/10.1007/978-3-032-06936-8_5

Introduction

The Golgi apparatus is a central organelle in the secretory pathway that is responsible for the modification and sorting of secretory proteins and lipids destined for various cellular locations [4, 8, 28, 53]. Its fundamental structural unit is the Golgi stack, which comprises several flattened membrane-bound compartments called cisternae, along with associated tubules and vesicles. In organisms like plants, *Caenorhabditis elegans*, and *Drosophila*, Golgi stacks are dispersed throughout the cytoplasm. However, in mammalian cells and also some other animal cells, multiple Golgi stacks are linked by lateral tubules to form a Golgi ribbon positioned near the microtubule-organizing center (MTOC) [2, 16, 29, 46, 69, 73, 83]. Golgi stacks are highly polarized structures with newly synthesized secretory proteins from the endoplasmic reticulum (ER) entering the *cis* face and progressing to the *trans* face through cisternal maturation [33, 38, 40, 47, 74, 75]. The *trans*-Golgi network (TGN), located on the *trans*-side of the Golgi stack, was initially considered the *trans*-most compartment of the Golgi stack and functions as a sorting center that directs newly synthesized cargo to distinct cellular destinations [27, 39]. However, it is now well established that the TGN also receives cargo from endosomes, making it a central hub for coordinating both endocytic and exocytic pathways.

Recycling endosomes (REs) were initially identified as perinuclear compartments responsible for slow recycling of transferrin internalized from the plasma membrane [41, 86]. Subsequently, it was discovered that the RE also receives newly synthesized exocytic cargoes from the Golgi apparatus and is thus recognized as a central hub for both the endocytic and exocytic pathways, similar to TGNs [1, 15, 37, 71]. In mammalian cells, REs are often closely associated with MTOC [20, 58, 79], particularly in COS1 cells, in which Golgi ribbons surround the circular RE and MTOC [36, 43, 44]. Although recycling endosomes are not specifically defined in plants, they appear to be scattered throughout the cell, similar to those observed in *Caenorhabditis elegans* and *Drosophila* cells [48, 56].

Recent advances in high-resolution and high-speed microscopy have revealed the dynamics of plant TGNs and animal REs [10, 23, 77]. Both organelles are highly mobile and exist in two distinct forms, a free form and a Golgi-attached form, with the ability to switch between these states. The remarkable similarity between plant TGNs and animal REs suggests that the mobile zones of plant TGNs are functionally equivalent to animal REs. In this chapter, I introduce our work related to this idea and discuss the nature and function of animal RE as a zone of the TGN. Additionally, I discuss the cell-wide arrangement of Golgi stacks in terms of Golgi/RE associations in a later part of the chapter.

Rab11 Is a Key Factor in the Post-Golgi Transport of Rhodopsin in Fly Photoreceptors

Drosophila photoreceptors are highly polarized, making them an excellent genetic model for studying the mechanisms of polarized transport [19, 24, 34, 63, 64, 72]. Using confocal microscopy of whole-mounted retinas, we simultaneously visualized three distinct plasma membrane domains across hundreds of photoreceptors [26, 82]. Notably, rhodopsin 1 (Rh1), a photosensitive protein, is selectively transported to the photosensitive rhabdomeres that form at the center of the apical plasma membrane of photoreceptors [18, 30]. Rh1 consists of an apo-protein, an opsin, and a chromophore, 11-*cis*-3-hydroxyretinal [65, 81]. In the absence of a chromophore, opsin cannot exit the ER, likely because of conformational problems. All-*trans*-3-hydroxyretinal accumulates in the eyes of flies containing only all-*trans*-retinal as the chromophore source, and upon blue light illumination of these flies, isomerization of all-*trans*-3-hydroxyretinal to 11-*cis*-3-hydroxyretinal immediately generates numerous Rh1 molecules that synchronously exit the ER (Fig. 5.1) [18, 22, 52, 68]. This allowed us to track Rh1 movement from the ER to rhabdomeres. We termed this method blue light-induced chromophore supply (BLICS) [24, 61–63].

Our investigation of the factors involved in the polarized transport of Rh1 to the rhabdomeres led to the identification of Rab11 as a key regulator of post-Golgi transport [62]. Rab11 localizes to the *trans*-Golgi network and Rh1-bearing post-Golgi vesicles at the base of the rhabdomeres in steady-state late pupal photoreceptors. Using BLICS to visualize Rh1 transport, we observed a transient but significant association between Rab11 and newly synthesized Rh1 in TGN and post-Golgi carriers (Fig. 5.1). Disruption of Rab11 function through deficiency, RNAi knockdown, or the expression of a dominant-negative Rab11 mutant prevents Rh1 accumulation in the rhabdomeres. Instead, Rh1 accumulated in the cytoplasm within abnormal vesicles. The precise nature of these vesicles remains unclear; however, they appear to be formed by the fusion of endocytic vesicles and post-Golgi carriers [87].

REs Attached to the *trans*-Side of Golgi Stacks in *Drosophila* Tissue

Rab11 is a well-established marker of recycling endosomes (REs) that regulates transferrin receptor (TfR) recycling in mammalian cells [15, 79]. Since REs are not well characterized in *Drosophila* photoreceptors, the identification of Rab11 as a key regulator of Rh1 post-Golgi transport, coupled with its localization on the *trans*-side of Golgi stacks, raises the question of whether REs serve as post-Golgi carriers, or if Rab11 operates independently of REs. To explore this, we examined the localization of other RE markers, human Vamp3 [42, 45, 88] and its fly ortholog Syb, in *Drosophila* photoreceptors [10, 57]. Both human Vamp3 and fly Syb were found to localize to the *trans*-side of Golgi stacks. Additionally, we observed Rab11

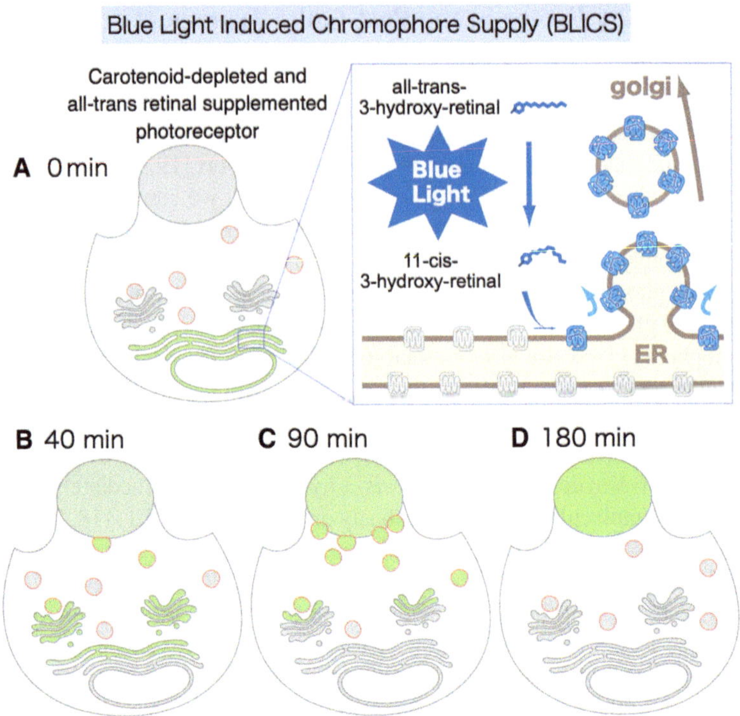

Fig. 5.1 Rh1 transport in *Drosophila* photoreceptors. (**a**) Schematic representation of the blue light-induced chromophore supply (BLICS) method, which synchronizes Rh1 release from the ER. Prior to blue light exposure (0 min), apo-protein opsins accumulate in the ER. (**b**–**d**) Schematic depictions of Rab11-positive vesicles (red-outlined) and Rh1 (green) localization within photoreceptor cells at 40, 90, and 180 min following blue light illumination

localization on the *trans*-side of Golgi stacks in wing discs and salivary glands [10]. These observations suggest that some REs associate with Golgi stacks and participate in post-Golgi transport, at least in *Drosophila*. Notably, a localization study using APEX2 fusion proteins with the RE marker Rab11GEF Pcs and the *trans*-Golgi cisternae marker GalT examined via electron microscopy revealed that REs are positioned directly adjacent to the *trans*-Golgi cisternae without any intervening membrane structures, presumably the *trans*-Golgi network (TGN) (Fig. 5.2) [10, 51].

REs Attached to the *trans*-Side of Golgi Stacks in Mammalian Cells and Sea Urchin Embryos

In mammalian cells, both Golgi stacks and REs assemble near centrosomes via dynein-dependent movement along microtubules, and Golgi stacks are connected laterally to form the Golgi ribbon [17, 20, 59, 60, 84, 85]. While the relationship

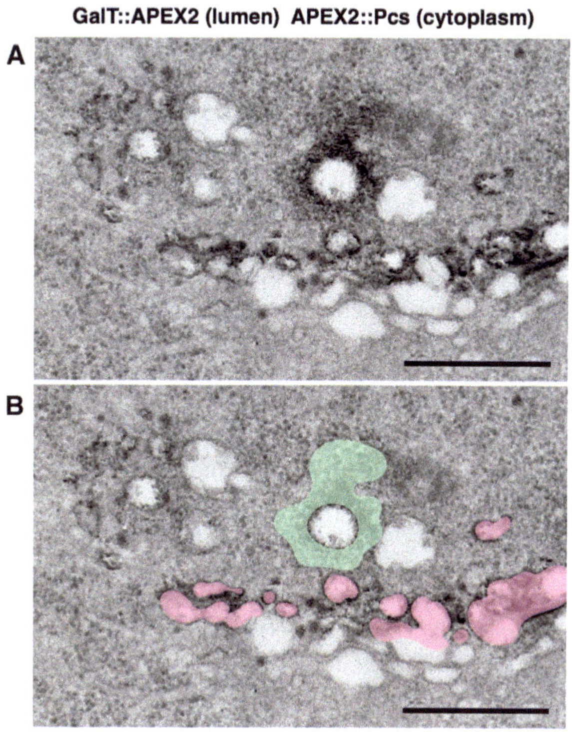

Fig. 5.2 RE attachment to the *trans*-cisternae of the Golgi stack. (**a**) Electron micrographs of S2 cells co-expressing GalT::APEX2 and APEX2::Pcs. APEX2 localization is visualized using osmium-enhanced DAB (diaminobenzidine) depositions. DAB depositions are observed within the *trans*-Golgi cisternae due to GalT::APEX2 and in the cytoplasm adjacent to recycling endosomes due to APEX2::Pcs. (**b**) Osmium-enhanced DAB depositions from GalT::APEX2 and APEX2::Pcs are pseudo-colored pink and green, respectively. Scale bars: 500 nm (**a**, **b**)

between the Golgi ribbon and REs is unclear in most cell types, COS1 cells display a remarkable arrangement: the REs cluster around the centrosome and the Golgi ribbon surrounds this cluster, with its *trans*-side facing the REs [36, 43, 44, 71]. In mammalian cells, the microtubule polymerization inhibitor nocodazole disperses Golgi stacks and REs throughout the cytoplasm [59, 60]. We found that more than 70% of these dispersed Golgi stacks were accompanied by REs on their *trans*-side, resembling the organization of *Drosophila* tissues [10]. A significant population of REs exists independently of Golgi stacks. These two forms are referred to as Golgi-associated REs (GA-REs) and Golgi-independent REs (free-REs). Both GA-REs and free-REs have been observed in nocodazole-treated sea urchin embryos at the blastula stage [12]. Thus, the presence of GA-REs and free-REs appears to be a common feature of eukaryotic cells.

GA-REs and Free-REs Are Interchangeable

The dynamics of GA-REs and free-REs in *Drosophila* S2 cells and nocodazole-treated HeLa cells were revealed by live imaging using super-resolution confocal live-imaging microscopy (SCLIM) [31–33, 74]. In both S2 cells and nocodazole-treated HeLa cells, the REs were in constant motion, repeatedly dividing and fusing with one another (Fig. 5.3a). In contrast, Golgi stacks are relatively static, with GA-REs being more stable than free-REs. The movement of these REs does not depend on microtubules or actin filaments, but requires the activity of ARFGEFs—Sec71 in *Drosophila* S2 cells [11] and BIG1/2 in HeLa cells (Takiguchi, unpublished data). In yeast, Golgi cisternae are highly mobile and their movement is inhibited in COPI mutants, suggesting that Golgi mobility may be at least partially driven by coat protein assembly [21].

Interestingly, GA-REs can detach to become free-REs, whereas free-REs can reattach to Golgi stacks to become GA-REs (Fig. 5.2a) [10]. This interchangeability highlights the dynamic nature of these endosomes. The detachment of GA-REs from the Golgi is inhibited by Brefeldin A (BFA) in S2 cells, while the attachment of free-REs to the Golgi remains unaffected. This inhibition leads to the formation of BFA bodies, which are aggregates of Golgi stacks centered around the REs in S2 cells (Fig. 5.4). Because TGN-localized ARFGEF Sec71 is the only BFA target responsible for BFA body formation in S2 cells, ARF activity is essential for the detachment of REs from the Golgi stacks [11]. Notably, Sec71 is a human ortholog of BIG1/2. Therefore, both RE mobility and the separation of REs from the TGN appear to depend on ARFGEF activity in the TGN.

Similar dynamics have been observed in other plant organelles. In *Arabidopsis thaliana*, two types of TGNs exist: Golgi-associated TGN (GA-TGN) and Golgi-independent TGN (GI-TGN) [23, 25, 77, 78, 80]. GA-TGNs can detach from Golgi stacks to become GI-TGNs, whereas GI-TGNs can reattach to the Golgi stacks and transform into GA-TGNs. Although REs are not well-characterized in plants, TGNs are known to function as endosomes [5, 6, 23, 35, 54, 80]. The mobility of Golgi stacks in plants depends on actin filaments [49, 50], but the mechanisms that facilitate the detachment or attachment of TGNs to Golgi stacks remain unclear. Most mature *trans*-cisternae peel off from the Golgi stacks and become GI-TGNs, suggesting that the progression of cargo transport or cisternal maturation may induce the release of TGNs from the Golgi stacks [25, 67].

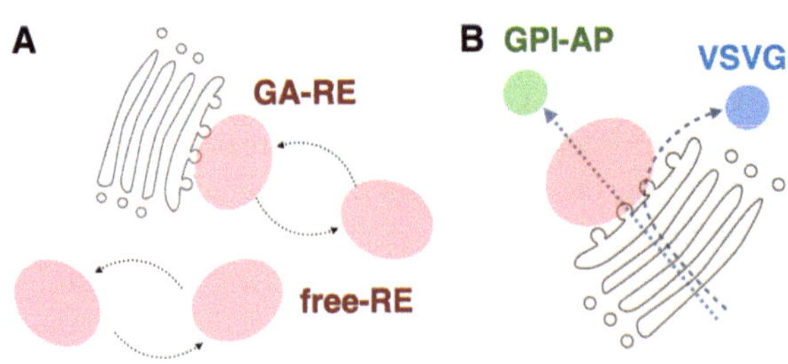

Fig. 5.3 Models illustrating the interaction between RE and Golgi Stack and their functional roles. (**a**) A proposed model depicting the interaction between REs and the Golgi stacks. The pink circle represents REs. GA-REs and free-REs are shown as interchangeable, with REs undergoing repeated cycles of fusion and separation. (**b**) A model of cargo transport mechanisms. Green represents GPI-anchored protein (GPI-AP), which are transported from the Golgi stack to GA-REs before being delivered to the plasma membrane. Blue represents VSV-G, which is transported directly from the Golgi stack to the plasma membrane

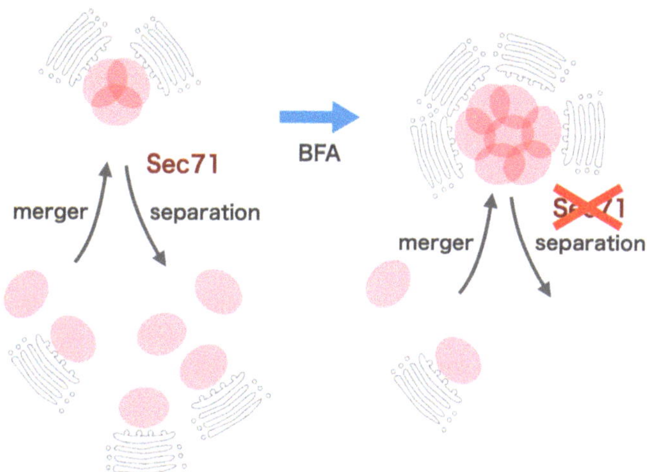

Fig. 5.4 Model of Golgi stack/RE unit merger and separation. Golgi stacks/RE units undergo continuous merging and separation cycles. In the presence of Brefeldin A (BFA), this separation is inhibited, leading to the formation of a BFA-induced structure, known as a BFA body

REs Receive Certain Types of Cargo from Golgi Stacks

The role of RE attachment to Golgi stacks has been elucidated by visualizing the localization of cargo after its release from the endoplasmic reticulum (ER) using the RUSH system, a widely used method for synchronous cargo release based on biotin-streptavidin technology [3]. Glycosylphosphatidylinositol-anchored proteins (GPI-APs) are predominantly localized in GA-REs following their exit from the *trans*-Golgi cisternae. Subsequently, GA-REs containing GPI-APs move away from the Golgi stacks and transition into free-REs that also carry GPI-APs. This behavior is characteristic of post-Golgi carriers. In contrast, vesicular stomatitis virus glycoprotein (VSV-G) and tumor necrosis factor-alpha (TNFα) are not localized to GA-REs but are found in close proximity to them (Fig. 5.3b) [10].

Our lab recently developed an additional cargo release method called RudLOV (Tago, unpublished: RudLOV method is published: PMID: 39658747). In RudLOV, the cargo fused to Zdk1 is retained in the endoplasmic reticulum (ER) by binding to LOV2, which is fused to Sec61b. Illumination induces a conformational change in LOV2, leading to the release of the Zdk1-cargo, allowing the cargo to exit the ER synchronously. The intensity, region, and timing of illumination are precisely controlled, enabling easy and clear visualization of cargo movements on the cisternae in the RudLOV. Using this method, we examined and compared the movements of GPI-AP, VSV-G, and TNFα (Tago, unpublished). GPI-AP translocated smoothly from the *trans*-Golgi cisternae to the GA-RE, whereas VSV-G and TNFα did not enter the GA-RE. Instead, they accumulated in the space between the *trans*-Golgi cisternae and GA-RE after exiting the *trans*-Golgi cisternae. Interestingly, VSV-G and TNFα appeared to push the GA-RE away, creating additional space for themselves in the gap between the *trans*-Golgi cisternae and the GA-RE. The nature of the membrane on which VSV-G and TNFα reside remains unknown, but it likely corresponds to a zone of the TGN. It is not yet clear whether this zone can detach from Golgi stacks.

Two distinct zones of GA-TGNs have been clearly visualized in plants using super-resolution confocal live imaging microscopy (SCLIM) [66, 67]. The first is the secretory trafficking zone, which is characterized by the localization of R-SNARE VAMP721, the adaptor protein AP-1, and the coat protein clathrin. The second is the vacuolar trafficking zone, where the R-SNARE VAMP727 and the adaptor protein AP-4 are found. Notably, the secretory trafficking zone can detach from Golgi stacks to form Golgi-independent TGNs (GI-TGNs) or free-TGNs, whereas the vacuolar trafficking zone does not exhibit this behavior.

REs Influence the Distribution and Arrangement of Golgi Stacks Within the Cell

As mentioned earlier, the kinetics of REs affect the distribution and arrangement of Golgi stacks within the cell. BFA inhibits the detachment of REs from Golgi stacks, leading to the formation of BFA bodies with central REs in S2 cells (Fig. 5.4) [10]. BFA bodies were first described in plants [7, 13, 14, 55]. In *Arabidopsis thaliana*, BFA body formation upon BFA administration results in an aggregate of Golgi stacks surrounding the *trans*-Golgi network (TGN), rather than REs, as observed in *Drosophila*. Therefore, the configuration of BFA bodies in plants and flies suggests that the properties of plant TGNs are analogous to those of animal REs.

The appearance of BFA bodies in *Drosophila* S2 and *Arabidopsis thaliana* cells resembles that of Golgi stacks and REs in COS1 cells [36, 43, 76]. This fried egg-like arrangement, in which the RE is the yolk, has also been clearly visualized in ascidian and sea urchin embryos after the blastula stage [69]. Golgi orientation, *trans*-side facing the RE, was clear in both COS1 cells and sea urchin embryos. These observations suggest that RE attachment to the Golgi stack and RE-RE association may contribute to the organization of a single gigantic Golgi apparatus.

The most crucial factor in the centralization of Golgi stacks is undoubtedly microtubules. Disruption of microtubules transforms Golgi ribbons into distinct Golgi stacks located in front of the ER exit sites in mammalian cells [9, 29]. These Golgi stacks are relatively static. Although the mechanism of ribbon formation must remain intact, the close proximity or collision of Golgi stacks is essential for this process. Consequently, the Golgi stacks in nocodazole-treated cells maintain their separate configurations. In contrast, in plants and fruit flies, the continuous movement of Golgi stacks facilitates collisions and docking, leading to the fusion or connection of these stacks [11, 77], although the counteracting strong separation activity of REs/TGNs prevents the centralization of Golgi stacks. Only when REs/TGNs separation is inhibited by BFA treatment do Golgi stacks dock and centralize, resulting in the formation of BFA bodies. We recently discovered that several Golgi stacks cluster at the center of REs when SNAREs required for COPI vesicle fusion are deficient in *Drosophila*. Surprisingly, some Golgi stacks exhibited lateral connections, suggesting that such connections may be suppressed by COPI [70].

Conclusion and Future Perspective

Both TGNs and REs are recognized as central hubs in the exocytic and endocytic pathways; however, their functional relationships remain largely unexplored. While the role of TGNs as early endosomes is well-established in plants and yeast, it is less clear in animal cells, even though TGNs eventually receive endocytic material through retrograde trafficking from endosomes. Our unexpected discovery of a close association between REs and the *trans*-side of Golgi stacks suggests that REs may be functionally equivalent to at least part of the TGN. Given that the TGN is now understood to consist of distinct zones, REs may represent the *trans*-most zones of the TGN. Further research is required to elucidate the exact nature of the TGN and RE.

In addition to cargo sorting, the TGNs/REs may influence the overall distribution of Golgi stacks within the cell. We propose that the dynamic association between Golgi stacks and TGNs/REs, regulated by cycles of merging and separation, plays a key role in governing the cell-wide arrangement of Golgi stacks, in addition to the organization of microtubules and the lateral-linking activity of Golgi stacks.

Acknowledgments This work was supported by grants from the Japan Society for the Promotion of Science (JSPS) (KAKENHI grant no. 22H02617); the Japan Science and Technology Agency (JST) (CREST grant no. JPMJCR22E2); Core Research for Organelle Diseases funding from Hiroshima University; Takeda Science Foundation, and Ohsumi Frontier Science Foundation (Grant Number JP22H04926). The author expresses gratitude to Dr. Koichi Ozaki for his insightful discussions on the chromophore of fly rhodopsin, and to Ms. Rina Satoh for her assistance in illustrating the model and coloring the EM images. The author thanks Editage (http://www.editage.com) for editing and reviewing the manuscript for English language.

Compliance with Ethical Standards The author declares that the research was conducted in the absence of any commercial or financial relationships that could be construed as a potential conflict of interest.

References

1. Ang AL, Taguchi T, Francis S, Fölsch H, Murrells LJ, Pypaert M, Warren G, Mellman I (2004) Recycling endosomes can serve as intermediates during transport from the Golgi to the plasma membrane of MDCK cells. J Cell Biol 167:531–543
2. Benvenuto G, Leone S, Astoricchio E, Bormke S, Jasek S, D'Aniello E, Kittelmann M, McDonald K, Hartenstein V, Baena V et al (2024) Evolution of the ribbon-like organization of the Golgi apparatus in animal cells. Cell Rep 43:113791

3. Boncompain G, Divoux S, Gareil N, de Forges H, Lescure A, Latreche L, Mercanti V, Jollivet F, Raposo G, Perez F (2012) Synchronization of secretory protein traffic in populations of cells. Nat Methods 9:493–498
4. Boncompain G, Weigel AV (2018) Transport and sorting in the Golgi complex: multiple mechanisms sort diverse cargo. Curr Opin Cell Biol 50:94–101
5. Chow CM, Neto H, Foucart C, Moore I (2008) Rab-A2 and Rab-A3 GTPases define a trans-golgi endosomal membrane domain in Arabidopsis that contributes substantially to the cell plate. Plant Cell 20:101–123
6. Dettmer J, Hong-Hermesdorf A, Stierhof YD, Schumacher K (2006) Vacuolar H+-ATPase activity is required for endocytic and secretory trafficking in Arabidopsis. Plant Cell 18:715–730
7. Driouich A, Zhang GF, Staehelin LA (1993) Effect of brefeldin A on the structure of the Golgi apparatus and on the synthesis and secretion of proteins and polysaccharides in sycamore maple (Acer pseudoplatanus) suspension-cultured cells. Plant Physiol 101:1363–1373
8. Farquhar MG, Palade GE (1998) The Golgi apparatus: 100 years of progress and controversy. Trends Cell Biol 8:2–10
9. Fourriere L, Divoux S, Roceri M, Perez F, Boncompain G (2016) Microtubule-independent secretion requires functional maturation of Golgi elements. J Cell Sci 129:3238–3250
10. Fujii S, Kurokawa K, Inaba R, Hiramatsu N, Tago T, Nakamura Y, Nakano A, Satoh T, Satoh AK (2020a) Recycling endosomes attach to the trans-side of Golgi stacks in Drosophila and mammalian cells. J Cell Sci 133(4):jcs236935
11. Fujii S, Kurokawa K, Tago T, Inaba R, Takiguchi A, Nakano A, Satoh T, Satoh AK (2020b) Sec71 separates Golgi stacks in *Drosophila* S2 cells. J Cell Sci 133:jcs245571
12. Fujii S, Tago T, Sakamoto N, Yamamoto T, Satoh T, Satoh AK (2020c) Recycling endosomes associate with Golgi stacks in sea urchin embryos. Commun Integr Biol 13:59–62
13. Geldner N, Anders N, Wolters H, Keicher J, Kornberger W, Muller P, Delbarre A, Ueda T, Nakano A, Jürgens G (2003) The Arabidopsis GNOM ARF-GEF mediates endosomal recycling, auxin transport, and auxin-dependent plant growth. Cell 112:219–230
14. Geldner N, Friml J, Stierhof YD, Jürgens G, Palme K (2001) Auxin transport inhibitors block PIN1 cycling and vesicle trafficking. Nature 413:425–428
15. Goldenring JR (2015) Recycling endosomes. Curr Opin Cell Biol 35:117–122
16. Gosavi P, Gleeson PA (2017) The function of the Golgi ribbon structure – an enduring mystery unfolds! BioEssays 39:1700063
17. Harada A, Takei Y, Kanai Y, Tanaka Y, Nonaka S, Hirokawa N (1998) Golgi vesiculation and lysosome dispersion in cells lacking cytoplasmic dynein. J Cell Biol 141:51–59
18. Harris WA, Ready DF, Lipson ED, Hudspeth AJ, Stark WS (1977) Vitamin A deprivation and Drosophila photopigments. Nature 266:648–650
19. Hebbar S, Schuhmann K, Shevchenko A, Knust E (2020) Hydroxylated sphingolipid biosynthesis regulates photoreceptor apical domain morphogenesis. J Cell Biol 219:e201911100
20. Hehnly H, Chen CT, Powers CM, Liu HL, Doxsey S (2012) The centrosome regulates the Rab11- dependent recycling endosome pathway at appendages of the mother centriole. Curr Biol 22:1944–1950
21. Ishii M, Suda Y, Kurokawa K, Nakano A (2016) COPI is essential for Golgi cisternal maturation and dynamics. J Cell Sci 129:3251–3261
22. Isono K, Tanimura T, Oda Y, Tsukahara Y (1988) Dependency on light and vitamin A derivatives of the biogenesis of 3-hydroxyretinal and visual pigment in the compound eyes of Drosophila melanogaster. J Gen Physiol 92:587–600
23. Ito Y, Uemura T (2022) Super resolution live imaging: the key for unveiling the true dynamics of membrane traffic around the Golgi apparatus in plant cells. Front Plant Sci 13:1100757
24. Iwanami N, Nakamura Y, Satoh T, Liu Z, Satoh AK (2016) Rab6 is required for multiple apical transport pathways but not the basolateral transport pathway in Drosophila photoreceptors. PLoS Genet 12:e1005828
25. Kang BH, Nielsen E, Preuss ML, Mastronarde D, Staehelin LA (2011) Electron tomography of RabA4b- and PI-4Kβ1-labeled trans Golgi network compartments in Arabidopsis. Traffic 12:313–329

26. Karagiosis SA, Ready DF (2004) Moesin contributes an essential structural role in Drosophila photoreceptor morphogenesis. Development 131:725–732
27. Kienzle C, von Blume J (2014) Secretory cargo sorting at the trans-Golgi network. Trends Cell Biol 24:584–593
28. Klumperman J (2011) Architecture of the mammalian Golgi. Cold Spring Harb Perspect Biol 3:a005181
29. Kondylis V, Rabouille C (2009) The Golgi apparatus: lessons from Drosophila. FEBS Lett 583:3827–3838
30. Kumar JP, Ready DF (1995) Rhodopsin plays an essential structural role in Drosophila photoreceptor development. Development 121:4359–4370
31. Kurokawa K, Ishii M, Suda Y, Ichihara A, Nakano A (2013) Live cell visualization of Golgi membrane dynamics by super-resolution confocal live imaging microscopy. Methods Cell Biol 118:235–242
32. Kurokawa K, Nakano A (2020) Live-cell imaging by Super-resolution Confocal Live Imaging Microscopy (SCLIM): simultaneous three-color and four-dimensional live cell imaging with high space and time resolution. Bio Protoc 10:e3732
33. Kurokawa K, Osakada H, Kojidani T, Waga M, Suda Y, Asakawa H, Haraguchi T, Nakano A (2019) Visualization of secretory cargo transport within the Golgi apparatus. J Cell Biol 218:1602–1618
34. Laffafian A, Tepass U (2019) Identification of genes required for apical protein trafficking in *Drosophila* photoreceptor cells. G3 (Bethesda) 9:4007–4017
35. Lam SK, Siu CL, Hillmer S, Jang S, An G, Robinson DG, Jiang L (2007) Rice SCAMP1 defines clathrin-coated, trans-golgi-located tubular-vesicular structures as an early endosome in tobacco BY-2 cells. Plant Cell 19:296–319
36. Lee S, Uchida Y, Wang J, Matsudaira T, Nakagawa T, Kishimoto T, Mukai K, Inaba T, Kobayashi T, Molday RS et al (2015) Transport through recycling endosomes requires EHD1 recruitment by a phosphatidylserine translocase. EMBO J 34:669–688
37. Lock JG, Stow JL (2005) Rab11 in recycling endosomes regulates the sorting and basolateral transport of E-cadherin. Mol Biol Cell 16:1744–1755
38. Losev E, Reinke CA, Jellen J, Strongin DE, Bevis BJ, Glick BS (2006) Golgi maturation visualized in living yeast. Nature 441:1002–1006
39. Luini A, Parashuraman S (2016) Signaling at the Golgi: sensing and controlling the membrane fluxes. Curr Opin Cell Biol 39:37–42
40. Matsuura-Tokita K, Takeuchi M, Ichihara A, Mikuriya K, Nakano A (2006) Live imaging of yeast Golgi cisternal maturation. Nature 441:1007–1010
41. Mayor S, Presley JF, Maxfield FR (1993) Sorting of membrane components from endosomes and subsequent recycling to the cell surface occurs by a bulk flow process. J Cell Biol 121:1257–1269
42. McMahon HT, Ushkaryov YA, Edelmann L, Link E, Binz T, Niemann H, Jahn R, Südhof TC (1993) Cellubrevin is a ubiquitous tetanus-toxin substrate homologous to a putative synaptic vesicle fusion protein. Nature 364:346–349
43. Misaki R, Morimatsu M, Uemura T, Waguri S, Miyoshi E, Taniguchi N, Matsuda M, Taguchi T (2010) Palmitoylated Ras proteins traffic through recycling endosomes to the plasma membrane during exocytosis. J Cell Biol 191:23–29
44. Misaki R, Nakagawa T, Fukuda M, Taniguchi N, Taguchi T (2007) Spatial segregation of degradation- and recycling-trafficking pathways in COS-1 cells. Biochem Biophys Res Commun 360:580–585
45. Murray RZ, Kay JG, Sangermani DG, Stow JL (2005) A role for the phagosome in cytokine secretion. Science 310:1492–1495
46. Nakano A (2022) The Golgi apparatus and its next-door neighbors. Front Cell Dev Biol 10:884360
47. Nakano A, Luini A (2010) Passage through the Golgi. Curr Opin Cell Biol 22:471–478

48. Naramoto S, Otegui MS, Kutsuna N, de Rycke R, Dainobu T, Karampelias M, Fujimoto M, Feraru E, Miki D, Fukuda H et al (2014) Insights into the localization and function of the membrane trafficking regulator GNOM ARF-GEF at the Golgi apparatus in Arabidopsis. Plant Cell 26:3062–3076
49. Nebenführ A, Gallagher LA, Dunahay TG, Frohlick JA, Mazurkiewicz AM, Meehl JB, Staehelin LA (1999) Stop-and-go movements of plant Golgi stacks are mediated by the acto-myosin system. Plant Physiol 121:1127–1142
50. Nebenführ A, Staehelin LA (2001) Mobile factories: Golgi dynamics in plant cells. Trends Plant Sci 6:160–167
51. Otsuka Y, Satoh T, Nakayama N, Inaba R, Yamashita H, Satoh AK (2019) Parcas is the predominant Rab11-GEF for rhodopsin transport in Drosophila photoreceptors. J Cell Sci 132:jcs231431
52. Ozaki K, Nagatani H, Ozaki M, Tokunaga F (1993) Maturation of major Drosophila rhodopsin, ninaE, requires chromophore 3-hydroxyretinal. Neuron 10:1113–1119
53. Papanikou E, Glick BS (2014) Golgi compartmentation and identity. Curr Opin Cell Biol 29:74–81
54. Richter S, Voss U, Jürgens G (2009) Post-Golgi traffic in plants. Traffic 10:819–828
55. Robinson DG, Langhans M, Saint-Jore-Dupas C, Hawes C (2008) BFA effects are tissue and not just plant specific. Trends Plant Sci 13:405–408
56. Rodriguez-Furlan C, Minina EA, Hicks GR (2019) Remove, recycle, degrade: regulating plasma membrane protein accumulation. Plant Cell 31:2833–2854
57. Rosenbaum EE, Vasiljevic E, Cleland SC, Flores C, Colley NJ (2014) The Gos28 SNARE protein mediates intra-Golgi transport of rhodopsin and is required for photoreceptor survival. J Biol Chem 289:32392–32409
58. Sakai T, Yamashina S, Ohnishi S (1991) Microtubule-disrupting drugs blocked delivery of endocytosed transferrin to the cytocenter, but did not affect return of transferrin to plasma membrane. J Biochem 109:528–533
59. Saraste J, Goud B (2007) Functional symmetry of endomembranes. Mol Biol Cell 18:1430–1436
60. Saraste J, Prydz K (2019) A new look at the functional organization of the Golgi ribbon. Front Cell Dev Biol 7:171
61. Satoh A, Tokunaga F, Kawamura S, Ozaki K (1997) In situ inhibition of vesicle transport and protein processing in the dominant negative Rab1 mutant of Drosophila. J Cell Sci 110(Pt 23):2943–2953
62. Satoh AK, O'Tousa JE, Ozaki K, Ready DF (2005) Rab11 mediates post-Golgi trafficking of rhodopsin to the photosensitive apical membrane of Drosophila photoreceptors. Development 132:1487–1497
63. Satoh T, Inagaki T, Liu Z, Watanabe R, Satoh AK (2013) GPI biosynthesis is essential for rhodopsin sorting at the trans-Golgi network in Drosophila photoreceptors. Development 140:385–394
64. Schopf K, Huber A (2017) Membrane protein trafficking in Drosophila photoreceptor cells. Eur J Cell Biol 96:391–401
65. Seki T, Fujishita S, Ito M, Matsuoka N, Kobayashi C, Tsukida K (1986) A fly, Drosophila melanogaster, forms 11-cis 3-hydroxyretinal in the dark. Vis Res 26:255–258
66. Shimizu Y, Takagi J, Ito E, Ito Y, Ebine K, Komatsu Y, Goto Y, Sato M, Toyooka K, Ueda T et al (2021) Cargo sorting zones in the trans-Golgi network visualized by super-resolution confocal live imaging microscopy in plants. Nat Commun 12:1901
67. Shimizu Y, Uemura T (2022) The sorting of cargo proteins in the plant. Front Plant Sci 13:957995
68. Stark WS, Schilly D, Christianson JS, Bone RA, Landrum JT (1990) Photoreceptor-specific efficiencies of beta-carotene, zeaxanthin and lutein for photopigment formation deduced from receptor mutant Drosophila melanogaster. J Comp Physiol A 166:429–436
69. Tago T, Fujii S, Sasaki S, Shirae-Kurabayashi M, Sakamoto N, Yamamoto T, Maeda M, Ueki T, Satoh T, Satoh AK (2024a) Cell-wide arrangement of Golgi/RE units depends on the microtubule organization. Cell Struct Funct 49:101

70. Tago T, Yamada Y, Goto Y, Toyooka K, Ochi Y, Satoh T, Satoh AK (2024b) Golgi clustering by the deficiency of COPI-SNARE in Drosophila photoreceptors. Front Cell Dev Biol 12:1442198
71. Taguchi T (2013) Emerging roles of recycling endosomes. J Biochem 153:505–510
72. Tepass U (2012) The apical polarity protein network in Drosophila epithelial cells: regulation of polarity, junctions, morphogenesis, cell growth, and survival. Annu Rev Cell Dev Biol 28:655–685
73. Terasaki M (2000) Dynamics of the endoplasmic reticulum and Golgi apparatus during early sea urchin development. Mol Biol Cell 11:897–914
74. Tojima T, Miyashiro D, Kosugi Y, Nakano A (2023) Super-resolution live imaging of cargo traffic through the Golgi apparatus in mammalian cells. Methods Mol Biol 2557:127–140
75. Tojima T, Suda Y, Jin N, Kurokawa K, Nakano A (2024) Spatiotemporal dissection of the Golgi apparatus and the ER-Golgi intermediate compartment in budding yeast. elife 13:e92900
76. Uchida Y, Hasegawa J, Chinnapen D, Inoue T, Okazaki S, Kato R, Wakatsuki S, Misaki R, Koike M, Uchiyama Y et al (2011) Intracellular phosphatidylserine is essential for retrograde membrane traffic through endosomes. Proc Natl Acad Sci USA 108:15846–15851
77. Uemura T, Nakano RT, Takagi J, Wang Y, Kramer K, Finkemeier I, Nakagami H, Tsuda K, Ueda T, Schulze-Lefert P et al (2019) A Golgi-released subpopulation of the trans-Golgi network mediates protein secretion in Arabidopsis. Plant Physiol 179:519–532
78. Uemura T, Ueda T, Ohniwa RL, Nakano A, Takeyasu K, Sato MH (2004) Systematic analysis of SNARE molecules in Arabidopsis: dissection of the post-Golgi network in plant cells. Cell Struct Funct 29:49–65
79. Ullrich O, Reinsch S, Urbé S, Zerial M, Parton RG (1996) Rab11 regulates recycling through the pericentriolar recycling endosome. J Cell Biol 135:913–924
80. Viotti C, Bubeck J, Stierhof YD, Krebs M, Langhans M, van den Berg W, van Dongen W, Richter S, Geldner N, Takano J et al (2010) Endocytic and secretory traffic in Arabidopsis merge in the trans-Golgi network/early endosome, an independent and highly dynamic organelle. Plant Cell 22:1344–1357
81. Vogt K, Kirschfeld K (1984) Chemical identity of the chromophores of fly visual pigment. Naturwissenschaften 77:211–213
82. Walther RF, Pichaud F (2006) Immunofluorescent staining and imaging of the pupal and adult Drosophila visual system. Nat Protoc 1:2635–2642
83. Wei JH, Seemann J (2017) Golgi ribbon disassembly during mitosis, differentiation and disease progression. Curr Opin Cell Biol 47:43–51
84. Yadav S, Linstedt AD (2011) Golgi positioning. Cold Spring Harb Perspect Biol 3:a005322
85. Yadav S, Puthenveedu MA, Linstedt AD (2012) Golgin160 recruits the dynein motor to position the Golgi apparatus. Dev Cell 23:153–165
86. Yamashiro DJ, Maxfield FR (1987) Acidification of morphologically distinct endosomes in mutant and wild-type Chinese hamster ovary cells. J Cell Biol 105:2723–2733
87. Yamashita H, Ochi Y, Yamada Y, Sasaki S, Tago T, Satoh T, Satoh AK (2022) Functions of neuronal Synaptobrevin in the post-Golgi transport of Rhodopsin in Drosophila photoreceptors. J Cell Sci 135:jcs260196
88. Yamazaki Y, Schönherr C, Varshney GK, Dogru M, Hallberg B, Palmer RH (2013) Goliath family E3 ligases regulate the recycling endosome pathway via VAMP3 ubiquitylation. EMBO J 32:524–537

Chapter 6
The Impact of Golgi Organization on Cell Cycle and Cancer Development

Antonino Colanzi and Inmaculada Ayala

Abstract The Golgi complex plays a pivotal role in the secretory pathway, cell signaling, and cytoskeleton organization. In mammalian cells, it is organized as a ribbon-like structure that can undergo dynamic reorganizations, which are crucial for various cellular processes, including mitosis and cell migration. During mitosis, the Golgi complex is subjected to a tightly regulated disassembly process, essential for proper cell division, while its reformation is critical for inheritance by daughter cells. Disruption of Golgi organization has been linked to pathological conditions, including cancer, as altered organization and function can contribute to tumor progression by impacting glycosylation, cell polarity, and extracellular matrix remodeling. This chapter explores the molecular mechanisms regulating Golgi disassembly during the cell cycle and its involvement in ensuring proper cell division. Additionally, we discuss the emerging understanding of the potential role of modifications of the Golgi organization in tumor development and metastasis. Collectively, the findings presented in this chapter underscore the physiological significance of the Golgi ribbon structure and the pathological implications of its alterations in cancer progression, which could offer novel therapeutic opportunities for cancer treatment.

Keywords Golgi complex · Ribbon · Centrosome · Mitosis · Checkpoint · Mitotic spindle

---

A. Colanzi · I. Ayala (✉)
Institute of Endotypes in Oncology, Metabolism, and Immunology "G. Salvatore" (IEOMI), National Research Council (CNR), Naples, Italy
e-mail: mariainmaculada.ayalagrande@cnr.it

A. Nakano, J. Saraste (eds.), *The Golgi Network, Volume I*, Subcellular Biochemistry 110, https://doi.org/10.1007/978-3-032-06936-8_6

Introduction

The Golgi complex (GC) plays a central role in the secretory pathway as it is responsible for the transport, processing, and sorting of proteins and lipids to their final destinations. It regulates the organization of a subset of microtubules (MTs) [1] and, importantly, also functions as a platform that integrates several signaling pathways [2].

In most vertebrate cells during interphase, the GC is organized in the form of polarized stacks of flattened cisternae connected by tubules, thus forming a continuous membrane system known as "ribbon." In contrast, the GC of plants [3] and *Drosophila melanogaster* [4] is organized into independent stacks, whereas in the yeast *Saccharomyces cerevisiae* it is composed of scattered cisternae [5]. The ribbon-like organization is not exclusive of vertebrate cells. A recent study suggested that the Golgi ribbon assembles early in development, indicating its potential involvement in cell differentiation and embryonic development processes [6]. Current evidence shows that the ribbon likely evolved from the common ancestor of cnidarians and bilaterians, thus proposing that it precedes the emergence of vertebrates [6].

However, in specific mammalian cells, the GC presents variations from the ribbon organization; for instance, in neurons, the GC is composed of outposts in dendrites [7]. Other examples of cell types with the GC arranged as dispersed stacks throughout the cytoplasm include uroepithelial cells of the urinary bladder [8], epithelial gastric parietal cells [9], oocytes [10], and differentiated muscle cells [11].

Furthermore, the structure and localization of the ribbon are dynamic and can vary depending on physiological processes. Notable events include the disassembly of the ribbon into stacks, or "unlinking," during the G2 phase of the cell cycle, which is essential for entry into mitosis [12, 13]. Moreover, the Golgi ribbon is transiently cleaved during cell migration to support its orientation toward the leading edge, thereby facilitating cell motility [14]. Importantly, alterations in the structure of the GC are associated with pathological manifestations [15, 16] such as infections [17], stress responses [18, 19], apoptosis [20], neurological disorders [21], aging-related processes [22], and cancer [23].

We will summarize the mechanisms involved in the maintenance of the ribbon and the machinery implicated in the dynamic modulation of its organization, with a focus on mitosis. Additionally, we will address the potential role of alterations of Golgi structure in cancer development.

Essential Regulators of the Structure of the Golgi Complex

The characteristic ribbon-like organization of the mammalian GC, positioned near the centrosome (CE), depends on a delicate balance achieved through the cooperation of several factors as briefly described below. For the purposes of this chapter, we will introduce here the main mechanisms, while certain aspects of specific interest will be covered in more detail in subsequent sections. For an in-depth exploration of specific topics, dedicated reviews are recommended.

The structure and function of the GC require a "Golgi matrix"; that is, a scaffold of proteins around this organelle that extends into the cytoplasm [24, 25]. The principal constituents are golgins and Golgi reassembly and stacking proteins (GRASP65 and GRASP55) [26]. Together they regulate the association of the cisternae into stacks and the stacks into a ribbon [27]. The golgin family consists of at least 11 proteins characterized by elongated coiled-coil domains, which facilitate their interaction with other golgins and Golgi-associated proteins, thus contributing to the formation of the matrix, which stabilizes the Golgi structure [28], and also act as tethers for transport vesicles and proteins involved in the regulation of the actin and microtubule (MT) cytoskeleton [16]. In addition, several golgins control the localization of Golgi enzymes [29, 30], thereby contributing to the ordered processing of proteins and lipids. The different golgins are associated with specific regions of the GC. For instance, GM130 is localized to the *cis*-Golgi, while Golgin-45 is found at medial-Golgi. Although some golgins are anchored at the GC via their transmembrane domains, the majority are recruited by Rab, Arf, and Arl GTPases [28, 31].

The GRASPs behave as membrane tethers that drive the stacking of cisternae and ribbon formation [32, 33]. They consist of two N-terminal PDZ-like domains (GRASP domain) and a C-terminal serine/proline-rich regulatory domain (SPR) [34]. GRASPs form homodimers in *trans* through the interaction of their GRASP domains, and homodimerization is inhibited by phosphorylation in response to various stimuli, including growth factors, oxidative stress, and cell division. In the SPR of GRASP65 and GRASP55, there are eight and five phosphorylation sites, respectively [35–37]. Their localization at the GC is mediated by the myristoylation of an N-terminal glycine and the interaction with the C-terminus of specific golgins. Precisely, GRASP65 binds GM130, while GRASP55 binds Golgin-45 [34]. Beyond their role in the architecture of the GC, GRASP65 and GRASP55 also regulate protein trafficking and processing within this organelle [38].

On the other hand, the maintenance of the structure of the GC relies on the cytoskeleton. For instance, disruption of actin filaments leads to the swelling of cisternae [39] and compaction of Golgi membranes around the CE, which acts as a microtubule-organizing center (MTOC). Conversely, actin-stabilizing agents induce extensive disassembly of the GC [40]. Thus, actin and actin-associated proteins control the morphology and rigidity of this organelle. In addition, MTs contribute to the organization and localization of the GC [41]. In most cell types, the CE directs the formation of radial MTs that, together with dynein motor proteins, cause the clustering of the GC near this MTOC. Conversely, depolymerization of MTs induces the dispersion of the GC in the form of ministacks throughout the cytoplasm [41].

Dynamic Modulation of the Structure of the Golgi Ribbon

As previously mentioned, the Golgi ribbon can undergo different levels of disassembly under various physiological and pathological conditions [42]. Given that the major structural reorganization involved in these processes is the transition from a ribbon to isolated stacks [43], the mechanisms controlling this aspect will be discussed in more detail.

At steady state, the structure of the ribbon results from a dynamic equilibrium between the formation and cleavage of membranes connecting the stacks (Fig. 6.1) [13]. In fact, live imaging studies have shown that the formation of the ribbon is mediated by the elongation of GRASP-coated membranous tubules from the cisternae of "donor" stacks [44, 45]. This elongation requires phospholipase A2 (PLA2) and is guided by dynein and MTs [46]. Ultimately, these tubules dock and fuse with homologous cisternae of "acceptor" stacks in a Syntaxin 5-dependent manner [44]. The reverse process, involving the cleavage of these connections, is driven by the fission-inducing protein BARS, although the exact mechanism and regulation remain unknown (Fig. 6.1) [13, 47]. The regulation in ribbon organization of golgins, actin, MTs, and cytoskeletal motors is less defined [16, 48–50].

Among the proteins involved in the formation of the ribbon, the GM130/GRASP65 complex plays a key role in coordinating different signals, interacting proteins, and membrane- or cytoskeleton-based mechanisms to form lateral connections between the stacks. Regarding GM130, it recruits not only GRASP65 but also the scaffold protein AKAP450, which drives the nucleation of MTs from the GC that are crucial for the clustering of the stacks, and the maintenance of the structural integrity of the organelle [51–53]. In addition, the nucleation of MTs from the GC is also required for the formation of a different class of MTs, with an asymmetric

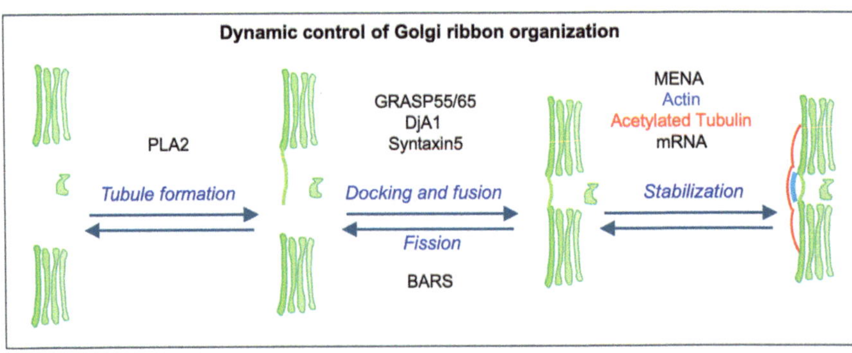

Fig. 6.1 Schematic representation of ribbon formation. The first step involves the formation of tubules from donor membranes, followed by their elongation, docking, and fusion with an acceptor membrane to establish a continuous compartment. The Golgi ribbon is further stabilized by actin, acetylated tubulin, the actin-binding protein MENA and mRNA. The reverse process, fission, is catalyzed by BARS. For simplicity, the role of Golgi-derived microtubules in inducing the clustering of stacks during the initial stages of ribbon formation is not shown

orientation, which are essential for the polarized delivery of cargoes to specific plasma membrane hotspots [51, 54], coincident with focal adhesions (FAs) [55]. They also facilitate the orientation of the organelle, the entire MT network, and the CE toward the leading edge during migration. Motility is particularly relevant for the ability of tumor cells to invade the extracellular matrix (ECM) [52, 56]. In addition, recent studies have shown that GM130 binds to RNA and forms RNA-dependent liquid-like condensates that are necessary for linking the stacks into a ribbon (Fig. 6.1). In particular, under cellular stress, the RNA is displaced from GM130 causing unlinking, which is reversed when the stress is overcome [57].

The essential role of GRASPs in ribbon formation is supported by evidence from siRNA-mediated depletion [58], and functional inactivation through kinase inhibitors and microinjection of blocking antibodies or recombinant proteins [13, 45, 59]. Although an acute silencing of GRASP65 or GRASP55, induces a prevalent loss of continuity at the *cis*-side, or at the medial/*trans* level, respectively [27], a more complex scenario emerges with the acute and concurrent depletion of both GRASPs, which results in unlinking but not in evident unstacking [60] in agreement with data based on the generation of a mouse lacking both GORASPs [61], thus suggesting that redundant mechanisms contribute to the process.

The *trans*-oligomerization of GRASP65, crucial for the tethering mechanism, is inhibited by various phosphorylation events under different physiological conditions [62]. Additional molecular mechanisms and binding partners fine-tune the role of GRASP65 in ribbon formation. For instance, GRASP65 interacts with MENA, a protein that promotes actin elongation, essential for linking the stacks into a ribbon [63], while the interaction with DjA1 promotes ribbon formation by stabilizing GRASP65 homodimerization [64] (Fig. 6.1). Another mechanism involves the Ajuba/PRMT5/Aurora-A complex, which regulates the activity of HURP, an MT-binding protein that favors Golgi ribbon organization by binding to GRASP65 [65]. Finally, GRASP65 can bind and stabilize Golgi-associated MTs, leading to their acetylation, a critical process for conferring structural resilience to the ribbon [66]. The processes of *trans*-oligomerization and MT stabilization are both inhibited by JNK/ERK-mediated phosphorylation of S274 (S277 in rat) during the G2 phase of the cell cycle or cell migration, resulting in unlinking [66]. In the latter case and as mentioned above, transient unlinking is also necessary for reorientating the GC and the CE toward the direction of migration [67].

The role of GRASP55 is primarily regulated by phosphorylation at T222/225 [44] and T249 [36], while molecular mechanisms and binding partners similar to those of GRASP65 have not yet been identified. Notably, GRASP55, but not GRASP65, is also involved in the unconventional protein secretion pathway, particularly under stress conditions [68], as well as the initiation and maturation of autophagosomes, especially during glucose deprivation, which is essential for cell survival [69]. Therefore, the two GRASPs, apart from their oligomerization capability, may have evolved different specialized functions.

Despite evidence that the GRASPs are major effectors of signaling pathways regulating the structure of the ribbon and fundamental cellular functions, important aspects regarding the ribbon organization remain poorly investigated. These include the "triggers" of the signaling acting on the GRASPs, the likely involvement of additional regulatory pathways, the role and regulation of golgins, the mechanism and regulation of BARS, and the cytoskeleton and membrane fusion-based processes [43]. A deeper investigation of these aspects is pivotal for a comprehensive understanding of the regulation and maintenance of the Golgi ribbon and for developing approaches to its modulation, which has potential implications for novel therapeutic strategies [43]. Another unclear area concerns the precise structure of the interstack connections. Although the formation of the ribbon seems to be driven by the formation and fusion of tubules linking the stacks [44], the ultrastructural studies performed to date suggest that the noncompact zones are composed of tubular-saccular membranes, indicating that the structure of the linkages is far more complex than simple tubules [70]. Therefore, we hope for a renewed interest in this topic, considering that it could be supported by the development of electron microscopy techniques characterized by improved resolution and automation (e.g., FIB-SEM, STEM, Array Tomography) [71–73], which now allow faster examination of the detailed ultrastructure of larger and deeper areas of the GC.

Gradual and Reversible Disassembly of the Golgi Complex During Mitosis

The most widely investigated mechanism of GC disassembly is related to the one observed during cell division, which can be described in four main phases. In the first, which occurs in G2, the ribbon separates into isolated stacks [43]. In the second, from prophase to metaphase, the stacks are completely disassembled into vesicles and tubular-vesicular clusters [74]. According to some hypotheses, some Golgi enzymes could relocate to the endoplasmic reticulum (ER) [75]. The third phase coincides with anaphase and involves the reformation of the stacks. The gradual reformation of the Golgi ribbon characterizes the fourth and final phase. During cytokinesis, two mini ribbons are located at the opposite poles of the nuclei of each daughter cell, one facing the mitotic furrow, and the other located in a pericentrosomal region at the other side of the nucleus. These mini ribbons gradually fuse into a single one (Fig. 6.2) [43, 74, 76].

The Cleavage of the Golgi Complex Is Essential for Entry into Mitosis

The continuity of the Golgi ribbon is disrupted during G2, but this cleavage is reversible, as functional inhibition of the GRASPs or BARS restores its integrity in G2-arrested cells [13, 45, 59]. The only regulatory mechanism known to date is the inhibition of the GRASPs-mediated tethering step. Regarding GRASP65,

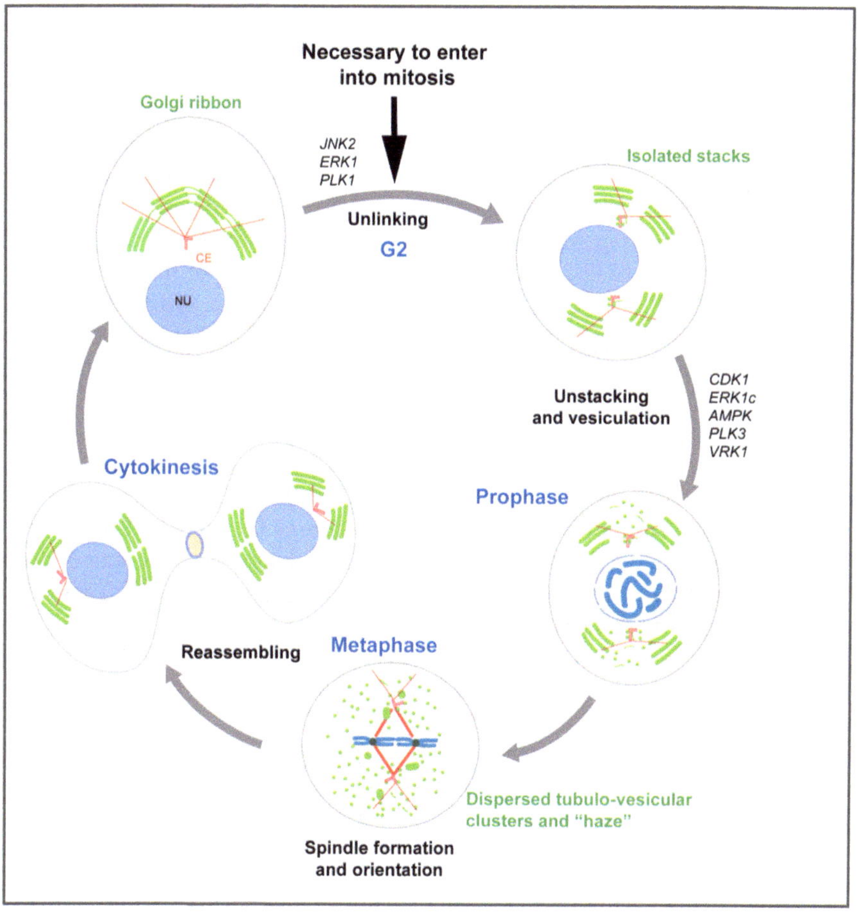

Fig. 6.2 Schematic representation of the main phases of Golgi disassembly and reformation during cell division. During the late G2 phase of the cell cycle, the Golgi ribbon disassembles into individual stacks, a process essential for entry into mitosis. After the onset of mitosis, the dissociation of cisternae within the stacks promotes increased vesiculation. Along with the inhibition of fusion, this leads to the dispersion of Golgi membranes into clusters and a Golgi haze, peaking at metaphase. Starting from anaphase, the processes that caused fragmentation and dispersion are reversed, with a gradual reformation of mini ribbons during cytokinesis. Once cytokinesis is complete, the mini ribbons reassemble into a single perinuclear one (*CE* centrosome, *NU* nucleus)

JNK2-mediated phosphorylation of S274 inhibits the *trans*-oligomerization and MT stabilization [66], facilitating the separation of the stacks. JNK inhibitors or JNK2 knockdown induce ribbon formation and cause a potent G2 arrest, which can be overridden by GRASP65 depletion, suggesting that the JNK2/GRASP65 signaling module plays a significant role in ribbon unlinking and the G2/M transition [45]. Moreover, the same residue is phosphorylated by ERK1 during stimulation of the cells with specific growth factors, during reorientation of the GC for migration [67,

77], during viral infections [78], or under cellular oxidative stress [79, 80]. In addition, PLK1 phosphorylates S189, located in the GRASP domain near a PDZ ligand that is essential for homodimerization, thereby leading to its inhibition [62]. Hence, GRASP65 integrates different stimuli to regulate Golgi structure and MT organization during cell division.

GRASP55 is also regulated throughout the cell cycle to facilitate the disassembly of the GC. During the G2/M transition, ERK1/2 [81], acting downstream of a signaling cascade involving MEK1 [82], RAF1 [83], and Protein Kinase D (PKD) [84], phosphorylates T222/225 and T249, inducing unlinking of the Golgi ribbon. Inhibition of MEK1 or expression of a phosphodefective GRASP55 mutant (T222/225A or T249A) prevents unlinking and delays the G2/M transition [36, 44].

Despite the significance of regulating ribbon formation and the studies conducted on the roles of GRASP65 and GRASP55, many aspects still need to be clarified [43]. These include the specific mechanisms by which phosphorylation of GRASPs in the regulatory domain affects their ability to form oligomers. Given that the stack/ribbon equilibrium is highly dynamic and unstable [13], it is unsurprising that numerous factors can influence it, some of which may be indirect. Actually, two proteomic studies identified about 200 potential regulators of the structure of the GC, including more than 30 mitogen-activated protein kinases (MAPKs) and other kinases, along with several dual specificity phosphatases (DUSPs), and regulators of phosphoinositide biosynthesis and actomyosin dynamics [85, 86]. This highlights the complexity and interconnected nature of the regulation of the arrangement of the GC.

However, regardless of the limited understanding of the mechanisms governing the structure of the GC, a key finding is that the inhibition of the Golgi unlinking (whether by using JNK inhibitors, microinjection of antibodies, or recombinant proteins that inhibit GRASP65, GRASP55, or BARS) results in a potent and prolonged G2 arrest [12, 13, 87]. This indicates that the segregation of the GC is essential for mitotic entry and is controlled by a "Golgi mitotic checkpoint" [88]. The primary mechanism by which the segregation of the Golgi ribbon affects the cell cycle will be described in section "Golgi unlinking and stack disassembly control key mitosis events".

After the Onset of Mitosis, the Stacks Are Extensively Disassembled

At the onset of mitosis, the GC undergoes a complete disassembly that is mediated by unstacking, extensive vesiculation, and the prevention of vesicle fusion with acceptor membranes, resulting in dispersed tubular-vesicular clusters and vesicles, referred to as the "Golgi haze" [74]. A fraction of the Golgi clusters, composed of Golgi matrix proteins and also the intermediate compartment (IC), associate with the spindle poles [89, 90], suggesting a role in the inheritance of Golgi "templates."

The unstacking is triggered by CyclinB-CDK1(CycB-CDK1)-mediated phosphorylation of the GRASP proteins. Regarding GRASP65, phosphorylation at S277 and S376 occurs from late G2 to telophase, while T220 and T224 are highly phosphorylated from prometaphase to anaphase, showing that these modifications are necessary for the full disassembly of the GC [74]. GRASP55 is phosphorylated at T222 and T225, and its *trans*-oligomerization is inhibited by acetylation at lysine 50 [91]. Additional pathways supporting unstacking involve AMPK, which phosphorylates the guanine-nucleotide exchange factor GBF1, causing its dissociation from Golgi membranes and inactivation of the GTPase Arf1, a key regulator of membrane traffic and Golgi structure [92]. Moreover, AMPK2 regulates Golgi fragmentation and the G2/M transition in mammalian cells by activating the CaMKKβ-AMPKα2 signaling complex [93]. Recently, a Gβγ isoform has been shown to regulate mitotic Golgi fragmentation and G2/M cell cycle progression [94].

The kinases PLK3, VRK1 [95], and ERK1c [96] have also been implicated in mitotic Golgi disassembly. PLK3 interacts with and specifically phosphorylates VRK1 at S342, inducing Golgi fragmentation [95]. ERK1c is a splice variant that is phosphorylated and activated by MEK1b [96]. The translocation of ERK1c to the GC during mitosis is triggered by its CDK1-mediated phosphorylation at S343, which induces the formation of a complex composed of PI4KIIIβ and 14-3-3γ, with its stability further regulated by PKD-mediated phosphorylation of PI4KIIIβ [97]. At the GC, ERK1c phosphorylates the MT-binding protein HOOK3, which is then phosphorylated by the mitotic kinase Aurora-A, leading to the dissociation of HOOK3 from the MTs and subsequent GC fragmentation [98]. Finally, an additional potential player is STK16, a kinase that directly interacts with actin and is necessary for GC integrity and cell cycle progression [99]. Still, its precise role in mitotic GC disassembly is unclear.

CycB-CDK1 also phosphorylates various golgins and molecular machinery involved in membrane fusion, which is regulated by the AAA ATPase p97/VCP (valosin-containing protein) and NSF (N-ethylmaleimide-sensitive factor) [100]. Specifically, p97/VCP forms a complex with the adaptor p37 and the deubiquitinating enzyme VCIP135, driving Golgi membrane fusion [101]. During mitosis, CDK1 phosphorylates p37 and VCIP135, inhibiting their binding to Golgi membranes, which blocks p97-mediated membrane fusion and contributes to Golgi disassembly by vesicle accumulation. NSF and α/β-SNAP (soluble NSF attachment proteins) regulate SNARE complex dissociation, blocking membrane fusion. This process is tightly controlled by phosphorylation and ubiquitination cycles. Specifically, CDK disrupts their interactions with tethering factors like p115, thus preventing NSF-mediated fusion. Moreover, the ubiquitin ligase HACE1 plays a role in ubiquitinating the Golgi t-SNARE Syntaxin 5 during early mitosis [102]. This ubiquitination disrupts the formation of the SNARE complex, blocking fusion events and promoting Golgi fragmentation [100, 101, 103, 104].

An indirect contribution to the cleavage of the GC may also arise from a block in ER exit of cargo. The ER exit sites are disassembled by casein kinase 1 (CK1)-mediated phosphorylation of TANGO1, resulting in a stop of secretion during mitosis that contributes to GC disassembly and potential retention of Golgi enzymes in the ER [105].

The Golgi Stacks Are Reformed During Anaphase

The definition of the mechanisms involved in the reformation of the stacks has been significantly advanced by the identification of proteins involved in membrane fusion, particularly through in vitro cisternal regrowth assays from mitotic Golgi fragments. Key factors include the ATPases NSF and p97, which function with specific co-factors: NSF-mediated fusion involves proteins such as SNAPs and p115, while the p97 pathway requires the formation of a complex with p47, VCIP135, and Syntaxin 5 [100, 103].

The reassembly process is triggered by the inhibition of CycB-CDK1 activity and the activation of phosphatases, mainly PP2A, which dephosphorylates NSF, p47, p37, and VCIP135. NSF plays a dual role in post-mitotic Golgi reassembly, where it first promotes a complex formation between GATE-16 and the v-SNARE GS28, preventing GS28 from interacting with the t-SNARE Syntaxin-5. Later, it facilitates the release of GATE-16, allowing SNARE pairing and membrane fusion. The other AAA ATPase, p97, along with its adaptor p47, is involved in promoting the regrowth of Golgi cisternae during post-mitotic reassembly. The p97/p47 complex specifically requires Syntaxin 5 for its function. Another p97 adaptor, p37, also plays a role in reassembly, requiring p115 and GS15 but not Syntaxin 5 [100, 106].

Ubiquitination plays a critical regulatory role in this process. The adaptor protein p47 contains a ubiquitin-associated (UBA) domain that binds mono-ubiquitin, and this interaction is vital for p97/p47-mediated membrane fusion. During mitosis, mono-ubiquitination of Golgi proteins, mediated by the E3 ligase HACE1, marks these proteins for recognition by the p97/p47 complex. The deubiquitinase VCIP135, a component of this complex, removes ubiquitin tags, facilitating the reassembly process. VCIP135 forms a quaternary complex with p97, p47, and Syntaxin 5, and its enzymatic activity is crucial for membrane fusion [100].

In summary, NSF and p97 play complementary roles in post-mitotic Golgi reassembly, with ubiquitination acting as a regulatory signal. NSF relies on the ubiquitin-like protein GATE-16 to regulate SNARE interactions, while p97 depends on the removal of ubiquitin tags by VCIP135 to facilitate membrane fusion [104].

Finally, the dephosphorylation of GRASP proteins, mainly by PP2A, promotes the stacking and linking of reformed Golgi cisternae, leading to the reassembly of the GC at the conclusion of mitosis [106]. Each daughter cell forms two distinct Golgi ribbons: a smaller one near the midbody and a larger one adjacent to the CE, at the opposite side of the nucleus. During cytokinesis, the smaller ribbon migrates to fuse with the larger one, suggesting that a functional Golgi ribbon is necessary for completing cell division [107].

Golgi Unlinking and Stack Disassembly Control Key Mitosis Events

The discovery of the "Golgi mitotic checkpoint" led to the identification of the mechanism through which the segregation of the GC controls the entry into mitosis. Specifically, the transformation from a ribbon to stacks, whether physiological or experimentally induced, results in the activation of Src at the GC [108]. Once activated, Src phosphorylates Aurora-A kinase at Y148, a site distinct from the canonical activation residue at T288 [108]. This phosphorylation enhances Aurora kinase activity, which then is recruited on the CE, where it activates the CycB-CDK1 complex, the master regulator of mitotic entry [109, 110], and promotes CE maturation, which involves the assembly of a protein network for the nucleation of MTs that form the mitotic spindle [111]. In agreement with a crucial physiological role, the inhibition of the ribbon unlinking prevents Src activation, consequently hindering Aurora-A recruitment and activation, leading to G2 arrest and reduced CE maturation [108, 111]. The Src-Aurora-A signaling axis is also essential for the formation of the spindle, as shown by our recent finding that the GC unlinking during G2 is necessary for the fidelity of subsequent mitotic events, such as proper bipolar spindle formation and cytokinesis (Fig. 6.3). Indeed, we have shown that when cells are forced to enter into mitosis during G2 with an intact Golgi ribbon, most of them develop multipolar spindles. The primary cause of these spindle defects is the insufficient activation of Aurora-A at the centrosomes, which results from the failure of Golgi unlinking [112] (Fig. 6.3). Therefore, defects in premitotic Golgi segregation could have a significant physiological impact, as mitotic spindle multipolarity, chromosomal instability, and cytokinesis failure that can cause transient tetraploidy, contributing to aneuploidy, a condition that promotes genomic instability, thus a critical risk factor for cell transformation and tumorigenesis [113, 114].

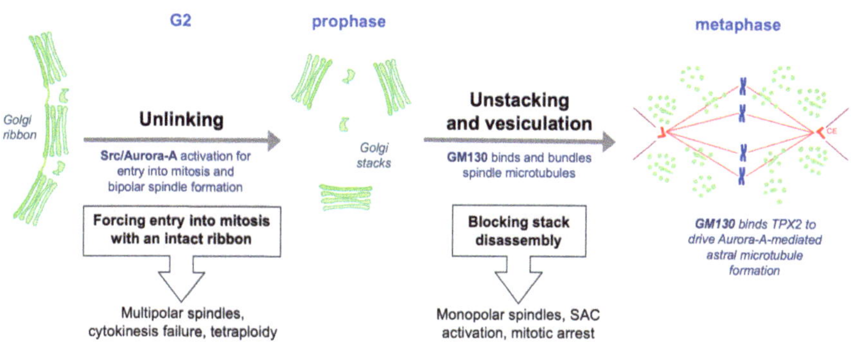

Fig. 6.3 Functional consequences of alterations in Golgi ribbon unlinking and stacks disassembly. Forcing the cells to enter mitosis with an intact ribbon results in division defects, such as spindle polarity issues, binucleation, and cytokinesis failure. In prophase, blocking the unstacking and vesiculation processes induces the spindle assembly checkpoint (SAC), leading to mitotic arrest

The mitosis-specific disassembly of the stacks is connected to cell cycle events. Arf1 dissociation from the mitotic GC membranes is necessary for the release of a set of proteins involved in spindle formation and cytokinesis [115]. The mitotic phosphorylation of GM130 leads to p115 dissociation [116], exposing a binding site for importin-α [117]. GM130 interacts with importin α via a classical nuclear localization signal (NLS). This sequestration of importin α by GM130 releases the spindle assembly factor TPX2, which activates Aurora-A kinase and stimulates MT nucleation. GM130 also directly binds to MTs, helping stabilize the spindle during mitosis (Fig. 6.3) [117]. Perturbing GM130 through the expression of an N-terminal sequence (aa 1-74) impairs spindle assembly, leading to defects in cell division, such as mitotic arrest and abnormal spindle formation. GM130 activates TPX2 on the Golgi membranes, contributing to astral MT growth and proper spindle orientation [118]. In fact, the inhibition of the disassembly of the stacks by experimentally induced polymerization of 3,3′-diaminobenzidine in the Golgi lumen causes prometaphase arrest with intact stacks clustered around abnormal spindles, and the activation of the spindle assembly checkpoint (SAC) [119, 120]. These findings highlight the importance of the G2-specific unlinking of the ribbon and mitotic disassembly of the stacks for cell cycle progression.

Why Is Golgi Unlinking Required for Entry into Mitosis?

The discovery of the role of the unlinking in triggering the Src/Aurora-A signaling axis does not explain why it is necessary for mitotic entry. Considering the close association of the GC with the CEs, one hypothesis is that the unlinking might remove a steric hindrance that could impede CEs separation or repositioning in G2. However, this seems unlikely since the GC dissociates from the CEs during this phase [121], and inhibition of the unlinking does not affect CEs separation [111] and positioning [122].

Alternative hypotheses arise from the temporal coincidence of the Golgi unlinking in G2 with an additional preparatory step for mitosis, the "Adhesion Checkpoint," that is based on the G2-specific expression of the scaffold protein DEPDC1B, which induces the selective disassembly of a subset of Focal Adhesions (FAs) [123], which is necessary for the retraction of the cell margins and formation of a rigid actomyosin cortex that guides the rounding of the cell in metaphase [124]. This process is crucial for the correct orientation and formation of a symmetric spindle, which in turn is essential for the equal segregation of homologous chromosomes [124]. The disassembly of FAs also allows scaffold proteins like HEF1 and Ajuba to relocate to the CEs, where they further contribute to the regulation of Aurora-A activity by protecting the phosphorylated T288 from the action of phosphatases [125].

Altogether, these findings suggest that Aurora-A acts as an integrator of different stimuli originating from the FAs dismantling and GC unlinking to drive proper spindle formation and CycB-CDK1 activation [126]. Notably, the downstream target of

Aurora-A, CycB-CDK1, also participates in FAs disassembly, suggesting the existence of signaling feedback loops coordinating the CEs and the FAs [127].

Therefore, what is the significant physiological event controlled by the Golgi checkpoint? As described in more detail below, we suggest that this checkpoint could control the equal distribution of Golgi proteins for their repurposing or, alternatively, the orientation of post-Golgi traffic during G2 to aid the cell polarity switch necessary for the process of cell rounding.

Repurposing of Golgi Proteins

This hypothesis is based on data showing that during mitosis a set of Golgi matrix proteins is relocated to the spindle, where they are repurposed for new mitosis-specific functions [128]. For instance, GM130 possesses a nuclear localization signal that is masked by the matrix protein p115 during interphase. As reported in section "Golgi unlinking and stack disassembly control key mitosis events", after the onset of mitosis, GM130 controls the formation of astral MTs of the spindle through the activation of TPX2 and the direct binding to MTs [117, 118]. In agreement with this role, GM130 depletion causes the formation of multipolar spindles [129].

Another Golgi protein involved in spindle formation is the vesicle tether p115, which during mitosis interacts with γ-tubulin via its armadillo-like fold and localizes at the CEs [130]. Depletion of p115 causes spindle abnormalities and chromosome segregation defects [130] (Fig. 6.4). Similarly, GRASP65 depletion leads to the formation of multipolar spindles [131], although its specific role in spindle formation has not yet been established. Notably, the depletion of GRASP55 does not affect the spindle, demonstrating a specific function of the p115-GM130-GRASP65 complex. The mitotic spindle is also involved in the inheritance of Golgi structural proteins, by partitioning "ribbon factors" necessary for post-mitotic Golgi

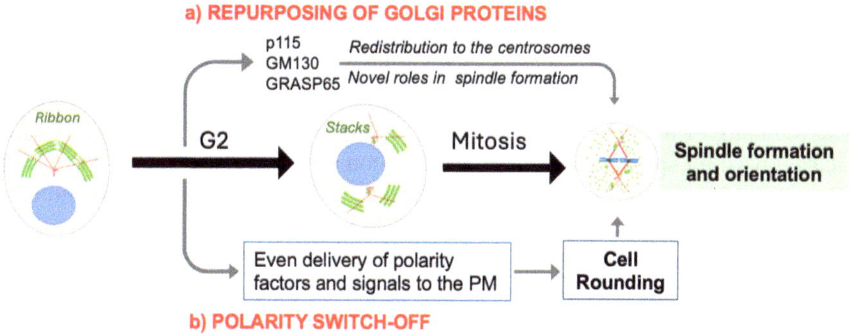

Fig. 6.4 Possible processes regulated by the Golgi mitotic checkpoint. The Golgi mitotic checkpoint may ensure the equal distribution of Golgi proteins for repurposing during mitosis (**a**). Alternatively, it could regulate the orientation of post-Golgi trafficking to support the cell polarity switch necessary for cell rounding (**b**). (*PM* plasma membrane)

reassembly [89, 132]. In particular, proteomic studies suggest that these "ribbon factors" include at least GM130, p115, and Golgin 160 [133].

Thus, the Golgi mitotic checkpoint could monitor the ribbon separation in G2 to oversee the formation of two equivalent pools of Golgi membranes as a prerequisite for the balanced redistribution of Golgi proteins at the spindle fibers. Supporting this hypothesis, and as mentioned earlier, the block of the disassembly of the stacks leads to metaphase arrest and monopolar spindles, probably because the intact Golgi stacks do not allow the correct distribution/function of Golgi proteins for the control of MT dynamics, which is crucial for the assembly of the spindle [119]. Moreover, when Arf1 is locked in its GTP-bound form, the Golgi stacks fail to disassemble during mitosis, leading to defective spindle formation and failed cytokinesis [115]. Also, live imaging studies indicate that the IC remains intact during mitosis, forming two membrane pools associated with the CEs and the spindles, suggesting that the IC may serve as a template for the reformation of a GC at mitotic exit, and that its equal partitioning could be critically important during cell division [90].

Importantly, to test this hypothesis would require determining whether the block of the unlinking hinders the redistribution of Golgi matrix proteins and the IC membranes to the separated CEs, a challenging task, as evidenced by the fact that the precision of their portioning is still an open question. Early studies on the accuracy of the distribution of Golgi membranes suggested that it is more precise than a stochastic mechanism [134]. Still, most of the Golgi markers are contained in fragments below the fluorescence microscopy resolution, complicating the quantitative assessment of their localization [135]. Therefore, we suggest a re-evaluation of this issue as modern high-resolution imaging techniques (e.g., STED, SIM, STORM) [136] provide an unprecedented opportunity for more reliable quantitative detection of the localization of Golgi and IC proteins during G2 and mitosis, as well as potential alterations in their localization when unlinking is blocked.

Orientation of Post-Golgi Traffic for Polarity Switch

The second hypothesis is that the unlinking in G2 may be required to modify the orientation of post-Golgi traffic, which could facilitate the process of cell rounding. The rounding of the cells is essential for spindle formation and cell division as it creates space for spindle assembly, ensures symmetric force distribution, increases cortical tension for proper spindle positioning, activates checkpoints for mitosis progression, and helps cells to overcome mechanical constraints from the environment for accurate chromosome segregation [137, 138]. Post-Golgi trafficking is essential for shaping the cell cortex by delivering proteins and lipids to specific membrane regions, which regulate actin dynamics, cortical tension, and polarity [52]. It also controls the transport of actin-regulating proteins like formin and Rho GTPases, guaranteeing proper actin polymerization and cortex stability (Fig. 6.4) [139].

This proposal is based on the widely accepted role of the Golgi ribbon in the polarized delivery of cargo to specific plasma membrane subdomains [52, 140]. Indeed, Golgi-originated non-centrosomal MT tracks typically direct post-Golgi transport, including integrins, lipids, and signaling proteins, to plasma membrane hotspots associated with a subset of FAs [55]. This process, regulated by Rab6 and the exocyst complex, modulates FAs dynamics and could generate asymmetry of the protein composition of the cell cortex and cell shape, even in nonpolarized cells. A relevant example is the formation of a leading edge during cell migration [141].

Therefore, the segregation of the ribbon into stacks during G2 could result in a more even secretion of cargo and likely of signals, which we hypothesize is necessary for transitioning from a cell axial symmetry, allowing the reshaping of the cell cortex to a spherical form, essential for cell rounding. As mentioned earlier, this is crucial for the symmetric spindle formation and the equal separation of genetic material [124]. To test this hypothesis, it would be necessary to employ dedicated live-imaging and high-resolution confocal experiments, to examine the effects of perturbations of Golgi unlinking on the directionality and distribution at the plasma membrane of post-Golgi carriers containing factors involved in cell polarity and cortex formation, as well as to determine the effects on FAs turnover in cell rounding.

A New Model for the Regulation of the G2/M Transition

The crucial and well-known regulatory step of the mitosis onset is the synthesis and accumulation of cyclin B, which forms a complex with the kinase CDK1 [109]. This complex is kept inactive by inhibitory phosphorylations at Y14 and Y15 of CDK1, carried out by the kinases Myt-1 and Wee1. The dephosphorylation of these sites is performed by members of the CDC25 phosphatase family, which are stimulated by Aurora-A. The balance between phosphorylation and dephosphorylation of these inhibitory residues on CDK1 is regulated by various physiological circumstances, primarily by the activation of the DNA replication/damage checkpoint, which acts as a brake to inhibit CDK1 until DNA replication is accurately completed [109].

The data discussed earlier indicate that CDK1 activation and mitosis onset are also under the control of the Golgi and adhesion checkpoints, which form signaling circuits that coordinate the disassembly of the GC and the dismantling of FAs with CE-based events [88, 123]. Thus, the FAs and Golgi checkpoints may ensure the correct execution of the preparatory steps necessary for the extensive morphological reorganizations necessary for cell rounding, spindle formation, and orientation. If these steps are successfully completed, they collectively activate CycB-CDK1 through Aurora-A stimulation, contributing to the initiation of mitosis [124]. CycB-CDK1, in turn, can also activate Aurora-A, establishing a positive feedback loop that functions as a bistable switch, making entry into mitosis irreversible (Fig. 6.5) [142].

Despite these potential pathological implications, interest in Golgi inheritance and cell division has declined over the past 15 years, as discussed in a recent commentary [43]. We urge renewed scientific interest in investigations of the

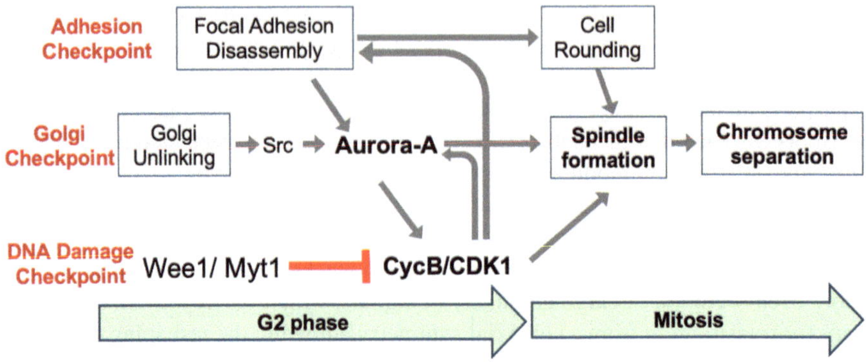

Fig. 6.5 Model of G2/M transition regulation. The Golgi and adhesion checkpoints coordinate the Golgi unlinking and focal adhesions to control cell rounding and spindle formation. In late G2, Aurora-A is activated and localizes at the centrosomes. During the G2/M transition, Aurora-A becomes fully activated by the CycB/CDK1 complex. (Adapted from Ref. [122])

fundamental mechanisms and regulation of mitotic Golgi disassembly, its role in mitotic processes, the effects of Golgi alterations in inheritance and on genomic stability. More morphofunctional studies of the GC structure in patient tissues and derived cells are needed to understand better the pathophysiological connection of alterations of GC organization. Defects in mitotic Golgi segregation could have significant physiological consequences, including mitotic spindle multipolarity, chromosomal instability, and cytokinesis failure, leading to transient tetraploidy. This condition may contribute to aneuploidy, promoting genomic instability, which is a significant risk factor for cellular transformation and tumorigenesis [113, 114].

Role of the Golgi Complex in Tumor Development

The reason for maintaining such a complex organization of the GC is not entirely clear, but precise structure and compartmentalization are likely necessary to perform critical functions in the transport and modification of proteins and lipids within the cell [76, 143]. Notably, the GC can undergo variable levels of disassembly under different pathological conditions [42]. In diseases such as Alzheimer's, Parkinson's, and amyotrophic lateral sclerosis (ALS), the GC often shows significant breakage, which may be linked to cellular stress and defects in intracellular transport processes [42, 144]. Moreover, certain viruses, such as hepatitis C and SARS-CoV2, can induce Golgi fragmentation to facilitate the assembly of viral particles [145, 146]. Oxidative stress and other forms of stress, such as those caused by DNA damage or metabolic alterations, can lead to structural changes in the GC [19]. The senescent-associated secretory phenotype (SASP), which is characterized by the release of pro-inflammatory cytokines, growth factors, and proteases, is also associated with Golgi ribbon cleavage, which could modify the traffic and secretion of

SASP factors, influencing the extracellular environment that senescent cells create [22, 147].

Regarding cancer development, many circumstances can lead to GC disassembly, including extracellular signals originating from the tumor microenvironment, oncogenes (such as kinases or GTPases), alterations in the expression levels or mutations of proteins directly involved in Golgi structure (e.g., Golgi matrix proteins), in membrane trafficking (e.g., GOLPH3, Arf, or Rab GTPases) or the cytoskeleton [23, 148, 149]. The changes in Golgi organization observed in cancer cells and tissues range from fragmentation to a honeycomb-like organization (Fig. 6.6a) [150, 151].

The main consequences of the Golgi structural alterations include modification of the secretion rate, disruption of polarized delivery of specific cargos, and variations in the glycosylation process (Fig. 6.6), which in turn can affect key cellular functions such as proliferation, survival/apoptosis, migration, and invasion [16, 31, 52]. However, comprehensive studies are still lacking, and a challenge arises from the multifunctional roles of several Golgi-associated proteins [143, 152]. This makes it difficult to determine whether the observed functional effects of alterations in a protein are a direct consequence of changes in GC structure or if they are solely due to the protein-specific functions (Fig. 6.6b). As a result, it remains unclear under which conditions the disassembly of the GC plays a causative role in cancer development or if it is simply a collateral effect of the altered protein function involved in GC structure. Furthermore, Golgi fragmentation has not been observed in all cancer cell models [148], but this is not necessarily problematic, as the fundamental goal is not to establish a universal role for the GC in tumor development but to identify specific tumors that may gain a competitive advantage from functional or structural alterations of the GC.

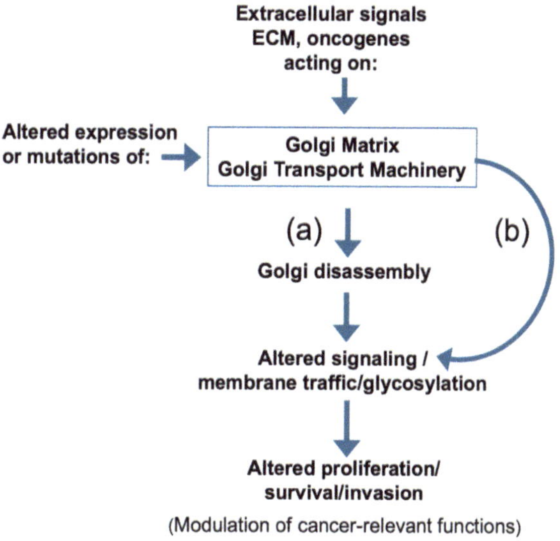

Fig. 6.6 Schematic representation of processes linking Golgi disassembly to cancer-related functions. Extracellular matrix cues, oncogenes, and altered expression or mutations in Golgi-associated proteins, trafficking mechanisms, and the cytoskeleton can induce Golgi disassembly (**a**). In turn, this disassembly can modify signaling, glycosylation, and membrane transport (**b**), leading to modulation of cancer-relevant processes

Following, we will first discuss the general potential consequences of morpho-functional modifications of the GC and associated proteins on processes related to the development of cancer. Then, we will examine in detail the potential relevance of each protein in this context.

Golgi-Related Functions That Impact Cancer-Promoting Processes

Since Golgi-associated proteins can affect multiple mechanisms potentially linked to cancer development, we will first present a general overview. A detailed description of the specific functional consequences linked with the alteration of each protein will follow in section "Contribution of proteins involved in Golgi structure and membrane trafficking to cancer-linked processes".

Signaling Regulation

The GC hosts numerous signaling proteins, including GTP-binding proteins and kinases, that are integral to signal transduction pathways controlling its dynamic organization and function. These pathways also regulate essential cellular processes such as proliferation, motility, differentiation, and stress responses, which are potentially linked to cancer [153–155]. This will be briefly addressed, while details are available in excellent recent reviews. For instance, the Src family kinases (SFKs) and PKD, which play crucial roles in Golgi-based membrane transport, are also fundamental in cancer progression [155]. SFKs regulate signal transduction during differentiation, adhesion, and migration and control Golgi fragmentation [156]. PKD, on the other hand, regulates protein secretion and is implicated in tumor development and progression [157, 158]. Additionally, protein kinase A (PKA) is involved in cAMP-dependent signaling and cancer progression, phosphorylating substrates such as CREB, which influences the expression of oncogenes like c-Jun and Cyclin D1 [159]. PKA also interacts with CDC42 interacting protein 4 (CIP4) and AKAP450 at the GC, coordinating membrane deformation, and actin polymerization to promote a more motile phenotype [160].

Another important aspect is the role that alterations of the structure of the GC play in cell signaling. Cellular responses commonly linked to Golgi unlinking, rather than the modification of a single Golgi protein, include the activation of the Src-Aurora-A signaling axis [108]. Notably, premature activation of Aurora-A can reduce the accuracy of the DNA damage checkpoint, thereby promoting tumorigenesis [161]. In addition, unlinking of the ribbon impairs polarized secretion, disrupting the targeted delivery not only of cargo but also of signaling proteins [140, 162]. Actually, multiple studies suggest that the GC becomes unlinked in response to the DNA damage checkpoint activation, potentially enhancing cell survival and resistance to DNA-damaging agents, thereby reducing the efficacy of chemotherapy [149].

In conclusion, although many studies indicate a connection between Golgi fragmentation and cellular responses, it is important to approach these findings with caution. It remains challenging to discern whether the observed functional effects are directly caused by the altered activity of specific proteins affecting Golgi structure or whether they are part of a larger signaling response to changes in Golgi architecture. Relevant examples of this will be discussed in section "Contribution of proteins involved in Golgi structure and membrane trafficking to cancer-linked processes".

Membrane Traffic and Glycosylation

One of the critical functions of the GC is glycosylation, a post-translational modification that involves the addition of sugar moieties to proteins and lipids, which is essential for their proper function and stability. Sequential glycosylation occurs as proteins and lipids transit across the Golgi cisternae. This pathway depends on glycosyltransferases, glycosidases, and the availability of sugar donors [163]. The oligosaccharides added to proteins and lipids are either N-linked or O-linked glycans. N-linked glycans are synthesized in the ER as lipid-linked precursors of a 14-sugar chain, from which three glucoses and one mannose are removed. Further processing begins in the *cis*-Golgi, leading to complex N-glycans through the addition and removal of sugars [164]. Conversely, O-glycans are entirely processed in the GC in the form of O-GalNAc (mucin-type) [165] and, in special cases, O-glycan processing starts with the addition of mannose [166].

Therefore, levels of expression [167], mislocalization of the enzymes [168], dysregulated chaperones [169], or altered nucleotide sugar transporters [163] can result in loss, gain, or abnormal glycosylation, which will impact physiological cell processes. Aberrant glycosylation is considered a hallmark of tumor development and progression as glycans regulate signal transduction, proliferation, invasion, metastasis, and angiogenesis [170]. Evidence suggests that tumor-associated truncated carbohydrates are linked to early carcinogenesis, whereas newly synthesized carbohydrates indicate an advanced stage of the tumor [171]. As a matter of fact, the most frequent glycosylation alterations in cancer include fucosylation, sialylation, N- and O-branching, and O-glycan truncation [172]. Of particular interest, O-GlcNAcylation modulates basic processes such as transcription, translation, and signal transduction [173].

Another cause of altered glycosylation is an abnormal intraluminal pH, which increases the expression of tumor-associated antigens [150, 174]. The distinct pH of each Golgi compartment is maintained by the vacuolar H^+-ATPase (V-ATPase). Glycosylation enzymes require a progressively more acidic pH across the *cis*- to *trans*-Golgi for proper activity, as well as for the transport and sorting of cargo through the cisternae [175]. Notably, in several tumor cell types, the pH of the GC is abnormally alkaline, causing the selective mislocalization of specific glycosyltransferases, which leads to defective glycosylation [176].

Therefore, proper glycosylation depends on a well-organized Golgi structure, accurate localization of glycosylation enzymes, and efficient intra-Golgi traffic. Disruptions in these processes, whether caused by mutations, mislocalization, or altered expression of specific Golgi proteins, can result in incomplete or defective glycosylation, which may affect the activity of kinases, integrins, binding to the ECM, or cell-to-cell interactions [164]. In addition, increasing evidence indicates that deregulation of intracellular transport and recycling of plasma membrane receptors or adhesion molecules contribute to the development of cancer [177].

Epithelial-to-Mesenchymal Transition

Most human cancers originate from epithelial tissues [178]. Epithelial cells have an apical membrane facing a luminal space and a basolateral membrane that connects to surrounding cells through cell junctions and to the underlying connective tissue. Cell polarization is crucial for the establishment and maintenance of asymmetry, which is essential for normal physiological function and tissue integrity. Epithelial–mesenchymal transition (EMT) is a process by which epithelial cells acquire mesenchymal, fibroblast-like characteristics, increasing their motility and invasiveness, thus playing a pivotal role in metastasis formation [179].

The hallmarks of EMT are represented by modifications of the cytoskeleton, the extracellular matrix (ECM), and plasma membrane-associated receptors, as well as the disruption of cell–cell adhesion. A critical regulator of cell–cell adhesion is E-cadherin, which is also a tumor suppressor and is frequently downregulated in malignant carcinomas [180]. This downregulation is associated with the activation of EMT transcription factors such as Snail, Slug, Twist, and Zeb, which subsequently upregulate N-cadherin, a protein linked to enhanced migratory ability [181]. Epithelial-derived tumors can also exhibit a phenotype capable of reestablishing epithelial features at distant sites through a process known as a mesenchymal–epithelial transition (MET). This trait is often associated with a more malignant phenotype, characterized by rapid tumor formation and growth and spread, frequently leading to a poor prognosis [182].

In this context, proper regulation of the formation of cargo carriers that transport E-cadherin from the GC to the plasma membrane is essential for maintaining cell–cell contacts, which are crucial for preserving epithelial cell polarity and preventing the EMT. Disruption of this pathway leads to the accumulation of cargo within a Golgi compartment [183]. Various microenvironmental signals known to trigger EMT such as growth factors (e.g., TGF-β, EGF, FGF), also affect Golgi integrity [77]. EMT-inducing transcription factors like Zeb1 can cause structural changes in the organization of the GC, leading to its compaction, a process mediated by the scaffolding protein PAQR11 [184]. Elevated PAQR11 levels are associated with EMT, enhancing vesicle transport and increasing metastatic potential. Conversely, PAQR11 depletion causes Golgi dispersal and impaired vesicle transport, highlighting its critical role in maintaining the integrity of the GC during EMT and promoting metastasis [184].

Thus, although the specific role of Golgi structural alterations in EMT is not yet fully understood, its function in modification, sorting, and packaging of proteins and lipids is likely crucial for this process.

Extracellular Matrix Remodeling and Cell Invasion

The ECM is a complex network of proteins, glycoproteins, and polysaccharides that provides structural and biochemical support to cells, influencing processes such as growth, differentiation, migration, and survival. The ECM plays a critical role in maintaining tissue integrity, facilitating cell communication, regulating biochemical signals, and supporting wound healing and tissue regeneration. There are two major types of ECM that differ in both localization and composition: the interstitial connective matrix, which offers structural support within tissues, and the basement membrane, which separates the epithelium from the stroma [185]. The ECM includes around 300 core proteins, including 43 collagen subunits, 36 proteoglycans, and about 200 complex glycoproteins [186]. The ECM is composed of two main macromolecule classes. The first comprises fibrous proteins such as collagen, elastin, fibronectin, and laminin, which provide elasticity, tensile strength, and adhesion sites that facilitate cell migration [187]. The second class consists of glycosaminoglycans (GAGs), which are linear polymers of disaccharides that bind to carboxyl, hydroxyl, or sulphate groups (e.g., chondroitin, keratan, and heparan sulphate). Except for hyaluronan, GAGs bind covalently to core proteins, forming a hydrated gel of proteoglycans that resists compressive forces [188]. Hyaluronan is synthesized at the plasma membrane and secreted directly, while chondroitin sulphate (CS) and heparan sulphate (HS) are synthesized in the GC before being transported and secreted. Disruption of the GC by depleting GRASPs increases global GAGs synthesis while reducing the sulfation and secretion of CS and HS [189], linking the GC to proteoglycans as key ECM components critical for cell–cell interaction, ligand-receptor binding, adhesion, maintenance of the homeostasis, and disease processes.

The ECM is dynamic, constantly modified through secretion and degradation, necessary for maintaining tissue homeostasis. Imbalances in ECM composition or stiffness can cause diseases like fibrosis and cancer. Cells sense the properties of the ECM and adjust the expression of components based on biochemical cues from the ECM. For instance, growth factor receptors and adhesion proteins on cell membranes are able to activate signaling pathways that induce differentiation, proliferation, migration, or apoptosis. The ECM also sequesters growth factors and bioactive molecules that, when exposed after proteolysis, can promote tumor angiogenesis [185], crucial for sustaining tumor growth and spread. The formation of the ECM or its remodeling is intimately linked to cell migration, which is a critical step in metastasis [187]. The acquisition of a highly motile phenotype is characteristic of metastatic cells and is essential in wound healing and tissue development, requiring ECM modifications.

In addition, the GC is pivotal in the remodeling of the ECM by secreting matrix metalloproteases (MMPs), zinc-dependent endopeptidases crucial in both physiological processes like wound repair and pathological conditions. MMPs cleave ECM components, cell surface proteins [190, 191], and intracellular proteins [192]. The MMP family has 23 members, many associated with tumor progression and metastasis [193]. Some MMPs are soluble, while others are anchored to the plasma membrane through a transmembrane domain or a GPI anchor [194]. MMPs are synthesized in the ER as zymogens (proenzymes), glycosylated in the GC, and activated by proteolytic cleavage through convertases like furin in the *trans*-Golgi network (TGN), by other MMPs or through autoactivation [195]. MMPs are secreted in a Golgi-dependent manner, emphasizing the role of this organelle in cancer progression.

On the other hand, invadopodia are actin-rich specialized protrusions of the plasma membrane of invasive cells in which proteases and integrins concentrate and are devoted to the focal degradation of the ECM [196]. Moreover, the Golgi position relative to the nucleus defines the front-rear polarity in migrating cells, with several Golgi proteins promoting or inhibiting cell migration.

In conclusion, alterations in Golgi structure play a critical role in ECM remodeling by regulating the modification, secretion, and activation of ECM components and enzymes, influencing cell migration, invasion, and tissue homeostasis.

Contribution of Proteins Involved in Golgi Structure and Membrane Trafficking to Cancer-Linked Processes

In this section, we outline the contribution of proteins involved in membrane trafficking and Golgi structure to cancer-related processes. A prominent example is the multiprotein complex COPI, which plays a role in the formation of vesicles transporting cargo within the GC and between the GC and the ER [143]. The $\zeta 1$ subunit of this complex is a potential cancer target as its depletion leads to Golgi fragmentation, inhibition of autophagy, and death of both dividing and nondividing tumor cells, while sparing normal cells [197]. COPB2, another subunit of the COPI coatomer complex, is overexpressed in various cancers, including prostate cancer, where it promotes cancer cell proliferation, survival, and metastasis. Conversely, knockdown of COPB2 in prostate cancer cells has been shown to inhibit cell proliferation and induce apoptosis, underscoring its role as an oncogene in several cancer types [198]. These findings suggest that targeting COPI subunits could offer a novel therapeutic approach to cancer treatment.

The Arf family of GTPases plays a crucial role in membrane trafficking by regulating COPI vesicle formation, cargo selection, lipid biosynthesis, and actin cytoskeleton assembly. Consequently, Arfs are essential for maintaining the structure and function of the GC [143]. Arf1, for instance, is overexpressed in invasive breast cancer cells and regulates key pathways involved in cell proliferation and invasion. Specifically, Arf1 controls the Rho/MLC pathway, which regulates MMP-9 activity, essential for degrading the ECM and facilitating cancer cell invasion. Additionally, Arf1 is involved in the regulation of the retinoblastoma protein (pRB), which is central in cell cycle progression and cancer cell proliferation [199, 200]. Arf3 also plays a role in cancer cell invasion by regulating N-cadherin levels, which are essential for cell–cell adhesion and the collective movement of cancer cells. In fact, Arf3 acts as a switch, determining whether cancer cells invade individually or as a collective group, thereby impacting metastasis [201].

The Conserved Oligomeric Golgi (COG) complex, a multisubunit protein complex, is involved in intra-Golgi trafficking, particularly in the retrograde transport of vesicles within the GC. It works in concert with COPI vesicles and Golgi SNAREs, such as Syntaxin 5, Syntaxin 6, and GS27, which are critical for vesicle tethering and fusion. These components collectively maintain Golgi homeostasis, ensure proper protein glycosylation, and facilitate the tethering and fusion of intra-Golgi transport vesicles [143]. Disruption of COG function can result in significant glycosylation defects, which are implicated in congenital disorders of glycosylation [202].

Golgi phosphoprotein 3 (GOLPH3), a protein mainly localized at the TGN, is involved in several critical cellular functions, including maintaining Golgi structure, vesicle trafficking, and glycosylation. GOLPH3 binds to phosphatidylinositol 4-phosphate (PI4P), which is crucial for its localization and function [203]. GOLPH3 regulates the intra-Golgi recycling of glycoenzymes during cisternal maturation, guaranteeing their proper localization within Golgi compartments and preventing their degradation or mislocalization [204]. The recycling of glycoenzymes sustains glycosphingolipid biosynthesis, which is essential for cell growth. In tumors, elevated GOLPH3 levels can disrupt glycosphingolipid synthesis and plasma membrane composition, leading to increased mitogenic signaling and cell proliferation [149, 203, 204]. Additionally, GOLPH3 links the GC to the actin cytoskeleton by interacting with MYO18A, which controls vesicular transport and Golgi organization [205]. GOLPH3 promotes tumor cell proliferation, survival, focal adhesion dynamics, migration, and invasion through mechanisms involving the PI3K/Akt/mTOR and Wnt/β-catenin signaling pathways. GOLPH3 is also phosphorylated by DNA-PK in response to DNA damage, enhancing cancer cell survival and resistance to therapy [205, 206]. Furthermore, GOLPH3 plays a role in EMT, promoting Golgi dispersal and enhancing vesicular trafficking, which are crucial for tumor cell migration and invasion [149]. As a result, GOLPH3 acts as an oncoprotein in several solid tumor types, including breast, lung, and colon cancers, where high levels correlate with advanced disease stages and reduced survival rates [207, 208].

Golgi membrane protein 1 (GOLM1), also known as GP73 or GOLPH2, is a type II transmembrane protein enriched in the *cis*- and medial-GC. It plays a significant role in various processes related to protein trafficking and cell signaling. GOLM1 has been directly associated with the progression, metastasis, and poor prognosis of various cancers [209–211].

Golgi matrix proteins are crucial not only for maintaining the organization and function of the GC but also for ensuring the proper localization of Golgi enzymes and the intra-Golgi transport of cargo [23, 30, 212]. They may also modulate Golgi-based signaling events. Functional alterations in these proteins can disrupt both Golgi structure and the processing of secretory proteins or signals, with potential implications for cancer-related processes [153]. For instance, depletion of GRASPs causes unstacking of the Golgi cisternae and accelerates trafficking through the GC, resulting in reduced N-linked glycosylation, decreased glycan complexity, altered glycolipid composition of the plasma membrane, and reduced $\alpha 5\beta 1$-integrin, which leads to decreased cell adhesion and migration, as well as faster growth and proliferation [38, 213]. Additionally, GRASP65 mediates programmed cell death via apoptosis [214], and it has been demonstrated that dihydromyricetin (DHM) significantly reduces ovarian cancer cell migration and invasion while inducing apoptosis by downregulating GRASP65, which is overexpressed in ovarian cancer tissues [215]. A potential role for GRASP65 has also been demonstrated in cellular models of prostate cancer [216].

GRASP55 is also involved in autophagy, a process involved in the degradation of proteins or damaged organelles under conditions that disrupt homeostasis, such as nutrient deprivation [217]. Autophagy involves the delivery of cytoplasmic components to an autophagosome, which fuses with a lysosome to form an autolysosome where degradation occurs [218]. GRASP55 has been shown to be O-GlcNAcylated under normal growth conditions. However, during glucose deprivation, it becomes de-O-GlcNAcylated and is directed to autophagosome-lysosome fusion sites, where it acts as a membrane tether by binding with LC3-II and LAMP2 [69]. GRASP55 is necessary for the growth of certain leukemic cells [219], while in preclinical models of lung adenocarcinoma, it has been identified as a key player in the secretion of pro-tumorigenic proteins [220, 221].

GM130, by binding to GRASP65, is responsible for the lateral fusion of homologous cisternae from different stacks to form the Golgi ribbon. Depletion of either protein leads to the formation of unlinked perinuclear mini-stacks, resulting in uneven distribution of processing enzymes and, consequently, increased anomalous post-translational modifications [222]. GM130 also plays a significant role in cell division and polarized cell migration as it activates kinases and GTPases essential for cell migration by targeting substrates such as 14-3-3ζ, which affects cell adhesion and polarity [223]. Furthermore, it is overexpressed in several cancer types, where it mediates EMT [224] (see below). Downregulation of GM130 suppresses tumorigenesis in lung cancer models [225], while in breast cancer cells, it has been linked to increased cell migration, invasion, and loss of cell polarity [226].

An isoform of the vesicle transport factor p115 is associated with worse prognostic outcomes in hepatocellular carcinoma patients. The isoform USO1-T lacks a domain involved in dimerization and tethering to GM130; thus, its expression induces unstacking of the GC, correlates with increased ERK1-mediated GRASP65 phosphorylation, and accelerates trafficking through the GC and plasma membrane [227].

The golgin giantin is linked to p115, which in turn binds to GM130, and GM130 subsequently binds to GRASP65 [143]. Loss of giantin function in prostate cancer causes a shift in the targeting of specific glycosyltransferases to the GM130-GRASP65 complex, leading to abnormally processed glycoproteins that alter the cell surface [216]. Downregulation of giantin correlates with poor survival in breast cancer patients. Additionally, the loss of giantin increases the number of intravasating cells, which are cells that enter the bloodstream and contribute to metastasis [228].

Golgin-97 is a matrix protein recruited to the TGN, which is the main delivery station for proteins and lipids to their final destinations, as well as the entry site for some endocytic cargo. Its depletion has been shown to prevent the exit of E-cadherin from the TGN [229], a process crucial for maintaining cell–cell adhesion and epithelial integrity [183]. Indeed, low expression of golgin-97 is associated with increased invasiveness in breast cancer and the expression of invasion-promoting genes [230].

The expression levels of the golgin GCC88 regulate both GC structure and mTORC1 activation. mTORC1 is a kinase that plays a critical role in regulating cell growth, proliferation, metabolism, and autophagy in response to growth factors, nutrients, and cellular energy status [231]. Proteins involved in mTORC1 activation such as Rab1A, GOLPH3, and Arf1 are located at the GC or the GC/ER interface, indicating that the GC is an important site for mTORC1 activation. Many of these proteins are identified as oncogenes and their overexpression is associated with enhanced mTORC1 signaling, contributing to tumorigenesis and cancer progression [232]. Further studies are needed to determine whether the reported effects are specifically related to GCC88 or a more general disassembly of the GC.

PITPNC1 is a TGN protein involved in phosphatidylinositol transfer that is overexpressed in several cancers and correlates with metastatic progression. PITPNC1 interacts with Rab1B to increase PI4P levels at the TGN, promoting the secretion of MMP1 and other pro-metastatic factors [233].

Finally, the KDEL receptors regulate membrane traffic by retrieving ER-resident proteins that have escaped to the GC [143]. KDELR induces Src activation at invadopodia, essential for invasive tumor growth and metastasis [234, 235]. Specifically, KDELR2, a subtype of the KDEL receptor, is highly expressed in glioblastoma tissues and its expression correlates with shorter survival rates in patients. Moreover, KDELR2 promotes tumorigenesis through the mTOR pathway [236] and its overexpression is associated with increased metastatic potential in non-small cell lung cancer and glioblastomas, promoting MMP secretion and cell invasion [237].

In summary, the dysregulation of various Golgi matrix proteins contributes to cancer progression by enhancing cell proliferation, metastasis, and resistance to therapy, making them potential therapeutic targets and prognostic biomarkers in multiple cancers.

Conclusions

The GC plays a critical and multifaceted role not only in the modification and delivery of cargo, but also in broader cellular functions, especially during mitosis and cell migration [143]. The structure and localization of the GC are highly dynamic, as shown by its ability to disassemble and reassemble during cell division. In particular, the process of Golgi unlinking, which facilitates mitotic entry by activating the Src-Aurora-A signaling axis, highlights a regulatory mechanism, the "Golgi mitotic checkpoint" that ensures a close coordination between Golgi reorganization and mitotic events, such as CE maturation and spindle formation, and, ultimately, the successful completion of cell division and accurate chromosome segregation [43]. Failures in these processes lead to significant cellular dysfunctions, including spindle multipolarity and chromosomal instability, which are hallmarks of cancerous transformation [122].

Accordingly, the disassembly of the GC during mitosis is not merely a structural phenomenon but is intrinsically linked to key cell cycle events [31, 120]. Moreover, morphofunctional alterations in the GC have profound implications in cancer biology due to its involvement in post-translational modifications as glycosylation, sorting, and membrane trafficking, therefore, with a direct impact on ECM remodeling, tumor progression, and metastasis [148]. Besides, the role of the GC in EMT further emphasizes its importance in metastasis by supporting the cellular changes required for detachment and invasion of surrounding tissues.

Importantly, the findings presented in this chapter suggest that targeting the molecular pathways that regulate the organization and function of the GC could offer new therapeutic strategies for diseases characterized by an abnormal function of this organelle. By inhibiting its disassembly or modulating its associated signaling pathways, it could be possible to disrupt the cellular processes that cancer cells exploit for migration or invasion, and thus metastasis [238].

In conclusion, the GC is much more than an organelle specialized in cargo processing, sorting, and delivery; it also serves as a central hub for activities essential to both normal cell function and disease progression. Understanding the intricate mechanisms that control its dynamics during the cell cycle and in pathological conditions opens new avenues for research. Future studies should focus on exploring the therapeutic potential of targeting Golgi disassembly, Golgi-associated proteins, or signaling pathways in cancer and other diseases characterized by an abnormal Golgi function, thereby improving outcomes for patients with cancer and other Golgi-related disorders.

Acknowledgments Research in AC and IA lab was supported by the Italian Association for Cancer Research (AIRC, Milan, Italy; IG 2017 id. 20095 to AC), and the Next-Generation EU, in the context of the National Recovery and Resilience Plan, Investment PE8—Project Age-It: "Ageing Well in an Ageing Society" (CUP: B83C22004880006). The authors apologize to all colleagues whose work has not been cited in this chapter due to space limitations.

References

1. Wilson C, Venditti R, Rega LR, Colanzi A, D'Angelo G, De Matteis MA (2011) The Golgi apparatus: an organelle with multiple complex functions. Biochem J 433(1):1–9. https://doi.org/10.1042/BJ20101058
2. Cancino J, Luini A (2013) Signaling circuits on the Golgi complex. Traffic 14(2):121–134. https://doi.org/10.1111/tra.12022
3. Nebenfuhr A, Staehelin LA (2001) Mobile factories: Golgi dynamics in plant cells. Trends Plant Sci 6(4):160–167
4. Kondylis V, Rabouille C (2009) The Golgi apparatus: lessons from Drosophila. FEBS Lett 583(23):3827–3838. https://doi.org/10.1016/j.febslet.2009.09.048
5. Preuss D, Mulholland J, Franzusoff A, Segev N, Botstein D (1992) Characterization of the Saccharomyces Golgi complex through the cell cycle by immunoelectron microscopy. Mol Biol Cell 3(7):789–803
6. Ferraro F, Kriston-Vizi J, Metcalf DJ, Martin-Martin B, Freeman J, Burden JJ, Westmoreland D, Dyer CE, Knight AE, Ketteler R, Cutler DF (2014) A two-tier Golgi-based control of organelle size underpins the functional plasticity of endothelial cells. Dev Cell 29(3):292–304. https://doi.org/10.1016/j.devcel.2014.03.021
7. Valenzuela JI, Perez F (2015) Diversifying the secretory routes in neurons. Front Neurosci 9:358. https://doi.org/10.3389/fnins.2015.00358
8. Kreft ME, Di Giandomenico D, Beznoussenko GV, Resnik N, Mironov AA, Jezernik K (2010) Golgi apparatus fragmentation as a mechanism responsible for uniform delivery of uroplakins to the apical plasma membrane of uroepithelial cells. Biol Cell 102(11):593–607. https://doi.org/10.1042/BC20100024
9. Gunn PA, Gliddon BL, Londrigan SL, Lew AM, van Driel IR, Gleeson PA (2011) The Golgi apparatus in the endomembrane-rich gastric parietal cells exist as functional stable mini-stacks dispersed throughout the cytoplasm. Biol Cell 103(12):559–572. https://doi.org/10.1042/BC20110074
10. Moreno RD, Schatten G, Ramalho-Santos J (2002) Golgi apparatus dynamics during mouse oocyte in vitro maturation: effect of the membrane trafficking inhibitor brefeldin A. Biol Reprod 66(5):1259–1266. https://doi.org/10.1095/biolreprod66.5.1259
11. Lu Z, Joseph D, Bugnard E, Zaal KJ, Ralston E (2001) Golgi complex reorganization during muscle differentiation: visualization in living cells and mechanism. Mol Biol Cell 12(4):795–808. https://doi.org/10.1091/mbc.12.4.795
12. Sutterlin C, Hsu P, Mallabiabarrena A, Malhotra V (2002) Fragmentation and dispersal of the pericentriolar Golgi complex is required for entry into mitosis in mammalian cells. Cell 109(3):359–369. doi:S0092867402007201 [pii]
13. Colanzi A, Hidalgo Carcedo C, Persico A, Cericola C, Turacchio G, Bonazzi M, Luini A, Corda D (2007) The Golgi mitotic checkpoint is controlled by BARS-dependent fission of the Golgi ribbon into separate stacks in G2. EMBO J 26(10):2465–2476. doi:7601686 [pii] 10.1038/sj.emboj.7601686
14. Millarte V, Farhan H (2012) The Golgi in cell migration: regulation by signal transduction and its implications for cancer cell metastasis. ScientificWorldJournal 2012:498278. https://doi.org/10.1100/2012/498278
15. Zappa F, Failli M, De Matteis MA (2018) The Golgi complex in disease and therapy. Curr Opin Cell Biol 50:102–116. https://doi.org/10.1016/j.ceb.2018.03.005
16. Kulkarni-Gosavi P, Makhoul C, Gleeson PA (2019) Form and function of the Golgi apparatus: scaffolds, cytoskeleton and signalling. FEBS Lett 593(17):2289–2305. https://doi.org/10.1002/1873-3468.13567
17. Gitsels A, Sanders N, Vanrompay D (2019) Chlamydial infection from outside to inside. Front Microbiol 10:2329. https://doi.org/10.3389/fmicb.2019.02329

18. Ji G, Song X, Wang L, Li Z, Wu H, Dong H (2019) Golgi apparatus fragmentation participates in oxidized low-density lipoprotein-induced endothelial cell injury. J Cell Biochem 120(11):18862–18870. https://doi.org/10.1002/jcb.29205
19. Machamer CE (2015) The Golgi complex in stress and death. Front Neurosci 9:421. https://doi.org/10.3389/fnins.2015.00421
20. Carrara G, Parsons M, Saraiva N, Smith GL (2017) Golgi anti-apoptotic protein: a tale of camels, calcium, channels and cancer. Open Biol 7(5). https://doi.org/10.1098/rsob.170045
21. Caracci MO, Fuentealba LM, Marzolo MP (2019) Golgi complex dynamics and its implication in prevalent neurological disorders. Front Cell Dev Biol 7:75. https://doi.org/10.3389/fcell.2019.00075
22. Despres J, Ramdani Y, di Giovanni M, Benard M, Zahid A, Montero-Hadjadje M, Yvergnaux F, Saguet T, Driouich A, Follet-Gueye ML (2019) Replicative senescence of human dermal fibroblasts affects structural and functional aspects of the Golgi apparatus. Exp Dermatol 28(8):922–932. https://doi.org/10.1111/exd.13886
23. Petrosyan A (2015) Onco-Golgi: Is fragmentation a gate to cancer progression? Biochem Mol Biol J 1(1). https://doi.org/10.21767/2471-8084.100006
24. Slusarewicz P, Nilsson T, Hui N, Watson R, Warren G (1994) Isolation of a matrix that binds medial Golgi enzymes. J Cell Biol 124(4):405–413. https://doi.org/10.1083/jcb.124.4.405
25. Lowe M (2011) Structural organization of the Golgi apparatus. Curr Opin Cell Biol 23(1):85–93. https://doi.org/10.1016/j.ceb.2010.10.004
26. Zhang X, Wang Y (2015) GRASPs in Golgi structure and function. Front Cell Dev Biol 3:84. https://doi.org/10.3389/fcell.2015.00084
27. Jarvela T, Linstedt AD (2014) Isoform-specific tethering links the Golgi ribbon to maintain compartmentalization. Mol Biol Cell 25(1):133–144. doi:mbc.E13-07-0395 [pii] 10.1091/mbc.E13-07-0395
28. Munro S (2011) The Golgin coiled-coil proteins of the Golgi apparatus. Cold Spring Harb Perspect Biol 3(6) doi:cshperspect.a005256 [pii] 10.1101/cshperspect.a005256
29. Petrosyan A, Ali MF, Cheng PW (2012) Glycosyltransferase-specific Golgi-targeting mechanisms. J Biol Chem 287(45):37621–37627. https://doi.org/10.1074/jbc.C112.403006
30. Pothukuchi P, Agliarulo I, Pirozzi M, Rizzo R, Russo D, Turacchio G, Nuchel J, Yang JS, Gehin C, Capolupo L, Hernandez-Corbacho MJ, Biswas A, Vanacore G, Dathan N, Nitta T, Henklein P, Thattai M, Inokuchi JI, Hsu VW, Plomann M, Obeid LM, Hannun YA, Luini A, D'Angelo G, Parashuraman S (2021) GRASP55 regulates intra-Golgi localization of glycosylation enzymes to control glycosphingolipid biosynthesis. EMBO J 40(20):e107766. https://doi.org/10.15252/embj.2021107766
31. Li J, Ahat E, Wang Y (2019) Golgi structure and function in health, stress, and diseases. Results Probl Cell Differ 67:441–485. https://doi.org/10.1007/978-3-030-23173-6_19
32. Barr FA, Puype M, Vandekerckhove J, Warren G (1997) GRASP65, a protein involved in the stacking of Golgi cisternae. Cell 91(2):253–262
33. Shorter J, Watson R, Giannakou ME, Clarke M, Warren G, Barr FA (1999) GRASP55, a second mammalian GRASP protein involved in the stacking of Golgi cisternae in a cell-free system. EMBO J 18(18):4949–4960. https://doi.org/10.1093/emboj/18.18.4949
34. Vinke FP, Grieve AG, Rabouille C (2011) The multiple facets of the Golgi reassembly stacking proteins. Biochem J 433(3):423–433. doi:BJ20101540 [pii] 10.1042/BJ20101540
35. Preisinger C, Korner R, Wind M, Lehmann WD, Kopajtich R, Barr FA (2005) Plk1 docking to GRASP65 phosphorylated by Cdk1 suggests a mechanism for Golgi checkpoint signalling. EMBO J 24(4):753–765. doi:7600569 [pii] 10.1038/sj.emboj.7600569
36. Duran JM, Kinseth M, Bossard C, Rose DW, Polishchuk R, Wu CC, Yates J, Zimmerman T, Malhotra V (2008) The role of GRASP55 in Golgi fragmentation and entry of cells into mitosis. Mol Biol Cell 19(6):2579–2587. doi:E07-10-0998 [pii] 10.1091/mbc.E07-10-0998
37. Tang D, Yuan H, Vielemeyer O, Perez F, Wang Y (2012) Sequential phosphorylation of GRASP65 during mitotic Golgi disassembly. Biol Open 1(12):1204–1214. https://doi.org/10.1242/bio.20122659

38. Xiang Y, Zhang X, Nix DB, Katoh T, Aoki K, Tiemeyer M, Wang Y (2013) Regulation of protein glycosylation and sorting by the Golgi matrix proteins GRASP55/65. Nat Commun 4:1659. doi:ncomms2669 [pii] 10.1038/ncomms2669
39. Valderrama F, Babia T, Ayala I, Kok JW, Renau-Piqueras J, Egea G (1998) Actin microfilaments are essential for the cytological positioning and morphology of the Golgi complex. Eur J Cell Biol 76(1):9–17. https://doi.org/10.1016/S0171-9335(98)80012-5
40. Egea G, Lazaro-Dieguez F, Vilella M (2006) Actin dynamics at the Golgi complex in mammalian cells. Curr Opin Cell Biol 18(2):168–178. https://doi.org/10.1016/j.ceb.2006.02.007
41. Thyberg J, Moskalewski S (1999) Role of microtubules in the organization of the Golgi complex. Exp Cell Res 246(2):263–279. doi:S0014-4827(98)94326-9 [pii] 10.1006/excr.1998.4326
42. Makhoul C, Gosavi P, Gleeson PA (2019) Golgi dynamics: the morphology of the mammalian Golgi apparatus in health and disease. Front Cell Dev Biol 7:112. https://doi.org/10.3389/fcell.2019.00112
43. Ayala I, Colanzi A (2022) Structural organization and function of the Golgi ribbon during cell division. Front Cell Dev Biol 10:925228. https://doi.org/10.3389/fcell.2022.925228
44. Feinstein TN, Linstedt AD (2008) GRASP55 regulates Golgi ribbon formation. Mol Biol Cell 19(7):2696–2707. doi:E07-11-1200 [pii] 10.1091/mbc.E07-11-1200
45. Cervigni RI, Bonavita R, Barretta ML, Spano D, Ayala I, Nakamura N, Corda D, Colanzi A (2015) JNK2 controls fragmentation of the Golgi complex and the G2/M transition through phosphorylation of GRASP65. J Cell Sci 128(12):2249–2260. https://doi.org/10.1242/jcs.164871
46. Bechler ME, de Figueiredo P, Brown WJ (2012) A PLA1-2 punch regulates the Golgi complex. Trends Cell Biol 22(2):116–124. https://doi.org/10.1016/j.tcb.2011.10.003
47. Hidalgo Carcedo C, Bonazzi M, Spano S, Turacchio G, Colanzi A, Luini A, Corda D (2004) Mitotic Golgi partitioning is driven by the membrane-fissioning protein CtBP3/BARS. Science 305(5680):93–96. doi:10.1126/science.1097775 305/5680/93 [pii]
48. Witkos TM, Lowe M (2015) The Golgin family of coiled-coil tethering proteins. Front Cell Dev Biol 3:86. https://doi.org/10.3389/fcell.2015.00086
49. Wu J, Akhmanova A (2017) Microtubule-organizing centers. Annu Rev Cell Dev Biol. https://doi.org/10.1146/annurev-cellbio-100616-060615
50. Yadav S, Linstedt AD (2011) Golgi positioning. Cold Spring Harb Perspect Biol 3(5). https://doi.org/10.1101/cshperspect.a005322
51. Sanders AA, Kaverina I (2015) Nucleation and dynamics of Golgi-derived microtubules. Front Neurosci 9:431. https://doi.org/10.3389/fnins.2015.00431
52. Ravichandran Y, Goud B, Manneville JB (2020) The Golgi apparatus and cell polarity: roles of the cytoskeleton, the Golgi matrix, and Golgi membranes. Curr Opin Cell Biol 62:104–113. https://doi.org/10.1016/j.ceb.2019.10.003
53. Rivero S, Cardenas J, Bornens M, Rios RM (2009) Microtubule nucleation at the cis-side of the Golgi apparatus requires AKAP450 and GM130. EMBO J 28(8):1016–1028. https://doi.org/10.1038/emboj.2009.47
54. Miller PM, Folkmann AW, Maia AR, Efimova N, Efimov A, Kaverina I (2009) Golgi-derived CLASP-dependent microtubules control Golgi organization and polarized trafficking in motile cells. Nat Cell Biol 11(9):1069–1080. https://doi.org/10.1038/ncb1920
55. Fourriere L, Kasri A, Gareil N, Bardin S, Hugo B, David P, Perez F, Goud B, Gaelle B, Miserey-Lenkei S (2019) RAB6 and microtubules restrict protein secretion to focal adhesions. J Cell Biol 218:2215–2231. https://doi.org/10.1083/jcb.201805002
56. Kaverina I, Straube A (2011) Regulation of cell migration by dynamic microtubules. Semin Cell Dev Biol 22(9):968–974. https://doi.org/10.1016/j.semcdb.2011.09.017
57. Zhang Y, Seemann J (2024) RNA scaffolds the Golgi ribbon by forming condensates with GM130. Nat Cell Biol 26(7):1139–1153. https://doi.org/10.1038/s41556-024-01447-2

58. Xiang Y, Wang Y (2010) GRASP55 and GRASP65 play complementary and essential roles in Golgi cisternal stacking. J Cell Biol 188(2):237–251. doi:jcb.200907132 [pii] 10.1083/jcb.200907132
59. Feinstein TN, Linstedt AD (2007) Mitogen-activated protein kinase kinase 1-dependent Golgi unlinking occurs in G2 phase and promotes the G2/M cell cycle transition. Mol Biol Cell 18(2):594–604. doi:E06-06-0530 [pii] 10.1091/mbc.E06-06-0530
60. Zhang Y, Seemann J (2021) Rapid degradation of GRASP55 and GRASP65 reveals their immediate impact on the Golgi structure. J Cell Biol 220(1). https://doi.org/10.1083/jcb.202007052
61. Grond R, Veenendaal T, Duran JM, Raote I, van Es JH, Corstjens S, Delfgou L, El Haddouti B, Malhotra V, Rabouille C (2020) The function of GORASPs in Golgi apparatus organization in vivo. J Cell Biol 219(9). https://doi.org/10.1083/jcb.202004191
62. Truschel ST, Zhang M, Bachert C, Macbeth MR, Linstedt AD (2012) Allosteric regulation of GRASP protein-dependent Golgi membrane tethering by mitotic phosphorylation. J Biol Chem 287(24):19870–19875. doi:M111.326256 [pii] 10.1074/jbc.M111.326256
63. Tang D, Zhang X, Huang S, Yuan H, Li J, Wang Y (2016) Mena-GRASP65 interaction couples actin polymerization to Golgi ribbon linking. Mol Biol Cell 27(1):137–152. https://doi.org/10.1091/mbc.E15-09-0650
64. Li J, Tang D, Ireland SC, Wang Y (2019) DjA1 maintains Golgi integrity via interaction with GRASP65. Mol Biol Cell 30(4):478–490. https://doi.org/10.1091/mbc.E18-10-0613
65. Chiu SC, Yang XT, Wei TW, Liao YA, Chen JM, Kuo YC, Liu CJ, Cheng CY, Huang YJ, Huang YJ, Wu HJ, Wan CX, Tsai JR, Yu CR (2023) The crescent-like Golgi ribbon is shaped by the Ajuba/PRMT5/Aurora-A complex-modified HURP. Cell Commun Signal 21(1):156. https://doi.org/10.1186/s12964-023-01167-4
66. Ayala I, Crispino R, Colanzi A (2019) GRASP65 controls Golgi position and structure during G2/M transition by regulating the stability of microtubules. Traffic 20(10):785–802. https://doi.org/10.1111/tra.12682
67. Bisel B, Wang Y, Wei JH, Xiang Y, Tang D, Miron-Mendoza M, Yoshimura S, Nakamura N, Seemann J (2008) ERK regulates Golgi and centrosome orientation towards the leading edge through GRASP65. J Cell Biol 182(5):837–843. https://doi.org/10.1083/jcb.200805045
68. Zhang X, Wang Y (2020) Nonredundant roles of GRASP55 and GRASP65 in the Golgi apparatus and beyond. Trends Biochem Sci 45(12):1065–1079. https://doi.org/10.1016/j.tibs.2020.08.001
69. Zhang X, Wang L, Lak B, Li J, Jokitalo E, Wang Y (2018) GRASP55 senses glucose deprivation through O-GlcNAcylation to promote autophagosome-lysosome fusion. Dev Cell 45(2):245–261 e246. https://doi.org/10.1016/j.devcel.2018.03.023
70. Saraste J, Prydz K (2019) A new look at the functional organization of the Golgi ribbon. Front Cell Dev Biol 7:171. https://doi.org/10.3389/fcell.2019.00171
71. Ohta K, Hirashima S, Miyazono Y, Togo A, Nakamura KI (2021) Correlation of organelle dynamics between light microscopic live imaging and electron microscopic 3D architecture using FIB-SEM. Microscopy (Oxf) 70(2):161–170. https://doi.org/10.1093/jmicro/dfaa071
72. Koga D, Ushiki T, Watanabe T (2017) Novel scanning electron microscopy methods for analyzing the 3D structure of the Golgi apparatus. Anat Sci Int 92(1):37–49. https://doi.org/10.1007/s12565-016-0380-8
73. Ferguson S, Steyer AM, Mayhew TM, Schwab Y, Lucocq JM (2017) Quantifying Golgi structure using EM: combining volume-SEM and stereology for higher throughput. Histochem Cell Biol 147(6):653–669. https://doi.org/10.1007/s00418-017-1564-6
74. Wang Y, Seemann J (2011) Golgi biogenesis. Cold Spring Harb Perspect Biol 3(10):a005330. https://doi.org/10.1101/cshperspect.a005330
75. Zaal KJ, Smith CL, Polishchuk RS, Altan N, Cole NB, Ellenberg J, Hirschberg K, Presley JF, Roberts TH, Siggia E, Phair RD, Lippincott-Schwartz J (1999) Golgi membranes are absorbed into and reemerge from the ER during mitosis. Cell 99(6):589–601. doi:S0092-8674(00)81548-2 [pii]

76. Nakamura N, Wei JH, Seemann J (2012) Modular organization of the mammalian Golgi apparatus. Curr Opin Cell Biol 24(4):467–474. https://doi.org/10.1016/j.ceb.2012.05.009
77. Yoshimura S, Yoshioka K, Barr FA, Lowe M, Nakayama K, Ohkuma S, Nakamura N (2005) Convergence of cell cycle regulation and growth factor signals on GRASP65. J Biol Chem 280(24):23048–23056. doi:M502442200 [pii] 10.1074/jbc.M502442200
78. Rebmann GM, Grabski R, Sanchez V, Britt WJ (2016) Phosphorylation of Golgi peripheral membrane protein Grasp65 is an integral step in the formation of the human cytomegalovirus cytoplasmic assembly compartment. MBio 7(5). https://doi.org/10.1128/mBio.01554-16
79. Zhu BL, Xie CL, Hu NN, Zhu XB, Liu CF (2018) Inhibiting of GRASP65 phosphorylation by DL-3-N-butylphthalide protects against cerebral ischemia-reperfusion injury via ERK signaling. Behav Neurol 2018:5701719. https://doi.org/10.1155/2018/5701719
80. Ji G, Zhang W, Quan M, Chen Y, Qu H, Hu Z (2016) The role of Golgi reassembly and stacking protein 65 phosphorylation in H2O2-induced cell death and Golgi morphological changes. Med Mol Morphol 49(4):217–223. https://doi.org/10.1007/s00795-016-0138-3
81. Acharya U, Mallabiabarrena A, Acharya JK, Malhotra V (1998) Signaling via mitogen-activated protein kinase kinase (MEK1) is required for Golgi fragmentation during mitosis. Cell 92(2):183–192. doi:S0092-8674(00)80913-7 [pii]
82. Colanzi A, Deerinck TJ, Ellisman MH, Malhotra V (2000) A specific activation of the mitogen-activated protein kinase kinase 1 (MEK1) is required for Golgi fragmentation during mitosis. J Cell Biol 149(2):331–339
83. Colanzi A, Sutterlin C, Malhotra V (2003) RAF1-activated MEK1 is found on the Golgi apparatus in late prophase and is required for Golgi complex fragmentation in mitosis. J Cell Biol 161(1):27–32. doi:10.1083/jcb.200208099 jcb.200208099 [pii]
84. Kienzle C, Eisler SA, Villeneuve J, Brummer T, Olayioye MA, Hausser A (2013) PKD controls mitotic Golgi complex fragmentation through a Raf-MEK1 pathway. Mol Biol Cell 24(3):222–233. doi:mbc.E12-03-0198 [pii] 10.1091/mbc.E12-03-0198
85. Chia J, Goh G, Racine V, Ng S, Kumar P, Bard F (2012) RNAi screening reveals a large signaling network controlling the Golgi apparatus in human cells. Mol Syst Biol 8:629. https://doi.org/10.1038/msb.2012.59
86. Farhan H, Wendeler MW, Mitrovic S, Fava E, Silberberg Y, Sharan R, Zerial M, Hauri HP (2010) MAPK signaling to the early secretory pathway revealed by kinase/phosphatase functional screening. J Cell Biol 189(6):997–1011. https://doi.org/10.1083/jcb.200912082
87. Hidalgo Carcedo C, Donazzi M, Spano S, Turacchio G, Colanzi A, Luini A, Corda D (2004) Mitotic Golgi partitioning is driven by the membrane-fissioning protein CtBP3/BARS. Science 305(5680):93–96. https://doi.org/10.1126/science.1097775
88. Corda D, Barretta ML, Cervigni RI, Colanzi A (2012) Golgi complex fragmentation in G2/M transition: an organelle-based cell-cycle checkpoint. IUBMB Life 64(8):661–670. https://doi.org/10.1002/iub.1054
89. Seemann J, Pypaert M, Taguchi T, Malsam J, Warren G (2002) Partitioning of the matrix fraction of the Golgi apparatus during mitosis in animal cells. Science 295(5556):848–851. doi:10.1126/science.1068064 295/5556/848 [pii]
90. Marie M, Dale HA, Kouprina N, Saraste J (2012) Division of the intermediate compartment at the onset of mitosis provides a mechanism for Golgi inheritance. J Cell Sci 125(Pt 22):5403–5416. https://doi.org/10.1242/jcs.108100
91. Zhang X, Brachner A, Kukolj E, Slade D, Wang Y (2019) SIRT2 deacetylates GRASP55 to facilitate post-mitotic Golgi assembly. J Cell Sci 132(21). https://doi.org/10.1242/jcs.232389
92. Mao L, Li N, Guo Y, Xu X, Gao L, Xu Y, Zhou L, Liu W (2013) AMPK phosphorylates GBF1 for mitotic Golgi disassembly. J Cell Sci 126(Pt 6):1498–1505. https://doi.org/10.1242/jcs.121954
93. Lee IJ, Lee CW, Lee JH (2015) CaMKKbeta-AMPKalpha2 signaling contributes to mitotic Golgi fragmentation and the G2/M transition in mammalian cells. Cell Cycle 14(4):598–611. https://doi.org/10.4161/15384101.2014.991557

94. Rajanala K, Klayman LM, Wedegaertner PB (2021) Gbetagamma regulates mitotic Golgi fragmentation and G2/M cell cycle progression. Mol Biol Cell 32(20):br2. https://doi.org/10.1091/mbc.E21-04-0175
95. Lopez-Sanchez I, Sanz-Garcia M, Lazo PA (2009) Plk3 interacts with and specifically phosphorylates VRK1 in Ser342, a downstream target in a pathway that induces Golgi fragmentation. Mol Cell Biol 29(5):1189–1201. doi:MCB.01341-08 [pii] 10.1128/MCB.01341-08
96. Shaul YD, Seger R (2006) ERK1c regulates Golgi fragmentation during mitosis. J Cell Biol 172(6):885–897. doi:jcb.200509063 [pii] 10.1083/jcb.200509063
97. Wortzel I, Hanoch T, Porat Z, Hausser A, Seger R (2015) Mitotic Golgi translocation of ERK1c is mediated by a PI4KIIIbeta-14-3-3gamma shuttling complex. J Cell Sci 128(22):4083–4095. https://doi.org/10.1242/jcs.170910
98. Wortzel I, Maik-Rachline G, Yadav SS, Hanoch T, Seger R (2021) Mitotic HOOK3 phosphorylation by ERK1c drives microtubule-dependent Golgi destabilization and fragmentation. iScience 24(6):102670. https://doi.org/10.1016/j.isci.2021.102670
99. Liu J, Yang X, Li B, Wang J, Wang W, Liu J, Liu Q, Zhang X (2017) STK16 regulates actin dynamics to control Golgi organization and cell cycle. Sci Rep 7:44607. https://doi.org/10.1038/srep44607
100. Tang D, Wang Y (2013) Cell cycle regulation of Golgi membrane dynamics. Trends Cell Biol 23(6):296–304. https://doi.org/10.1016/j.tcb.2013.01.008
101. Totsukawa G, Matsuo A, Kubota A, Taguchi Y, Kondo H (2013) Mitotic phosphorylation of VCIP135 blocks p97ATPase-mediated Golgi membrane fusion. Biochem Biophys Res Commun 433(2):237–242. doi:S0006-291X(13)00375-6 [pii] 10.1016/j.bbrc.2013.02.090
102. Tang D, Xiang Y, De Renzis S, Rink J, Zheng G, Zerial M, Wang Y (2011) The ubiquitin ligase HACE1 regulates Golgi membrane dynamics during the cell cycle. Nat Commun 2:501. https://doi.org/10.1038/ncomms1509
103. Huang S, Tang D, Wang Y (2016) Monoubiquitination of syntaxin 5 regulates Golgi membrane dynamics during the cell cycle. Dev Cell 38(1):73–85. https://doi.org/10.1016/j.devcel.2016.06.001
104. Zhang X, Wang Y (2015) Cell cycle regulation of VCIP135 deubiquitinase activity and function in p97/p47-mediated Golgi reassembly. Mol Biol Cell 26(12):2242–2251. doi:mbc.E15-01-0041 [pii] 10.1091/mbc.E15-01-0041
105. Maeda M, Komatsu Y, Saito K (2020) Mitotic ER exit site disassembly and reassembly are regulated by the phosphorylation status of TANGO1. Dev Cell 55(2):237–250 e235. https://doi.org/10.1016/j.devcel.2020.07.017
106. Tang D, Mar K, Warren G, Wang Y (2008) Molecular mechanism of mitotic Golgi disassembly and reassembly revealed by a defined reconstitution assay. J Biol Chem 283(10):6085–6094. https://doi.org/10.1074/jbc.M707715200
107. Goss JW, Toomre DK (2008) Both daughter cells traffic and exocytose membrane at the cleavage furrow during mammalian cytokinesis. J Cell Biol 181(7):1047–1054. doi:jcb.200712137 [pii] 10.1083/jcb.200712137
108. Barretta ML, Spano D, D'Ambrosio C, Cervigni RI, Scaloni A, Corda D, Colanzi A (2016) Aurora-A recruitment and centrosomal maturation are regulated by a Golgi-activated pool of Src during G2. Nat Commun 7:11727. https://doi.org/10.1038/ncomms11727
109. Nigg EA (2001) Mitotic kinases as regulators of cell division and its checkpoints. Nat Rev Mol Cell Biol 2(1):21–32. doi:10.1038/35048096 35048096 [pii]
110. Marumoto T, Hirota T, Morisaki T, Kunitoku N, Zhang D, Ichikawa Y, Sasayama T, Kuninaka S, Mimori T, Tamaki N, Kimura M, Okano Y, Saya H (2002) Roles of aurora-A kinase in mitotic entry and G2 checkpoint in mammalian cells. Genes Cells 7(11):1173–1182. doi:592 [pii]
111. Persico A, Cervigni RI, Barretta ML, Corda D, Colanzi A (2010) Golgi partitioning controls mitotic entry through Aurora-A kinase. Mol Biol Cell 21(21):3708–3721. doi:E10-03-0243 [pii] 10.1091/mbc.E10-03-0243

112. Mascanzoni F, Ayala I, Iannitti R, Luini A, Colanzi A (2024) The Golgi checkpoint: Golgi unlinking during G2 is necessary for correct spindle formation and cytokinesis. Life Sci Alliance (In press)
113. Cohen-Sharir Y, McFarland JM, Abdusamad M, Marquis C, Bernhard SV, Kazachkova M, Tang H, Ippolito MR, Laue K, Zerbib J, Malaby HLH, Jones A, Stautmeister LM, Bockaj I, Wardenaar R, Lyons N, Nagaraja A, Bass AJ, Spierings DCJ, Foijer F, Beroukhim R, Santaguida S, Golub TR, Stumpff J, Storchova Z, Ben-David U (2021) Aneuploidy renders cancer cells vulnerable to mitotic checkpoint inhibition. Nature 590(7846):486–491. https://doi.org/10.1038/s41586-020-03114-6
114. Kuznetsova AY, Seget K, Moeller GK, de Pagter MS, de Roos JA, Durrbaum M, Kuffer C, Muller S, Zaman GJ, Kloosterman WP, Storchova Z (2015) Chromosomal instability, tolerance of mitotic errors and multidrug resistance are promoted by tetraploidization in human cells. Cell Cycle 14(17):2810–2820. https://doi.org/10.1080/15384101.2015.1068482
115. Altan-Bonnet N, Phair RD, Polishchuk RS, Weigert R, Lippincott-Schwartz J (2003) A role for Arf1 in mitotic Golgi disassembly, chromosome segregation, and cytokinesis. Proc Natl Acad Sci USA 100(23):13314–13319. doi:10.1073/pnas.2234055100 2234055100 [pii]
116. Nakamura N (2010) Emerging new roles of GM130, a cis-Golgi matrix protein, in higher order cell functions. J Pharmacol Sci 112(3):255–264
117. Wei JH, Zhang ZC, Wynn RM, Seemann J (2015) GM130 regulates Golgi-derived spindle assembly by activating TPX2 and capturing microtubules. Cell 162(2):287–299. doi:S0092-8674(15)00697-2 [pii] 10.1016/j.cell.2015.06.014
118. Guo H, Wei JH, Zhang Y, Seemann J (2021) Importin alpha phosphorylation promotes TPX2 activation by GM130 to control astral microtubules and spindle orientation. J Cell Sci. https://doi.org/10.1242/jcs.258356
119. Guizzunti G, Seemann J (2016) Mitotic Golgi disassembly is required for bipolar spindle formation and mitotic progression. Proc Natl Acad Sci USA 113(43):E6590–E6599. https://doi.org/10.1073/pnas.1610844113
120. Wei JH, Seemann J (2017) Golgi ribbon disassembly during mitosis, differentiation and disease progression. Curr Opin Cell Biol 47:43–51. https://doi.org/10.1016/j.ceb.2017.03.008
121. Frye K, Renda F, Fomicheva M, Zhu X, Gong L, Khodjakov A, Kaverina I (2020) Cell cycle-dependent dynamics of the Golgi-centrosome association in motile cells. Cells 9(5). https://doi.org/10.3390/cells9051069
122. Mascanzoni F, Ayala I, Iannitti R, Luini A, Colanzi A (2024) The Golgi checkpoint: Golgi unlinking during G2 is necessary for spindle formation and cytokinesis. Life Sci Alliance 7(5). https://doi.org/10.26508/lsa.202302469
123. Marchesi S, Montani F, Deflorian G, D'Antuono R, Cuomo A, Bologna S, Mazzoccoli C, Bonaldi T, Di Fiore PP, Nicassio F (2014) DEPDC1B coordinates de-adhesion events and cell-cycle progression at mitosis. Dev Cell 31(4):420–433. https://doi.org/10.1016/j.devcel.2014.09.009
124. Champion L, Linder MI, Kutay U (2017) Cellular reorganization during mitotic entry. Trends Cell Biol 27(1):26–41. https://doi.org/10.1016/j.tcb.2016.07.004
125. Pugacheva EN, Golemis EA (2006) HEF1-aurora A interactions: points of dialog between the cell cycle and cell attachment signaling networks. Cell Cycle 5(4):384–391. https://doi.org/10.4161/cc.5.4.2439
126. Tavernier N, Sicheri F, Pintard L (2021) Aurora A kinase activation: different means to different ends. J Cell Biol 220(9). https://doi.org/10.1083/jcb.202106128
127. Chen NP, Aretz J, Fassler R (2022) CDK1-cyclin-B1-induced kindlin degradation drives focal adhesion disassembly at mitotic entry. Nat Cell Biol 24(5):723–736. https://doi.org/10.1038/s41556-022-00886-z
128. Mascanzoni F, Ayala I, Colanzi A (2019) Organelle inheritance control of mitotic entry and progression: implications for tissue homeostasis and disease. Front Cell Dev Biol 7:133. https://doi.org/10.3389/fcell.2019.00133

129. Kodani A, Sutterlin C (2008) The Golgi protein GM130 regulates centrosome morphology and function. Mol Biol Cell 19(2):745–753. https://doi.org/10.1091/mbc.E07-08-0847
130. Radulescu AE, Mukherjee S, Shields D (2011) The Golgi protein p115 associates with gamma-tubulin and plays a role in Golgi structure and mitosis progression. J Biol Chem 286(24):21915–21926. doi:M110.209460 [pii] 10.1074/jbc.M110.209460
131. Sutterlin C, Polishchuk R, Pecot M, Malhotra V (2005) The Golgi-associated protein GRASP65 regulates spindle dynamics and is essential for cell division. Mol Biol Cell 16(7):3211–3222. doi:E04-12-1065 [pii] 10.1091/mbc.E04-12-1065
132. Wei JH, Seemann J (2009) Spindle-dependent partitioning of the Golgi ribbon. Commun Integr Biol 2(5):406–407
133. Sauer G, Korner R, Hanisch A, Ries A, Nigg EA, Sillje HH (2005) Proteome analysis of the human mitotic spindle. Mol Cell Proteomics 4(1):35–43. https://doi.org/10.1074/mcp.M400158-MCP200
134. Shima DT, Cabrera-Poch N, Pepperkok R, Warren G (1998) An ordered inheritance strategy for the Golgi apparatus: visualization of mitotic disassembly reveals a role for the mitotic spindle. J Cell Biol 141(4):955–966
135. Puri S, Telfer H, Velliste M, Murphy RF, Linstedt AD (2004) Dispersal of Golgi matrix proteins during mitotic Golgi disassembly. J Cell Sci 117(Pt 3):451–456. doi:10.1242/jcs.00863 jcs.00863 [pii]
136. Feng H, Wang X, Xu Z, Zhang X, Gao Y (2018) Super-resolution fluorescence microscopy for single cell imaging. Adv Exp Med Biol 1068:59–71. https://doi.org/10.1007/978-981-13-0502-3_6
137. Lancaster OM, Le Berre M, Dimitracopoulos A, Bonazzi D, Zlotek-Zlotkiewicz E, Picone R, Duke T, Piel M, Baum B (2013) Mitotic rounding alters cell geometry to ensure efficient bipolar spindle formation. Dev Cell 25(3):270–283. https://doi.org/10.1016/j.devcel.2013.03.014
138. Thery M, Bornens M (2008) Get round and stiff for mitosis. HFSP J 2(2):65–71. https://doi.org/10.2976/1.2895661
139. Fukata M, Nakagawa M, Kaibuchi K (2003) Roles of Rho-family GTPases in cell polarisation and directional migration. Curr Opin Cell Biol 15(5):590–597. https://doi.org/10.1016/s0955-0674(03)00097-8
140. Yadav S, Puri S, Linstedt AD (2009) A primary role for Golgi positioning in directed secretion, cell polarity, and wound healing. Mol Biol Cell 20(6):1728–1736. https://doi.org/10.1091/mbc.E08-10-1077
141. Rong Y, Yang W, Hao H, Wang W, Lin S, Shi P, Huang Y, Li B, Sun Y, Liu Z, Wu C (2021) The Golgi microtubules regulate single cell durotaxis. EMBO Rep 22(3):e51094. https://doi.org/10.15252/embr.202051094
142. Van Horn RD, Chu S, Fan L, Yin T, Du J, Beckmann R, Mader M, Zhu G, Toth J, Blanchard K, Ye XS (2010) Cdk1 activity is required for mitotic activation of aurora A during G2/M transition of human cells. J Biol Chem 285(28):21849–21857. https://doi.org/10.1074/jbc.M110.141010
143. Arab M, Chen T, Lowe M (2024) Mechanisms governing vesicle traffic at the Golgi apparatus. Curr Opin Cell Biol 88:102365. https://doi.org/10.1016/j.ceb.2024.102365
144. Shirai R, Yamauchi J (2024) Emerging evidence of Golgi stress signaling for neuropathies. Neurol Int 16(2):334–348. https://doi.org/10.3390/neurolint16020024
145. Zhang J, Kennedy A, de Melo Jorge DM, Xing L, Reid W, Bui S, Joppich J, Rose M, Ercan S, Tang Q, Tai AW, Wang Y (2024) SARS-CoV-2 remodels the Golgi apparatus to facilitate viral assembly and secretion. bioRxiv. https://doi.org/10.1101/2022.03.04.483074
146. Hansen MD, Johnsen IB, Stiberg KA, Sherstova T, Wakita T, Richard GM, Kandasamy RK, Meurs EF, Anthonsen MW (2017) Hepatitis C virus triggers Golgi fragmentation and autophagy through the immunity-related GTPase M. Proc Natl Acad Sci USA 114(17):E3462–E3471. https://doi.org/10.1073/pnas.1616683114

147. Cho JH, Saini DK, Karunarathne WK, Kalyanaraman V, Gautam N (2011) Alteration of Golgi structure in senescent cells and its regulation by a G protein gamma subunit. Cell Signal 23(5):785–793. https://doi.org/10.1016/j.cellsig.2011.01.001
148. Bui S, Mejia I, Diaz B, Wang Y (2021) Adaptation of the Golgi apparatus in cancer cell invasion and metastasis. Front Cell Dev Biol 9:806482. https://doi.org/10.3389/fcell.2021.806482
149. Sechi S, Frappaolo A, Karimpour-Ghahnavieh A, Piergentili R, Giansanti MG (2020) Oncogenic roles of GOLPH3 in the physiopathology of cancer. Int J Mol Sci 21(3). https://doi.org/10.3390/ijms21030933
150. Kellokumpu S, Sormunen R, Kellokumpu I (2002) Abnormal glycosylation and altered Golgi structure in colorectal cancer: dependence on intra-Golgi pH. FEBS Lett 516(1-3):217–224. https://doi.org/10.1016/s0014-5793(02)02535-8
151. Sano T, Kovacs K, Asa SL, Yamada S, Sanno N, Yokoyama S, Takami H (2002) Pituitary adenoma with “honeycomb Golgi” appearance showing a phenotypic change at recurrence from clinically nonfunctioning to typical Cushing disease. Endocr Pathol 13(2):125–130. https://doi.org/10.1385/ep:13:2:125
152. Rabouille C, Linstedt AD (2016) GRASP: a multitasking tether. Front Cell Dev Biol 4:1. https://doi.org/10.3389/fcell.2016.00001
153. Makhoul C, Gosavi P, Gleeson PA (2018) The Golgi architecture and cell sensing. Biochem Soc Trans 46(5):1063–1072. https://doi.org/10.1042/BST20180323
154. Spano D, Colanzi A (2022) Golgi complex: a signaling hub in cancer. Cells 11(13). https://doi.org/10.3390/cells11131990
155. Del Giudice S, De Luca V, Parizadeh S, Russo D, Luini A, Di Martino R (2022) Endogenous and exogenous regulatory signaling in the secretory pathway: role of Golgi signaling molecules in cancer. Front Cell Dev Biol 10:833663. https://doi.org/10.3389/fcell.2022.833663
156. Poh AR, Ernst M (2023) Functional roles of SRC signaling in pancreatic cancer: recent insights provide novel therapeutic opportunities. Oncogene 42(22):1786–1801. https://doi.org/10.1038/s41388-023-02701-x
157. Azoitei N, Cobbaut M, Becher A, Van Lint J, Seufferlein T (2018) Protein kinase D2: a versatile player in cancer biology. Oncogene 37(10):1263–1278. https://doi.org/10.1038/s41388-017-0052-8
158. Zhang X, Connelly J, Chao Y, Wang QJ (2021) Multifaceted functions of protein kinase D in pathological processes and human diseases. Biomolecules 11(3). https://doi.org/10.3390/biom11030483
159. Mavillard F, Hidalgo J, Megias D, Levitsky KL, Velasco A (2010) PKA-mediated Golgi remodeling during cAMP signal transmission. Traffic 11(1):90–109. https://doi.org/10.1111/j.1600-0854.2009.01007.x
160. Larocca MC, Shanks RA, Tian L, Nelson DL, Stewart DM, Goldenring JR (2004) AKAP350 interaction with cdc42 interacting protein 4 at the Golgi apparatus. Mol Biol Cell 15(6):2771–2781. https://doi.org/10.1091/mbc.e03-10-0757
161. Zheng D, Li J, Yan H, Zhang G, Li W, Chu E, Wei N (2023) Emerging roles of Aurora-A kinase in cancer therapy resistance. Acta Pharm Sin B 13(7):2826–2843. https://doi.org/10.1016/j.apsb.2023.03.013
162. Baschieri F, Farhan H (2015) Endomembrane control of cell polarity: relevance to cancer. Small GTPases 6(2):104–107. https://doi.org/10.1080/21541248.2015.1018402
163. Kumamoto K, Goto Y, Sekikawa K, Takenoshita S, Ishida N, Kawakita M, Kannagi R (2001) Increased expression of UDP-galactose transporter messenger RNA in human colon cancer tissues and its implication in synthesis of Thomsen-Friedenreich antigen and sialyl Lewis A/X determinants. Cancer Res 61(11):4620–4627
164. Zhang X (2021) Alterations of Golgi structural proteins and glycosylation defects in cancer. Front Cell Dev Biol 9:665289. https://doi.org/10.3389/fcell.2021.665289
165. Bennett EP, Mandel U, Clausen H, Gerken TA, Fritz TA, Tabak LA (2012) Control of mucin-type O-glycosylation: a classification of the polypeptide GalNAc-transferase gene family. Glycobiology 22(6):736–756. https://doi.org/10.1093/glycob/cwr182

166. Pothukuchi P, Agliarulo I, Russo D, Rizzo R, Russo F, Parashuraman S (2019) Translation of genome to glycome: role of the Golgi apparatus. FEBS Lett. 593(17):2390-2411. https://doi.org/10.1002/1873-3468.13541
167. Marcos NT, Bennett EP, Gomes J, Magalhaes A, Gomes C, David L, Dar I, Jeanneau C, DeFrees S, Krustrup D, Vogel LK, Kure EH, Burchell J, Taylor-Papadimitriou J, Clausen H, Mandel U, Reis CA (2011) ST6GalNAc-I controls expression of sialyl-Tn antigen in gastrointestinal tissues. Front Biosci (Elite Ed) 3(4):1443–1455. https://doi.org/10.2741/e345
168. Gill DJ, Chia J, Senewiratne J, Bard F (2010) Regulation of O-glycosylation through Golgi-to-ER relocation of initiation enzymes. J Cell Biol 189(5):843–858. https://doi.org/10.1083/jcb.201003055
169. Wang Y, Ju T, Ding X, Xia B, Wang W, Xia L, He M, Cummings RD (2010) Cosmc is an essential chaperone for correct protein O-glycosylation. Proc Natl Acad Sci USA 107(20):9228–9233. https://doi.org/10.1073/pnas.0914004107
170. Pinho SS, Reis CA (2015) Glycosylation in cancer: mechanisms and clinical implications. Nat Rev Cancer 15(9):540–555. https://doi.org/10.1038/nrc3982
171. Hakomori S, Kannagi R (1983) Glycosphingolipids as tumor-associated and differentiation markers. J Natl Cancer Inst 71(2):231–251
172. Christiansen MN, Chik J, Lee L, Anugraham M, Abrahams JL, Packer NH (2014) Cell surface protein glycosylation in cancer. Proteomics 14(4–5):525–546. https://doi.org/10.1002/pmic.201300387
173. Groves JA, Lee A, Yildirir G, Zachara NE (2013) Dynamic O-GlcNAcylation and its roles in the cellular stress response and homeostasis. Cell Stress Chaperones 18(5):535–558. https://doi.org/10.1007/s12192-013-0426-y
174. Axelsson MA, Karlsson NG, Steel DM, Ouwendijk J, Nilsson T, Hansson GC (2001) Neutralization of pH in the Golgi apparatus causes redistribution of glycosyltransferases and changes in the O-glycosylation of mucins. Glycobiology 11(8):633–644. https://doi.org/10.1093/glycob/11.8.633
175. Weisz OA (2003) Organelle acidification and disease. Traffic 4(2):57–64. https://doi.org/10.1034/j.1600-0854.2003.40201.x
176. Rivinoja A, Hassinen A, Kokkonen N, Kauppila A, Kellokumpu S (2009) Elevated Golgi pH impairs terminal N-glycosylation by inducing mislocalization of Golgi glycosyltransferases. J Cell Physiol 220(1):144–154. https://doi.org/10.1002/jcp.21744
177. Goldenring JR (2013) A central role for vesicle trafficking in epithelial neoplasia: intracellular highways to carcinogenesis. Nat Rev Cancer 13(11):813–820. https://doi.org/10.1038/nrc3601
178. McCaffrey LM, Macara IG (2011) Epithelial organization, cell polarity and tumorigenesis. Trends Cell Biol 21(12):727–735. https://doi.org/10.1016/j.tcb.2011.06.005
179. Nieto MA, Huang RY, Jackson RA, Thiery JP (2016) Emt: 2016. Cell 166(1):21–45. https://doi.org/10.1016/j.cell.2016.06.028
180. Rosso M, Majem B, Devis L, Lapyckyj L, Besso MJ, Llaurado M, Abascal MF, Matos ML, Lanau L, Castellvi J, Sanchez JL, Perez Benavente A, Gil-Moreno A, Reventos J, Santamaria Margalef A, Rigau M, Vazquez-Levin MH (2017) E-cadherin: a determinant molecule associated with ovarian cancer progression, dissemination and aggressiveness. PLoS One 12(9):e0184439. https://doi.org/10.1371/journal.pone.0184439
181. Araki K, Shimura T, Suzuki H, Tsutsumi S, Wada W, Yajima T, Kobayahi T, Kubo N, Kuwano H (2011) E/N-cadherin switch mediates cancer progression via TGF-beta-induced epithelial-to-mesenchymal transition in extrahepatic cholangiocarcinoma. Br J Cancer 105(12):1885–1893. https://doi.org/10.1038/bjc.2011.452
182. Jolly MK, Tripathi SC, Jia D, Mooney SM, Celiktas M, Hanash SM, Mani SA, Pienta KJ, Ben-Jacob E, Levine H (2016) Stability of the hybrid epithelial/mesenchymal phenotype. Oncotarget 7(19):27067–27084. https://doi.org/10.18632/oncotarget.8166
183. Grimaldi G, Filograna A, Schembri L, Lo Monte M, Di Martino R, Pirozzi M, Spano D, Beccari AR, Parashuraman S, Luini A, Valente C, Corda D (2022) PKD-dependent

PARP12-catalyzed mono-ADP-ribosylation of Golgin-97 is required for E-cadherin transport from Golgi to plasma membrane. Proc Natl Acad Sci USA 119(1). https://doi.org/10.1073/pnas.2026494119
184. Tan X, Banerjee P, Guo HF, Ireland S, Pankova D, Ahn YH, Nikolaidis IM, Liu X, Zhao Y, Xue Y, Burns AR, Roybal J, Gibbons DL, Zal T, Creighton CJ, Ungar D, Wang Y, Kurie JM (2017) Epithelial-to-mesenchymal transition drives a pro-metastatic Golgi compaction process through scaffolding protein PAQR11. J Clin Invest 127(1):117–131. https://doi.org/10.1172/JCI88736
185. Bonnans C, Chou J, Werb Z (2014) Remodelling the extracellular matrix in development and disease. Nat Rev Mol Cell Biol 15(12):786–801. https://doi.org/10.1038/nrm3904
186. Hynes RO, Naba A (2012) Overview of the matrisome—an inventory of extracellular matrix constituents and functions. Cold Spring Harb Perspect Biol 4(1):a004903. https://doi.org/10.1101/cshperspect.a004903
187. Rozario T, DeSimone DW (2010) The extracellular matrix in development and morphogenesis: a dynamic view. Dev Biol 341(1):126–140. https://doi.org/10.1016/j.ydbio.2009.10.026
188. Frantz C, Stewart KM, Weaver VM (2010) The extracellular matrix at a glance. J Cell Sci 123(Pt 24):4195–4200. https://doi.org/10.1242/jcs.023820
189. Ahat E, Xiang Y, Zhang X, Bekier ME 2nd, Wang Y (2019) GRASP depletion-mediated Golgi destruction decreases cell adhesion and migration via the reduction of alpha5beta1 integrin. Mol Biol Cell 30(6):766–777. https://doi.org/10.1091/mbc.E18-07-0462
190. Ohuchi E, Imai K, Fujii Y, Sato H, Seiki M, Okada Y (1997) Membrane type 1 matrix metalloproteinase digests interstitial collagens and other extracellular matrix macromolecules. J Biol Chem 272(4):2446–2451. https://doi.org/10.1074/jbc.272.4.2446
191. Koshikawa N, Giannelli G, Cirulli V, Miyazaki K, Quaranta V (2000) Role of cell surface metalloprotease MT1-MMP in epithelial cell migration over laminin-5. J Cell Biol 148(3):615–624. https://doi.org/10.1083/jcb.148.3.615
192. Golubkov VS, Chekanov AV, Doxsey SJ, Strongin AY (2005) Centrosomal pericentrin is a direct cleavage target of membrane type-1 matrix metalloproteinase in humans but not in mice: potential implications for tumorigenesis. J Biol Chem 280(51):42237–42241. https://doi.org/10.1074/jbc.M510139200
193. Deryugina EI, Quigley JP (2006) Matrix metalloproteinases and tumor metastasis. Cancer Metastasis Rev 25(1):9–34. https://doi.org/10.1007/s10555-006-7886-9
194. Ra HJ, Parks WC (2007) Control of matrix metalloproteinase catalytic activity. Matrix Biol 26(8):587–596. https://doi.org/10.1016/j.matbio.2007.07.001
195. Kinoshita T, Sato H, Okada A, Ohuchi E, Imai K, Okada Y, Seiki M (1998) TIMP-2 promotes activation of progelatinase A by membrane-type 1 matrix metalloproteinase immobilized on agarose beads. J Biol Chem 273(26):16098–16103. https://doi.org/10.1074/jbc.273.26.16098
196. Ayala I, Baldassarre M, Caldieri G, Buccione R (2006) Invadopodia: a guided tour. Eur J Cell Biol 85(3–4):159–164. https://doi.org/10.1016/j.ejcb.2005.09.005
197. Shtutman M, Roninson IB (2011) A subunit of coatomer protein complex offers a novel tumor-specific target through a surprising mechanism. Autophagy 7(12):1551–1552. https://doi.org/10.4161/auto.7.12.17659
198. Feng Y, Lei X, Zhang L, Wan H, Pan H, Wu J, Zou M, Zhu L, Mi Y (2021) COPB2: a transport protein with multifaceted roles in cancer development and progression. Clin Transl Oncol 23(11):2195–2205. https://doi.org/10.1007/s12094-021-02630-9
199. Schlienger S, Ramirez RA, Claing A (2015) ARF1 regulates adhesion of MDA-MB-231 invasive breast cancer cells through formation of focal adhesions. Cell Signal 27(3):403–415. https://doi.org/10.1016/j.cellsig.2014.11.032
200. Boulay PL, Schlienger S, Lewis-Saravalli S, Vitale N, Ferbeyre G, Claing A (2011) ARF1 controls proliferation of breast cancer cells by regulating the retinoblastoma protein. Oncogene 30(36):3846–3861. https://doi.org/10.1038/onc.2011.100
201. Sandilands E, Freckmann EC, Cumming EM, Roman-Fernandez A, McGarry L, Anand J, Galbraith L, Mason S, Patel R, Nixon C, Cartwright J, Leung HY, Blyth K, Bryant DM

(2023) The small GTPase ARF3 controls invasion modality and metastasis by regulating N-cadherin levels. J Cell Biol 222(4). https://doi.org/10.1083/jcb.202206115
202. Shestakova A, Zolov S, Lupashin V (2006) COG complex-mediated recycling of Golgi glycosyltransferases is essential for normal protein glycosylation. Traffic 7(2):191–204. https://doi.org/10.1111/j.1600-0854.2005.00376.x
203. Kuna RS, Field SJ (2019) GOLPH3: a Golgi phosphatidylinositol(4)phosphate effector that directs vesicle trafficking and drives cancer. J Lipid Res 60(2):269–275. https://doi.org/10.1194/jlr.R088328
204. Rizzo R, Parashuraman S, D'Angelo G, Luini A (2017) GOLPH3 and oncogenesis: what is the molecular link? Tissue Cell 49(2 Pt A):170–174. https://doi.org/10.1016/j.tice.2016.06.008
205. Buschman MD, Rahajeng J, Field SJ (2015) GOLPH3 links the Golgi, DNA damage, and cancer. Cancer Res 75(4):624–627. https://doi.org/10.1158/0008-5472.CAN-14-3081
206. Farber-Katz SE, Dippold HC, Buschman MD, Peterman MC, Xing M, Noakes CJ, Tat J, Ng MM, Rahajeng J, Cowan DM, Fuchs GJ, Zhou H, Field SJ (2014) DNA damage triggers Golgi dispersal via DNA-PK and GOLPH3. Cell 156(3):413–427. https://doi.org/10.1016/j.cell.2013.12.023
207. Zeng Z, Lin H, Zhao X, Liu G, Wang X, Xu R, Chen K, Li J, Song L (2012) Overexpression of GOLPH3 promotes proliferation and tumorigenicity in breast cancer via suppression of the FOXO1 transcription factor. Clin Cancer Res 18(15):4059–4069. https://doi.org/10.1158/1078-0432.CCR-11-3156
208. Wang R, Ke ZF, Wang F, Zhang WH, Wang YF, Li SH, Wang LT (2015) GOLPH3 overexpression is closely correlated with poor prognosis in human non-small cell lung cancer and mediates its metastasis through upregulating MMP-2 and MMP-9. Cell Physiol Biochem 35(3):969–982. https://doi.org/10.1159/000369753
209. Song Q, He X, Xiong Y, Wang J, Zhang L, Leung EL, Li G (2021) The functional landscape of Golgi membrane protein 1 (GOLM1) phosphoproteome reveal GOLM1 regulating P53 that promotes malignancy. Cell Death Discov 7(1):42. https://doi.org/10.1038/s41420-021-00422-2
210. Varambally S, Laxman B, Mehra R, Cao Q, Dhanasekaran SM, Tomlins SA, Granger J, Vellaichamy A, Sreekumar A, Yu J, Gu W, Shen R, Ghosh D, Wright LM, Kladney RD, Kuefer R, Rubin MA, Fimmel CJ, Chinnaiyan AM (2008) Golgi protein GOLM1 is a tissue and urine biomarker of prostate cancer. Neoplasia 10(11):1285–1294. https://doi.org/10.1593/neo.08922
211. Mao Z, Wu Y, Yao P, Xing C (2023) GOLM1 facilitates human colorectal cancer progression and metastasis via activating the AKT/GSK3beta/EMT axis. Neoplasma 70(1):136–144. https://doi.org/10.4149/neo_2023_220816N835
212. Lowe M (2019) The physiological functions of the Golgin vesicle tethering proteins. Front Cell Dev Biol 7:94. https://doi.org/10.3389/fcell.2019.00094
213. Bekier ME 2nd, Wang L, Li J, Huang H, Tang D, Zhang X, Wang Y (2017) Knockout of the Golgi stacking proteins GRASP55 and GRASP65 impairs Golgi structure and function. Mol Biol Cell 28(21):2833–2842. https://doi.org/10.1091/mbc.E17-02-0112
214. Lane JD, Lucocq J, Pryde J, Barr FA, Woodman PG, Allan VJ, Lowe M (2002) Caspase-mediated cleavage of the stacking protein GRASP65 is required for Golgi fragmentation during apoptosis. J Cell Biol 156(3):495–509. https://doi.org/10.1083/jcb.200110007
215. Wang F, Chen X, Yuan D, Yi Y, Luo Y (2019) Golgi reassembly and stacking protein 65 downregulation is required for the anti-cancer effect of dihydromyricetin on human ovarian cancer cells. PLoS One 14(11):e0225450. https://doi.org/10.1371/journal.pone.0225450
216. Bhat G, Hothpet VR, Lin MF, Cheng PW (2017) Shifted Golgi targeting of glycosyltransferases and alpha-mannosidase IA from giantin to GM130-GRASP65 results in formation of high mannose N-glycans in aggressive prostate cancer cells. Biochim Biophys Acta Gen Subj 1861(11 Pt A):2891–2901. https://doi.org/10.1016/j.bbagen.2017.08.006
217. Galluzzi L, Pietrocola F, Levine B, Kroemer G (2014) Metabolic control of autophagy. Cell 159(6):1263–1276. https://doi.org/10.1016/j.cell.2014.11.006

218. Debnath J, Gammoh N, Ryan KM (2023) Autophagy and autophagy-related pathways in cancer. Nat Rev Mol Cell Biol 24(8):560–575. https://doi.org/10.1038/s41580-023-00585-z
219. Bailly AL, Grenier JMP, Cartier-Michaud A, Bardin F, Balzano M, Goubard A, Lissitzky JC, De Grandis M, Mancini SJC, Serge A, Aurrand-Lions M (2020) GRASP55 is dispensable for normal hematopoiesis but necessary for Myc-dependent leukemic growth. J Immunol 204(10):2685–2696. https://doi.org/10.4049/jimmunol.1901124
220. Sehgal K, Barbie DA (2021) Targeting the mutant p53 secretome. J Clin Invest 131(1). https://doi.org/10.1172/JCI143945
221. Tan X, Shi L, Banerjee P, Liu X, Guo HF, Yu J, Bota-Rabassedas N, Rodriguez BL, Gibbons DL, Russell WK, Creighton CJ, Kurie JM (2021) A protumorigenic secretory pathway activated by p53 deficiency in lung adenocarcinoma. J Clin Invest 131(1). https://doi.org/10.1172/JCI137186
222. Puthenveedu MA, Bachert C, Puri S, Lanni F, Linstedt AD (2006) GM130 and GRASP65-dependent lateral cisternal fusion allows uniform Golgi-enzyme distribution. Nat Cell Biol 8(3):238–248. doi:ncb1366 [pii] 10.1038/ncb1366
223. Baschieri F, Confalonieri S, Bertalot G, Di Fiore PP, Dietmaier W, Leist M, Crespo P, Macara IG, Farhan H (2014) Spatial control of Cdc42 signalling by a GM130-RasGRF complex regulates polarity and tumorigenesis. Nat Commun 5:4839. https://doi.org/10.1038/ncomms5839
224. Zhao J, Yang C, Guo S, Wu Y (2015) GM130 regulates epithelial-to-mesenchymal transition and invasion of gastric cancer cells via snail. Int J Clin Exp Pathol 8(9):10784–10791
225. Chang SH, Hong SH, Jiang HL, Minai-Tehrani A, Yu KN, Lee JH, Kim JE, Shin JY, Kang B, Park S, Han K, Chae C, Cho MH (2012) GOLGA2/GM130, cis-Golgi matrix protein, is a novel target of anticancer gene therapy. Mol Ther 20(11):2052–2063. https://doi.org/10.1038/mt.2012.125
226. Baschieri F, Uetz-von Allmen E, Legler DF, Farhan H (2015) Loss of GM130 in breast cancer cells and its effects on cell migration, invasion and polarity. Cell Cycle 14(8):1139–1147. https://doi.org/10.1080/15384101.2015.1007771
227. Yoon S, Choi JH, Shah M, Kwon SM, Yang J, Park YN, Wang HJ, Woo HG (2021) USO1 isoforms differentially promote liver cancer progression by dysregulating the ER-Golgi network. Carcinogenesis 42(9):1208–1220. https://doi.org/10.1093/carcin/bgab067
228. Ghannoum S, Fantini D, Zahoor M, Reiterer V, Phuyal S, Leoncio Netto W, Sorensen O, Iyer A, Sengupta D, Prasmickaite L, Maelandsmo GM, Kohn-Luque A, Farhan H (2023) A combined experimental-computational approach uncovers a role for the Golgi matrix protein Giantin in breast cancer progression. PLoS Comput Biol 19(4):e1010995. https://doi.org/10.1371/journal.pcbi.1010995
229. Lock JG, Hammond LA, Houghton F, Gleeson PA, Stow JL (2005) E-cadherin transport from the trans-Golgi network in tubulovesicular carriers is selectively regulated by golgin-97. Traffic 6(12):1142–1156. https://doi.org/10.1111/j.1600-0854.2005.00349.x
230. Hsu RM, Zhong CY, Wang CL, Liao WC, Yang C, Lin SY, Lin JW, Cheng HY, Li PY, Yu CJ (2018) Golgi tethering factor golgin-97 suppresses breast cancer cell invasiveness by modulating NF-kappaB activity. Cell Commun Signal 16(1):19. https://doi.org/10.1186/s12964-018-0230-5
231. Luke MR, Kjer-Nielsen L, Brown DL, Stow JL, Gleeson PA (2003) GRIP domain-mediated targeting of two new coiled-coil proteins, GCC88 and GCC185, to subcompartments of the trans-Golgi network. J Biol Chem 278(6):4216–4226. https://doi.org/10.1074/jbc.M210387200
232. Makhoul C, Gleeson PA (2021) Regulation of mTORC1 activity by the Golgi apparatus. Fac Rev 10:50. https://doi.org/10.12703/r/10-50
233. Halberg N, Sengelaub CA, Navrazhina K, Molina H, Uryu K, Tavazoie SF (2016) PITPNC1 recruits RAB1B to the Golgi network to drive malignant secretion. Cancer Cell 29(3):339–353. https://doi.org/10.1016/j.ccell.2016.02.013

234. Ruggiero C, Grossi M, Fragassi G, Di Campli A, Di Ilio C, Luini A, Sallese M (2018) The KDEL receptor signalling cascade targets focal adhesion kinase on focal adhesions and invadopodia. Oncotarget 9(12):10228–10246. https://doi.org/10.18632/oncotarget.23421
235. Ruggiero C, Fragassi G, Grossi M, Picciani B, Di Martino R, Capitani M, Buccione R, Luini A, Sallese M (2015) A Golgi-based KDELR-dependent signalling pathway controls extracellular matrix degradation. Oncotarget 6(5):3375–3393. https://doi.org/10.18632/oncotarget.3270
236. Liao Z, She C, Ma L, Sun Z, Li P, Zhang X, Wang P, Li W (2019) KDELR2 promotes glioblastoma tumorigenesis targeted by HIF1a via mTOR signaling pathway. Cell Mol Neurobiol 39(8):1207–1215. https://doi.org/10.1007/s10571-019-00715-2
237. Bajaj R, Kundu ST, Grzeskowiak CL, Fradette JJ, Scott KL, Creighton CJ, Gibbons DL (2020) IMPAD1 and KDELR2 drive invasion and metastasis by enhancing Golgi-mediated secretion. Oncogene 39(37):5979–5994. https://doi.org/10.1038/s41388-020-01410-z
238. Colanzi A, Parashuraman S, Reis CA, Ungar D (2022) Editorial: does the Golgi complex enable oncogenesis? Front Cell Dev Biol 10:1000946. https://doi.org/10.3389/fcell.2022.1000946

Chapter 7
The Neuronal Golgi Network and Its Acolytes

Alexandre Varangot and Cyril Hanus

Abstract Like other cells, neurons synthesize, chemically modify, and traffic secreted and surface membrane proteins in the specialized biosynthetic membrane network of the secretory pathway. Unlike other cells, neurons develop and maintain very long protrusions with ramified geometries that engage in stable synaptic connections with thousands of other cells, hence dealing with unique logistic constraints to synthesize and bring secretory proteins where they are needed. In this chapter, we discuss how the neuronal Golgi apparatus and associated membranes evolved to achieve this daunting task.

Keywords Neurons · Endoplasmic reticulum · ER-Golgi Intermediate compartment · Golgi apparatus and satelitte systems · Unconventionnal secretory trafficking

Introduction

Neuron development and function require numerous membrane and secreted proteins, including growth factors, surface receptors, adhesion molecules, and ion channels, and thus heavily rely on the *secretory pathway* where these proteins are synthesized, chemically modified, and trafficked where they are needed [85].

With their long neurites of varying shapes and branching patterns, neurons are the largest and morphologically most complex cells found in metazoans. The surface of the neuronal plasma membrane is up to 10,000 times larger than that of a small globular cell such as a lymphocyte, putting a large demand and unique spatial constraints on protein homeostasis [68]. Consequently, neurons have evolved specialized mechanisms to upscale their biosynthetic capacities and deliver nascent

A. Varangot · C. Hanus (✉)
Institute of Psychiatry and Neurosciences of Paris, INSERM UMR1266, Paris-Cité University, Paris, France
e-mail: cyril.hanus@inserm.fr

A. Nakano, J. Saraste (eds.), *The Golgi Network, Volume I*, Subcellular Biochemistry 110, https://doi.org/10.1007/978-3-032-06936-8_7

secretory proteins[1] up to millimeters away from their cell body, a few meters in some instances if one thinks of motoneurons in large mammals [58, 63, 68].

Neurons are also extraordinarily compartmentalized. First, because of the polarity of synaptic signaling. In the most frequent case, synaptic transmission requires the spatial segregation of the neurotransmitter-releasing and neurotransmitter-signaling machinery in distinct neurites: the "emitter" compartment of chemical synaptic signaling, the axon, and the "receiver" compartment formed by the soma and its multiple dendritic protrusions. Second and perhaps on a more impressive scale, the somatodendritic surface is compartmentalized by the tens of thousands of synaptic contacts that impinge on the somatodendritic surface of a typical mammalian neuron. Synapses compartmentalize the plasma membrane in a multitude of distinct microdomains with specific molecular composition and signaling efficacy, requiring efficient mechanisms to synthesize, sort, and recruit specific membrane proteins on an exquisitely local scale [63, 85].

Whether it is the individual synapse, clusters of synapses, or segments of dendritic branches, what constitutes the elementary unit of synaptic signaling is still debated [68]. Yet, it is clear that synapse composition and signaling are tuned on a synapse-by-synapse basis. Local protein synthesis plays an important role in this context, enabling targeted functionalization of specific synapses during learning [68]. Numerous proteins have been shown to be synthesized locally in dendrites and axon terminals [54, 58, 80, 85, 153]. These include secretory proteins and notably neurotransmitter receptors. To avoid any ambiguity, let us be reminded that, in contrast to cytoplasmic proteins, the synthesis and turnover of membrane proteins require coordination of the multiple membrane compartments of the secretory and endosomal networks where these proteins are made, recycled, or degraded. Underlying mechanisms thus lay at the core of our brain's ability to maintain synaptic homeostasis while, at the same time, modify specific synapses during memory formation and learning. These mechanisms are altered in numerous debilitating human pathologies and are the focus of intense scrutiny [186].

In this chapter, we provide a critical overview of current research on the morphology and function of the neuronal Golgi apparatus (GA) and associated membrane networks, and their role in neuronal differentiation and plasticity.

Generic and Neuron-Specific Organization of the Secretory Network

In vertebrate cells and as described in other chapters of the present book, the GA constitutes a central hub and sorting station for membrane trafficking and stands close to the nucleus at the intersection of biosynthetic and recycling membrane

[1] The terms *secretory proteins* and *secretory cargo* are used thereafter as synonymous to describe all the nascent membrane and secreted proteins that are synthesized in and trafficked through the secretory pathway.

trafficking pathways (see chapters by AK Satoh; J. Saraste and K. Prydz; B. Glick and D.Majaran and L. Lu; A. Colanzi and I. Ayala). The GA plays major functions in the chemical modification of lipids and proteins that are important for their subsequent sorting and their activity. It also plays important roles in the control of actin and microtubule assembly (see chapter by R.M. Rios and M.P.Gavilán) and is the point of convergence of multiple signaling pathways originating from the cell surface (see chapters by L Lu / D Mahajan and A Colanzi).

Although its shape and size vary significantly between cell types and species, the GA, or more strictly speaking Golgi dictyosomes, always display the characteristic and well-delineated ribbon ultrastructure that has mesmerized so many cell biologists. Golgi dictyosomes are encased within pleomorphic endomembrane networks[2] whose boundaries are less strictly defined and that blend with tubulo-vesicular transport systems. These mediate "anterograde" and "retrograde" membrane flux between two opposing poles defined by, on one side, the endoplasmic reticulum (ER)—where secretory proteins and membrane components are synthesized—and, on the other side, the plasma membrane where these components are sent and from which they can be retrieved and recycled (Fig. 7.1a).

As shown primarily in non-neuronal cells, the *cis* cisternae of Golgi dictyosomes interfaces the *cis*-Golgi network (CGN) that blends with the endoplasmic reticulum–Golgi intermediate compartment (ERGIC, hereafter called the intermediate compartment, IC, Fig. 7.1a). As its name suggests, the IC enables bidirectional membrane exchanges between the GA and the ER. The *trans*-Golgi interfaces the *trans*-Golgi network (TGN), which is an integral part of the endosomal network (EN) and mediates bidirectional membrane exchanges with the plasma membrane and lysosomes. As shown by temperature block experiments and inhibition of clathrin and COPI function in plant and animal cells, the CGN and TGN are also interfacing semi-autonomous trafficking hubs, respectively, the biosynthetic recycling compartment (BRC) and the endosomal recycling compartment (ERC) [137, 138]. These mirror compartments normally exist in close proximity to Golgi dictyosomes, typically relatively close to the centrosome, and develop from the IC and endosomal network in the noncompact regions surrounding and connecting Golgi dictyosomes (Fig. 7.1a). In typical globular vertebrate cells, the BRC and ERC are thus usually not easily recognizable from the CGN and the TGN. Yet in some instances, for example, in plants and migrating animal cells, these separate from the Golgi ribbon and engage in polarized exchanges with specific subdomains of the plasma membrane [137, 138].

In neurons, the morphology of the GA varies depending on neuron development and geometry. As we will discuss, Golgi dictyosomes are, at least in mammals, found only in the soma and the most proximal section of some large dendrites and are largely excluded from neuronal protrusions [48, 122] (Fig. 7.1b). In contrast, the ER, IC, and EN are found throughout the neuroplasm, often in close apposition to

[2] The term *endomembrane* is used here in a general sense to designate all intracellular membrane cisternal systems other than mitochondria, and thus not just membranes of the endosomal network.

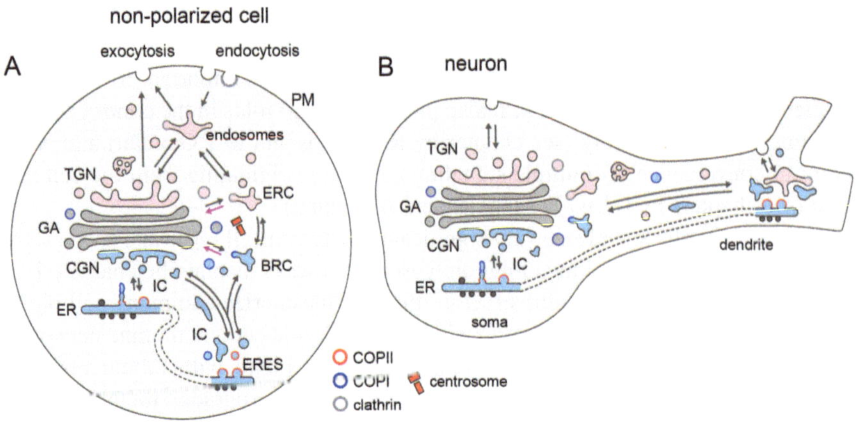

Fig. 7.1 Models of early biosynthetic and endocytic trafficking in nonpolarized cells and neurons. (**a**) In nonpolarized cells, membranes move along microtubules either inward toward the cell center (centripetal) or outward toward the surface (centrifugal). This organization sorts trafficking compartments into "early/peripheral" or "late/central," based on their position, age, and content. Near the centrosome (red), tubular structures of the intermediate compartment (IC) and endosomes gather to form two stable hubs: the biosynthetic recycling compartment (BRC), which returns material to the ER, and the endosomal recycling compartment(ERC), which returns material to the plasma membrane. These hubs are also important for maintaining the central Golgi. Pink arrows highlight transport steps regulated by Arf1 and blocked by brefeldin A. (**b**) In neurons (and other polarized cells), these tubular hubs spread out and are redirected to support polarized trafficking, enabling local exchanges between the IC, endosomes, and the plasma membrane. Abbreviations: PM, plasma membrane; ER, endoplasmic reticulum; ERES, ER exit sites; IC, intermediate compartment; CGN, *cis*-Golgi network; GA, Golgi apparatus; TGN, *trans*-Golgi network; BRC, biosynthetic recycling compartment; ERC, endosomal recycling compartment. Adapted from [137, 138]

one another (Fig. 7.1b). Owing to neuron's gigantic size, we argue in this chapter the presence and close proximity of these compartments at a large distance from the GA favors direct exchanges between the IC, endosomes, and related endomembranes providing multiple routes for nascent proteins to reach the plasma membrane. Although this question is still a matter of debate, evidence suggests that, while seconding the somatic GA in regulating protein homeostasis, these satellite systems have a distinct and specific protein composition and engage in Golgi-independent trafficking, thus in effect acting as rogue Golgi acolytes.

In neurons, as in any eukaryotic cell, membrane trafficking requires molecular motors and cytoskeleton-modifying proteins. Commensurate with their unrivaled size and morphological complexity, neurons are bestowed with cytoplasmic specializations that impact how microtubule-based transport is regulated and where secretory proteins are delivered to the cell surface [1]. Let us first briefly touch on the cytoskeleton before dealing with the neuronal Golgi and associated endomembranes more specifically.

Neuron-Specific Features of the Cytoskeleton and Impact on Membrane Trafficking

The overall morphology of a typical compact and globular interphasic cell is supported and constrained by the radial polarization of the microtubule network (Fig. 7.2a). This "plus-end out" polarization enables centripetal collection and centrifugal redistribution of material between extended membrane systems distributed throughout the cell (e.g., the ER, IC, and EN) and more compact organelles, most saliently the Golgi apparatus, that are located in the perinuclear area in close association with the centrosome as mentioned in the previous section. In contrast, differentiated neurons seem to lack a dominant active centrosomal microtubule-organizing center [34, 152]. Thus, and more prominently than our typical interphasic cell, neurons use diverse microtubule nucleation sites—including the plasma membrane, the Golgi apparatus, and endosomes—and display a mixed microtubule polarity in specific neurite segments ([1], Fig. 7.2b). In the most frequent case in vertebrates, microtubules show a preferential centrifugal ("plus-end out") orientation in the axon and the distal tips of dendrites, while more proximal segments of dendrites contain microtubules with a mixed orientation. Consequently, plus-end and minus-end-directed microtubule motors (i.e., kinesins and dynein) can both be used for membrane transport toward or away from the neuronal cell body and the GA [1], hence expanding the repertoire of possible membrane trafficking itineraries controlled by generic motor proteins.

Another important feature of membrane trafficking in neurons directly results from the tubular morphology of neurites. There, microtubule tracks available for membrane transport run parallel and a few tens of nanometers away from the plasma membrane and its associated subcortical actin cytoskeleton [1, 122]. In our typical interphasic cell, secretory processing requires anterograde and retrograde flux of membranes along axes that are orthogonal to the plane of the plasma membrane. Spontaneous or regulated pauses of membrane transport between distinct processing stations will thus in most cases stall membrane cargo at a distance to the next protein processing station or the cell surface. In neurites in contrast, transport shuttles scan and may hence pause in the immediate vicinity of specific synapses (Fig. 7.2c), thus facilitating local membrane exchanges with the plasma membrane through transport by actin-based transport [175] or phase separation [127].

In synergy with the emergence of these specializations of the neuronal cytoskeleton, the localization and abundance of endomembranes pertaining to the BRC and ERC networks in dendrites and axons likely favored and evolved through strong local interconnections between the two systems.

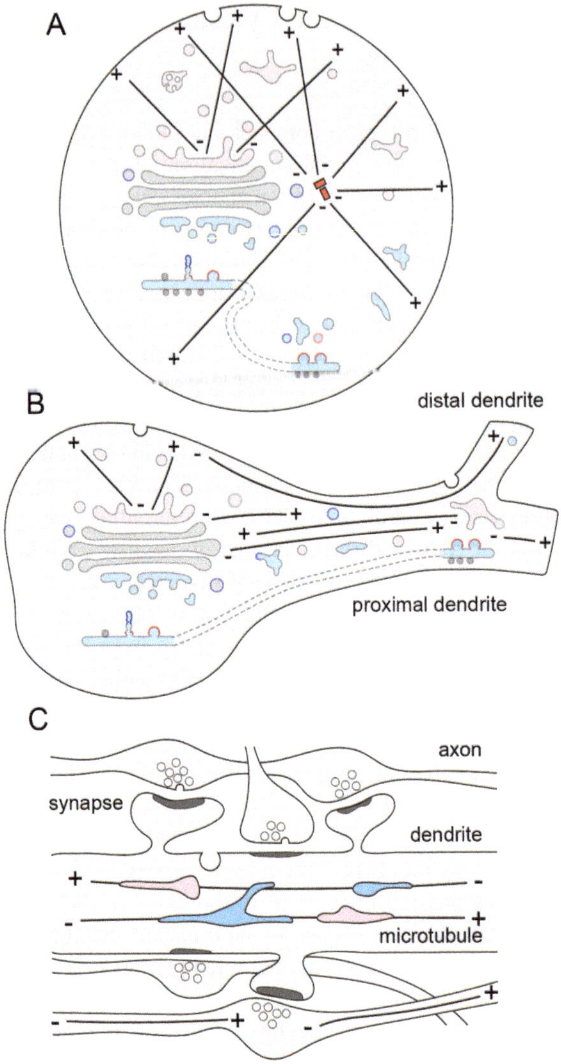

Fig. 7.2 Models of microtubule organization in nonpolarized cells and neurons. (**a**) In nonpolarized cells, most microtubules grow from the centrosome or the Golgi. They extend "plus-end out" toward the cell surface, creating a clear transport system: kinesins move cargo outward (plus-end directed), while dynein (with p150/glued) moves cargo inward (minus-end directed). (**b**) In neurons, the centrosome is no longer the main source of microtubules. Instead, new microtubules can form at multiple sites, including the plasma membrane and internal membranes. Axons (not shown) and distal dendrites mostly have "plus-end-out" microtubules, but proximal dendritic regions show mixed orientations, allowing both kinesins and dynein to operate in opposite directions. (**c**) In dendrites, microtubules align with the main dendritic axis, running just beneath the plasma membrane. This layout allows them to "scan" the neuronal surface and support trafficking near specific synapses

The Endoplasmic Reticulum

Morphology and Neuron-Specific Features

The ER is one of the major sites of protein and lipid synthesis and related biochemical processes such as protein lipidation and N-glycosylation [3, 47]. It is directly involved in a range of other cellular processes, including calcium homeostasis and signaling, mRNA sorting and autophagy, and maintains direct membrane contacts to numerous other organelles throughout the cell, most prominently the neuronal plasma membrane, the GA, endosomes, and mitochondria.

The ER occupies more than a third of the neuroplasm [149, 163, 183] and is the largest neuronal organelle. As shown by 3D electron microscopy (3D-EM) reconstructions in brain slices and photobleaching experiments in live neurons, it constitutes a continuous dynamic membrane network that distributes throughout the entire somatodendritic compartment and axons, including presynaptic terminals and some dendritic spines (Fig. 7.3a, g) [29, 149, 174, 183]. As seen in other cell types, the neuronal ER mainly consists of elongated tubule networks with varying degree of anastomosis and multiple types of flattened cisternae, collectively called ER sheets, that are often fenestrated with ~100-nm-wide nanoholes [142] and are usually studded with ER-bound ribosomes [122]. This ultrastructure is determined by membrane tension [194] and controlled by resident transmembrane proteins and interactions with the cytoskeleton. These either promote membrane flattening through *cis-* and *trans-* interactions in the ER lumen (e.g., CLIMP63 or SigmaR1) or, at the opposite, bend the ER membrane and induce the formation of ER tubules or nanoholes (e.g., reticulons and REEP protein families) or favor tubule branching (Atlastins and Sey1p) [142, 146, 169]. The ER membrane is very dynamic and is continuously reshaped by fenestration, fragmentation, and membrane fusion but, in effect, constitutes a single membrane that delimitates a single continuous internal lumen throughout the neuron.

In the soma, the ER consists of the nuclear envelope and extended ER sheets that are decorated with ER-bound ribosomes and connected by reticulated tubules [122, 183]. ER contacts with the plasma membrane are particularly prominent in the soma and are detected in excess of 10% of the cell body surface area [183].

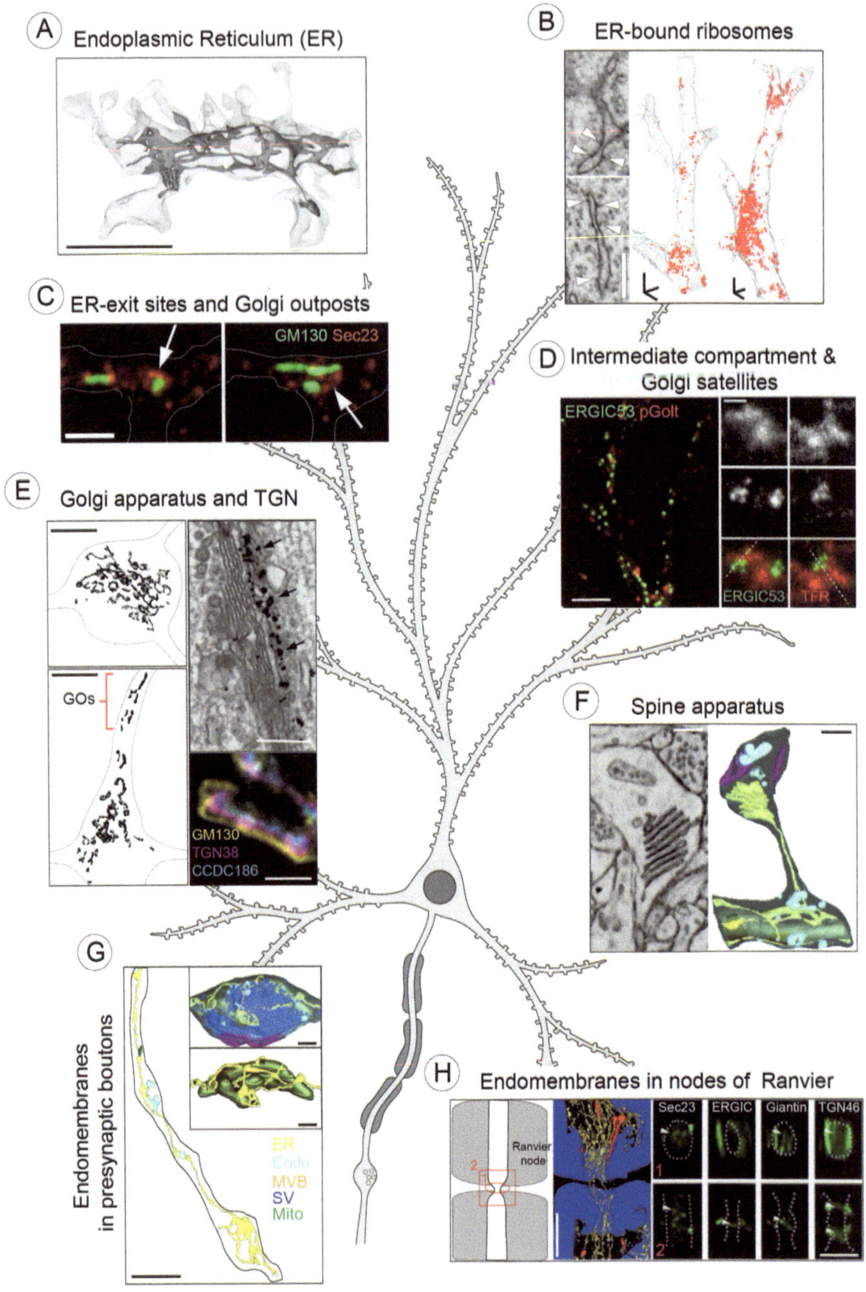

Fig. 7.3 Neuronal Golgi and associated membrane systems – key features. (a, b) Dendritic ER & ribosomes: 3D EM shows ER structure (**a**) and ribosome enrichment at branch points and dendritic spine bases (**b**). Scale bar, 1 μm. (**c**) Golgi outposts & ER-exit sites: Fluorescence images of GM130 (green) and Sec23 (red) highlight dendritic branch points (arrows). Scale bar, 2.5 μm. (**d**) (continued)

In dendrites, the ER consists of long elongated tubules running along the dendritic shaft, which connect specialized hot spots of increased membrane branching and bunching, and in some instances small ER sheets, that are located at dendritic branch points and at the base of dendritic spines [93, 149, 174] (Fig. 7.3a). From these specialized *complexity hot spots*, the ER occasionally protrudes from the dendritic shaft into the head of dendritic spines, sometimes extending in direct proximity to the postsynaptic signaling machinery [29, 149, 174, 183]. In hippocampal CA1 pyramidal neurons, ER membranes are found in approximately half of dendritic spines at any given time but the ER can be seen entering 70% of active dendritic spines over a spine life span [121].

In axons, the ER has a simpler and more monotonous morphology than in dendrites and typically consists of a single or a few elongated tubules with occasional anastomosis and small ER sheets [183], in particular at axonal branch points [117] (Fig. 7.3g). ER anastomosis is regulated by reticulons and seems more frequent and regularly spaced in developing than in mature axons [190]. Finally, an ER specialization with a morphology reminiscent to that of the spine apparatus in dendritic spines (see the next paragraph), the *cisternal organelle*, can be observed at the axonal initial segment [10, 123], the physical barrier that marks the boundary between the axon and the somatodendritic compartments and where axon potentials are generated.

The Spine Apparatus

In the majority of large mature dendritic spines ("mushroom" spines), the ER forms a specialized array of stacked sheets called the spine apparatus [44, 60, 149] (Fig. 7.3f). The genesis and maintenance of the spine apparatus are controlled by the concerted action of actin binding/regulatory proteins, notably MyosinV, synaptopodin, and caldendrin [88, 89]. As seen in EM micrographs, the membrane of the spine apparatus is devoid of ER-bound ribosomes, and it seems that the main function of

Fig 7.3 (continued) Intermediate compartments & Golgi satellites: ERGIC53 (green), pGolt (red), and recycling endosomes (TfR, red) mark local trafficking hubs. Scale bars: left, 5 μm; right, 400 nm. (**e**) Golgi & TGN: GM130 (Golgi) and TGN38/CCDC186 (TGN) visualized by fluorescence and EM. Scale bars: left panels, 10 μm; top-right, 250 nm; bottom-right, 2 μm. (**f**) Spine apparatus: 3D reconstruction shows ER (yellow), mitochondria (green), vesicles (cyan), plasma membrane (transparent green), and synaptic interface (magenta). Scale bars: left, 240 nm; right, 400 nm. (**g**) Axonal ER & presynaptic varicosities: Same color code as (f), with multivesicular bodies (orange) and synaptic vesicles (dark blue). Scale bars: axon, 800 nm; varicosity, 320 nm. (**h**) Nodes of Ranvier: EM/fluorescence images show ER (yellow), mitochondria (red), myelin (blue), ER-exit sites (Sec23), IC (ERGIC), and Golgi/TGN markers (Giantin, TGN-46). Axon outlines are dotted. Scale bars: left panels, 2.5 μm; right panels, 5 μm. Adapted from: (a, b) [29174] with permission from the Society for Neuroscience, copyright 2002 and from Elsevier, copyright 2012; (d) [18,111] with permission from Elsevier, copyright 2016; (e) [72] with permission from Elsevier, copyright 2005 and [164], https://creativecommons.org/licenses/by/4.0/

this organelle is to regulate calcium signaling and incidentally the local trafficking of neurotransmitter receptors during synaptic plasticity [85, 145].

The morphology of the spine apparatus and Golgi-like appearance has long suggested a role in local secretory trafficking [85, 145]. Consistent with this possibility, early EM immunogold labeling studies showed that the spine apparatus was immunoreactive for a number of IC, GA, and TGN proteins, namely ERGIC53, Rab1b, α-mannosidase II, Giantin, Rab6, and TGN38 [126]. However, subsequent studies based on fluorescence-based immunolabeling or expression of tagged versions of some of these proteins have generally failed to observe localization within dendritic spines. For example, in neuronal cultures, TGN38 staining is sparse within dendrites, with the majority of the signal confined to the cell body or Golgi outposts (GOs, see Sect. 5) located within proximal regions of the apical dendrite [18, 24, 43, 67, 72, 73, 78, 79, 101, 103, 111, 129]. Consistently, little, if any, signal was observed for ERGIC53 in dendritic spines either by overexpressing GFP-tagged ERGIC53 or by immunolabeling endogenous ERGIC53 [18, 67]. Thus, it remains controversial which, if any, of the defining factors for the IC or the GA consistently localize to dendritic spines.

As observed by 3D-EM, the majority of spines that harbor a spine apparatus also contain distinct endosomal membranes [93, 149, 174, 183]. Using fluorescent imaging, Bowen et al. [18] observed that spines that contain a spine apparatus always contain recycling endosomes (REs, [18]). It is, however, not clear yet whether these endosomes represent newly formed vesicular structures that arose from the SA or whether they formed independently, either through endocytosis of the spine membrane or RE translocation from the dendritic shaft [29, 85].

Because REs can act as trafficking intermediaries between the GA and the PM (and, for that matter the PM, IC, and other endomembranes as discussed in Sects. 5 and 6), spine REs, regardless of their origin, may receive secretory cargo from the spine apparatus for sorting and delivery to the spine PM. Thus, a definitive role for the spine apparatus in forward biosynthetic trafficking remains to be demonstrated.

Engagement of the Dendritic and Axonal ER in the Local Synthesis of Secretory Proteins

While it has long been thought that the bulk of secretory protein synthesis occurred in the soma, components of the translocon complex and ER-bound ribosomes are also found in distal dendrites [94, 149, 160, 174], indicating ongoing membrane protein synthesis at these locations (Fig. 7.3b). Similarly, ER-exit sites (ERES) are more abundant in the soma but are distributed throughout dendrites, where they are enriched at ER complexity hot spots [6, 174]. ERES are largely excluded from axons but are occasionally observed in presynaptic terminals, suggesting local ER-exit of nascent presynaptic membrane proteins [102]. As indicated by experiments performed in isolated neurites and purified synapses, membrane proteins can indeed be locally synthesized and trafficked to the surface in presynaptic terminals and dendrites [54, 58, 80, 153]. Surprisingly, EM studies have to this day failed to detect ER-bound ribosomes in presynaptic terminals [102, 183]. In contrast, ER-bound ribosomes were described on small ER sheets in 3D-EM reconstructions of axonal branch points [117]. However, recent super-resolution fluorescent imaging indicates that the axonal ER (i.e., mostly ER tubules) bind ribosomes thanks to the ER integral protein P180/RRBP1 [37, 90]. Corresponding "contact sites" (here dubbed so to avoid confusion with canonical ER-bound ribosomes) are distributed along the axon, including presynaptic terminals and axonal branch points, consistently with aforementioned EM data [117]. As shown by protein puromycilation and protein knockdown experiments, these ER-associated ribosomes appear to be sites of protein synthesis, seemingly accounting for up to 30% of total protein synthesis occurring in the axon in cultured neurons [90]. However, it is unclear yet whether these ribosomes correspond to proper ER-bound ribosomes, that is, are indeed actively engaged in the synthesis of secretory proteins and translocation in the ER lumen.

As mentioned in the previous section, the spine apparatus is devoid of ER-bound ribosomes. Nevertheless, EM studies have described the presence of polysomes in close proximity of the dendritic ER and the spine apparatus at the base of dendritic spines [151]. However and as we have alluded above, the cytoplasmic face of the ER membrane is known to be an important platform for the sorting and translation of mRNAs encoding nonsecretory proteins [74, 81, 104, 133]. The presence of polysomes in close proximity to the ER and the spine apparatus thus does not equate to an engagement in a local synthesis for proteins to enter the ER.

Compartmentalized Protein Processing in the Neuronal ER

While the ER is a continuous membrane, it does exhibit some degree of compartmentalization. Prior to ER-exit, membrane and luminal proteins can explore large fractions of the neuronal ER by lateral diffusion [59, 174]. This free diffusion is, however, counteracted by binding of nascent proteins to the cytoskeleton and by confinement by the topology of the EM membrane [162, 174]. For example, as shown in dendrites, ER complexity hotspots act as diffusion traps and increase local dwell time of proteins at dendritic branch points and near synapses [174]. This confined diffusion opposes long-range ER-cargo exchanges between the soma and dendrites [18, 174] and favor local ER-exit, as was shown by local enrichment of COPII protein components upon acute increase in nascent secretory protein levels [174]. Interestingly, as mentioned above, ER-bound ribosomes are also frequent at ER complexity hot spots, suggesting synergy between local protein synthesis and local ER-exit [174]. Moreover, these sites of ER-bound ribosomes and ERES enrichment are spatially coordinated with clusters of IC and recycling endosomes where membrane cargo accumulates after exiting the ER [18, 29, 149, 174, 183]. As seen for the

SNARE protein VAMP2 [174] and other secretory proteins [87], insertion into the dendritic PM occurs preferentially at branch points, suggesting that the entire secretory network can be locally organized.

As mentioned above, ER ultrastructure morphs over time scales of seconds to minutes through regulation by proteins that control ER membrane curvature, fission, branching, and interactions with the cytoskeleton. All these factors may strongly affect nascent protein dynamics and large-scale exchanges within the neuronal ER. As shown, for example, for CLIMP63-dependent binding of the ER to microtubules or interactions of nascent neurotransmitter receptors in the ER to cytoplasmic motors [162, 174], multiple signaling pathways control ER morphology and nascent protein dynamics in this organelle, enabling compartmentalized processing of nascent proteins to satisfy local demand. Nascent proteins that have correctly folded and passed quality control then exit the ER at ERES from which they are then transferred to the intermediate compartment (IC).

ER–Golgi Intermediate Compartment

The IC consists of a dynamic array of pleomorphic tubulovesicular elements that sorts and bidirectionally shuttles lipids and proteins between the ER and the GA (see [138] for an excellent review). The IC is an essential recycling platform that allows the retrieval of ER proteins thanks to the KDEL receptor. It is also directly involved in a range of other processes, including ER-exit, folding and quality control of nascent proteins, O- and N-glycosylation, glycosaminoglycans, and lipid synthesis and autophagy [138].

At the interface with the ER, the IC receives nascent proteins and lipids exiting the ER. As recently shown by EM in cell lines, COPI coats, in addition to COPII, are detected in 60–75% of ERES [178]. Upon synchronized release of cargo from the ER, the two coat systems assembled on distinct subdomains of interwoven membrane structures forming at ERES, which sometimes extend over several micrometers thanks to interaction with Dynactin/dynein [178]. These membrane intermediates are then released by membrane fission to generate COPI-positive transport intermediates that traffic toward the centrosome and the CGN [178]. As shown in a variety of cell types in the presence or absence of synchronized cargo release from the ER, IC elements generated at ERES may also coalesce and form relatively stable structures, very close (see notes on the BRC in the introduction), and sometimes further away from the centrosome [137, 138]. Interestingly, while ER-exit seems to directly generate or recruit available IC membranes, COPII assembly is highly plastic and distinct modes of COPII assembly have been described for larger cargo such as collagen [132, 140]. This thus indicates that local ER-exit of distinct cargo informs the formation of specific and spatially segregated ERES and associated IC elements.

At the interface with the GA, IC membranes are incorporated in the CGN and the *cis*-Golgi thanks to interaction between Rab1, bound to IC membranes, and Golgi tethering proteins such as p115 and GM130. Conversely, as part of the CGN, the IC receives membranes from the *cis*-Golgi, enabling retrograde trafficking and recycling of lipids and proteins back to the ER [138].

In neurons and as shown by the distribution of some of its core components—that is, Rab1, Rab2, COPI, GS27, and ERGIC53—the neuronal IC is distributed throughout dendrites and axons (Fig. 7.3d, h) [18, 36, 57, 67, 91, 110]. In the dendritic shaft, the IC comprises small tubulovesicular carriers that rapidly move along microtubules and larger and more stable structures that rapidly morph but persist over tens of minutes at specific locations, most notably at dendritic branch points [36, 67].

In dendrites, IC seems to be exclusively found in the dendritic shaft and is excluded from dendritic spines. In 3D-EM micrographs, the dendritic IC closely resembles endosomes [183], which makes it difficult to unambiguously distinguish the two types of membranes [183]. The xx is further complicated by the fact that, as shown by STORM,[3] membranes pertaining to these two systems are often found in very close proximity, potentially engaging in direct transfer of lipids and proteins (Fig. 7.3d) ([18]; see Sect. 7).

As documented by thermosensitive, pharmacologically or optogenetically synchronized release of secretory proteins (see also Sect. 8), dendritic proteins exiting the ER accumulate in clusters of IC membranes that are spatially coordinated with hot spots of ER complexity [18, 25, 67, 174]. From there, nascent cargo may be processed locally or distributed over large segments of the somatodendritic compartment through microtubule-dependent transport [67, 71]. Interestingly, this transport is directly regulated by synaptic activity through control of microtubule motors. For example, as described for Kif17, increased synaptic activity induces a CaMKII-dependent phosphorylation of the kinesin and prevents its binding to IC membranes, releasing cargo from microtubule tracks. Synaptic activity (and elevated cytoplasmic calcium concentration) hence prevents long-range transport along dendrites but does not prevent the exocytosis of nascent proteins to the dendritic surface [67] (see also [42]). Increased synaptic activity in specific segments of dendrites may hence regulate local surface trafficking of nascent proteins from stable IC clusters.

As mentioned above, some dendritic IC membranes are found in very close apposition to REs [18]. Expression of dominant-negative variants of specific rab GTPases required for anterograde trafficking from the TGN (Rab8), recycling to the cell surface via REs (Rab11), indicates that nascent proteins in the IC access the neuronal surface via REs. Interestingly, while both Rab8 and Rab11 dominant negative mutants inhibit cargo exocytosis at relatively long time intervals after synchronized ER-release (4 h), only Rab11 inhibits exocytosis at short time scales (2 hours). Consistently, transport to the cell surface at these early time points is not impaired after Golgi disruption with brefelfin A (BFA) [18]. These experiments thus indicate

[3] STochastic Optical Reconstruction Microscopy. A fluorescence imaging technique based on the detection of single fluorophore emission allowing their location with a xy resolution close to 20 nm.

that nascent proteins in the IC directly access the neuronal surface via REs and hence bypass the GA.

Fewer data on the axonal IC are available. One of the most compelling examples of a specialized axonal differentiation of the IC comes from studies of nodes of Ranvier in dorsal root ganglion (DRG) neurons [57]. In these neurons, two long axons emanate from the cell body and project to the body periphery and the spinal cord, giving their archetypical pseudo-unipolar morphology to these cells. As documented by immunofluorescence (IF) labeling and overexpression of fluorescently tagged proteins, both nodal and internodal regions of the axonal shaft contain discrete ER accumulations of ER membranes immuno-positive for KDEL and ERP72 [57]. In contrast, COPII components (Sec23), p58/ERGIC53 and Golgi markers (but see next section), are detected only in nodal regions, suggesting a local engagement in the local trafficking of nascent membrane proteins. Consistently, ligand-controllable chimera of the NAV channels β2 subunit were shown to access the axonal surface after release of ER-exit blockade in DRG axons isolated from primary cultures [57].

Golgi Apparatus, Golgi Outposts, and Related Satellite Systems

Morphology and Distribution of Golgi Membranes in Neurons

The GA is predominantly localized close to the nucleus in the neuronal cell body. GA morphology varies with that of the dendritic tree but the GA is usually polarized toward the largest dendrites [72, 129](Fig. 7.3c, d and e). In the very asymmetric dendritic tree of pyramidal neurons, it often extends prominently in the proximal section of the apical dendrite. There, it occasionally forms smaller (0.5–2 micrometers long) units termed Golgi outposts (GOs) that can be found up to 30 micrometers away from the basis of the dendrite (Fig. 7.3d, e; see also later sections), a number that is surprisingly constant across neuron types and studies. However, as documented by classic EM studies, the most proximal segment of apical dendrites shares common ultrastructural features with the cell body [48], making it challenging to unambiguously assign the end of the soma and the beginning of the dendrite without corresponding EM micrographs.

As shown in classic studies in plant cells transduced with viruses and overexpressing membrane proteins, the volume of the GA is modulated by and directly reflects cell secretory load [61]. In the hippocampus and depending on neuron types, the volume of the GA seems to vary proportionally to total dendritic length, as seen, for example, in granule cells of the dentate gyrus and CA3 pyramidal neurons [68]. In contrast, however, GA volume in CA1 pyramidal neurons is much smaller than dendritic size would predict [68]. In other terms, only some neuron types seem to upscale the volume of the GA to match the size of their dendritic tree. It is thus likely that distinct dendritic lengths impose commensurate demands on secretory functions but affect Golgi morphology in a cell type-specific manner. Distinct ratios of GA volume to dendritic length may thus reflect a varying need for processing in the GA and, conversely, a varying reliance on dendritic processing pathways bypassing the somatic Golgi [68].

Dendritic Golgi Outposts and Related Satellite Systems in Vertebrates

As reported by a number of fluorescence imaging studies in untransfected neurons (see Table 7.1), the majority of dendrites of central neurons lack detectable Golgi membranes, at least defined by enrichment of classical GA and TGN markers (Table 7.1), in particular GM130 (*cis*-Golgi), mannosidase II (*med-trans*), GRASP55 (medial-*trans*), giantin (medial) and TGN38 (TGN), all of which colocalize with GM130 and are mostly detected in or in very close proximity to the somatic GA. As shown by EM, these dendritic GOs present the canonical ultrastructure of dictyosomes [72]. As quantified after GM130 staining in rat cortical and hippocampal neuron cultures and in the brain [33, 72], GOs are found in only ~5% of dendrites in basal conditions (i.e., ~20% of hippocampal neurons contain GOs in one of their approximately four to five central dendritic processes), typically within the first 30 micrometers of the apical dendrite as mentioned above [33, 72].

In rare instances, GOs have been observed after labeling of GM130 in more distal dendritic segments of some CA1 hippocampal neurons in close proximity of hotspots of mRNAs encoding subunits of the AMPA receptor subunits (one of the primary glutamate receptors throughout the brain)[33]. This thus indicates that genuine dendritic GOs may be more frequent in specific populations of pyramidal neurons where dendritic synthesis of AMPA receptors is also upregulated [153].

Of note, some studies have reported detection of low levels of GM130 immunoreactivity in discrete punctate structures dispersed in dendrites (e.g., [111]) and axon growth cones [110]. However, the persistence of this faint immunoreactivity despite severe reduction of GM130 expression by RNAi [111] questions the physiological relevance of this signal.

Consistent with these fluorescence imaging studies, EM data—including 3D reconstructions of large segments of dendrites in primary cultures or in the brain [149, 174, 183]—have insofar failed to detect Golgi dictyosomes. Large datasets of serial 3D EM imaging of volumes of hundreds of micrometers of cross section in the mouse cortex are now publicly available at a relatively high spatial resolution [83, 157, 170]. Assisted with machine learning, it should now be possible to rigorously assess the occurrence of Golgi dictyosomes in large numbers of neurons and dendrites. We anticipate that these analyses will confirm that, in mammals at least, generic Golgi dictyosomes are largely excluded from dendrites and axons.

Table 7.1 Target organelle and distribution of "Golgi" markers in neurons

protein		target organelle	overexpression		endogenous expression		sample type / species : drosophila, chicken, rodents, hIPS			reference
full name	short		Tag (XFP)	EM	IF	EM	primary culture	*in vivo*	present in distal dendrites	
Small COPII coat GTPase	SAR1A	ERES, IC			x		x		x	1
Coat Protein Complex II	COPII	ERES, IC			x		x		x	1
Golgi-specific brefeldin A-resistance guanine nucleotide exchange 1	GBF1	ERES, IC, *cis*-GA	x				x		x	2
ADP-ribosylation factor 1	ARF1	ERES, IC, *cis*-GA	x				x	x	x x	3,4,5
Ras-related protein Rab-1A	Rab1a	IC, *cis*-GA	x			x	x	x	x	6
Golgi reassembly-stacking protein of 65 kDa	GRASP65	*cis*-GA	x				x			7
cis-Golgi matrix protein 1	GM130	*cis*-GA	x					x	x	8, 9
cis-Golgi matrix protein 1	GM130	*cis*-GA			x (#)			x		5, 10, 11, 12
cis-Golgi matrix protein 1	GM130	*cis*-GA			x	x	x	x		6, 7, 13, 14, 15, 16, 17,18, 19, 20,21
Golgin subfamily B member 1	Giantin	*cis*-medial-GA			x	x		x		22, 23
Golgin subfamily B member 1	Giantin	*cis*-medial-GA			x		x			21, 24, 25
CTR433		*cis*-medial-GA				x		x		26
α-mannosidase II	ManII	medial-GA	x				x x	x x	x x	5, 6, 8, 9, 10, 11, 21, 27, 28, 29, 30, 31, 32, 33, 34, 35, 36, 37, 38, 39, 40, 41, 42, 43
α-mannosidase II	ManII	medial-GA				x		x		22, 44
galactosyl-transferase	GalT	*trans*-GA	x				x	x x	x x	4, 8, 13, 18, 25, 32, 36, 41, 42, 45, 46
galactosyl-transferase 2	GalT2	*trans*-GA	x				x			16, 45
N-acetylgalactosaminyltransferase 2	GalNacT2	*trans*-GA	x				x	x	x	40, 47
pGolgi*	pGolgi*	*trans*-GA	x				x	x	x x	3, 6, 11, 45
Golgin subfamily B member 1	Golgin97	*trans*-GA	x				x		x	48
Lactosylceramide alpha-2,3-sialyltransferase	ST3GAL5	*trans*-GA	x				x		x	6
ST8 alpha-N-acetyl-neuraminide alpha-2,8-sialyltransferase 2	ST8SIA2	*trans*-GA	x				x		x	21
pGolt*	pGolt*	TGN	x	x			x		x	2, 6, 19, 21
Trans-Golgi network integral membrane protein 38	TGN38	TGN			x		x	x		4, 6, 15,16, 17, 18, 19, 24, 49, 50
Trans-Golgi network integral membrane protein 38	TGN38	TGN				x		x	x	26,51
Trans-Golgi network integral membrane protein 38	TGN38	TGN	x				x		x	2
Trans-Golgi network integral membrane protein 46	TGN46	TGN			x			x		23
Calcium-binding protein 7	Caln 2	TGN	x				x		x	6

#, in cells over-expressing membrane protein reporters

References: 1, Aridor et al. [6]; 2, Cornejo et al. [32]; 3, Hasdemir et al. [69]; 4, (Jeyifous et al. [78]; 5, Wang et al. [173]; 6, Mikhaylova et al. [111]; 7, Horton et al. [72]; 8, Zhou et al. [193]; 9, Cheng et al. [26]; 10, Mukherjee et al. [115]; 11, Li et al. [97]; 12, Mukherjee et al. [114] 13, Horton and Ehlers [71]; 14, Chua and Tang [27]; 15, Merianda et al. [110]; 16, Quassollo et al. [129]; 17, Islam et al. [73]; 18, Bowen et al. [18]; 19, Fabbretti et al. [43]; 20, Bollmann et al. [14]; 21, Andres-Alonso et al.[4]; 22, Pierce et al. [126]; 23, González et al. [57]; 24, Chen et al. [24]; 25, Jones et al. [79]; 26, Gardiol et al. [52]; 27, Ye et al. [187]; 28, Zheng et al. [192]; 29, Satoh et al. [139]; 30, Chandramouli Iyer et al. [23]; 31, Ori-McKenney et al. [119]; 32, Nguyen et al. [118]; 33, Copf [30]; 34, Arthur et al. [7]; 35, Lin et al. [99]; 36, Yalgin et al. [184]; 37, [28]; 38, Kelliher et al. [84]; 39, Fu et al. [50]; 40, Yang and Wildonger [185]; 41, Park et al. [120]; 42, Du et al. [40]; 43, Bhattacharjee et al. [11]; 44, Torre and Steward [160]; 45, Valenzuela et al. [162]; 46, Tozer et al. [161]; 47, Mitchell and Wildonger [112]; 48, Wang et al. [171]; 49, Lowenstein et al. [101]; 50, Ma and Morris [103]; 51, [99]

Differing from this generic picture in cortical and hippocampal neurons, long dendritic protrusions of the GA have been reported in some motoneurons of the rat spinal cord [52]. These protrusions have been described after immunolabeling of TGN38 and CTR433 stainings in motoneurons of the spinal cord and extend up to at least 70 micrometers away from the nucleus [52]. Consequently as shown by EM, canonical dictyosomes (as well as ribosomes-studded ER membranes and membrane protein mRNAs) can hence be found in these dendritic sections, sometimes just a few hundred nanometers away from inhibitory synapses [52].

In contrast to what is seen in untransfected neurons, studies of mammalian neurons expressing chimeric or intact Golgi proteins tagged with fluorescent reporters have reported their distribution throughout the dendritic arbor and axons (see Table 7.1). In a comparable manner, dendritic GOs containing GM130 (the endogenous protein) are also more frequently observed after neuron transfection with exogenous membrane proteins used as cargo reporters [72, 174]. As discussed for the IC, GOs also engage in the local trafficking of cargo after ER-exit [67, 72, 174].

As shown in Sect. 5, GOs observed with exogenous fluorescent proteins seem to be more frequent in invertebrates, notably in flies, but, as we will discuss, their identity is still debated. Indeed, evidence suggests that they may in fact in some instances (if not all) correspond to endosomal structures.

However, a significant caveat to these observations is that exogenous proteins may localize to dendritic organelles due to interference of fluorescent protein tags with normal trafficking or high expression levels exceeding the capacity of target organelles. For example, some Golgi reporters including the widely used pGolgi-GFP chimera are only detected in dendrites at high expression levels [4, 69, 111, 162]. These results have thus been difficult to reconcile with the very low or the absence of staining with antibodies against numerous endogenous GA organizing proteins (see Table 7.1). However, it is possible that dendritic Golgi or functionally related membranes may lack the defining features of the somatic GA, including *cis* and *trans* stacked-membrane structures, and thus may not require many of the structural proteins that have long been used as generic GA markers.

Alternatively, a distinct organelle may fulfill the role of the GA at dendritic sites. For example, as described in Sect. 5, the IC is present in dendrites and nascent secretory proteins released from the ER accumulates in ICs with similar kinetics as the somatic GA [18, 67]. Because the GA houses key enzymes involved in refining the glycosylation patterns of membrane proteins, it will be important to understand whether these refinements can occur in dendrites lacking GA. So far and as discussed below, only a few studies have addressed that question. Nevertheless, interesting insights on this topic have come from the description of synthetic probes derived from TGN proteins as we will now discuss in the next sections.

Dendritic Golgi Satellites

In mammalian neurons, overexpression of GFP-tagged versions of the TGN protein calneuron1 revealed 0.5–1-micrometer-wide structures reminiscent of GOs distributed throughout the entire dendritic tree [111]. As the expression of this tagged-TGN reporter impairs surface expression of glutamate receptors in co-transfection experiments, a synthetic version of the related protein calneuron2 was generated to encompass only its TGN targeting sequence, which was fused to an efficient ER-exit sequence and a mCherry tag. As seen in vivo and in cultured neurons, the resulting chimera, pGolT, localizes to pleomorphic membranes centered on the TGN but have a broader distribution in the cell body [4, 111].

In dendrites, pGolT labels membranes distributed throughout the dendritic shaft, including distal dendritic segments (Fig. 7.3d), and that are presumably found also in the axon [70]. These structures, called Golgi satellites (GSs), correspond to mobile and more stable elements and are observed even at relatively low expression levels of the reporter protein. In dendrites, GSs are observed in the vicinity but do not colocalize with IC markers (i.e., Rab1 and p58). In contrast, they extensively colocalize with clathrin and Vps35 (vacuolar protein sorting ortholog 35), a component of the endosomal retromer complex that plays important roles in regulating retrograde trafficking from sorting endosomes to the TGN or recycling to the plasma membrane [144, 180]. Dendritic GSs also partly colocalize with membranes containing Rab6, a rab protein primarily involved in anterograde trafficking from the TGN but that also controls retrograde transport from endosomes [49, 106]. Consistently, GS engage in the anterograde trafficking of ligand-controlled cargo after synchronized ER-release [4], thus indicating that GS (akin to the IC and REs) are also visited by nascent membrane proteins on their way to the plasma membrane [111].

Correlative light electron microscopy experiments in transgenic mouse lines in vivo showed that GSs correspond to pleomorphic arrays of elongated saccules and tubules, an ultrastructure reminiscent of the TGN and endosomes [4]. Indeed, dendritic GSs also engage in the retrograde trafficking of CDMPR (cation-dependent and cation-independent mannose-6-phosphate receptors) from the neuronal surface [4]. Thus, consistent with the detection of the retromer and clathrin on GSs, these structures are likely involved in cargo recycling after endocytosis. As discussed in Sect. 5, REs are closely associated with IC membranes and appear to be directly involved in local secretory exocytosis after ER-exit [18]. Interestingly, GSs also closely associate with REs as indicated by live cell imaging of Rab11-GFP [111]. Yet, colocalization of GSs and REs is only transient and is only seen for a fraction of the dendritic Rab11-positive tubulovesicular network. Also worth mentioning, GSs or related TGN-derived structures engage in the local recycling of ECM components after endocytosis, as was shown by localization of internalized Tenascin R in dendritic endomembranes immuno-reactive for TGN38 [35].

The exact sequence of secretory cargo progression through dendritic IC, GS, and REs is insofar not clear. In particular, more work will be needed to decipher which (if not all) of these compartments are directly involved in local membrane exocytosis. Yet, altogether these results suggest that GSs correspond to trafficking hubs, endosomal in nature because of their affiliation to the TGN, but that sit at the interface between the BRC and the ERC networks [137, 138]. One may thus suggest that *TGN satellites* would be a more fitting name than Golgi satellites. Time will tell.

Interestingly, GSs present a strong reactivity for specific lectins and contain high levels of PSA-NCAM [111], a heavily sialylated adhesion molecule with important functions in the maturation of neuronal circuits. Interestingly, St8Sia2, a sialyl-transferase responsible for the poly-sialylation of NCAM, also localizes to GSs, strongly suggesting that GSs engage in N-glycan processing [111]. Consistently, imaging after bioorthogonal metabolic labeling with sialic acid precursors has shown that GSs contain high levels of sialylated glycoproteins (i.e., proteins at terminal stages of O- or N-glycosylation). However, the incubation times that were used to label neurons with tagged precursors were relatively long (24 hours) [4], precluding a clear resolution of their site of synthesis. Thus, it is possible that detected "nascent" sialylated glycoproteins found in GSs may have been synthesized in other compartments before residence in GSs.

Similarly, fluorescently tagged version of St3Gal5, a sialyl-transferase specifically involved in the synthesis of glycolipids, GM3 gangliosides [165] perfectly colocalizes with pGolT [111] and has also been used as a GS reporter both in dendrites [111] and axons [70].

Extra-Golgi Localization of the Golgi Glycosylation Machinery

The modification of lipids and proteins through N- and O-glycosylation is one of the most important functions of the GA (e.g., see chapter by H Yagi). As shown in Table 7.1, enzymes of the N-glycosylation pathway and, in particular, Mannosidases-2 and galactosyl-transferases that are involved in the maturation of N-glycoproteins after ER-exit [165], are commonly used as markers of Golgi membranes. Thanks to the work of Jürgen Roth [135], Kelley Moremen [113], and many other groups, the localization and role of these enzymes in the maturation of glycoproteins in the GA have been solidly established. Yet, for the most part this was demonstrated primarily in cell lines [113, 156, 165]. Interestingly, seminal studies in the 1990s and more recent work have also reported that these enzymes can in multiple instances be found in other organelles as well, most prominently in the ER, IC, endosomes, and the plasma membrane [12]. For example, *N*-acetylgalactosaminyl transferases (GalNac-Ts), an enzyme involved in the synthesis of mucin O-glycan, are known to redistribute from the GA to the RE upon cell exposure to growth factors [53]. Along the same line, initiation of O-glycosylation of some viral proteins has been suggested to take place in the IC, making it strongly reactive for cognate lectins [91, 138, 159]. Similarly, some galactosyl transferases that are involved in N-glycan maturation (GalNAcT2) cycle between the Golgi and the IC [76]. As shown for type I and II Golgi mannosidases, these enzymes are primarily localized in the medial and *trans* cisternae of the GA in cell lines. Nevertheless, their localization within the GA varies significantly depending on cell type and both Man I and Man II have also been detected within secretory granules and at the cell surface of enterocytes, pancreatic acinar cells, and goblet cells [166]. These results thus indicate that the distribution of Man I and Man II within the GA overlaps considerably but that their distribution is more variable and less compartmentalized than generally assumed [166]. Intriguingly, ectopic localization of glycosylation enzymes appears to be particularly salient in primary cell cultures, cells obtained ex vivo, and in tissue sections [12]. Of note and to close the present point, the endogenous form of St3Gal5, the GM3 synthase that localizes to dendritic GSs, was previously shown to localize to dendrites and axons of cultured neurons and PC12 cells protrusions upon differentiation by growth factors [150]. In conclusion, what is commonly inferred about the distribution of the glycosylation machinery in neurons, more specifically its strict localization to Golgi membranes, may thus not always be true.

Axonal Golgi Membranes

As we have already mentioned, a number of membrane and secreted proteins are thought to be locally synthesized in dendrites and axons. Yet, the bulk of neuronal secretory proteins seems to traffic through the somatic GA and TGN from where secretory cargo is distributed to dendrites and to the axon. From a cytoarchitectural point of view, no physical barrier separates the dendrites from the soma and only the axon is physically isolated from the somatodendritic compartment. Once loaded in vesicles budding from the TGN, secretory cargo destined to the axon follow a specific path and are assigned to specific motor proteins for long-range transport in a specialized transitional area at the beginning of the axon called the pre-axonal-exclusion zone (PAEZ)[45]. After crossing the PAEZ, transport vesicles arrive at a second sorting zone known as the axon initial segment (AIS). The AIS literally functions as a selective barrier preventing entry of somatodendritic cargos due to their selective association with specific kinesins [77, 96, 148]. However, there are a few exceptions to this rule as, for example, neuroligins, postsynaptic cell adhesion molecules, were shown to be inserted in the plasma membrane of the AIS [17]. This odd behavior suggests that the actual boundary between the axonal and somatodendritic compartments lies at the distal end of the AIS rather than at the PAEZ for particular cargos.

Another more recent immunofluorescence (IF) study showed the presence of Golgi markers in small axonal organelles in vivo and in vitro. Indeed, it was shown that nodes of Ranvier of DRG neurons contain discrete membrane structures immuno-positive for the *cis*/medial-Golgi (Giantin) and TGN (TGN38) markers were found at the nodal and internodal level [57](Fig. 7.3h). This study also showed the presence of markers of ER and ERES (KDEL, ERP72), ERGIC (pERGIC-53/p58) in the immediate vicinity of these presumed Golgi membranes [57]. These findings thus suggest that the nodal and paranodal regions of some myelinated axons contain an autonomous machinery for secretory trafficking. The same team also reported the colocalization of TRPM8—an ion channel involved in cold sensations—with axonal GSs labeled with pGolt (see Table 7.1) [32, 111]. Interestingly, TRPM8 also colocalizes extensively with the lysosomial protein LAMP1 and both proteins are thought to visit GSs before being sent out for unconventional secretion from lysosome-related structures [32].

Also indicative of a genuine continuity between the Golgi network and the axonal endo-lysosomal system, a recent study in vivo and in cultured neurons showed that generic mature lysosomes are restricted from axons, likely through the action of minus-end-directed microtubule motors ([98]). In contrast, nondegradative TGN transport carriers (TCs) containing lysosomal components but not fully acidified are observed in axons where they engage in bidirectional transport. These structures, which are upregulated in dystrophic axons in mouse AD models, thus likely represent bidirectional axonal TCs in axons previously thought to be lysosomes [99]. These observations thus support the notion that TCs, rather than lysosomes, mediate anterograde transport of lysosomial components in axons [99]. From a larger perspective, these results document that axonal endomembrane display hybrid identities, in the present case affiliating these structures to TGN, late endosomes, and lysosomes. This phenomenon also likely occurs in dendrites where degradative compartments are confined to proximal dendrites [98).

To conclude this section, discrete membrane structures containing Golgi markers seem to be present in axons and could potentially perform post-translational modification and secretion functions locally [32, 57, 58]. However and as noted for dendrites, Golgi dictyosomes seem to be absent from axons. More studies will be needed to clarify the actual structure and functionality of axonal Golgi-related structures. Finally, there remains the question of what the term GA actually encompasses. It is well established that the so-called classical GA, which in neuron lies in the soma, consists of a set of one or more dictyosomes encased in zones of vesicle budding and fusion. However, it is much less clear whether this term should encompass the TGN, which represents a pleomorphic network of tubulovesicular structures at the *trans*-side of Golgi dictyosomes, in a similar way that the IC mediates communications between the ER and the *cis*-side of Golgi dictyosomes. This nuance is worth highlighting as the TGN is a compartment in permanent exchange with the endo-lysosomal system, potentially leading to some confusion in the interpretation of experimental data.

Polarization of the Golgi Apparatus and Satellite Systems during Neuronal Morphogenesis

Polarization of Golgi Membranes During Neuronal Morphogenesis

The acquisition of neuronal polarity and the growth of long and ramified axons and dendrites result in the expansion of the surface area of the neuronal membrane over up to 4 orders of magnitude and hence puts a colossal demand on the secretory system [68]. Accordingly, neuronal morphogenesis is accompanied by a reorganization of the distribution and morphology of secretory organelles where membrane components are synthesized and trafficked. This reorganization is particularly salient for the GA but also occurs for other endomembranes that may experience even more dramatic reorganization during neuronal morphogenesis.

As discussed in Sect. 2, the centriole and the GA are important microtubule-organizing/nucleating centers from which microtubules radiate from the cell center toward the cell periphery, enabling optimal membrane collection from the ER and IC and redistribution throughout the cell [1]. The centrosome-Golgi vector (CGV) thus constitutes a core axis that structures cellular polarization and whose orientation varies with asymmetric cell growth. Such variations in CGV orientation are particularly important when neurons acquire their axon-dendrite polarity, with first the specification and elongation of a single axon and then the growth of multiple dendrites [9, 136, 155]. In the developing cortex in vivo, the CGV reorients itself as developing pyramidal neurons migrate into the cortical plate [20]. Initially oriented toward the basal pole (rear) of the cell, the CGV subsequently reorients toward the opposite pole (front) to favor asymmetric growth of the leading process used for radial migration that subsequently becomes the apical dendrite [9, 72, 136]. In cultured neurons—that is, in the absence of the strong orientation constraints existing in the developing brain—the orientation of the CGV is regulated in a cell-autonomous manner and determines the site of axon outgrowth [20].

In neuronal cultures and in the brain, the polarization of the somatic GA precedes asymmetric dendritic growth and does not occur in neurons with symmetrical dendritic trees, for example, GABAergic interneurons [72]. This polarization is associated with a bias in post-Golgi trafficking toward the apical dendrite [72]. Experimental fragmentation of the GA by overexpression of the Golgi stacking protein GRASP65, and resulting lack of dendritic polarization of the GA, prevents the preferential growth of the apical dendrite and does so without diminishing the overall rate and extent of dendritic outgrowth [72]. Similar results documenting a close relationship between Golgi polarization and asymmetric dendritic growth were obtained in multiple model systems and neuron types [154]. Thus, Golgi polarization and, presumably, resulting vectorial secretory trafficking seem required to sustain asymmetrical neuron growth.

A comparable polarization process was described downstream of synaptic transmission in spiny stellate neurons of the mouse barrel cortex[4] [116]. Synaptic activity is a key determinant of dendritic morphology during the formation and refinement of neuronal circuits in the developing brain [182]. In specific layers of the barrel cortex, the GA of spiny stellate neurons loses its original apical positioning after the initial growth of the apical dendrite and redistributes laterally to promote asymmetric dendritic growth toward a single cortical barrel [116]. As shown by sparse and

[4] Barrel cortex. These structures represent specific cytoarchitectonic units visible in specific layers (i.e., layer 4) or the somatosensory cortex where thalamocortical projections are organized with the same topology than individual units of peripheral sensory organs from which they originate. They are particularly salient in the cortical receptive fields of individual vibrissae.

cell type-specific knockout of GluN1 (an obligatory subunit of the NMDA glutamate receptor) or fragmentation of the GA by overexpression of GRASP65, this lateralization of the GA is controlled by synaptic activity and disrupts asymmetric dendritic projection and, consequently, alters whisker-specific response of spiny neurons [116].

A direct link between Golgi morphology and neuronal morphogenesis was also uncovered downstream of Reelin signaling. Reelin, a large secreted glycoprotein, is released by specialized neurons and plays essential roles in controlling neuron position and dendritic growth during brain development [109]. In pyramidal hippocampal neurons, Reelin induces the deployment of the Golgi apparatus in the apical dendrites [109], presumably promoting asymmetric dendritic growth. Interestingly, this effect antagonizes that of Stk25, a kinase known to dampen Reelin/Dab1 signaling that interacts with the LKB1 kinase and GM130. When overexpressed, Stk25 induces the compaction of the GA and promotes the formation of multiple axons [109]. At the opposite, Stk25 knockdown by RNAi impairs axon formation and maturation. Mirroring the effect of its overexpression, downregulation of Stk25 fragments the Golgi and increases the diameter of basal dendrites [109]. These results thus indicate that well-known signaling pathways that control neuronal morphology do so, at least in part, by regulating the polarization of the Golgi apparatus with important consequences on neuronal circuit formation and function.

More recently, Andres-Alonso and colleagues showed in Calneuron1 KO mice that loss of this TGN protein halves the number of CA1 pyramidal neurons showing an expansion of the GA in the apical dendrite, concomitantly decreasing the amount of GSs and PSA-NCAM expression (but not that of NCAM) in distal dendrites [4]. This thus indicates that GS may be generated by fragmentation from the TGN in the cell body (see also the next section) and that GS (and/or the TGN) are required for the trafficking of PSA-NCAM to distal dendritic segments.

One may argue, however, that membrane processing at the GA and cytoskeleton dynamics are so intricately interdependent that it is often challenging to decipher the exact underlying cause of cellular phenotypes resulting from redistribution of the GA. As described in previous sections, the GA (and presumably related membranes) is a major hub for microtubule nucleation (see the chapter by R.M. Rios and M.P. Gavilán), actin remodeling, and the integration of signaling pathways that may not be directly linked to secretory trafficking [106]. The polarization of the GA toward growing axons and dendrites may thus reflect a local remodeling of the cytoskeleton rather than a need for vectoral delivery of new lipids and proteins. Indeed, in young neurons, local stabilization of microtubules is both necessary and sufficient for the specification of the axon [19, 181]. Interestingly, after this step and when the GA is no longer strongly polarized toward the emerging axon, membrane cargo is still efficiently delivered to this compartment [176]. In hippocampal neurons, for instance, dendritic proteins that exit the GA are equally likely to enter dendrites and axons but are sorted by motor proteins at the AIS to ensure that only axonal proteins are allowed to enter the axon while dendritic proteins are rerouted to the somatodendritic compartment [2, 92]. A comparable mechanism has been described upon cargo entry intro dendrites where the microtubule-associated septin

SEPT9 favors the entry and distal transport of dendritic proteins while impeding entry and favoring retrograde transport of cargo destined to the axon [82].

Regardless of the actual importance of direct vectorial trafficking, the polarization of the somatic GA is accompanied by important changes in how and to which extent axons and dendrites rely on post-Golgi trafficking. As shown in mammals and insects, axons and dendrites show striking differences in their sensitivity to disruptions of the secretory pathway. For example, in rat neurons after axonal polarization, disrupting COPI and Golgi function with BFA or expression of inhibitory mutants of Arf1 (the primary switch controlling COPI assembly), and protein kinase D (PKD), a protein regulating membrane fission at the TGN, all prevent dendritic growth and maintenance but have no measurable effect on axonal outgrowth [72]. Similarly, in flies, mutagenesis screens identified mutations in the genes encoding Sar1 and Sec23 (ERES/IC) and Rab1 (IC), three proteins of the ER–Golgi interface that are involved in forward secretory trafficking, selectively disrupt dendritic growth or maintenance in mechanosensory neurons but, as seen in rat neurons, do not alter axons [187]. Intriguingly, disruption of secretory trafficking earlier during cellular development prevents axonal polarization [187] or, if occurring shortly after axon specification, prevents axonal growth [75].

Along the same line, Sar1 and active ERES are enriched in newly specified axons [5], indicating a role for local secretory trafficking in supporting axon generation and growth. Consistently, inhibiting Sar1 prevents axonal growth in cultured neurons in their first day in vitro but, interestingly, has no measurable effect on the length of axons in neurons cultured for 2 or 3 days [5]. As discussed above, Golgi disruption with BFA prevents the growth of developing dendrites but, interestingly, has only a moderate effect on the maintenance of mature dendrites as indicated by ~20% reduction of dendritic length in mature hippocampal neurons [72]. This thus indicates that neuronal reliance on conventional, Golgi-dependent, secretory trafficking may strongly vary during development and maturation.

A distinct sensitivity of axons and dendrites to disrupted secretory trafficking during development has also been observed after disruption of PKD function. As mentioned earlier, inhibition of PKD1 in hippocampal neurons prevents dendritic growth and maintenance but does not affect the axon [72]. In contrast, similar manipulations in cultured neurons at an earlier stage induce the formation of multiple axons [189]. At this stage, PKD1 inhibition prevents the asymmetric growth of a single neurite while preserving total neurite outgrowth. This thus indicates that PKD determines the polarization rather than the bulk level of secretory trafficking required to support growth. This effect may be due to a neuron-specific regulation of PKD function in post-GA trafficking. Indeed, some studies have shown that PKD1 regulates the sorting of cargo to be delivered in axons or dendrites in the TGN, rather than membrane fission for bulk export to the cell surface as seen in fibroblasts and cancer cells [13, 105]. Contrasting with PKD1, knockdown of LIM kinase 1 (LIMK1)—another kinase controlling Golgi morphology and post-Golgi trafficking—inhibits both axon and dendrite growth, thus in this case likely controlling bulk GA membrane export but not the polarity of membrane trafficking [134].

Genesis of Golgi Outposts and Role in Dendritic Development

As discussed earlier, dendritic GOs are rare in mammalian hippocampal and cortical neurons. They are, however, occasionally observed in the most proximal segment of major dendrites, in particular in the apical dendrite of pyramidal neurons where they form by fission of the somatic GA [129, 187]. This fragmentation is controlled by RhoA through Rock, PKD1, LIMK1, cofilin, and dynamin [129], thus sharing multiple mechanistic features with GA dispersal and reformation during and after mitosis [28]. An opposite mechanism has been described for Citron-N, a brain-specific Rho binding protein, which promotes integrity of the GA through recruitment of RockII and profilin [21].

As shown in multiple instances of cell polarization during embryonic development, these actin regulatory proteins are important effectors of polarity protein complexes (notably PAR proteins). While it is not clear to what extent GA fragmentation itself impacts dendritic morphogenesis and function in vertebrates, these mechanisms indicate that polarity proteins controlling neuronal polarization directly impact the morphology of the neuronal GA. Importantly, similar mechanisms seem to occur in neurodegenerative disease models where the GA is often fragmented [56, 108]. This topic is described in the chapters by J Hutkin and P Gleeson and will not be discussed here.

A particularly interesting case is that of neurons derived from human iPSC cells. A recent study identified distinct Golgi structures and behavior during the differentiation of hiPSC-derived neurons and in mature human neurons [171]. These neurons contain multi-cisternae GOs in the proximal and distal regions of the dendrites and reveal major differences from rodent and *Drosophila* models [171, 172]. These GOs contain *cis*- (GM130) and *trans*-Golgi markers (p230/golgin245) and likely correspond to intact Golgi mini-stacks in human neuron cultures and in organoids [171]. Using live cell imaging, it was shown that early in neuronal development the entire somatic GA transiently translocates from the cell body into a single neurite facing the direction of the cell migration [171]. In addition, late in development, small Golgi structures containing *cis* and *trans* elements are deployed from the soma to distant regions of dendrites while the somatic Golgi remains in place [171]. These dendritic GOs present highly dynamic features and are transported bidirectionally. Finally, they do engage in the trafficking of nascent secretory proteins as was shown with controllable secretory cargo, demonstrating that they can receive proteins after ER-exit [171]. At steady state, 16% of iPSC-derived neurons contained dendritic Golgi membranes; however, given the relatively rapid movements of these Golgi structures, a much larger percentage of neurons probably present dendritic GOs over extended time periods. This work has revealed unique features of the secretory system in neurons derived from human iPSCs where GOs are highly dynamic and constantly moving throughout the dendritic network.

In contrast to vertebrates, dendritic GOs appear to be abundant and regulate dendritic morphogenesis in invertebrates, most notably in class I and class IV "da" neurons of the fly larvae body wall. These superficial mechanosensory neurons, devoid of synaptic input, have a planar and stereotyped dendritic patterning that has been extensively used as a genetic model of neurite morphogenesis. In these neurons, heterologous expression of fluorescently tagged Golgi reporters, mostly Mannosidase II (Man II, see Table 7.1), has shown a widespread distribution of GOs throughout the dendritic arbor [99, 119, 187, 193]. As observed for other endomembranes [139], *Drosophila* GOs are preferentially positioned at branch points and are thus ideally positioned to regulate cargo trafficking between multiple dendritic protrusions [187]. Indeed, altering GO localization by disrupting the interaction of Golgi membranes with microtubule motors (see hereafter) reduces dendrite branching and results in aberrant dendritic morphology, indicating a role for GOs in driving dendritic growth and branching [99, 187].

To our knowledge, the ultrastructure of dendritic GOs in *Drosophila* da neurons has not been investigated by EM yet. It is thus not known whether these GOs display the canonical ultrastructure of dictyosomes observed in hippocampal dendrites [72]. Evidence suggests that they may not and probably encompass a larger structural and biochemical diversity than initially anticipated. As shown by Ye and colleagues, whereas ManII-GFP and GalT-TagRFP are both found in more than 90% of Golgi elements in the soma, only half of dendritic GOs contain both proteins [193]. Smaller dendritic GOs containing only one of these proteins are typically found in unbranched segments of dendrites and are much more mobile than GOs located at branch points. Interestingly, akin to vertebrates, the fly homolog of GM130 is required for the maintenance of GA structure and is found only in ring-shaped Golgi units in the soma and stable dendritic GOs found at branch points [193].

Localization of GOs in the dendrites of da neurons is determined by dynein and the Golgi adaptor protein Lava lamp [99, 187]. GO interaction with dynein is dynamic and is inhibited by phosphorylation of Lava lamp by Lrkk, which promotes GO stability and dendritic branching [99]. Both loss of function of dynein [192] or gain of function of conventional kinesins result in the redistribution of GOs in proximal segments of dendrites and the axon [84, 99, 187]. This thus indicates that, akin to other organelles, GO positioning is determined by balancing the antagonist action of dynein and kinesins.

Consistent with a role in dendritic morphogenesis, the presence and escape of GOs from specific branch points are correlated with the extension but also the retraction of nearby terminal branches [187]. Dendritic remodeling seems equally frequent at branch points with or without GOs. However, local disruption by laser irradiation selectively diminishes both extension and retraction of terminal dendritic branches when GOs are present [187]. This thus indicates a causal relationship between GO and branching dynamics. The correlation between GO stability and branch elongation supports the notion that vectorial anterograde trafficking favors neurite extension. Conversely, correlation between GO and branch retraction indicates a role for retrograde membrane trafficking in redirecting membranes retrieved by endocytosis or, nonalternatively, Golgi functions not directly related to membrane trafficking [85].

Studies of microtubule dynamics have suggested that, in flies, dendritic GOs regulate dendritic morphology by nucleating microtubules thanks to recruitment of the gamma-tubulin ring complex (γ-TuRC) by orthologs of AKAP450 and CDK5RAP [119]. However, subsequent work showed that elimination of these two proteins does not affect dendritic microtubule polarity in the same neuron type [185]. Moreover, depletion of ManII-containing dendritic GOs by interfering with microtubule motor function does not deplete γ-tubulin from dendrite branch points [118]. These data thus indicate that GO-independent mechanisms determine γ-TuRC localization and function in fly dendrites. Consistently, more recent experiments that were partly based on endogenous tagging of endosomal proteins indicate that dendritic Rab5-positive endosomes, rather than GOs, are an important site of microtubule nucleation/stabilization [179]. Interestingly, a plausible explanation for this discrepancy may be due to the propensity of ManII (and potentially other glycosylation enzymes) to localize to endomembranes other than the GA, at least when overexpressed [179] (see comments on the glycosylation machinery in Sect. 6). Thus, structures assumed to be GOs in other studies may actually be early endosomes into which overexpressed ManII had leaked. The involvement of endosomes in nucleating microtubules has been described in other models [1] and is thus not specific to insects. As discussed in the next section, endosome distribution and function also are a major determinant of dendritic morphogenesis.

Polarization of Pre-Golgi Membranes and Endosomes During Neuronal Morphogenesis

As we have seen, the polarization of the somatic GA varies during neuronal development and reflects important changes in the organization of the cytoskeleton and the directionality of vectorial membrane trafficking. Similarly, and perhaps on a more impressive scale, pre-Golgi membranes and endosomes also undergo significant reorganization during neuronal growth.

For example, the ER rapidly extends in growing neurites thanks to microtubule-binding proteins and motors. Knockdown of the ER microtubule-interacting protein p600 impairs proper neuronal migration and neurite differentiation [147]. In growing dendrites, the appearance of local zones of ER complexity is spatially and temporally correlated with dendritic branching. Increasing or decreasing ER complexity by interfering with CLIMP63 phospho-dependent interaction with microtubules augments or diminishes dendritic growth and complexity, respectively [174]. This thus indicates that ER complexity and resulting compartmentalization of ER cargo processing are limiting for dendritogenesis.

In a consistent manner, ERES are initially abundant throughout the axon during early cellular development but then become progressively more abundant in dendrites as dendritogenesis advances [5, 6]. Multiple proteins influencing ER morphology, including REEP, reticulons, and atlastin, have been linked to dendritic and axonal outgrowth, further strengthening the notion that the morphology and dynamics of the neuronal ER strongly impact neuronal growth and maturation ([59], Liu et al. 2019).

As mentioned in Sect. 4, the dendritic and axonal ER forms specialized contact zones with endosomes and the PM [183]. As shown for the SNARE Sec22b, ER-PM contact sites are structured by non-fusogenic SNAREs and are required for proper neurite outgrowth, likely contributing to membrane expansion by nonvesicular lipid transfer [51, 124].

ER contacts with endosomes also regulate neurite extension [150]. As described for the ER protein Protrudin, ER contacts with late endosome (LE) enable the transfer of Kinesin1 from Protrudin to LEs, hence favoring LE translocation and exocytosis in growing neurites [131].

Endosomes are abundant in neuronal processes [95] and play a major role in trafficking membranes to sustain the growth, maintenance, and plasticity of dendritic spines [22, 41, 86]. Interestingly, spine remodeling after induction of long-term synaptic potentiation (LTP), a well-characterized form of synaptic plasticity, is accompanied by co-variations of the morphology and abundance of ER and endosomes that vary depending on spine types [93] (see also the description of the spine apparatus in Sect. 4).

Endosomes play a comparable role in developing dendrites. In *Drosophila* da neurons, as we have already alluded to, endosomes are distributed throughout the dendritic tree through interactions with Dynein and Kinesin 1 and their dynamics closely match those of growing terminal branches [139]. Similarly to what was described for GOs, dendritic endosomes redistribute from distal to proximal dendritic segments upon dynein disruption [139] and disrupting endosome function by expression of dominant-negative mutants of Rab5 prevents this proximal branching phenotype [139]. Endosomes thus likely engage in vectorial trafficking of membranes in dendritic branching and local growth. This redistribution may thus also explain the increased branching that occurs in proximal dendrites in the dynein mutants described in GO studies (see also the comments on the leakage of Golgi reporters into endosomes in the previous section).

Diversity of Secretory Routes in Neurons

The biogenesis and secretory trafficking of key neuronal secretory proteins, including neurotransmitter receptors, adhesion molecules, secreted growth factors, and synaptic vesicle proteins, is an area of intense scrutiny and is too broad a topic to be summarized and discussed here (for recent reviews, see [38, 39, 58, 62, 85, 125, 143, 168]). In this final section, we discuss emerging tools and general concepts that have brought novel insights into how and where secretory proteins are sent to the neuronal surface.

Let us first point out that, perhaps ironically given that the GA was one of the first cell organelles to be described and was first observed in neurons [46, 55], we still know much less about the trafficking of neuronal proteins in secretory membranes than in endosomes. This discrepancy is for a large part due to technical limitations. After labeling at the cell surface at a given time, proteins can be directly tracked and monitored following endocytosis, which has greatly facilitated investigations of the role of endosomal trafficking in controlling the homeostasis of the neuronal surface. Consequently, as we have alluded to at multiple occasions throughout this chapter, it is well established that endosomes are a major source of membranes in neuronal processes, most particularly at axonal boutons and the postsynapse [87]. In contrast, labeling specific neuronal proteins at specific steps of their secretory processing is much more challenging.

Controllable Reporters for Synchronized Monitoring of Secretory Cargo

The development of molecular systems enabling synchronized release of specific tagged candidate proteins from specific secretory organelles has thus been a significant breakthrough in this context. In brief, these strategies include the use of protein aggregation domains suitable for ligand-controlled deaggregation, enabling inducible release of virtually any tagged protein of interest from the ER [16, 18, 57]. Coupled to local photoactivation approaches [18], this strategy has been successfully used to monitor exchanges of nascent proteins between the soma and dendrites. Importantly, this approach was used to demonstrate that nascent proteins released from the dendritic ER can directly access the dendritic surface via IC (and probably GSs) and REs without passing through the GA in the cell body [18].

More recently, optogenetics approaches based on photocleavable linkers [64] were developed for local control of ER-exit by removal of a KDEL sequence [17]. Using this strategy, Bourke and colleagues have reported that molecular identity rather than subcellular location dictates the trafficking route of dendritic proteins [17]. For example, significant differences were observed between NL1, a synaptic adherence molecule and GluA1, an abundant subunit of glutamate AMPA receptors [17], two proteins that localize to postsynaptic excitatory membranes throughout the dendritic tree. When targeted for ER-release from the same compartment (i.e., soma vs dendrites), the two proteins display differing rates of accumulation in the somatic GA and distinct sensitivity to manipulations of neuronal activity in regard to the extent and spatial distribution of their exocytosis and subsequent diffusion at the neuronal surface [17].

Finally, numerous studies have successfully used the RUSH system developed by Franck Perez and colleagues to monitor the trafficking of secretory neuronal proteins [15]. In this approach, fusion of streptavidin to various anchors for luminal presentation in specific endomembranes enables a biotin-dependent release of candidate proteins from these compartments. In neurons, this approach has been used primarily to monitor cargo trafficking following ER-exit and release from the somatic GA [4, 42, 171, 176].

In complement to these approaches, techniques enabling imaging of endogenous proteins and glycans after metabolic labeling with bioorthogonal tagged precursors have provided valuable insights into the site of synthesis and redistribution of bulk secretory cargo [111] or specific candidate proteins [158].

Together with studies of endogenous or conventional fluorescently tagged secretory proteins, experiments performed with synchronizable-secretory cargo all point to the existence of distinct secretory trafficking pathways for diverse neuronal proteins to specific axonal or dendritic compartments [17, 18, 58, 63, 65, 78](Fig. 7.4).

Numerous important questions still need to be addressed regarding the exact itinerary of nascent proteins between the ER and the plasma membrane and interconnections between the IC, GS, and REs. Yet, and as discussed throughout this chapter, it is clear that secretory trafficking in dendrites and axons, involves a local interplay between biosynthetic membranes and other endomembranes.

Atypical Prevalence of Golgi-Independent Secretory Trafficking in Neurons and Consequences for Protein Glycosylation

Given the lack of canonical Golgi membranes in neurites, how can all or a subset of membrane and secreted proteins synthesized in dendrites and axons be trafficked to the neuronal PM without first visiting the somatic GA? As discussed throughout this

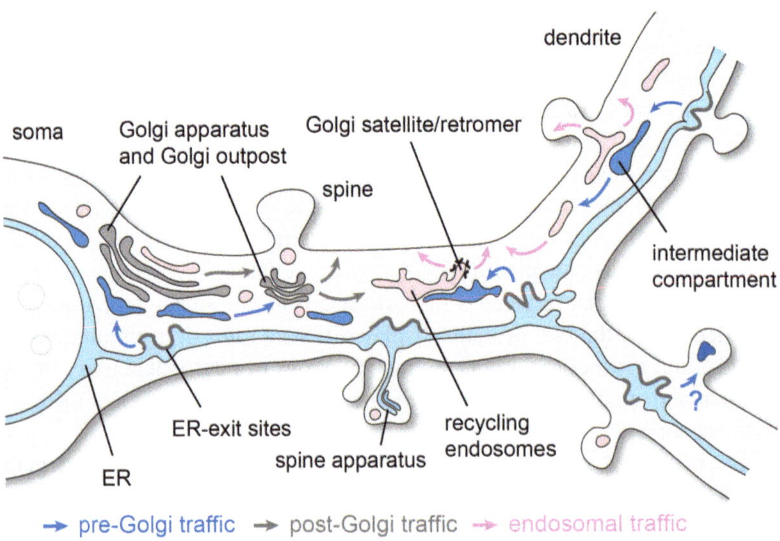

Fig. 7.4 Golgi-dependent and Golgi-independent secretory routes in dendrites. Nascent secretory proteins leave the ER at ER exit sites (ERES) located at hotspots of complex ER and accumulate in intermediate compartments (IC) throughout the soma and dendrites. From the IC, proteins can either travel to the somatic Golgi and nearby Golgi outposts in proximal dendrites (Golgi-dependent pathway) or bypass the Golgi and move directly to recycling endosomes to reach the plasma membrane (Golgi-independent pathway)

chapter, studies of candidate proteins have documented multiple instances of Golgi bypass in dendrites and axons [58, 85]. The exact itinerary of cargo that follow these Golgi-independent secretory routes is not completely understood yet. However, it most likely involves mixed identity organelles sharing molecular and functional features of biosynthetic (i.e., ER, IC) endosomal (GSs, REs) and lysosomal compartments.

More generally speaking, processing in the GA was generally thought to be a requisite step for trafficking most integral membrane and secreted proteins in non-neuronal cells. However, the list of exceptions keeps expanding. Just to name a few, cystic fibrosis transmembrane conductance regulator (CFTR), CD45, Connexins 26 and 30, the *Drosophila* integrin αPS1 and, in the axons of mammalian neurons, the voltage-gated channel Kv2.1, can be delivered to the cell surface after disrupting the GA with BFA, Syntaxin 5 knockout or expression of Arf1 mutants [8, 107, 128, 141]. It is now well established that GA-independent trafficking occurs in most cell types in basal conditions and is further enhanced by various forms of cellular stress [130].

As discussed in Sect. 7, GA disruption with BFA blocks dendritic growth in developing neurons but has only a moderate effect on mature neurons [72], thus suggesting that Golgi-independent trafficking becomes more prominent as neurons mature. Consistently, metabolic labeling of endogenous proteins combined with cell fractionation revealed that a surprisingly large fraction of newly synthesized proteins reach the neuronal surface despite GA disruption with BFA, in stark contrast to COS7 cells where this drug largely disrupts nascent protein expression at the PM [66].

Further support for GA bypass comes from analyses of the N-glycosylation of integral membrane proteins, which all indicate that the most abundant N-glycans in human and mouse brains, the neuronal surface and purified synapses correspond to N-glycans that have passed quality control in the ER but have bypassed or have not been fully processed in the GA [18, 66, 191]. Interestingly, the sensitivity or lack of sensitivity of the surface expression of candidate neuronal proteins to Golgi disruption seems to correlate with their decoration with high-mannose N-glycans [18, 66], thus suggesting that the core-glycans that are expressed at atypically high levels at the neuronal surface may not originate from the GA but may be processed in other endomembranes.

Consistent with N-glycosylation's role in controlling protein folding and trafficking to the cell surface [113], pharmacological inhibition of N-glycosylation in the ER with tunicamycin reduces neuronal growth and induces neuron death [66]. In contrast and consistent with the moderate effect of BFA on mature neurons, inhibition of the maturation of core-glycans after processing in the ER appears largely dispensable for dendritic growth and synaptic transmission in cultured hippocampal neurons [66]. This thus indicates that glycan processing in the GA is not required for the biogenesis of functional proteins. However, conditional MGAT1 knockout mice have shown that neuron-specific disruption of N-glycan maturation into hybrid and complex N-glycans in the GA does induce apoptosis in specific neuronal populations, resulting in severe locomotion deficits and animal life span shortening

[188]. Surprisingly, however, the density of neurons and the overall anatomy of the brain, notably in the hippocampus and the cortex, appear largely unaffected by loss of mature N-glycans [188]. It is thus likely that MGAT1 is required for animal survival because of its requirement in specific neuron types and development/maturation stages. But, overall, it seems that N-glycan maturation in the GA is not needed for cell viability in all neuron types. Indeed, it was recently shown that targeted knockout of MGAT1 in hippocampal neurons in adult mice slightly diminishes synaptic density but is not required for cell viability [191]. Consistent with this notion, studies of conditional GM130 knockout mice have shown that neuron-specific disruption of the GA has severe deleterious effects on the cerebellum but has little or no effect on the cortex and the hippocampus [100]. Thus, multiple lines of evidence indicate that Golgi-independent trafficking, and consequently protein core-glycosylation, is atypically prominent across multiple neuron types.

Unresolved questions remain as to whether some early steps of N-glycan processing that were classically thought to happen in the GA may actually also happen in neurites. Indeed, as we have seen at the end of Sect. 6, numerous studies have reported that Golgi glycosylation enzymes can be in multiple instances localized and probably also function in extra-Golgi compartments. It will thus be important for future studies to determine the exact distribution and site of function of N-glycan processing enzymes in neurons.

Finally, let us close this chapter by elaborating on what could be the functional consequences of bypassing the somatic Golgi from the perspective of N-glycosylation. As shown for surface glutamate AMPA receptors, core-glycosylated proteins are much less stable than mature glycoforms [66], suggesting a causal link between glycosylation and protein lifetime. Glycans can also influence the signaling properties of these receptors and most notably their desensitization [31, 167, 177, 191]. It is thus likely that the subcellular location of protein synthesis and subsequent processing in neurites or the soma could dictate the longevity and biophysical properties of the same protein. For example, long-lived somatic GA-processed proteins with mature N-linked glycans could be tailored to withstand long-range transport to distant locations in the axonal or dendritic arbor. Conversely, shorter-lived and locally processed core-glycosylated proteins could be better suited for transient modifications of specific synapses.

Acknowledgments We thank Lisa Kochen and Diana Zala for their careful reading of the manuscript. Work in the laboratory of CH is supported by the *Agence Nationale de la Recherche* and the *Fondation Recherche Alzheimer*.

References

1. Akhmanova A, Kapitein LC (2022) Mechanisms of microtubule organization in differentiated animal cells. Nat Rev Mol Cell Biol 23(8):541–558. https://doi.org/10.1038/s41580-022-00473-y

2. Al-Bassam S, Xu M, Wandless TJ, Arnold DB (2012) Differential trafficking of transport vesicles contributes to the localization of dendritic proteins. Cell Rep 2(1):89–100. https://doi.org/10.1016/j.celrep.2012.05.018
3. Alberts B, Johnson A, Lewis J, Raff M, Roberts K, Walter P (2002) Molecular biology of the cell, 4th edn. Garland Science
4. Andres-Alonso M, Borgmeyer M, Mirzapourdelavar H, Lormann J, Klein K, Schweizer M, Hoffmeister-Ullerich S, Oelschlegel AM, Dityatev A, Kreutz MR (2023) Golgi satellites are essential for polysialylation of NCAM and expression of LTP at distal synapses. Cell Rep 42(7):112692. https://doi.org/10.1016/j.celrep.2023.112692
5. Aridor M, Fish KN (2009) Selective targeting of ER exit sites supports axon development. Traffic 10(11):1669–1684. https://doi.org/10.1111/j.1600-0854.2009.00974.x
6. Aridor M, Guzik AK, Bielli A, Fish KN (2004) Endoplasmic reticulum export site formation and function in dendrites. J Neurosci 24(15):3770–3776. https://doi.org/10.1523/JNEUROSCI.4775-03.2004
7. Arthur AL, Yang SZ, Abellaneda AM, Wildonger J (2015) Dendrite arborization requires the dynein cofactor Nud E. J Cell Sci 128:2191–2201. https://doi.org/10.1242/jcs.170316
8. Baldwin TA, Ostergaard HL (2002) The protein-tyrosine phosphatase CD45 reaches the cell surface via Golgi-dependent and -independent pathways *. J Biol Chem 277(52):50333–50340. https://doi.org/10.1074/jbc.M209075200
9. Barnes AP, Polleux F (2009) Establishment of axon-dendrite polarity in developing neurons. Annual review of neuroscience 32(volume 32, 2009):347–381. https://doi.org/10.1146/annurev.neuro.31.060407.125536
10. Bas Orth C, Schultz C, Müller C, Frotscher M, Deller T (2007) Loss of the cisternal organelle in the axon initial segment of cortical neurons in synaptopodin-deficient mice. J Comp Neurol 504(5). https://doi.org/10.1002/cne.21445
11. Bhattacharjee S, Lottes EN, Nanda S, Golshir A, Patel AA, Ascoli GA, Cox DN (2022) PP2A phosphatase regulates cell-type specific cytoskeletal organization to drive dendrite diversity. Front Mol Neurosci 15. https://doi.org/10.3389/fnmol.2022.926567
12. Berger EG (2002) Ectopic localizations of Golgi glycosyltransferases. Glycobiology 12(2):29R–36R. https://doi.org/10.1093/glycob/12.2.29r
13. Bisbal M, Conde C, Donoso M, Bollati F, Sesma J, Quiroga S, Añel AD, Malhotra V, Marzolo MP, Cáceres A (2008) Protein kinase D regulates trafficking of dendritic membrane proteins in developing neurons. J Neurosci 28(37):9297–9308. https://doi.org/10.1523/JNEUROSCI.1879-08.2008
14. Bollmann C, Schöning S, Kotschnew K, Grosse J, Heitzig N, Fischer von Mollard G (2022) Primary neurons lacking the SNAREs vti 1a and vti1b show altered neuronal development. Neural Dev 17:12. https://doi.org/10.1186/s13064-022-00168-2
15. Boncompain G, Divoux S, Gareil N, de Forges H, Lescure A, Latreche L, Mercanti V, Jollivet F, Raposo G, Perez F (2012) Synchronization of secretory protein traffic in populations of cells. Nat Methods 9(5):493–498. https://doi.org/10.1038/nmeth.1928
16. Bonnet C, Charpentier J, Retailleau N, Choquet D, Coussen F (2023) Regulation of different phases of AMPA receptor intracellular transport by 4.1N and SAP97. eLife 12:e85609. doi:https://doi.org/10.7554/eLife.85609
17. Bourke AM, Schwartz SL, Bowen AB, Kleinjan MS, Winborn CS, Kareemo DJ, Gutnick A, Schwarz TL, Kennedy MJ (2021) zapERtrap: a light-regulated ER release system reveals unexpected neuronal trafficking pathways. J Cell Biol 220(9):e202103186. https://doi.org/10.1083/jcb.202103186
18. Bowen AB, Bourke AM, Hiester BG, Hanus C, Kennedy MJ (2017) Golgi-independent secretory trafficking through recycling endosomes in neuronal dendrites and spines. In: eLife. https://elifesciences.org/articles/27362. Accessed 10 Oct 2024
19. Burute M, Jansen KI, Mihajlovic M, Vermonden T, Kapitein LC (2022) Local changes in microtubule network mobility instruct neuronal polarization and axon specification. Sci Adv 8(44):eabo2343. https://doi.org/10.1126/sciadv.abo2343

20. Calderon de Anda F, Gärtner A, Tsai L-H, Dotti CG (2008) Pyramidal neuron polarity axis is defined at the bipolar stage. J Cell Sci 121(2):178–185. https://doi.org/10.1242/jcs.023143
21. Camera P, Da Silva JS, Griffiths G, Giuffrida MG, Ferrara L, Schubert V, Imarisio S, Silengo L, Dotti CG, Di Cunto F (2003) Citron-N is a neuronal rho-associated protein involved in Golgi organization through actin cytoskeleton regulation. Nat Cell Biol 5(12):1071–1078. https://doi.org/10.1038/ncb1064
22. Chambers JP, Tsin AT, Raymond NY, Aldape FG, Rodriguez KA (1986) Effects of glycosylation inhibitors on the frog retina. Brain Res Bull 17(2):259–263. https://doi.org/10.1016/0361-9230(86)90124-3
23. Chandramouli Iyer S, Ramachandran Iyer EP, Meduri R, Rubaharan M, Kuntimaddi A, Karamsetty M, Cox DN (2013) Cut via CrebA transcriptionally regulates the COPII secretory pathway to direct dendrite development in drosophila. J Cell Sci jcs.131144. https://doi.org/10.1242/jcs.131144
24. Chen A, Göβling EK, Witkowski L, Bhindi A, Bauch C, Roussy G, Sarret P, Kreienkamp H-J, Stroh T (2012) Regional and subcellular distribution of the receptor-targeting protein PIST in the rat central nervous system. J Comp Neurol 520(5):889–913. https://doi.org/10.1002/cne.22774
25. Chen D, Gibson ES, Kennedy MJ (2013) A light-triggered protein secretion system. J Cell Biol 201(4):631–640. https://doi.org/10.1083/jcb.201210119
26. Cheng G, Chang J, Gong H, Zhou W (2023) A distinct Golgi-targeting mechanism of dGM130 in drosophila neurons. Front Mol Neurosci 16. https://doi.org/10.3389/fnmol.2023.1206219
27. Chua CEL, Tang BL (2008) Syntaxin 16 is enriched in neuronal dendrites and may have a role in neurite outgrowth. Mol Membr Biol 25:35–45. https://doi.org/10.1080/09687680701504649
28. Chung CG, Kwon MJ, Jeon KH, Hyeon DY, Han MH, Park JH, Cha IJ, Cho JH, Kim K, Rho S, Kim GR, Jeong H,Lee JW, Kim T, Kim K, Kim KP, Ehlers MD, Hwang D, Lee SB (2017) Golgi Outpost Synthesis Impaired by ToxicPolyglutamine Proteins Contributes to Dendritic Pathology in Neurons. Cell Reports 20:356–369.https://doi.org/10.1016/j.celrep.2017.06.059
29. Colanzi A, Suetterlin C, Malhotra V (2003) Cell-cycle-specific Golgi fragmentation: how and why? Curr Opin Cell Biol 15(4):462–467. https://doi.org/10.1016/S0955-0674(03)00067-X
30. Cooney J, Hurlburt J, Selig D, Harris K, Fiala J (2002) Endosomal compartments serve multiple hippocampal dendritic spines from a widespread rather than a local store of recycling membrane. The journal of neuroscience : the official journal of the society for. Neuroscience 22(6). https://doi.org/10.1523/JNEUROSCI.22-06-02215.2002
31. Copf T (2014) Developmental shaping of dendritic arbors in drosophila relies on tightly regulated intra-neuronal activity of protein kinase a (PKA). Dev Biol 393:282–297. https://doi.org/10.1016/j.ydbio.2014.07.002
32. Copits BA, Vernon CG, Sakai R, Swanson GT (2014) Modulation of ionotropic glutamate receptor function by vertebrate galectins. J Physiol 592(10):2079–2096. https://doi.org/10.1113/jphysiol.2013.269597
33. Cornejo VH, González C, Campos M, Vargas-Saturno L, de los Juricic MÁ, Miserey-Lenkei S, Pertusa M, Madrid R, Couve A (2020) Non-conventional axonal organelles control TRPM8 Ion Channel trafficking and peripheral cold sensing. Cell Rep 30(13):4505–4517.e5. https://doi.org/10.1016/j.celrep.2020.03.017
34. Cox DJ, Racca C (2013) Differential dendritic targeting of AMPA receptor subunit mRNAs in adult rat hippocampal principal neurons and interneurons. J Comp Neurol 521(9):1954–2007. https://doi.org/10.1002/cne.23292
35. Cunha-Ferreira I, Chazeau A, Buijs RR, Stucchi R, Will L, Pan X, Adolfs Y, van der Meer C, Wolthuis JC, Kahn OI, Schätzle P, Altelaar M, Pasterkamp RJ, Kapitein LC, Hoogenraad CC (2018) The HAUS complex is a key regulator of non-centrosomal microtubule organization during neuronal development. Cell Rep 24(4):791–800. https://doi.org/10.1016/j.celrep.2018.06.093

36. Dankovich TM, Kaushik R, Olsthoorn LHM, Petersen GC, Giro PE, Kluever V, Agüi-Gonzalez P, Grewe K, Bao G, Beuermann S, Hadi HA, Doeren J, Klöppner S, Cooper BH, Dityatev A, Rizzoli SO (2021) Extracellular matrix remodeling through endocytosis and resurfacing of tenascin-R. Nat Commun 12(1):7129. https://doi.org/10.1038/s41467-021-27462-7
37. de los Ángeles Juricic Urzúa M, Rojas JG, Correa AC, Cerda M, Gründler SH, González-Silva C (2022) The dendritic ERGIC: microtubule and actin cytoskeletons participate in stop-and-go movement of mobile carriers between stable structures. Traffic 23(3):174–187. https://doi.org/10.1111/tra.12832
38. Deng C, Moradi M, Reinhard S, Ji C, Jablonka S, Hennlein L, Lüningschrör P, Doose S, Sauer M, Sendtner M (2021) Dynamic remodeling of ribosomes and endoplasmic reticulum in axon terminals of motoneurons. J Cell Sci 134(22):jcs258785. https://doi.org/10.1242/jcs.258785
39. Devine MJ, Kittler JT (2018) Mitochondria at the neuronal presynapse in health and disease. Nat Rev Neurosci 19(2):63–80. https://doi.org/10.1038/nrn.2017.170
40. Diering GH, Huganir RL (2018) The AMPA receptor code of synaptic plasticity. Neuron 100(2):314–329. https://doi.org/10.1016/j.neuron.2018.10.018
41. Du Q, Chang J, Cheng G, Zhao Y, Zhou W (2021) Sunday driver mediates multi-compartment Golgi outposts defects induced by amyloid precursor protein. Front Neurosci 15. https://doi.org/10.3389/fnins.2021.673684
42. Esteves da Silva M, Adrian M, Schätzle P, Lipka J, Watanabe T, Cho S, Futai K, Wierenga CJ, Kapitein LC, Hoogenraad CC (2015) Positioning of AMPA receptor-containing endosomes regulates synapse architecture. Cell Rep 13(5):933–943. https://doi.org/10.1016/j.celrep.2015.09.062
43. Evans AJ, Gurung S, Wilkinson KA, Stephens DJ, Henley JM (2017) Assembly, secretory pathway trafficking, and surface delivery of Kainate receptors is regulated by neuronal activity. Cell Rep 19(12):2613–2626. https://doi.org/10.1016/j.celrep.2017.06.001
44. Fabbretti E, Antognolli G, Tongiorgi E (2021) Amyloid-β impairs dendritic trafficking of Golgi-like organelles in the early phase preceding neurite atrophy: rescue by mirtazapine. Front Mol Neurosci 14. https://doi.org/10.3389/fnmol.2021.661728
45. Falahati H, Wu Y, De Camilli P (2024) Ectopic reconstitution of a spine-apparatus like structure provides insight into mechanisms underlying its formation. bioRxiv:2024.04.16.589782. https://doi.org/10.1101/2024.04.16.589782
46. Farías GG, Guardia CM, Britt DJ, Guo X, Bonifacino JS (2015) Sorting of dendritic and axonal vesicles at the pre-axonal exclusion zone. Cell Rep 13(6):1221–1232. https://doi.org/10.1016/j.celrep.2015.09.074
47. Farquhar MG, Palade GE (1998) The Golgi apparatus: 100 years of progress and controversy. Trends Cell Biol 8(1):2–10. https://doi.org/10.1016/s0962-8924(97)01187-2
48. Ferro-Novick S, Rapoport T, Schekman R (2023) The endoplasmic reticulum. In: Cold Spring Harbor perspectives in biology collection, 2nd edn. Cold Spring Harbor Laboratory Press, Cold Spring Harbor (NY)
49. Fiala JC, Spacek J, Harris KM (2007) Dendrite structure. In: Stuart G, Spruston N, Häusser M (eds) Dendrites. Oxford University Press, p 0
50. Fourriere L, Kasri A, Gareil N, Bardin S, Bousquet H, Pereira D, Perez F, Goud B, Boncompain G, Miserey-Lenkei S (2019) RAB6 and microtubules restrict protein secretion to focal adhesions. J Cell Biol 218(7):2215–2231. https://doi.org/10.1083/jcb.201805002
51. Fu M, McAlear TS, Nguyen H, Oses-Prieto JA, Valenzuela A, Shi RD, Perrino JJ, Huang T-T, Burlingame AL, Bechstedt S, Barres BA (2019) The Golgi outpost protein TPPP nucleates microtubules and is critical for myelination. Cell 179:132–146.e14. https://doi.org/10.1016/j.cell.2019.08.025
52. Gallo A, Vannier C, Galli T (2016) Endoplasmic reticulum–plasma membrane associations: structures and functions. Annu Rev Cell Dev Biol 32:279–301. https://doi.org/10.1146/annurev-cellbio-111315-125024

53. Gardiol A, Racca C, Triller A (1999) Dendritic and postsynaptic protein synthetic machinery. J Neurosci 19(1):168–179. https://doi.org/10.1523/JNEUROSCI.19-01-00168.1999
54. Gill DJ, Chia J, Senewiratne J, Bard F (2010) Regulation of O-glycosylation through Golgi-to-ER relocation of initiation enzymes. J Cell Biol 189(5):843–858. https://doi.org/10.1083/jcb.201003055
55. Glock C, Biever A, Tushev G, Nassim-Assir B, Kao A, Bartnik I, Tom Dieck S, Schuman EM (2021) The translatome of neuronal cell bodies, dendrites, and axons. Proc Natl Acad Sci USA 118(43):e2113929118. https://doi.org/10.1073/pnas.2113929118
56. Golgi C, Lipsky NG (1989) On the structure of nerve cells. J Microsc 155(1):3–7. https://doi.org/10.1111/j.1365-2818.1989.tb04294.x
57. Gonatas NK, Stieber A, Gonatas JO (2006) Fragmentation of the Golgi apparatus in neurodegenerative diseases and cell death. J Neurol Sci 246(1):21–30. https://doi.org/10.1016/j.jns.2006.01.019
58. González C, Cánovas J, Fresno J, Couve E, Court FA, Couve A (2016) Axons provide the secretory machinery for trafficking of voltage-gated sodium channels in peripheral nerve. Proc Natl Acad Sci 113(7):1823–1828. https://doi.org/10.1073/pnas.1514943113
59. González C, Cornejo VH, Couve A (2018) Golgi bypass for local delivery of axonal proteins, fact or fiction? Curr Opin Cell Biol 53:9–14. https://doi.org/10.1016/j.ceb.2018.03.010
60. Goyal U, Blackstone C (2013) Untangling the web: mechanisms underlying ER network formation. Biochimica et Biophysica Acta (BBA)—Molecular Cell Research 1833(11):2492–2498. https://doi.org/10.1016/j.bbamcr.2013.04.009
61. Gray E (1959) Axo-somatic and axo-dendritic synapses of the cerebral cortex: an electron microscope study. J Anat 93(Pt 4)
62. Griffiths G, Fuller SD, Back R, Hollinshead M, Pfeiffer S, Simons K (1989) The dynamic nature of the Golgi complex. J Cell Biol 108(2):277–297. https://doi.org/10.1083/jcb.108.2.277
63. Groc L, Choquet D (2020) Linking glutamate receptor movements and synapse function. Science 368(6496):eaay4631. https://doi.org/10.1126/science.aay4631
64. Grochowska KM, Andres-Alonso M, Karpova A, Kreutz MR (2022) The needs of a synapse—how local organelles serve synaptic proteostasis. EMBO J 41(7):e110057. https://doi.org/10.15252/embj.2021110057
65. Gutnick A, Banghart MR, West ER, Schwarz TL (2019) The light-sensitive dimerizer zapalog reveals distinct modes of immobilization for axonal mitochondria. Nat Cell Biol 21(6):768–777. https://doi.org/10.1038/s41556-019-0317-2
66. Hanus C, Ehlers MD (2016) Specialization of biosynthetic membrane trafficking for neuronal form and function. Curr Opin Neurobiol 39:8–16. https://doi.org/10.1016/j.conb.2016.03.004
67. Hanus C, Geptin H, Tushev G, Garg S, Alvarez-Castelao B, Sambandan S, Kochen L, Hafner A-S, Langer JD, Schuman EM (2016) Unconventional secretory processing diversifies neuronal ion channel properties. elife 5:e20609. https://doi.org/10.7554/eLife.20609
68. Hanus C, Kochen L, Tom Dieck S, Racine V, Sibarita J-B, Schuman EM, Ehlers MD (2014) Synaptic control of secretory trafficking in dendrites. Cell Rep 7(6):1771–1778. https://doi.org/10.1016/j.celrep.2014.05.028
69. Hanus C, Schuman EM (2013) Proteostasis in complex dendrites. Nat Rev Neurosci 14(9):638–648. https://doi.org/10.1038/nrn3546
70. Hasdemir B, Fitzgerald DJ, Prior IA, Tepikin AV, Burgoyne RD (2005) Traffic of Kv4 K+ channels mediated by KChIP1 is via a novel post-ER vesicular pathway. J Cell Biol 171(3):459–469. https://doi.org/10.1083/jcb.200506005
71. Hertrich N, Doil N, Konietzny A, Mikhaylova M (2024) Myosin VI controls localization of Golgi satellites at active presynaptic boutons. 2024.05.08.593268
72. Horton AC, Ehlers MD (2003) Dual modes of endoplasmic reticulum-to-Golgi transport in dendrites revealed by live-cell imaging. J Neurosci 23(15):6188–6199. https://doi.org/10.1523/JNEUROSCI.23-15-06188.2003

73. Horton AC, Rácz B, Monson EE, Lin AL, Weinberg RJ, Ehlers MD (2005) Polarized secretory trafficking directs cargo for asymmetric dendrite growth and morphogenesis. Neuron 48(5):757–771. https://doi.org/10.1016/j.neuron.2005.11.005
74. Islam MA, Sharif SR, Lee H, Seog D-H, Moon IS (2015) N-acetyl-D-glucosamine kinase interacts with dynein light-chain roadblock type 1 at Golgi outposts in neuronal dendritic branch points. Exp Mol Med 47(8):e177–e177. https://doi.org/10.1038/emm.2015.48
75. Jan CH, Williams CC, Weissman JS (2014) Principles of ER cotranslational translocation revealed by proximity-specific ribosome profiling. Science 346(6210):1257521. https://doi.org/10.1126/science.1257521
76. Jareb M, Banker G (1997) Inhibition of axonal growth by Brefeldin a in hippocampal neurons in culture. J Neurosci 17(23):8955–8963. https://doi.org/10.1523/JNEUROSCI.17-23-08955.1997
77. Jarvela T, Linstedt AD (2012) Irradiation-induced protein inactivation reveals Golgi enzyme cycling to cell periphery. J Cell Sci 125(Pt 4):973–980. https://doi.org/10.1242/jcs.094441
78. Jenkins B, Decker H, Bentley M, Luisi J, Banker G (2012) A novel split kinesin assay identifies motor proteins that interact with distinct vesicle populations. J Cell Biol 198(4):749–761. https://doi.org/10.1083/jcb.201205070
79. Jeyifous O, Waites CL, Specht CG, Fujisawa S, Schubert M, Lin EI, Marshall J, Aoki C, de Silva T, Montgomery JM, Garner CC, Green WN (2009) SAP97 and CASK mediate sorting of NMDA receptors through a previously unknown secretory pathway. Nat Neurosci 12(8):1011–1019. https://doi.org/10.1038/nn.2362
80. Jones KA, Eng AG, Raval P, Srivastava DP, Penzes P (2014) Scaffold protein X11α interacts with Kalirin-7 in dendrites and recruits it to Golgi outposts. J Biol Chem 289(51):35517–35529. https://doi.org/10.1074/jbc.M114.587709
81. Jung H, Gkogkas CG, Sonenberg N, Holt CE (2014) Remote control of gene function by local translation. Cell 157(1):26–40. https://doi.org/10.1016/j.cell.2014.03.005
82. Kaewsapsak P, Shechner DM, Mallard W, Rinn JL, Ting AY (2017) Live-cell mapping of organelle-associated RNAs via proximity biotinylation combined with protein-RNA cross-linking. elife 6:e29224. https://doi.org/10.7554/eLife.29224
83. Karasmanis EP, Phan C-T, Angelis D, Kesisova IA, Hoogenraad CC, McKenney RJ, Spiliotis ET (2018) Polarity of neuronal membrane traffic requires sorting of kinesin motor cargo during entry into dendrites by a microtubule-associated Septin. Dev Cell 46(2):204–218.e7. https://doi.org/10.1016/j.devcel.2018.06.013
84. Kasthuri N, Hayworth KJ, Berger DR, Schalek RL, Conchello JA, Knowles-Barley S, Lee D, Vázquez-Reina A, Kaynig V, Jones TR, Roberts M, Morgan JL, Tapia JC, Seung HS, Roncal WG, Vogelstein JT, Burns R, Sussman DL, Priebe CE, Pfister H, Lichtman JW (2015) Saturated reconstruction of a volume of neocortex. Cell 162(3):648–661. https://doi.org/10.1016/j.cell.2015.06.054
85. Kelliher MT, Yue Y, Ng A, Kamiyama D, Huang B, Verhey KJ, Wildonger J (2018) Autoinhibition of kinesin-1 is essential to the dendrite-specific localization of Golgi outposts. J Cell Biol 217(7):2531–2547. https://doi.org/10.1083/jcb.201708096
86. Kennedy M, Hanus C (2019) Architecture and dynamics of the neuronal secretory network. Annu Rev Cell Dev Biol 35. https://doi.org/10.1146/annurev-cellbio-100818-125418
87. Kennedy MJ, Davison IG, Robinson CG, Ehlers MD (2010) Syntaxin-4 defines a domain for activity-dependent exocytosis in dendritic spines. Cell 141(3):524–535. https://doi.org/10.1016/j.cell.2010.02.042
88. Kennedy MJ, Ehlers MD (2006) Organelles and trafficking machinery for postsynaptic plasticity. Annu Rev Neurosci 29:325–362. https://doi.org/10.1146/annurev.neuro.29.051605.112808
89. Konietzny A, Grendel J, Kadek A, Bucher M, Han Y, Hertrich N, Dekkers DHW, Demmers JAA, Grünewald K, Uetrecht C, Mikhaylova M (2022) Caldendrin and myosin V regulate synaptic spine apparatus localization via ER stabilization in dendritic spines. EMBO J 41(4):e106523. https://doi.org/10.15252/embj.2020106523

90. Konietzny A, Wegmann S, Mikhaylova M (2023) The endoplasmic reticulum puts a new spin on synaptic tagging. Trends Neurosci 46(1):32–44. https://doi.org/10.1016/j.tins.2022.10.012
91. Koppers M, Özkan N, Nguyen HH, Jurriens D, McCaughey J, Nguyen DTM, Li CH, Stucchi R, Altelaar M, MacGillavry HD, Kapitein LC, Hoogenraad CC, Farías GG (2024) Axonal endoplasmic reticulum tubules control local translation via P180/RRBP1-mediated ribosome interactions. Dev Cell 59(16):2053–2068.e9. https://doi.org/10.1016/j.devcel.2024.05.005
92. Krijnse-Locker J, Parton RG, Fuller SD, Griffiths G, Dotti CG (1995) The organization of the endoplasmic reticulum and the intermediate compartment in cultured rat hippocampal neurons. MBoC 6(10):1315–1332. https://doi.org/10.1091/mbc.6.10.1315
93. Kuijpers M, van de Willige D, Freal A, Chazeau A, Franker MA, Hofenk J, Rodrigues RJC, Kapitein LC, Akhmanova A, Jaarsma D, Hoogenraad CC (2016) Dynein regulator NDEL1 controls polarized cargo transport at the axon initial segment. Neuron 89(3):461–471. https://doi.org/10.1016/j.neuron.2016.01.022
94. Kulik YD, Watson DJ, Cao G, Kuwajima M, Harris KM (2019) Structural plasticity of dendritic secretory compartments during LTP-induced synaptogenesis. elife 8:e46356. https://doi.org/10.7554/eLife.46356
95. Land R, Fetter R, Liang X, Tzeng CP, Taylor CA, Shen K (2023) Endoplasmic reticulum exit sites scale with somato-dendritic size in neurons. MBoC 34(11):ar106. https://doi.org/10.1091/mbc.E23-03-0090
96. Lasiecka ZM, Winckler B (2011) Mechanisms of polarized membrane trafficking in neurons—focusing in on endosomes. Mol Cell Neurosci 48(4):278–287. https://doi.org/10.1016/j.mcn.2011.06.013
97. Leterrier C (2018) The axon initial segment: an updated viewpoint. J Neurosci 38(9):2135–2145. https://doi.org/10.1523/JNEUROSCI.1922-17.2018
98. Li H, Sung H-H, Huang Y-C, Cheng Y-J, Yeh H-F, Pi H, Giniger E, Chien C-T (2022) Fringe-positive Golgi outposts unite temporal Furin 2 convertase activity and spatial Delta signal to promote dendritic branch retraction. Cell Rep 40:111372. https://doi.org/10.1016/j.celrep.2022.111372
99. Lie PPY, Yang D-S, Stavrides P, Goulbourne CN, Zheng P, Mohan PS, Cataldo AM, Nixon RA (2021) Post-Golgi carriers, not lysosomes, confer lysosomal properties to pre-degradative organelles in normal and dystrophic axons. Cell Rep 35(4). https://doi.org/10.1016/j.celrep.2021.109034
100. Lin C-H, Li H, Lee Y-N, Cheng Y-J, Wu R-M, Chien C-T (2015) Lrrk regulates the dynamic profile of dendritic Golgi outposts through the golgin lava lamp. J Cell Biol 210(3):471–483. https://doi.org/10.1083/jcb.201411033
101. Liu C, Mei M, Li Q, Roboti P, Pang Q, Ying Z, Gao F, Lowe M, Bao S (2017) Loss of the golgin GM130 causes Golgi disruption, Purkinje neuron loss, and ataxia in mice. Proc Natl Acad Sci USA 114(2):346–351. https://doi.org/10.1073/pnas.1608576114
102. Lowenstein PR, Morrison EE, Bain D, Shering AF, Banting G, Douglas P, Castro MG (1994) Polarized distribution of the trans-Golgi network marker TGN38 during the in vitro development of neocortical neurons: effects of Nocodazole and Brefeldin a. Eur J Neurosci 6(9):1453–1465. https://doi.org/10.1111/j.1460-9568.1994.tb01007.x
103. Luarte A, Cornejo VH, Bertin F, Gallardo J, Couve A (2018) The axonal endoplasmic reticulum: one organelle—many functions in development, maintenance, and plasticity. Dev Neurobiol 78(3):181–208. https://doi.org/10.1002/dneu.22560
104. Ma D, Morris JF (2002) Protein synthetic machinery in the dendrites of the magnocellular neurosecretory neurons of wild-type Long-Evans and homozygous Brattleboro rats. J Chem Neuroanat 23(3):171–186. https://doi.org/10.1016/S0891-0618(01)00158-2
105. Ma W, Mayr C (2018) A Membraneless organelle associated with the endoplasmic reticulum enables 3′UTR-mediated protein-protein interactions. Cell 175(6):1492–1506.e19. https://doi.org/10.1016/j.cell.2018.10.007
106. Makhoul C, Gosavi P, Gleeson PA (2018) The Golgi architecture and cell sensing. Biochem Soc Trans 46:1063–1072. https://doi.org/10.1042/BST20180323

107. Malhotra V, Campelo F (2011) PKD regulates membrane fission to generate TGN to cell surface transport carriers. Cold Spring Harb Perspect Biol 3(2):a005280. https://doi.org/10.1101/cshperspect.a005280
108. Mallard F, Tang BL, Galli T, Tenza D, Saint-Pol A, Yue X, Antony C, Hong W, Goud B, Johannes L (2002) Early/recycling endosomes-to-TGN transport involves two SNARE complexes and a Rab6 isoform. J Cell Biol 156(4):653–664. https://doi.org/10.1083/jcb.200110081
109. Martin PEM, Blundell G, Ahmad S, Errington RJ, Evans WH (2001) Multiple pathways in the trafficking and assembly of connexin 26, 32 and 43 into gap junction intercellular communication channels. J Cell Sci 114(21):3845–3855. https://doi.org/10.1242/jcs.114.21.3845
110. Martínez-Menárguez JÁ, Tomás M, Martínez-Martínez N, Martínez-Alonso E (2019) Golgi fragmentation in neurodegenerative diseases: is there a common cause? Cells 8(7):748. https://doi.org/10.3390/cells8070748
111. Matsuki T, Matthews RT, Cooper JA, van der Brug MP, Cookson MR, Hardy JA, Olson EC, Howell BW (2010) Reelin and stk25 have opposing roles in neuronal polarization and dendritic Golgi deployment. Cell 143(5):826–836. https://doi.org/10.1016/j.cell.2010.10.029
112. Merianda TT, Lin AC, Lam JSY, Vuppalanchi D, Willis DE, Karin N, Holt CE, Twiss JL (2009) A functional equivalent of endoplasmic reticulum and Golgi in axons for secretion of locally synthesized proteins. Mol Cell Neurosci 40(2):128–142. https://doi.org/10.1016/j.mcn.2008.09.008
113. Mikhaylova M, Bera S, Kobler O, Frischknecht R, Kreutz MR (2016) A Dendritic Golgi Satellite between ERGIC and Retromer. Cell Rep 14(2):189–199. https://doi.org/10.1016/j.celrep.2015.12.024
114. Mitchell JW, Wildonger J (2023) Live imaging of Golgi outposts in drosophila dendritic arbors. Methods Mol Biol 2557:635–644. https://doi.org/10.1007/978-1-0716-2639-9_38
115. Moremen KW, Tiemeyer M, Nairn AV (2012) Vertebrate protein glycosylation: diversity, synthesis and function. Nat Rev Mol Cell Biol 13(7):448–462. https://doi.org/10.1038/nrm3383
116. Mukherjee A, Andrés Jeske Y, Becam I, Taïeb A, Brooks P, Aouad J, Monguillon C, Conduit PT (2024) γ-TuRCs and the augmin complex are required for the development of highly branched dendritic arbors in drosophila. J Cell Sci 137. https://doi.org/10.1242/jcs.261534
117. Mukherjee A, Brooks PS, Bernard F, Guichet A, Conduit PT (2020) Microtubules originate asymmetrically at the somatic golgi and are guided via Kinesin2 to maintain polarity within neurons. eLife 9:e58943. doi:https://doi.org/10.7554/eLife.58943
118. Nakagawa N, Iwasato T (2023) Golgi polarity shift instructs dendritic refinement in the neonatal cortex by mediating NMDA receptor signaling. Cell Rep 42(8). https://doi.org/10.1016/j.celrep.2023.112843
119. Nedozralova H, Basnet N, Ibiricu I, Bodakuntla S, Biertümpfel C, Mizuno N (2022) In situ cryo-electron tomography reveals local cellular machineries for axon branch development. J Cell Biol 221(4):e202106086. https://doi.org/10.1083/jcb.202106086
120. Nguyen MM, McCracken CJ, Milner ES, Goetschius DJ, Weiner AT, Long MK, Michael NL, Munro S, Rolls MM (2014) Γ-tubulin controls neuronal microtubule polarity independently of Golgi outposts. Mol Biol Cell 25(13):2039–2050. https://doi.org/10.1091/mbc.E13-09-0515
121. Ori-McKenney KM, Jan LY, Jan Y-N (2012) Golgi outposts shape dendrite morphology by functioning as sites of Acentrosomal microtubule nucleation in neurons. Neuron 76(5):921–930. https://doi.org/10.1016/j.neuron.2012.10.008
122. Park JH, Chung CG, Seo J, Lee B-H, Lee Y-S, Kweon JH, Lee SB (2020) C9orf72-associated arginine-rich dipeptide repeat proteins reduce the number of Golgi outposts and dendritic branches in drosophila neurons. Mol Cells 43:821–830. https://doi.org/10.14348/molcells.2020.0130
123. Perez-Alvarez A, Yin S, Schulze C, Hammer JA, Wagner W, Oertner TG (2020) Endoplasmic reticulum visits highly active spines and prevents runaway potentiation of synapses. Nat Commun 11(1):5083. https://doi.org/10.1038/s41467-020-18889-5

124. Peters A, Palay SL, Webster H deF (1991) Fine structure of the nervous system: neurons and their supporting cells, 3rd edition. Oxford University Press, New York Oxford
125. Peters A, Proskauer CC, Kaiserman-Abramof IR (1968) The small pyramidal neuron of the rat cerebral cortex : the axon hillock and initial segment. J Cell Biol 39(3):604–619. https://doi.org/10.1083/jcb.39.3.604
126. Petkovic M, Jemaiel A, Daste F, Specht CG, Izeddin I, Vorkel D, Verbavatz J-M, Darzacq X, Triller A, Pfenninger KH, Tareste D, Jackson CL, Galli T (2014) The SNARE Sec22b has a non-fusogenic function in plasma membrane expansion. Nat Cell Biol 16(5):434–444. https://doi.org/10.1038/ncb2937
127. Pick JE, Ziff EB (2018) Regulation of AMPA receptor trafficking and exit from the endoplasmic reticulum. Mol Cell Neurosci 91:3–9. https://doi.org/10.1016/j.mcn.2018.03.004
128. Pierce JP, Mayer T, McCarthy JB (2001) Evidence for a satellite secretory pathway in neuronal dendritic spines. Curr Biol 11(5):351–355. https://doi.org/10.1016/S0960-9822(01)00077-X
129. Qiu H, Wu X, Ma X, Li S, Cai Q, Ganzella M, Ge L, Zhang H, Zhang M (2024) Short-distance vesicle transport via phase separation. Cell 187(9):2175–2193.e21. https://doi.org/10.1016/j.cell.2024.03.003
130. Qu C, Gardner P, Schrijver I (2009) The role of the cytoskeleton in the formation of gap junctions by Connexin 30. Exp Cell Res 315(10):1683–1692. https://doi.org/10.1016/j.yexcr.2009.03.001
131. Quassollo G, Wojnacki J, Salas DA, Gastaldi L, Marzolo MP, Conde C, Bisbal M, Couve A, Cáceres A (2015) A RhoA signaling pathway regulates dendritic Golgi outpost formation. Curr Biol 25(8):971–982. https://doi.org/10.1016/j.cub.2015.01.075
132. Rabouille C (2017) Pathways of unconventional protein secretion. Trends Cell Biol 27(3):230–240. https://doi.org/10.1016/j.tcb.2016.11.007
133. Raiborg C, Wenzel EM, Pedersen NM, Olsvik H, Schink KO, Schultz SW, Vietri M, Nisi V, Bucci C, Brech A, Johansen T, Stenmark H (2015) Repeated ER–endosome contacts promote endosome translocation and neurite outgrowth. Nature 520(7546):234–238. https://doi.org/10.1038/nature14359
134. Raote I, Ortega-Bellido M, Santos AJ, Foresti O, Zhang C, Garcia-Parajo MF, Campelo F, Malhotra V (2018) TANGO1 builds a machine for collagen export by recruiting and spatially organizing COPII, tethers and membranes. eLife 7:e32723. doi:https://doi.org/10.7554/eLife.32723
135. Reid DW, Nicchitta CV (2015) Diversity and selectivity in mRNA translation on the endoplasmic reticulum. Nat Rev Mol Cell Biol 16(4):221–231. https://doi.org/10.1038/nrm3958
136. Rosso S, Bollati F, Bisbal M, Peretti D, Sumi T, Nakamura T, Quiroga S, Ferreira A, Cáceres A (2004) LIMK1 regulates Golgi dynamics, traffic of Golgi-derived vesicles, and process extension in primary cultured neurons. Mol Biol Cell 15(7):3433–3449. https://doi.org/10.1091/mbc.e03-05-0328
137. Roth J (2002) Protein N-glycosylation along the secretory pathway: relationship to organelle topography and function, protein quality control, and cell interactions. Chem Rev 102(2):285–303. https://doi.org/10.1021/cr000423j
138. Sakakibara A, Hatanaka Y (2015) Neuronal polarization in the developing cerebral cortex. Front Neurosci 9. https://doi.org/10.3389/fnins.2015.00116
139. Saraste J, Goud B (2007) Functional symmetry of endomembranes. Mol Biol Cell 18(4):1430–1436. https://doi.org/10.1091/mbc.e06-10-0933
140. Saraste J, Marie M (2018) Intermediate compartment (IC): from pre-Golgi vacuoles to a semi-autonomous membrane system. Histochem Cell Biol 150(5):407–430. https://doi.org/10.1007/s00418-018-1717-2
141. Satoh D, Sato D, Tsuyama T, Saito M, Ohkura H, Rolls MM, Ishikawa F, Uemura T (2008) Spatial control of branching within dendritic arbors by dynein-dependent transport of Rab5-endosomes. Nat Cell Biol 10(10):1164–1171. https://doi.org/10.1038/ncb1776

142. Saxena S, Foresti O, Liu A, Androulaki S, Rodriguez MP, Raote I, Aridor M, Cui B, Malhotra V (2024) Endoplasmic reticulum exit sites are segregated for secretion based on cargo size. Dev Cell 59(19):2593–2608.e6. https://doi.org/10.1016/j.devcel.2024.06.009
143. Schotman H, Karhinen L, Rabouille C (2008) dGRASP-mediated noncanonical integrin secretion is required for drosophila epithelial remodeling. Dev Cell 14(2):171–182. https://doi.org/10.1016/j.devcel.2007.12.006
144. Schroeder L, Barentine A, Merta H, Schweighofer S, Zhang Y, Baddeley D, Bewersdorf J, Bahmanyar S (2019) Dynamic nanoscale morphology of the ER surveyed by STED microscopy. J Cell Biol 218(1). https://doi.org/10.1083/jcb.201809107
145. Schwenk J, Fakler B (2021) Building of AMPA-type glutamate receptors in the endoplasmic reticulum and its implication for excitatory neurotransmission. J Physiol 599(10):2639–2653. https://doi.org/10.1113/JP279025
146. Seaman MNJ (2012) The retromer complex—endosomal protein recycling and beyond. J Cell Sci 125(Pt 20):4693–4702. https://doi.org/10.1242/jcs.103440
147. Segal M, Vlachos A, Korkotian E (2010) The spine apparatus, Synaptopodin, and dendritic spine plasticity. Neuroscientist 16(2):125–131. https://doi.org/10.1177/1073858409355829
148. Shibata Y, Shemesh T, Prinz WA, Palazzo AF, Kozlov MM, Rapoport TA (2010) Mechanisms determining the morphology of the peripheral ER. Cell 143(5):774–788. https://doi.org/10.1016/j.cell.2010.11.007
149. Shim SY, Wang J, Asada N, Neumayer G, Tran HC, Ishiguro K, Sanada K, Nakatani Y, Nguyen MD (2008) Protein 600 is a microtubule/endoplasmic reticulum-associated protein in CNS neurons. J Neurosci 28(14):3604–3614. https://doi.org/10.1523/JNEUROSCI.5278-07.2008
150. Shirane M, Nakayama KI (2006) Protrudin Induces Neurite Formation by Directional Membrane Trafficking.Science 314:818–821. https://doi.org/10.1126/science.1134027
151. Song A, Wang D, Chen G, Li Y, Luo J, Duan S, Poo M (2009) A selective filter for cytoplasmic transport at the axon initial segment. Cell 136(6):1148–1160. https://doi.org/10.1016/j.cell.2009.01.016
152. Spacek J, Harris K (1997) Three-dimensional organization of smooth endoplasmic reticulum in hippocampal CA1 dendrites and dendritic spines of the immature and mature rat. The journal of neuroscience : the official journal of the society for. Neuroscience 17(1). https://doi.org/10.1523/JNEUROSCI.17-01-00190.1997
153. Stern CA, Tiemeyer M (2001) A ganglioside-specific Sialyltransferase localizes to axons and non-Golgi structures in neurons. J Neurosci 21(5):1434–1443. https://doi.org/10.1523/JNEUROSCI.21-05-01434.2001
154. Steward O, Reeves TM (1988) Protein-synthetic machinery beneath postsynaptic sites on CNS neurons: association between polyribosomes and other organelles at the synaptic site. J Neurosci 8(1):176–184. https://doi.org/10.1523/JNEUROSCI.08-01-00176.1988
155. Stiess M, Maghelli N, Kapitein LC, Gomis-Rüth S, Wilsch-Bräuninger M, Hoogenraad CC, Tolić-Nørrelykke IM, Bradke F (2010) Axon extension occurs independently of centrosomal microtubule nucleation. Science 327(5966):704–707. https://doi.org/10.1126/science.1182179
156. Sun C, Schuman E (2023) A multi-omics view of neuronal subcellular protein synthesis. Curr Opin Neurobiol 80:102705. https://doi.org/10.1016/j.conb.2023.102705
157. Tanabe K, Kani S, Shimizu T, Bae Y-K, Abe T, Hibi M (2010) Atypical protein kinase C regulates primary dendrite specification of cerebellar Purkinje cells by localizing Golgi apparatus. J Neurosci 30(50):16983–16992. https://doi.org/10.1523/JNEUROSCI.3352-10.2010
158. Taverna E, Huttner WB (2019) The Golgi apparatus in polarized Neuroepithelial stem cells and their progeny: canonical and noncanonical features. In: Kloc M (ed) The Golgi apparatus and centriole: functions, interactions and role in disease. Springer, Cham, pp 359–375
159. Tie HC, Ludwig A, Sandin S, Lu L (2018) The spatial separation of processing and transport functions to the interior and periphery of the Golgi stack. elife 7:e41301. https://doi.org/10.7554/eLife.41301

160. Titze B, Genoud C (2016) Volume scanning electron microscopy for imaging biological ultrastructure. Biol Cell 108(11):307–323. https://doi.org/10.1111/boc.201600024
161. Tom Dieck S, Kochen L, Hanus C, Heumüller M, Bartnik I, Nassim-Assir B, Merk K, Mosler T, Garg S, Bunse S, Tirrell DA, Schuman EM (2015) Direct visualization of newly synthesized target proteins in situ. Nat Methods 12(5):411–414. https://doi.org/10.1038/nmeth.3319
162. Tooze SA, Tooze J, Warren G (1988) Site of addition of N-acetyl-galactosamine to the E1 glycoprotein of mouse hepatitis virus-A59. J Cell Biol 106:1475–1487. https://doi.org/10.1083/jcb.106.5.1475
163. Torre ER, Steward O (1996) Protein synthesis within dendrites: glycosylation of newly synthesized proteins in dendrites of hippocampal neurons in culture. J Neurosci 16(19):5967–5978. https://doi.org/10.1523/JNEUROSCI.16-19-05967.1996
164. Tozer S, Baek C, Fischer E, Goiame R, Morin X (2017) Differential routing of Mindbomb1 via Centriolar satellites regulates asymmetric divisions of neural progenitors. Neuron 93:542–551.e4. https://doi.org/10.1016/j.neuron.2016.12.042
165. Valenzuela JI, Jaureguiberry-Bravo M, Salas DA, Ramírez OA, Cornejo VH, Lu HE, Blanpied TA, Couve A (2014) Transport along the dendritic endoplasmic reticulum defines the trafficking modality for GABAB receptors. J Cell Sci jcs.151092. https://doi.org/10.1242/jcs.151092
166. Valm AM, Cohen S, Legant WR, Melunis J, Hershberg U, Wait E, Cohen AR, Davidson MW, Betzig E, Lippincott-Schwartz J (2017) Applying systems-level spectral imaging and analysis to reveal the organelle interactome. Nature 546(7656):162–167. https://doi.org/10.1038/nature22369
167. van Bommel DM, Toonen RF, Verhage M (2023) Mapping localization of 21 endogenous proteins in the Golgi apparatus of rodent neurons. Sci Rep 13:2871. https://doi.org/10.1038/s41598-023-29998-8
168. Varki A, Cummings RD, Esko JD, Stanley P, Hart GW, Aebi M, Mohnen D, Kinoshita T, Packer NH, Prestegard JH, Schnaar RL, Seeberger PH (eds) (2022) Essentials of Glycobiology, 4th edn. Cold Spring Harbor Laboratory Press, Cold Spring Harbor (NY)
169. Velasco A, Hendricks L, Moremen KW, Tulsiani DR, Touster O, Farquhar MG (1993) Cell type-dependent variations in the subcellular distribution of alpha-mannosidase I and II. J Cell Biol 122(1):39–51. https://doi.org/10.1083/jcb.122.1.39
170. Vernon CG, Copits BA, Stolz JR, Guzmán YF, Swanson GT (2017) N-glycan content modulates kainate receptor functional properties. J Physiol 595(17):5913–5930. https://doi.org/10.1113/JP274790
171. Vieira MM, Jeong J, Roche KW (2021) The role of NMDA receptor and neuroligin rare variants in synaptic dysfunction underlying neurodevelopmental disorders. Curr Opin Neurobiol 69:93–104. https://doi.org/10.1016/j.conb.2021.03.001
172. Voeltz GK, Prinz WA, Shibata Y, Rist JM, Rapoport TA (2006) A class of membrane proteins shaping the tubular endoplasmic reticulum. Cell 124(3):573–586. https://doi.org/10.1016/j.cell.2005.11.047
173. Vogelstein JT, Perlman E, Falk B, Baden A, Gray Roncal W, Chandrashekhar V, Collman F, Seshamani S, Patsolic JL, Lillaney K, Kazhdan M, Hider R, Pryor D, Matelsky J, Gion T, Manavalan P, Wester B, Chevillet M, Trautman ET, Khairy K, Bridgeford E, Kleissas DM, Tward DJ, Crow AK, Hsueh B, Wright MA, Miller MI, Smith SJ, Vogelstein RJ, Deisseroth K, Burns R (2018) A community-developed open-source computational ecosystem for big neuro data. Nat Methods 15(11):846–847. https://doi.org/10.1038/s41592-018-0181-1
174. Wang J, Daniszewski M, Hao MM, Hernández D, Pébay A, Gleeson PA, Fourriere L (2023) Organelle mapping in dendrites of human iPSC-derived neurons reveals dynamic functional dendritic Golgi structures. Cell Rep 42(7):112709. https://doi.org/10.1016/j.celrep.2023.112709
175. Wang J, Fourriere L, Gleeson PA (2020) Local secretory trafficking pathways in neurons and the role of dendritic Golgi outposts in different cell models. Front Mol Neurosci 13. https://doi.org/10.3389/fnmol.2020.597391

176. Wang Q, Wang Y, Yu F (2018) Yif1 associates with Yip1 on Golgi and regulates dendrite pruning in sensory neurons during drosophila metamorphosis. Development dev.164475. https://doi.org/10.1242/dev.164475
177. Wang T, Hanus C, Cui T, Helton T, Bourne J, Watson D, Harris KM, Ehlers MD (2012) Local zones of endoplasmic reticulum complexity confine cargo in neuronal dendrites. Cell 148(1–2). https://doi.org/10.1016/j.cell.2011.11.056
178. Wang Z, Edwards JG, Riley N, Provance DW, Karcher R, Li X, Davison IG, Ikebe M, Mercer JA, Kauer JA, Ehlers MD (2008) Myosin Vb mobilizes recycling endosomes and AMPA receptors for postsynaptic plasticity. Cell 135(3):535–548. https://doi.org/10.1016/j.cell.2008.09.057
179. Watson ET, Pauers MM, Seibert MJ, Vevea JD, Chapman ER (2023) Synaptic vesicle proteins are selectively delivered to axons in mammalian neurons. elife 12:e82568. https://doi.org/10.7554/eLife.82568
180. Weigand E, Keller BU (1998) Functional diversity of synaptic AMPA/KA receptors from rat as revealed by subtype-specific antagonists. Eur J Neurosci 10(1):64–70. https://doi.org/10.1046/j.1460-9568.1998.00014.x
181. Weigel AV, Chang C-L, Shtengel G, Xu CS, Hoffman DP, Freeman M, Iyer N, Aaron J, Khuon S, Bogovic J, Qiu W, Hess HF, Lippincott-Schwartz J (2021) ER-to-Golgi protein delivery through an interwoven, tubular network extending from ER. Cell 184(9):2412–2429.e16. https://doi.org/10.1016/j.cell.2021.03.035
182. Weiner AT, Seebold DY, Torres-Gutierrez P, Folker C, Swope RD, Kothe GO, Stoltz JG, Zalenski MK, Kozlowski C, Barbera DJ, Patel MA, Thyagarajan P, Shorey M, Nye DMR, Keegan M, Behari K, Song S, Axelrod JD, Rolls MM (2020) Endosomal Wnt signaling proteins control microtubule nucleation in dendrites. PLoS Biol 18(3):e3000647. https://doi.org/10.1371/journal.pbio.3000647
183. Williams ET, Chen X, Otero PA, Moore DJ (2022) Understanding the contributions of VPS35 and the retromer in neurodegenerative disease. Neurobiol Dis 170:105768. https://doi.org/10.1016/j.nbd.2022.105768
184. Witte H, Neukirchen D, Bradke F (2008) Microtubule stabilization specifies initial neuronal polarization. J Cell Biol 180(3):619–632. https://doi.org/10.1083/jcb.200707042
185. Wong ROL, Ghosh A (2002) Activity-dependent regulation of dendritic growth and patterning. Nat Rev Neurosci 3(10):803–812. https://doi.org/10.1038/nrn941
186. Wu Y, Whiteus C, Xu C, Hayworth K, Weinberg R, Hess H, De Camilli P (2017) Contacts between the endoplasmic reticulum and other membranes in neurons. Proc Natl Acad Sci USA 114(24). https://doi.org/10.1073/pnas.1701078114
187. Yalgin C, Ebrahimi S, Delandre C, Yoong LF, Akimoto S, Tran H, Amikura R, Spokony R, Torben-Nielsen B, White KP, Moore AW (2015) Centrosomin represses dendrite branching by orienting microtubule nucleation. Nat Neurosci 18:1437–1445. https://doi.org/10.1038/nn.4099
188. Yang SZ, Wildonger J (2020) Golgi outposts locally regulate microtubule orientation in neurons but are not required for the overall polarity of the dendritic cytoskeleton. Genetics 215(2):435–447. https://doi.org/10.1534/genetics.119.302979
189. Yarwood R, Hellicar J, Woodman PG, Lowe M (2020) Membrane trafficking in health and disease. Dis Model Mech 13(4):dmm043448. https://doi.org/10.1242/dmm.043448
190. Ye B, Zhang Y, Song W, Younger SH, Jan LY, Jan YN (2007) Growing dendrites and axons differ in their reliance on the secretory pathway. Cell 130(4):717–729. https://doi.org/10.1016/j.cell.2007.06.032
191. Ye Z, Marth JD (2004) N-glycan branching requirement in neuronal and postnatal viability. Glycobiology 14(6):547–558. https://doi.org/10.1093/glycob/cwh069
192. Yin D-M, Huang Y-H, Zhu Y-B, Wang Y (2008) Both the establishment and maintenance of neuronal polarity require the activity of protein kinase D in the Golgi apparatus. J Neurosci 28(35):8832–8843. https://doi.org/10.1523/JNEUROSCI.1291-08.2008

193. Zamponi E, Meehl JB, Voeltz GK (2022) The ER ladder is a unique morphological feature of developing mammalian axons. Dev Cell 57(11):1369–1382.e6. https://doi.org/10.1016/j.devcel.2022.05.002
194. Zhang C-L, Moutoussamy C, Tuffery M, Varangot A, Piskorowski R, Hanus C (2024) Core-N-glycans are atypically abundant at the neuronal surface and regulate glutamate receptor signaling:2024.03.25.586577
195. Zheng Y, Wildonger J, Ye B, Zhang Y, Kita A, Younger SH, Zimmerman S, Jan LY, Jan YN (2008) Dynein is required for polarized dendritic transport and uniform microtubule orientation in axons. Nat Cell Biol 10(10):1172–1180. https://doi.org/10.1038/ncb1777
196. Zhou W, Chang J, Wang X, Savelieff MG, Zhao Y, Ke S, Ye B (2014) GM130 is required for compartmental Organization of Dendritic Golgi Outposts. Curr Biol 24(11):1227–1233. https://doi.org/10.1016/j.cub.2014.04.008
197. Zucker B, Kozlov MM (2022) Mechanism of shaping membrane nanostructures of endoplasmic reticulum. Proc Natl Acad Sci 119(1):e2116142119. https://doi.org/10.1073/pnas.2116142119

Chapter 8
Models of Intracellular Transport: Contradictions and Current Understanding

Alexander A. Mironov and Galina V. Beznoussenko

Abstract Intracellular transport is one of the most complex and debated topics in cell biology. Numerous models and their combinations have been proposed to explain experimental data on this process. This article describes our current understanding of the secretory pathway by analyzing different transport models. To identify the most realistic model, we identified observations that cannot be fully explained by the vesicular, diffusion, or cisterna maturation–progression models. The kiss-and-run model appears as the most powerful. From this perspective, vesicles at the ER exit sites are considered to be COPI-dependent. In cells with the central Golgi, the movement of carriers between the Golgi and peripheral structures occurs along microtubules via a bolus-like mechanism. In cells lacking microtubules, free ER-to-Golgi and Golgi-to-plasmalemma (or Golgi-to-endosome) carriers are extremely rare. Intra-Golgi transport relies heavily on temporary connections between cisternae, which subsequently break down. Cisternal perforations play a crucial role in this process. Cargo moves directionally through the Golgi stack, reaching the last two medial-cisternae and/or the *trans*-most cisterna, which function as the Golgi exit site. At this point, the dynamics of transport shifts from linear cargo movement to an exponential decay pattern. Post-Golgi carriers have to replace their own SNAREs with SNAREs that are able to facilitate their fusion with the plasmalemma.

Keywords Intracellular · Diffusion model · Kiss-and-run model · Cisterna maturation-progression model · Vesicular (dissociation) model

A. A. Mironov (✉) · G. V. Beznoussenko
Department of Cell Biology, IFOM ETS—The AIRC Institute of Molecular Oncology, Milan, Italy
e-mail: alexandre.mironov@external.ifom.eu; galina.beznusenko@ifom.eu

A. Nakano, J. Saraste (eds.), *The Golgi Network, Volume I*, Subcellular Biochemistry 110, https://doi.org/10.1007/978-3-032-06936-8_8

Abbreviations

COPI	coatomer I
COPII	coatomer II
CMPM	cisternal maturation and progression model
DM	diffusion model
ER	endoplasmic reticulum
KARM	kiss-and-run model
TGN	*trans*-Golgi network
VLDL	very low-density lipoprotein
VM	vesicular model

Introduction

In most cells, the Golgi plays a crucial role. Numerous studies have explored the connection between defects in intracellular transport and human pathologies. Moreover, most of the molecular machinery involved in intracellular transport, as well as its interaction with secretory compartments, has now been elucidated (reviewed by [115, 167, 176], Mironov et al. 2022, [66, 102, 119, 127]). Despite these advancements, many issues in the field of secretory trafficking remain unclear.

The main challenge in the field of intracellular transport is the ongoing competition between the vesicular model, the cisterna maturation–progression model, the diffusion model, and the kiss-and-run model for dominance as the prevailing paradigm. Despite this, the predominant view on intracellular transport remains largely restricted to vesicle-based models, often overlooking the wealth of information accumulated during recent years. For example, Koike and Jahn [84] write in their abstract: " … transport vesicles contain targeting signals such as Rab GTPases and polyphosphoinositides that are recognized by tethering factors in the cytoplasm and that connect the vesicles with their respective destination compartment." Similarly, Mogre et al. [123] assert that "a recycling pathway depends on the motor-driven transport of coat protein I (COPI)-coated vesicles, which shuttle transport receptors and leaked ER-resident proteins back into the ER, thereby maintaining proteostasis within the organelles."

In 2028, we will mark the 130th anniversary of the discovery of the Golgi. This year commemorates the 50th anniversary of Palade's [138] introduction of the ideas explaining intracellular transport, the 45th anniversary of the publication of the vesicular model [154], and the 35th anniversary of the presentation of the seminal paper by Kaiser and Schekman [75]. Additionally, nearly 60 years has passed since the first description of intracellular transport by Palade and coworkers [138]. However, in recent years, a substantial amount of data has accumulated that cannot be explained by the intracellular transport model based solely on transport vesicles. The history of various models describing intracellular transport is summarized in

Table 8.1 History of the models of intracellular transport

Year	Formulation of a model	References
1980	The vesicular model (VM) based on the recycling of resident proteins by clathrin vesicles	[154]
1981	Improvement of the VM	[46]
1989	Progression model	[92]
1997	Cisterna maturation–progression model (CMPM) based on COPI vesicles	[9, 122, 60, 161]
1997	CMPM based on retrograde diffusion along connections between compartments	[122]
1998	The synthetic model of intra-Golgi traffic	[111]
2000	Percolating vesicles	[134]
2000	Mega-vesicles as carries	[180]
2001	Maturation of cargo domain (or the asymmetric variant of the kiss-and-run model (KARM) of intra-Golgi transport)	[113]
2001	The concentration by exclusion	[131]
2002	Comparison of the models	[17]
2005	Mega-vesicles near the ER exit sites	[54]
2008	KARM at the level of intra-Golgi transport	[112]
2008	KARM for regulated protein secretion	[111]
2009	KARM at the level of post-Golgi transport	[72]
2010	The cisternal progenitor model	[142]
2014	KARM at the level of ER–Golgi transport	[88]
2019	Improved KARM at the ER–Golgi transport	[115]
2019	Channels for the exit of mega-cargo from the ER	[151]

Table 8.1. Therefore, re-evaluation of these transport models is a crucial task for modern cell biology.

Structure of the Secretory Pathway

The structure of the ER–Golgi interface and the Golgi is well known ([8, 120, 143]; Mironov and Pavelka 2008; [78, 121]). Briefly, ER exit sites appear as distinct, long-lived puncta scattered throughout both the perinuclear and peripheral ER network. In routinely used mammalian cells, which typically have a diameter of approximately 40 μm, there are around 200 ER exit sites [64, 123, 168, 181].

ER exit sites consist of coatomer II (COPII)-coated buds on the granular ER, COPI-coated buds on smooth tubules, a three-dimensional tubular network with partial COPII coat, and free uncoated vesicles with a mean diameter of about 50 nm [8, 120]. ER exit sites are in direct continuity with the ER [35, 68, 87, 109, 120, 122, 170, 189]. It is important to understand that ER–Golgi and post-Golgi transport take place within a dense aqueous medium where the motion of any particle necessitates flow or rearrangement of the surrounding fluid.

ER exit sites are not found in absorptive enterocytes or in the endothelial cells of the aorta [162, 163]. However, in endothelial cells, these structures are present in cells that are in the G2-phase [163]. In the acinar epithelial cells of the pancreas and salivary glands, ER exit sites are large and located only near the Golgi [121]. A similar situation is observed in neurons (see the book by [141]) and in Goblet cells, where ER exit sites are small [117].

The Golgi forms stacks composed of 3–8 (rarely more) flat cisternae with perforation along their rims [16, 91, 121]. Generally, stacks can be connected by tubules, forming a ribbon. In most cells within tissues, the Golgi ribbon is formed primarily by the cisternae themselves [80 83], whereas in cell culture Golgi structure can be connected by tubules. The ribbon can also be formed by cisternae arranged in a spiral-like structure [175]. The Golgi ribbon is absent in cells lacking microtubules [121]. In *S. cerevisiae* and the microsporidia *Paranosema (Antonospora) grylli*, Golgi stacks are absent.

Transporting stacks additionally contain the *cis*-most and the *trans*-most Golgi cisternae. These cisternae are absent in resting stacks [103, 121, 177]. The rims of Golgi cisternae contain COPI-coated buds. Near cisternal rims, 50–52-nm uncoated vesicles are observed. The number of the 52-nm vesicles is higher in resting stacks, whereas in transporting stacks, inter-cisternal connections can be detected [121].

Importantly, in some organisms, coatomer I-dependent vesicles are either absent or very few in number. This has been observed in the microsporidia *Paranosema (Antonospora) grylli* and *Paranosema locustae* (see Fig. 8.4c–e presented in [15]); the alga *Ostreococcus tauri* [67]; the parasite *Plasmodium falciparum* [69]; and *Tripanosoma cruzi* (see movies and Fig. 8.2i presented in [59]).

Clermont et al. [36] described the *trans*-Golgi network (TGN) as a tubular network connected to the rims of the two last Golgi cisternae and various types of endosomes. Clathrin-dependent uncoated vesicles with diameters of 60, 70, and 80 nm are visible within the post-Golgi zone. Some of these free vesicles are clathrin-coated, while others have already shed their coats [121].

Certain apical proteins, such as the raft-associated apical marker influenza hemagglutinin, have also been found to move through endosomes on their way to the plasmalemma [38]. A *trans*-endosomal route is known for several newly synthesized plasma membrane proteins [27, 56, 62], including the asialoglycoprotein receptor [93], polymeric immunoglobulin receptor (pIgR) [136], and E-cadherin [97]. This route has also been observed for the G protein of the vesicular stomatitis virus [3].

In cells where secretion is regulated, secretory granules are visible. These granules can form in the center of the cell near the Golgi and may then be transported to the basolateral plasmalemma, either already as mature or still undergoing maturation. In some cells, secretory granules form when the cargo is outside the Golgi [36, 111, 150].

Problems and Contradictions of Different Models

Schemes of the most popular models of intracellular transport and the requirements for different models are presented in Figs. 8.1 and 8.2 and in Table 8.2.

The Vesicular (Dissociation) Model (VM)

In general, the vesicular model proposes that small membrane spheres (vesicles) form within proximal secretory compartments, move toward their target destination, and transport cargo to the distal compartment. In most cases, vesicle formation is mediated by protein coat machineries such as coatomer and clathrin, while vacuoles may form through other mechanisms (Fig. 8.1a).

However, many immunolabelling studies have shown that the concentration of most cargos in COPI-dependent vesicles is lower than in precursor compartments [16, 18, 105, 113, 115, 122]. Additionally, the existence of isolated mega-vesicles has not been conclusively demonstrated. Moreover, vesicles isolated by Adolf et al. [1] containing COPII subunits do not contain albumin, indicating that albumin is not transported by COPII vesicles.

An important problem of the vesicular model is the absence of a mechanism capable of accelerating the diffusion of 52-nm vesicles, which presumably function as carriers. In traditionally used cells, the diffusion of 50-nm particles is nearly impossible [99, 100]. However, in cells, active transport can be achieved through advection, where particles are carried along by the flow of intracellular fluids. This flow could drive small vesicles toward the cell periphery, but vesicle strings prevent movement [133]. Importantly, in neurons, motor-driven transport takes approximately 10 days to deliver particles from the cell body to the tips of these long cellular processes [124]. It remains unclear why VSVG, a routinely used cargo, leaves the ER at 40 °C under conditions where COPI is non-functional, either due to the addition of brefeldin A [160] or after incubation with AlF_4 [65]. Experiments with brefeldin A suggest that COPI is the main mechanism responsible for the exit of Golgi enzymes from the ER [10, 121], whereas COPII-dependent vesicles are not necessary for the transport of Golgi enzymes from the ER to the Golgi [40, 76].

We believe that chylomicrons—and presumably very low-density lipoproteins (VLDL)—move through the smooth tubes of the ER [162] as a result of peristalsis. However, when the size of mega-cargos becomes excessive, the autophagy pathway is activated. Autophagosomes derived from ER protrusions filled with protein aggregates have been described ([52, 53, 130]; see also [49]). We suggest that the mega-vesicles observed by Gorur et al. [61] and the large procollagen-positive immobile dots observed by Omari et al. [130] and McCaughey et al. ([108]: see Figure 8.2 in their paper) represent the ER-derived autophagosomes (also called reticulophagosomes; see [49, 52, 53, 130]).

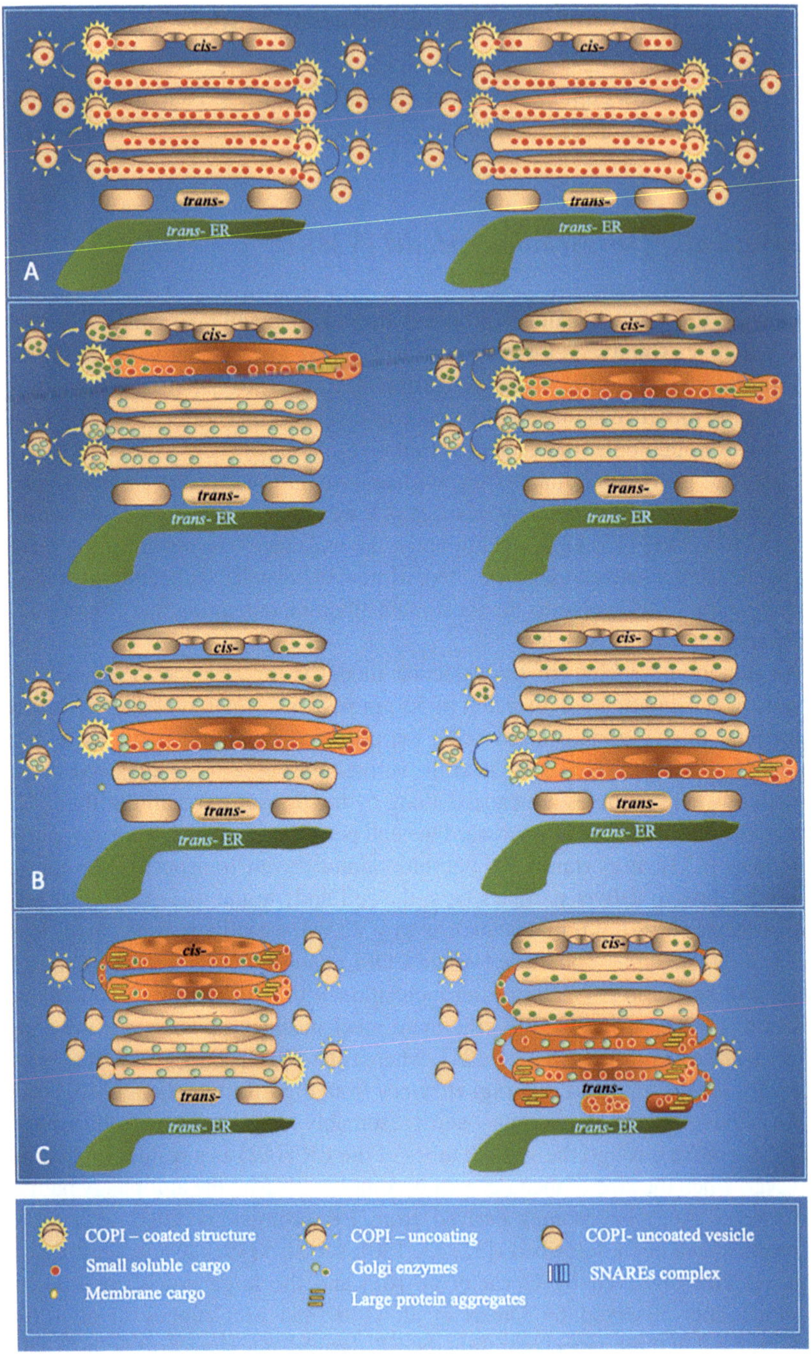

Fig. 8.1 Scheme of the main principles and different models of the intracellular transport. (**a**) The Vesicular (Dissociation) Model (VM). This model is based on the formation of a small membrane

(continued)

Within the post-Golgi interface, the delivery of post-Golgi carriers to the apical or basolateral plasmalemma depends on the localization of SNAREs, Rabs, and clathrin adaptors ([27, 86]; [41, 62]). It has been proven that vesicle fusion does not contribute to the formation of post-Golgi carriers [146].

Albumin follows a completely different mechanism of post-Golgi transport that does not rely on carriers containing concentrated cargo. Albumin is highly enriched in the lumen of vacuoles derived from the *trans*-most cisterna/TGN. These vacuoles are particularly evident in hepatocytes and are connected to the plasmalemma by thin tubular structures [14, 16, 18].

Tubules and dots positive for the apically directed protein p75 move from the TGN toward the apical plasmalemma along microtubules and actin bundles [41, 85]. However, their 3D structures remain unknown. The apical plasmalemma contains microvilli and a dense actin network beneath them, making transport rather difficult. Until now, no one has demonstrated these carriers passing across this actin barrier [111].

Attempts to modify the vesicular model by assuming the existence of so-called mega-buds and mega-vesicles were not convincing because COPII-coat was not observed on procollagen I (PCI)- or VLDL-containing ER protrusions and post-Golgi carriers [115]. Additionally, according to the vesicular model, the putative COPII vesicles should contain Sec22, which binds to Sec24, but this was also not observed [126, 152].

The Cisterna Maturation–Progression Model (CMPM)

Initially, the cisterna maturation–progression model (CMPM) existed in the form of a simple progression model describing how a large compartment detaches from the ER or the Golgi and directly fuses with the Golgi or plasmalemma, respectively. However, this model supposes that Golgi resident proteins should be present on the plasmalemma, which is not the case. As a result, a new model combining cisternal progression and recycling of the residents was proposed in two forms, namely, the model depicting COPI-vesicle-dependent recycling (Fig. 8.1b) and another one

Fig. 8.1 (continued) bud on the proximal donor compartment with the help of a protein coat. Following its detachment, the bud undergoes fission and subsequent uncoating. The resulting coat-dependent vesicle then moves toward the distal target compartment, where it is captured by a tethering system. Finally, with the aid of SNAREs, the vesicle fuses with the distal compartment. Cargo moves through the secretory pathway via these transport vesicles. (**b**) The Cisterna Maturation–Progression Model (CMPM). In this model, cisternae themselves move from the distal pole of the Golgi stack toward the proximal one, traveling from ER exit sites to the Golgi, through the Golgi, and toward the endosome or plasmalemma. Simultaneously, small vesicles enriched in resident proteins (Golgi enzymes) must detach from the distal compartment, be transported to the proximal compartment, and deliver their resident proteins there. (**c**) The Diffusion Model (DM). The cargo diffuses passively along the membrane continuity (continuous tubular networks connecting compartments) between a proximal compartment and distal compartment

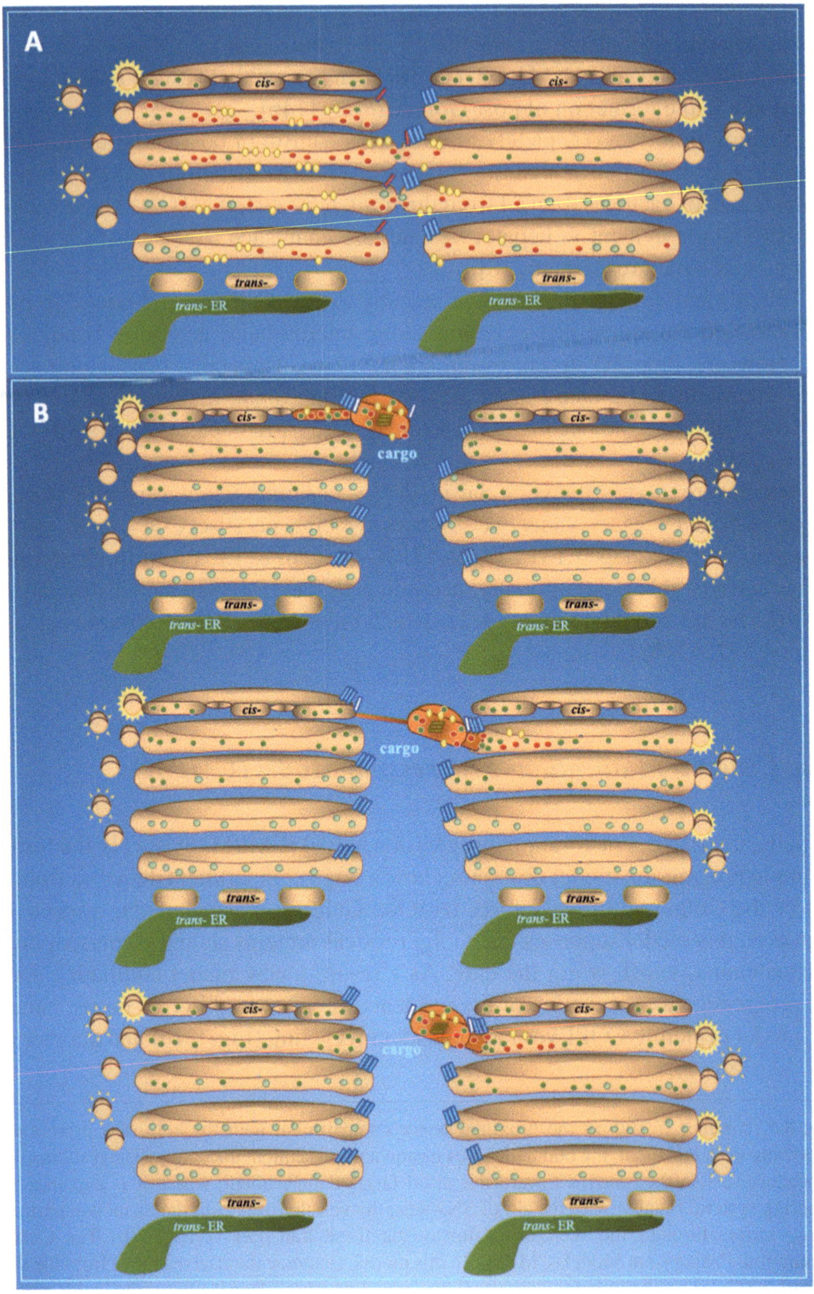

Fig. 8.2 Scheme of the Kiss-and-Run Model (KARM). This model proposes that two compartments temporarily fuse. To facilitate fusion, SNAREs must be concentrated on the distal surface of the cargo domain and the proximal surface of the distal target compartment. While the two compartments are connected, cargo and resident proteins can diffuse in both directions. Subsequently,

(continued)

based on the recycling of residents along membrane continuities [9, 60, 122]. Currently, the CMPM is often considered as a dogma [4, 101, 102].

At the ER–Golgi interface, the modern CMPM suggests that within the ER exit site, COPII forms a protrusion on the ER filled with cargo. In the next step, COPI-coated buds form on the surface of this protrusion. These buds then undergo fission, leading to the formation of COPI-coated vesicles enriched in resident proteins or retrograde cargo. Subsequently, the COPI coat detaches, allowing the vesicle to fuse with the ER or ER exit site. This process facilitates the recycling of resident proteins.

At the level of the Golgi stack, the CMPM proposes that the ER–Golgi carriers arrive at the *cis*-side of the stack and form a new *cis*-Golgi cisterna. Simultaneously, the last cisterna on the *trans*-side of the stack is transformed into post-Golgi carriers. Then, COPI-coated vesicles enriched in Golgi resident proteins are formed in each Golgi cisterna, undergo fission, and subsequently lose their coats. Finally, the vesicles formed on the distal cisterna fuse with the proximal cisterna. This process occurs at all levels of the Golgi stacks, with each pair composed of proximal and distal cisternae. After this sequence of events, the newly formed *cis*-Golgi cisterna acquires resident proteins from the medial-Golgi cisterna, which, in turn, received resident proteins from *trans*-Golgi cisterna, and so on. This cycle is repeated multiple times. According to this model, the Golgi cisternae are not stable but are instead subject to continuous renewal [115].

Similarly, at the level of the Golgi exit sites, the protrusion of the *trans*-Golgi cisterna is transformed into a post-Golgi carrier through the recycling of resident proteins via clathrin- or COPI-dependent vesicles. These vesicles form, undergo uncoating, and subsequently fuse with the proximal cisterna. According to mathematical modeling, the concentration of resident proteins in COPI- or clathrin-dependent vesicles must be higher than in the compartments where the vesicles are formed [60]. However, there is no convincing evidence supporting the role of COPI vesicles as the retrograde carriers. For instance, Gilchrist et al. [58] demonstrated that four retrograde cargos are depleted in COPI vesicles. However, COPI-dependent vesicles have a very high surface area-to-volume ratio. As a result, they would remove more surface area than volume from immature ER-to-Golgi carriers. This could lead to a situation where the surface area is finished while some volume remains.

Fig. 8.2 (continued) the proximal site of the cargo domain undergoes fission. This process involves the fission of thin tubules or bridges between the pores. (A) The symmetric KARM assumes that two compartments temporarily fuse. During the period when they are connected, cargo proteins and resident proteins can diffuse in both directions. Subsequently, the site of fusion undergoes fission. (B) The asymmetric KARM. Within this framework, the cargo domain and the proximal compartment are connected by a thin tubule or a row of pores. The distal surface of the cargo domain then fuses with the proximal surface of the distal compartment. After a certain period, fission occurs, leading to the separation of the thin tubule or the bars between the pores. Thus, fusion occurs at one site, while fission takes place at another site, specifically where the thin tubule or pores are localized. To facilitate fusion, it is important to concentrate SNAREs at the distal site

Table 8.2 Requirements of models of intracellular transport

	Models of transport			
Necessary features	VM	CMPM	KARM	DM
1. COPII vesicles	Necessary	Not necessary	Not necessary	Not necessary
2. Golgi or ERES resident proteins in COPI vesicles	Not necessary	Necessary	Not necessary	Not necessary
3. Temporal fusion of the distal compartment with the proximal one	Not necessary	Not necessary	Necessary	Not necessary
4. Presence of GOS1 or GS27 or GS28 in 50-nm vesicles	Not necessary	Not necessary	Necessary	Not necessary
5. Presence of Sec22 in the 50-nm vesicles	Necessary	Not necessary	Not necessary	Not necessary
6. Mega-buds and mega-vesicles with COPII coat	Necessary	Not necessary	Not necessary	Not necessary
7. Presence of many SNAREs	Necessary	Necessary	Necessary	Not necessary
8. Total membrane continuity	Not necessary	Not necessary	Not necessary	Necessary

Moreover, a simultaneous increase in amylase concentration was only 3.7-fold, suggesting that several types of COPI vesicles (presumably for different cargo) may be operating. However, this has not yet been demonstrated. At the ER-to-Golgi level, Oprins et al. [131] observed a significant 57.6-fold enrichment of chymotrypsinogen at ER exit sites and proposed a model based on "concentration by exclusion." On the other hand, fluorescent spots carrying membrane cargo toward the Golgi do not exhibit an increase in fluorescence intensity during centripetal movement [149, 159, 169].

Importantly, in most studies performed in intact cells, Golgi vesicles are depleted of Golgi resident proteins [37, 55, 90, 132]. During mitosis, COPI-dependent vesicles are also depleted in Golgi enzymes ([57, 74]: [116]). Similarly, in *S. cerevisiae,* Golgi resident proteins are absent from COPI vesicles [20]. Moreover, the membranes of COPI-dependent vesicles are thinner than other membranes along the secretory pathway [135]. It is important to note that the Golgi contains not only Golgi glycosylation enzymes but also nucleotide sugar transporters, which have 10 transmembrane domains and form dimers and oligomers with a width of 20 nm [26, 31, 32, 45, 95, 106]. Thus, it is highly unlikely that these transporters—in their intact molecular form—could be enriched in membranes of COPI vesicles.

These data appear to contradict data published by Martinez-Menarguez et al. [105], Gilchrist et al. [58], Rizzo et al. [153], and Pothukuchi et al. [148]. However, upon detailed analysis, we demonstrated that these findings can also be interpreted with the framework of the kiss-and-run model (KARM; [115]). For instance, Martinez-Menarguez et al. [105] assumed that COPI-coated round profiles near the Golgi cisternae represent free vesicles. However, three-dimensional analysis was not performed, and the presence of Golgi enzymes in the 52-nm COPI-dependent vesicles was not confirmed. Moreover, genuine COPI-dependent vesicles are uncoated [25, 104]. On the other hand, the vesicles described by Munro's group [185, 186] could be clathrin-dependent.

Next, the augmentation of the numerical density of the molecules contained in cargo aggregates in the Golgi cisternae, moving from *cis*-to-*trans* side of the stack, does not align with the essence of the CMPM.

Finally, the CMPM cannot explain the observation that in *S. cerevisiae*, when cargo exit from the ER is blocked, the Golgi disappears [5, 124]. Notably, the elimination of COPI vesicles using a temperature-sensitive mutant of one of the COPI subunits does not block the maturation process [107]. However, maturation becomes slower. These findings suggest that while COPI plays an important role, another mechanism (presumably the KARM) also contributes to maturation. Additionally, the maturation model based on vesicle-mediated recycling of resident proteins is not suitable for the post-Golgi transport step, as there are no COPI-coated buds on the *trans*-most cisterna or the TGN.

Diffusion Model (DM)

The diffusion model (Fig. 8.1c) suggests that all compartments are interconnected, allowing cargo to diffuse along tubules from the proximal compartment to the distal one. Examples of such proximal-distal compartment pairs include ER exit sites and the Golgi, a proximal and a distal Golgi cisterna (or medial-cisterna and the *trans*-most cisterna), and the Golgi exit site and the TGN/endosome or plasma membrane.

The diffusion model is supported by several observations. First, cargo transport-dependent connections between Golgi cisternae have been demonstrated [177]. These connections could facilitate cargo diffusion [18]. Additionally, Pagano et al. [137] showed that in cells fixed with aldehyde (but not with OsO_4, which damages lipid bilayers), lipid markers added externally can reach the nuclear envelope. Furthermore, the regression line describing the dynamics of fluorescent cargo within the Golgi follows an exponential decay (or negative exponential) pattern [14, 140].

Within the framework of the diffusion model, ER-to-Golgi transport occurs via diffusion along persistent connections between the ER and the Golgi. Direct membrane continuities between organelles have been described multiple times [22, 24, 34, 47, 48, 51, 70, 87, 94, 98, 125, 129, 158, 164, 178, 184]), and have been reviewed by Mironov et al. [122] [121]).

These observations were based on various imaging and analytical techniques, including high-voltage electron microscopy [94], scanning electron microscopy [175], three-dimensional reconstruction of serial sections [164], functional transport analysis [87], and electron tomography [91]. Notably, connections between the ER and pre-Golgi carriers have been observed even after quick-freezing procedures [91, 120]. Ladinsky et al. [91] described a connection between the ER and a small cisterna that exhibits all the characteristic features of Golgi cisternae, including buds and small pores (see Figure 3 in [91]). They interpreted this structure as "the specialized domain of the ER." However, such small pores have never been observed in ER cisternae [91].

Analysis of the Golgi in Goblet cells and in other cells with regulated secretion raises new questions about the role of cargo diffusion [111, 117]. In secretory acinar cells of the pancreas and salivary glands, the Golgi forms a spiral structure [80–83, 117, 121] similar to that demonstrated by Tanaka et al. [174, 175]. In Goblet cells and other mucus-producing or secretory protein-producing cells, cargo diffusion serves as the primary mechanism for intra-Golgi transport. However, in Goblet cells, the diffusion model is only partially valid due to interruptions along the pathway between the ER and the apical plasmalemma. These interruptions occur at three key points: (1) between the ER exit site and one of the first three cisternal layers of the Golgi, (2) between the cisternal distension with the last cisternal layer and secretory granules, and (3) between the secretory granule and the apical plasmalemma [117].

Additionally, other regulated secretory proteins become progressively concentrated during their transport. Their concentration continues to increase in the Golgi and further in cells where these proteins are efficiently packaged into secretory granules [166]. While prolactin can be up to 200 times more concentrated in dense-core secretory granules compared to the lumen of the ER, amylase concentration in exocrine secretory granules is more moderate, involving small concentration steps at both the *cis*- or *trans*-sides of the Golgi stack [111]. Patches of clathrin have been observed on secretory granules [2, 111, 155], although their function is unknown. Therefore, in cells that secrete proteins via secretory granules, the most effective transport mechanism is likely to involve a combination of the KARM and DM models, with diffusion occurring only within the Golgi.

Additionally, the discovery of the ultrafast movement of the Golgi [66] has revived some older models that were previously considered impossible, such as the peristaltic movement proposed by Griffiths [63]. However, the DM alone cannot account for the necessity of SNAREs in intracellular transport or the concentration of cargo at different stages of transport.

Kiss-and-Run Model (KARM)

Initially, the kiss-and-run model was proposed to explain phenomena related to the function of synaptic vesicles by Ceccarelli et al. [29]. Later, this model was applied to fusion between endosomes and lysosomes [23], secretory granules and the plasma membrane in neuroendocrine cells (reviewed by [179]), and post-Golgi carriers and the plasma membrane [72]. Subsequently, we demonstrated the significant potential of this model for intracellular transport [112, 114]. Indeed, the formation of large dense-core vesicles in neurons [21, 79, 157], adrenal chromaffin cells [30, 187], and insulin-containing dense-core secretory vesicles in pancreatic ß-cells [156], as well as the secretion of synaptic vesicle contents in neurons [39, 192], and lysosome biogenesis [171] can be explained using KARM. At the level of the ER–Golgi interface, this mechanism has been proposed in yeast [50, 88]. Thus, to explain various biological transport processes, scientists have started to use KARM more frequently. Over time, KARM has gradually replaced other models.

The fast recycling of internalized receptors from early endosomes to the plasma membrane occurs via the kiss-and-run process [188]. Fusion between early and late endosomes follows the KARM [11, 12, 71]. Additionally, the contents of endosomes and lysosomes are mixed via two different pathways: kiss-and-run and direct fusion [73]. The mechanisms of the KARM are described in detail in our previous reviews [112, 114, 115].

The main principle of the KARM is the temporal fusion between proximal and distal compartments ([112, 114, 115]; Fig. 8.2). Within the framework of the symmetrical KARM, cargo diffuses through a narrow connection from the proximal to the distal compartment, followed by the concentration and retention of soluble cargo in the distal compartment. This type of model primarily applies to mobile soluble cargo, such as albumin. In contrast, the asymmetrical KARM applies to cargo with a well-defined cargo domain and limited mobility within the membrane compartment, such as procollagen (PCI). In this case, fusion occurs only between the distal part of the cargo domain and the proximal edges of the distal cisternal compartments, while fission takes place within the proximal compartment, where rows of pores or thin tubules connect the cargo domain to the rest of the proximal compartment. After fusion–fission events, the distal compartment increases in volume, as the cargo-containing domain becomes part of the distal compartment.

Within the framework of the asymmetric KARM, membrane cargo and megacargo should be organized into domains, as large cargo domains are more efficient for transport. SNAREs must be concentrated over cargo domains to enhance the efficiency of intracellular transport. These cargo domains should be separated from the rest of a proximal compartment by pores or thin tubules [16]. During the transport, these thin tubules must be broken down with the help of fission machinery [183, 190, 191]. These pores are important only for the KARM. In cells with a fragmented Golgi, ER–Golgi and post-Golgi carriers must at all times remain connected to other membranes [144–147]. If the distance between proximal and distal compartments is long, i.e., within the ER-to-Golgi or Golgi-to-plasmalemma interfaces, the distal compartment must extend a protrusion toward the proximal one and fuse with the cargo domain [28]. Free ER–Golgi and Golgi-to-plasmalemma carriers do not exist [112, 115]. The regression line describing the emptying of the Golgi zone would resemble an exponential decay. The arrival of an ER-to-Golgi carrier at the *cis*-Golgi induces a flux of Ca^{2+} from ER exit sites [110], stimulating the fusion of the carrier with the medial-Golgi. Notably, the KARM does not contradict the maturation mechanism [115].

The KARM explains many experimental findings that remain unclear within the framework of other models. For instance, the so-called "land-locked" experiments by Dunlop et al. [44], in which small cargos pass through a Golgi that is fragmented and synthetically "glued" to mitochondria, challenge not only the CMPM but also the KARM model. However, a careful analysis of the images presented in this study reveals that, under these conditions, mitochondria formed aggregates where Golgi cisternae were positioned close to one another, all grouped at the center [44]. Consequently, the proximal and distal compartments were in close proximity, making direct contact between them possible. Due to this arrangement, the transport processes could occur according to KARM.

Recently, we experimentally tested which model is most powerful [14, 16]. To this end, we compared the CMPM with the KARM in one paper, and the DM with the KARM in another paper. In both cases, the KARM proved to be much more powerful, at least at the level of the Golgi. The CMPM, DM, and VM cannot explain why any cargo domain is separated from the Golgi cisternae of the proximal compartment by a row of pores (within the Golgi stack), or a thin tubule (for instance, within the ER–Golgi interface), whereas the KARM easily explains this phenomenon [16].

Additionally, it is important to address the contradictions between the KARM and the findings described in the cornerstone papers by Kaiser and Schekman [75] and Barlowe et al. [13]. Kaiser and Schekman [75] used strains of *S. cerevisiae* containing temperature-sensitive mutations in several secretory machinery proteins, specifically Sec17 and Sec18 (analogs of mammalian NSF and αSNAP accessory proteins, which are essential for SNARE function), as well as Sec23, a subunit of COPII. In their experiments, the temperature-dependent inhibition of SNARE function led to the accumulation of 50-nm vesicles. The authors concluded that these vesicles are COPII-dependent. However, the size of these vesicles does not correspond to the size of COPII-coated buds [6, 21]. Therefore, an alternative interpretation is possible. We already published a detailed explanation of how the results presented in these two papers could be interpreted from the point of view of the KARM [115].

Another significant challenge for the KARM is the conclusion that COPI-dependent vesicles are enriched in Golgi enzymes. This conclusion is primarily based on the isolation of vesicular fractions. For example, Gilchrist et al. [58] demonstrated that the vesicular fraction of Golgi membranes, obtained after incubating isolated Golgi membranes with cytosol and GTP, was depleted of secretory cargos but enriched in Golgi enzymes. Specifically, they reported a 2.7-fold enrichment of Man-I, a 4.5-fold enrichment of GalNAc-T1 (Galactosyl N-Acetyl Transferase type 1), and a 1.4-fold enrichment of Man-II, compared to total Golgi membranes. Electron microscopic analysis further revealed that this vesicular fraction predominantly consisted of 50-nm vesicles. However, a careful analysis of the study reveals that the authors employed an additional purification step. Specifically, they further separated the components of the already isolated vesicular fraction by subjecting the samples to an additional round of centrifugation between sucrose cushion gradients.

The next issue challenging the kiss-and-run model is the observed leakage of Golgi enzymes, particularly Mannosidase II, into lysosomes in cells lacking COG or GOLPH3.

Considering that in cells lacking COG, both the *cis*-most cisterna and the *trans*-most cisterna are absent [77, 96, 172, 173], we hypothesized that this phenomenon could be responsible for this leakage of some Golgi enzymes into endosomes/

lysosomes. Indeed, the *cis*-most cisterna and the *trans*-most cisterna are absent in Golgi stacks that lack cargo. These cisternae form upon cargo arrival and simultaneous release of Ca^{2+} [110], which induces fusion between Golgi compartments [177]. The state of Golgi hyper-fusion is induced after the arrival of membrane cargos or mega-cargos. The phenomenon is based on the leakage of calcium from the ER, triggered by the cargo arrival [110] and followed by the subsequent fusion of COPI-dependent vesicles enriched in GS27/membrin and GS28/GOS28 [55, 177] with the Golgi cisternae. This fusion makes rims of these cisternae competent for fusion with each other, as they, under these conditions, contain all the components of SNARE complexes. Before cargo arrival, GS27 and GS28 are eliminated from the rims by COPI-dependent vesicles, preventing cisterna fusion.

We also propose that Golgins, COG, and GOLPH3 could play a role in the formation of vesicular string, a form of tethering described by Orci et al. [133], which prevents the diffusion of 52-nm vesicles out of the Golgi. The mechanisms could also involve several Golgins, which have binding sites for Rabs and Arls [33, 185]. Importantly, in the vesicular fraction isolated by Gilchrist et al. [58], Rab1A, Rab1B, Rab2a, and Rab14 are enriched in vesicles. Thus, Golgins, COG, and GOLPH3 could bind to Golgi membranes enriched in several Rabs and Arls on one hand, and to Golgi vesicles enriched in other Rabs on the other hand. These interactions could contribute to the formation of tethering strings, facilitating vesicle retention within the Golgi apparatus.

We also hypothesize that, within the framework of the KARM, COPI-dependent 50-nm vesicles play essential roles in intracellular transport and Golgi organization. These vesicles could serve multiple functions:

1. Facilitating COPI uncoating: The formation of COPI-coated vesicles may accelerate the dissociation of COPI subunits from membranes along the secretory pathway, ensuring proper membrane recycling and trafficking dynamics.
2. Preventing hyper-fusion: COPI vesicles might regulate membrane fusion events by eliminating Qb SNARE (GS27/membrin and GS28/GOS28) from Golgi cisternae [55, 177].
3. Protecting cisternal rims: Scattered COPI coats, which are typically observed as a dense layer only on COPI-coated buds [182], may help stabilize cisternal rims. This protection could prevent undesired fusion events with endosomes, other cisternae, or the ER.
4. Regulating membrane curvature: 50-nm vesicles may play a role in eliminating excessive membrane curvature [19, 118]. This function would help maintain proper Golgi architecture.
5. Mediating fission event: COPI may drive the fission of tubular connections at the ER–Golgi interface and within the Golgi stack [35, 169]. The presence of COPI-positive collars is observed during ER–Golgi transport [120], further supporting this role.

Experiments with brefeldin A suggest that COPI plays a role in extracting Golgi enzymes from the ER and may be involved in their initial retention and retrieval from the ER [43, 128]. Park et al. [139] demonstrated that COPI can sort anterograde cargoes into COPI-dependent tubules. Notably, in absorptive enterocytes, the exit of chylomicrons occurs independently of COPII [42, 165]. COPII is also responsible for concentrating SNAREs and membrane cargoes [7, 115].

The kiss-and-run mechanism explains the observations that, in *S. cerevisiae,* when cargo exit from the ER is blocked, the Golgi disappears [5, 124]. It accounts for the concentration of albumin, VSVG, mega-cargoes, and regulated secretory proteins as cargo moves from *cis*- to *trans*-side of the Golgi. Additionally, it explains the process of domain maturation described by Casler et al. [28] and Kurokawa et al. [89], as illustrated in Fig. 10 presented in the paper by Mironov and Beznoussenko [115].

Furthermore, the findings of Pagano et al. [137] became more understandable in this context. They demonstrated that fluorescent lipids diffuse along the secretory pathway even in aldehyde-fixed cells, but this ability disappears when cells are treated with OsO_4. This phenomenon can be explained by the continuous formation and subsequent breakdown of connections, leading to residual membrane connectivity along the secretory pathway. If a sufficient number of these connections exist at a given moment, a transient membrane pathway could potentially connect the ER to the plasmalemma.

Conclusions

The absence of free ER-to-Golgi and Golgi-to-plasmalemma (or endosome) carriers, including mega-vesicles, supports both the DM and the KARM models. However, the presence of a well-developed system of SNAREs at all levels of secretory transport contradicts the diffusion model. Additionally, it has been shown that the concentration of anterograde and retrograde cargo in small vesicles is lower than in precursor secretory compartments, which argues against the CMPM and VM models. Furthermore, neither the DM nor the CMPM can explain the increase in cargo concentration observed during its movements along certain segments of the secretory pathway. Vesicles within ER exit sites are COPI-dependent, while COPII plays a crucial role in enhancing cargo concentration at these sites. Cargo moves vectorially across the Golgi stack and reaches the *trans*-most cisterna, which is connected to the TGN. After this, the linear dynamics of cargo movement across the Golgi converts into a regression pattern resembling exponential delay at different stages of transport. The TGN acts in a rate-limiting step, leading to the accumulation of cargo-containing carriers in the post-Golgi zone. The KARM explains the accumulated experimental observations, including the disappearance of Golgi when ER exit is blocked, cargo concentration changes, and domain maturation within Golgi, as well as the role of COPI in cargo sorting and membrane remodeling. Thus, the KARM provides the most comprehensive explanation of intracellular transport of the currently known proteins, resolving contradictions that arise in other models. By integrating key findings from various studies, this model explains cargo dynamics, vesicle formation, and membrane connectivity along the secretory pathway more realistically than the alternative models.

Author Contributions All authors have conceptualized, written, read, and agreed to the published version of the manuscript.

Conflicts of Interest The authors declare no conflicts of interest.

Funding This research received no external funding.

References

1. Adolf F, Herrmann A, Hellwig A, Beck R, Brügger B, Wieland FT (2013) Scission of COPI and COPII vesicles is independent of GTP hydrolysis. Traffic 14(8):922–932. https://doi.org/10.1111/tra.12084
2. Andresen JM, Moore HP (2001) Biogenesis of processing-competent secretory organelles in vitro. Biochemistry 40(43):13020–13030. https://doi.org/10.1021/bi0112762
3. Ang AL, Taguchi T, Francis S, Folsch H, Murrells LJ, Pypaert M, Warren G, Mellman I (2004) Recycling endosomes can serve as intermediates during transport from the Golgi to the plasma membrane of MDCK cells. J Cell Biol 167:531–543
4. Arab M, Chen T, Lowe M (2024) Mechanisms governing vesicle traffic at the Golgi apparatus. Curr Opin Cell Biol 4(88):102365. https://doi.org/10.1016/j.ceb.2024.102365
5. Ayscough K, Warren G (1994) Inhibition of protein synthesis disrupts the Golgi apparatus in the fission yeast. Schizosaccharomyces pombe Yeast 10:1–11. https://doi.org/10.1002/yea.320100102
6. Bacia K, Futai E, Prinz S, Meister A, Daum S, Glatte D, Briggs JA, Schekman R (2011) Multibudded tubules formed by COPII on artificial liposomes. Sci Rep 1:17. https://doi.org/10.1038/srep00017
7. Bannykh S, Aridor M, Plutner H, Rowe T, Balch WE (1995) Regulated export of cargoes from the endoplasmic reticulum of mammalian cells. Cold Spring Harb Symp Quant Biol 60:127–137
8. Bannykh SI, Rowe T, Balch WE (1996) The organization of endoplasmic reticulum export complexes. J Cell Biol 135:19–35. https://doi.org/10.1083/jcb.135.1.19
9. Bannykh SI, Balch WE (1997) Membrane dynamics at the endoplasmic reticulum-Golgi interface. J Cell Biol 138:1–4. https://doi.org/10.1083/jcb.138.1.1
10. Bannykh SI, Plutner H, Matteson J, Balch WE (2005) The role of ARF1 and Rab GTPases in polarization of the Golgi stack. Traffic 6(9):803–819
11. Barbieri MA, Li G, Mayorga LS, Stahl PD (1996a) Characterization of Rab5:Q79L-stimulated endosome fusion. Arch Biochem Biophys 326(1):64–72. https://doi.org/10.1006/abbi.1996.0047
12. Barbieri MA, Roberts RL, Mukhopadhyay A, Stahl PD (1996b) Rab5 regulates the dynamics of early endosome fusion. Biocell 20(3):331–338
13. Barlowe C, Orci L, Yeung T, Hosobuchi M, Hamamoto S, Salama N, Rexach MF, Ravazzola M, Amherdt M, Schekman R (1994) COPII: a membrane coat formed by sec proteins that drive vesicle budding from the endoplasmic reticulum. Cell 77:895–907. https://doi.org/10.1016/0092-8674(94)90138-4
14. Beznoussenko GV, Bejan AI, Parashuraman S, Luini A, Kweon HS, Mironov AA (2023) The diffusion model of intra-Golgi transport has limited power. Int J Mol Sci 24(2):1375. https://doi.org/10.3390/ijms24021375
15. Beznoussenko GV, Dolgikh VV, Seliverstova EV, Semenov PB, Tokarev YS, Trucco A, Micaroni M, Di Giandomenico D, Auinger P, Senderskiy IV, Skarlato SO, Snigirevskaya ES, Komissarchik YY, Pavelka M, De Matteis MA, Luini A, Sokolova YY, Mironov AA (2007)

Analogs of the Golgi complex in microsporidia: structure and avesicular mechanisms of function. J Cell Sci 120:1288–1298. https://doi.org/10.1242/jcs.03402
16. Beznoussenko GV, Kweon HS, Sesorova IS, Mironov AA (2022) Comparison of the cisterna maturation-progression model with the kiss-and-run model of intra-Golgi transport: role of Cisternal pores and cargo domains. Int J Mol Sci 23:3590. https://doi.org/10.3390/ijms23073590
17. Beznoussenko GV, Mironov AA (2002) Models of intracellular transport and evolution of the Golgi complex. Anat Rec 268:226–238. https://doi.org/10.1002/ar.10157
18. Beznoussenko GV, Parashuraman S, Rizzo R, Polishchuk R, Martella O, Di Giandomenico D, Fusella A, Spaar A, Sallese M, Capestrano MG, Pavelka M, Vos MR, Rikers YG, Helms V, Mironov AA, Luini A (2014) Transport of soluble proteins through the Golgi occurs by diffusion via continuities across cisternae. elife 27(3):e02009. https://doi.org/10.7554/eLife.02009
19. Beznoussenko GV, Pilyugin SS, Geerts WJ, Kozlov MM, Burger KN, Luini A, Derganc J, Mironov AA (2015) Trans-membrane area asymmetry controls the shape of cellular organelles. Int J Mol Sci 16(3):5299–5333. https://doi.org/10.3390/ijms16035299
20. Beznoussenko GV, Ragnini-Wilson A, Wilson C, Mironov AA (2016) Three-dimensional and immune electron microscopic analysis of the secretory pathway in Saccharomyces cerevisiae. Histochem Cell Biol 146:515–527. https://doi.org/10.1007/s00418-016-1483-y
21. Bost A, Shaib AH, Schwarz Y, Niemeyer BA, Becherer U (2017) Large dense-core vesicle exocytosis from mouse dorsal root ganglion neurons is regulated by neuropeptide Y. Neuroscience 27(346):1–13. https://doi.org/10.1016/j.neuroscience.2017.01.006
22. Bracker C, Grove SN, Heintz CE, Morre DJ (1971) Continuity between endomembrane components in hyphae of Pythium spp. Cytobiologie 4:1–8
23. Bright NA, Gratian MJ, Luzio JP (2005) Endocytic delivery to lysosomes mediated by concurrent fusion and kissing events in living cells. Curr Biol 15(4):360–365
24. Broadwell RD, Cataldo AM (1983) The neuronal endoplasmic reticulum: its cytochemistry and contribution to the endomembrane system. I. Cell bodies and dendrities. J Histochem Cytochem 31:1077–1088. https://doi.org/10.1177/31.9.6309951
25. Bykov YS, Schaffer M, Dodonova SO, Albert S, Plitzko JM, Baumeister W, Engel BD, Briggs JA (2017) The structure of the COPI coat determined within the cell. elife 6:pii: e32493. https://doi.org/10.7554/eLife.32493
26. Caffaro CE, Hirschberg CB (2006) Nucleotide sugar transporters of the Golgi apparatus: from basic science to diseases. Acc Chem Res 39:805–812
27. Cancino J, Torrealba C, Soza A, Yuseff MI, Gravotta D, Henklein P, Rodriguez-Boulan E, González A (2007) Antibody to AP1B adaptor blocks biosynthetic and recycling routes of basolateral proteins at recycling endosomes. Mol Biol Cell 18(12):4872–4884
28. Casler JC, Papanikou E, Barrero JJ, Glick BS (2019) Maturation-driven transport and AP-1-dependent recycling of a secretory cargo in the Golgi. J Cell Biol 218:1582–1601. https://doi.org/10.1083/jcb.201807195
29. Ceccarelli B, Hurlbut WP, Mauro A (1972) Depletion of vesicles from frog neuromuscular junctions by prolonged tetanic stimulation. J Cell Biol 54:30–38
30. Chang M, Wu S-Z, Ryken SE et al (2021) The COPII components Sec23 and Sec24 form isoform specific subcomplexes with Sec23D/E and Sec24C/D essential for tip growth. bioRxiv. https://doi.org/10.1101/2020.06.22.165100
31. Chaudhari BY, Pradhan AG, Joshi RS (2025) Metabolic gatekeepers: Dynamic roles of sugar transporters in insect metabolism and physiology. Insect Mol Biol 34(1):1–18. https://doi.org/10.1111/imb.12963
32. Chen LQ, Cheung LS, Feng L, Tanner W, Frommer WB (2015) Transport of sugars. Annu Rev Biochem 84:865–894. https://doi.org/10.1146/annurev-biochem-060614-033904
33. Claas C, Mironov AA, Starkuviene V (2008) Chapter 2.2: Rabs. In: Mironov AA, Pavelka M (eds) Golgi apparatus. Springer, Wien, pp 66–77

34. Claude A (1970) Growth and differentiation of cytoplasmic membranes in the course of lipoprotein granule synthesis in the hepatic cell. I. Elaboration of elements of the Golgi complex. J Cell Biol 47:745–766. https://doi.org/10.1083/jcb.47.3.745
35. Clermont Y, Rambourg A, Hermo L (1994) Connections between the various elements of the cis- and mid-compartments of the Golgi apparatus of early rat spermatids. Anat Rec 240(4):469–480
36. Clermont Y, Rambourg A, Hermo L (1995) Trans-Golgi network (TGN) of different cell types: three-dimensional structural characteristics and variability. Anat Rec 242(3):289–301
37. Cosson P, Amherdt M, Rothman JE, Orci L (2002) A resident Golgi protein is excluded from peri-Golgi vesicles in NRK cells. Proc Natl Acad Sci USA 99(20):12831–12834. https://doi.org/10.1073/pnas.192460999
38. Cresawn KO, Potter BA, Oztan A, Guerriero CJ, Ihrke G, Goldenring JR, Apodaca G, Weisz OA (2007) Differential involvement of endocytic compartments in the biosynthetic traffic of apical proteins. EMBO J 26(16):3737–3748. https://doi.org/10.1038/sj.emboj.7601813
39. Crivellato E, Nico B, Ribatti D (2005) Ultrastructural evidence of piecemeal degranulation in large dense-core vesicles of brain neurons. Anat Embryol (Berl) 210(1):25–34. https://doi.org/10.1007/s00429-005-0002-z
40. Cutrona MB, Beznoussenko GV, Fusella A, Martella O, Moral P, Mironov AA (2013) Silencing of the mammalian Sar1 isoforms reveals COPII-independent protein sorting and transport. Traffic 14:691–708. https://doi.org/10.1111/tra.12060
41. Deborde S, Gravotta D, Lakkaraju A, Rodriguez-Boulan E (2008) Golgi apparatus and epithelial cell polarity: lessons from 3 decades of MDCK cells. In: The Golgi apparatus: state of the art 110 years after Camillo Golgi's discovery, pp 563–579
42. Denisova GN, Dimov ID, Zaitseva AV, Artiux LJ, Mironov AA, Karelina NR (2021) Overloading of differentiated Caco-2 cells during lipid transcytosis induces glycosylation mistakes in the Golgi complex. Biocell; Mendoza 45(3):773–783. https://doi.org/10.32604/biocell.2021.014233
43. Dominguez M, Dejgaard K, Füllekrug J, Dahan S, Fazel A, Paccaud JP, Thomas DY, Bergeron JJ, Nilsson T (1998) gp25L/emp24/p24 protein family members of the cis-Golgi network bind both COP I and II coatomer. J Cell Biol 140(4):751–765. https://doi.org/10.1083/jcb.140.4.751
44. Dunlop MH, Ernst AM, Schroeder LK, Toomre DK, Lavieu G, Rothman JE (2017) Land-locked mammalian Golgi reveals cargo transport between stable cisternae. Nat Commun 2017(8):432. https://doi.org/10.1038/s41467-017-00570-z
45. Ebert B, Orellana A (2025) Nucleotide sugar transporters: orchestrating luminal glycosylation in plants. Annu Rev Plant Biol. https://doi.org/10.1146/annurev-arplant-083123-075017
46. Farquhar MG, Palade GE (1981) The Golgi apparatus (complex)-(1954–1981)-from artifact to center stage. J Cell Biol 91(3 Pt 2):77s–103s
47. Flickinger CJ (1969) The development of Golgi complexes and their dependence upon the nucleus in amoebae. J Cell Biol 49:250–262. https://doi.org/10.1083/jcb.43.2.250
48. Flickinger CL (1973) Maintenance and regeneration of cytoplasmic organelles in hybrid amoebae formed by nuclear transplantation. Exp Cell Res 80:31–46. https://doi.org/10.1016/0014-4827(73)90272-3
49. Forrester A, De Leonibus C, Grumati P, Fasana E, Piemontese M, Staiano L et al (2019) A selective ER-phagy exerts procollagen quality control via a Calnexin-FAM134B complex. EMBO J 38:e99847. https://doi.org/10.15252/embj.201899847
50. Fougère L, Grison M, Laquel P, Montrazi M, Cordelières F, Fernández-Monreal M, Poujol C, Uemura T, Nakano A, Ito Y, Boutté Y (2025) ER-to-Golgi trafficking through a dynamic intermediate cis-Golgi tubular network in Arabidopsis. Nat Cell Biol 27(3):424–437. https://doi.org/10.1038/s41556-025-01624-x
51. Franke WW, Kartenbeck J (1976) Some principles of membrane differentiation. In: Mueller-Berat N (ed) Progress in differentiation research. Elsevier-North Holland Publishing Co, Amsterdam, pp 213–243

52. Fregno I, Molinari M (2018) Endoplasmic reticulum turnover: ER-phagy and other flavors in selective and non-selective ER clearance. F1000Res 7:454. https://doi.org/10.12688/f1000research.13968.1
53. Fregno I, Fasana E, Bergmann TJ, Raimondi A, Loi M, Soldà T et al (2018) ER-to-lysosome-associated degradation of proteasome-resistant ATZ polymers occurs via receptor-mediated vesicular transport. EMBO J 37:e99259. https://doi.org/10.15252/embj.201899259
54. Fromme JC, Schekman R (2005) COPII-coated vesicles: flexible enough for large cargo? Curr Opin Cell Biol 17:345–352. https://doi.org/10.1016/j.ceb.2005.06.004
55. Fusella A, Micaroni M, Di Giandomenico D, Mironov AA, Beznoussenko GV (2013) Segregation of the Qb-SNAREs GS27 and GS28 into Golgi vesicles regulates intra-Golgi transport. Traffic 2–13 14:568–584. https://doi.org/10.1111/tra.12055
56. Futter CE, Connolly CN, Cutler DF, Hopkins CR (1995) Newly synthesized transferrin receptors can be detected in the endosome before they appear on the cell surface. J Biol Chem 270:10999–11003
57. Gaietta GM, Giepmans BN, Deerinck TJ, Smith WB, Ngan L, Llopis J, Adams SR, Tsien RY, Ellisman MH (2006) Golgi twins in late mitosis revealed by genetically encoded tags for live cell imaging and correlated electron microscopy. Proc Natl Acad Sci USA 103(47):17777–17782. https://doi.org/10.1073/pnas.0608509103
58. Gilchrist A, Au CE, Hiding J, Bell AW, Fernandez-Rodriguez J, Lesimple S et al (2006) Quantitative proteomics analysis of the secretory pathway. Cell 127:1265–1281. https://doi.org/10.1016/j.cell.2006.10.036
59. Girard-Dias W, Alcântara CL, Cunha-E-Silva N, de Souza W, Miranda K (2012) On the ultrastructural organization of Trypanosoma cruzi using cryopreparation methods and electron tomography. Histochem Cell Biol 138:821–831. https://doi.org/10.1007/s00418-012-1002-8
60. Glick BS, lston E.T., G. Oster G. (1997) A cisternal maturation mechanism can explain the asymmetry of the Golgi stack. FEBS Lett 414:177–181
61. Gorur A, Yuan L, Kenny SJ, Baba S, Xu K, Schekman R (2017) COPII-coated membranes function as transport carriers of intracellular procollagen I. J Cell Biol 216(6):1745–1759. https://doi.org/10.1083/jcb.201702135
62. Gravotta D, Deora A, Perret E, Oyanadel C, Soza A, Schreiner R, Gonzalez A, Rodriguez-Boulan E (2007) AP1B sorts basolateral proteins in recycling and biosynthetic routes of MDCK cells. Proc Natl Acad Sci USA 104(5):1564–1569
63. Griffiths G (2000) Gut thoughts on the Golgi complex. Traffic 1:738–745
64. Hammond AT, Glick BS (2000) Dynamics of transitional endoplasmic reticulum sites in vertebrate cells. Mol Biol Cell 11(9):3013–3030. https://doi.org/10.1091/mbc.11.9.3013
65. Hammond C, Helenius A (1994) Folding of VSV G protein: sequential interaction with BiP and calnexin. Science 266:456–458
66. Harada A, Kunii M, Kurokawa K, Sumi T, Kanda S, Zhang Y, Nadanaka S, Hirosawa KM, Tokunaga K, Tojima T, Taniguchi M, Moriwaki K, Yoshimura SI, Yamamoto-Hino M, Goto S, Katagiri T, Kume S, Hayashi-Nishino M, Nakano M, Miyoshi E, Suzuki KGN, Kitagawa H, Nakano A (2024) Dynamic movement of the Golgi unit and its glycosylation enzyme zones. Nat Commun 15(1):4514. https://doi.org/10.1038/s41467-024-48901-1
67. Henderson, G.P., Gan, L., and Jensen, G.J. 2007.3-D Ultrastructure of O. Tauri: electron cryotomography of an entire eukaryotic cell. PLoS One 2(8):e749
68. Hermo L, Smith CE (1998) 1998. The structure of the Golgi apparatus: a sperm's eye view in principal epithelial cells of the rat epididymis. Histochem Cell Biol 109(5–6):431–447
69. Hohmann-Marriott MF, Sousa AA, Azari AA, Glushakova S, Zhang G, Zimmerberg J et al (2009) Nanoscale 3D cellular imaging by axial scanning transmission electron tomography. Nat Methods 6:729–731. https://doi.org/10.1038/nmeth.1367
70. Holzman E (1971) Cytochemical studies of protein transport in the nervous system. Philos Trans R Soc Lond 261:407–421. https://doi.org/10.1098/rstb.1971.0075

71. Jahraus A, Storrie B, Griffiths G, Desjardins M (1994) Evidence for retrograde traffic between terminal lysosomes and the prelysosomal/late endosome compartment. J Cell Sci 107(Pt 1):145–157. https://doi.org/10.1242/jcs.107.1.145
72. Jaiswal JK, Rivera VM, Simon SM (2009) Exocytosis of post-Golgi vesicles is regulated by components of the endocytic machinery. Cell 137(7):1308–1319. https://doi.org/10.1016/j.cell.2009.04.064
73. Jeger JL (2020) Endosomes, lysosomes, and the role of endosomal and lysosomal biogenesis in cancer development. Mol Biol Rep 47(12):9801–9810. https://doi.org/10.1007/s11033-020-05993-4
74. Jokitalo E, Cabrera-Poch N, Warren G, Shima DT (2001) Golgi clusters and vesicles mediate mitotic inheritance independently of the endoplasmic reticulum. J Cell Biol 154(2):317–330. https://doi.org/10.1083/jcb.200104073
75. Kaiser CA, Schekman R (1990) Distinct sets of SEC genes govern transport vesicle formation and fusion early in the secretory pathway. Cell 61:723–733. https://doi.org/10.1016/0092-8674(90)90483-U
76. Kasberg W, Luong P, Hanna MG, Minushkin K, Tsao A, Shankar R, Block S, Audhya A (2023) The Sar1 GTPase is dispensable for COPII-dependent cargo export from the ER. Cell Rep 42(6):112635. https://doi.org/10.1016/j.celrep.2023.112635
77. Khakurel A, Kudlyk T, Bonifacino JS, Lupashin VV (2021) The Golgi-associated retrograde protein (GARP) complex plays an essential role in the maintenance of the Golgi glycosylation machinery. Mol Biol Cell 32(17):1594–1610. https://doi.org/10.1091/mbc.E21-04-0169
78. Klumperman J (2011) Architecture of the mammalian Golgi. Cold Spring Harb Perspect Biol 3:a005181. https://doi.org/10.1101/cshperspect.a005181
79. Klyachko VA, Jackson MB (2002) Capacitance steps and fusion pores of small and large-dense-core vesicles in nerve terminals. Nature 418(6893):89–92. https://doi.org/10.1038/nature00852
80. Koga D, Kusumi S, Ushiki T (2016) Three-dimensional shape of the Golgi apparatus in different cell types: serial section scanning electron microscopy of the osmium-impregnated Golgi apparatus. Microscopy (Oxf) 65(2):145–157. https://doi.org/10.1093/jmicro/dfv360
81. Koga D, Ushiki T, Watanabe T (2017a) Novel scanning electron microscopy methods for analyzing the 3D structure of the Golgi apparatus. Anat Sci Int 92(1):37–49. https://doi.org/10.1007/s12565-016-0380-8
82. Koga D, Bochimoto H, Kusumi S, Ushiki T, Watanabe T (2017b) Changes in the three-dimensional ultrastructure of membranous organelles in male rat pituitary gonadotropes after castration. Biomed Res 38:1–18. https://doi.org/10.2220/biomedres.38.1
83. Koga D, Ushiki T (2006) Three-dimensional ultrastructure of the Golgi apparatus in different cells: high-resolution scanning electron microscopy of osmium-macerated tissues. Arch Histol Cytol 69:357–374. https://doi.org/10.1679/aohc.69.357
84. Koike S, Jahn R (2022) SNARE proteins: zip codes in vesicle targeting? Biochem J 479(3):273–288. https://doi.org/10.1042/BCJ20210719
85. Kreitzer G, Marmorstein A, Okamoto P, Vallee R, Rodriguez-Boulan E (2000) Kinesin and dynamin are required for post-Golgi transport of a plasma-membrane protein. Nat Cell Biol 2(2):125–127
86. Kreitzer G, Schmoranzer J, Low SH, Li X, Gan Y, Weimbs T, Simon SM, Rodriguez-Boulan E (2003) Three-dimensional analysis of post-Golgi carrier exocytosis in epithelial cells. Nat Cell Biol 5(2):126–136
87. Krijnse-Locker J, Ericsson M, Rottier PJ, Griffiths G (1994) Characterization of the budding compartment of mouse hepatitis virus: evidence that transport from the RER to the Golgi complex requires only one vesicular transport step. J Cell Biol 124:55–70. https://doi.org/10.1083/jcb.124.1.55
88. Kurokawa K, Okamoto M, Nakano A (2014) Contact of cis-Golgi with ER exit sites executes cargo capture and delivery from the ER. Nat Commun 14(5):3653. https://doi.org/10.1038/ncomms4653

89. Kurokawa K, Osakada H, Kojidani T, Waga M, Suda Y, Asakawa H et al (2019) Visualization of secretory cargo transport within the Golgi apparatus. J Cell Biol 218:1602–1618. https://doi.org/10.1083/jcb.201807194
90. Kweon HS, Beznoussenko GV, Micaroni M, Polishchuk RS, Trucco A, Martella O, Di Giandomenico D, Marra P, Fusella A, Di Pentima A, Berger EG, Geerts WJ, Koster AJ, Burger KN, Luini A, Mironov AA (2004) Golgi enzymes are enriched in perforated zones of golgi cisternae but are depleted in COPI vesicles. Mol Biol Cell 15:4710–4724. https://doi.org/10.1091/mbc.e03-12-0881
91. Ladinsky MS, Mastronarde DN, McIntosh JR, Howell KE, Staehelin LA (1999) Golgi structure in three dimensions, functional insights from the normal rat kidney cell. J Cell Biol 144:1135–1149. https://doi.org/10.1083/jcb.144.6.1135
92. Leblond CP (1989) Synthesis and secretion of collagen by cells of connective tissue, bone, and dentin. Anat Rec 224(2):123–138
93. Leitinger B, Hille-Rehfeld A, Spiess M (1995) Biosynthetic transport of the asialoglycoprotein receptor H1 to the cell surface occurs via endosomes. Proc Natl Acad Sci USA 92(22):10109–10113
94. Lindsey JD, Ellisman MH (1985) The neuronal endomembrane system. I. Direct links between rough endoplasmic reticulum and the cis-element of the Golgi apparatus. J Neurosci 5:3111–3123. https://doi.org/10.1523/JNEUROSCI.05-12-03111.1985
95. Liu W, Jiang H, Zeng F (2025) The sugar transporter proteins in plants: an elaborate and widespread regulation network-a review. Int J Biol Macromol 294:139252. https://doi.org/10.1016/j.ijbiomac.2024.139252
96. Lock JG, Hammond LA, Houghton F, Gleeson PA, Stow JL (2005) E-cadherin transport from the trans-Golgi network in tubulovesicular carriers is selectively regulated by golgin-97. Traffic 6(12):1142–1156. https://doi.org/10.1111/j.1600-0854.2005.00349.x
97. Lock JG, Stow JL (2005) Rab 11 in recycling endosomes regulates the sorting and basolateral transport of E-cadherin. Mol Biol Cell 16:1744–1755
98. Lockhausen J, Kristen U, Menhardt W, Dallas W (1990) Three- dimensional reconstruction of a plant dictyosome from series of ultrathin sections using computer image processing. J Microsc 158:197–205. https://doi.org/10.1111/j.1365-2818.1990.tb02993.x
99. Luby-Phelps K, Castle PE, Taylor DL, Lanni F (1987) Hindered diffusion of inert tracer particles in the cytoplasm of mouse 3T3 cells. Proc Natl Acad Sci USA 84(14):4910–4913
100. Luby-Phelps K (1994) Physical properties of cytoplasm. Curr Opin Cell Biol 6:3–9. https://doi.org/10.1016/0955-0674(94)90109-0
101. Malhotra V (2024) Tailored assemblies of COPII proteins in secretion. J Cell Biol 223(8):e202404013. https://doi.org/10.1083/jcb.202404013
102. Malhotra V (2025) The pathways of secretory cargo export at the endoplasmic reticulum. Nat Commun 16(1):2138. https://doi.org/10.1038/s41467-025-57408-2
103. Marra P, Salvatore L, Mironov A Jr, Di Campli A, Di Tullio G, Trucco A et al (2007) The biogenesis of the Golgi ribbon: the roles of membrane input from the ER and of GM130. Mol Biol Cell 18:1595–1608. https://doi.org/10.1091/mbc.E06-10-0886
104. Marsh BJ, Mastronarde DN, Buttle KF, Howell KE, McIntosh JR (2001) Organellar relationships in the Golgi region of pancreatic beta cell line, HIT-T15, visualized by high resolution electron tomography. Proc Natl Acad Sci USA 98:2399–2406. https://doi.org/10.1073/pnas.051631998
105. Martinez-Menárguez JA, Prekeris R, Oorschot VM, Scheller R, Slot JW, Geuze HJ et al (2001) Peri-Golgi vesicles contain retrograde but not anterograde proteins consistent with the cisternal progression model of intra-Golgi transport. J Cell Biol 155:1213–1224. https://doi.org/10.1083/jcb.200108029
106. Maszczak-Seneczko D, Wiktor M, Skurska E, Wiertelak W, Olczak M (2022) Delivery of nucleotide sugars to the mammalian Golgi: a very well (un)explained story. Int J Mol Sci 23(15):8648. https://doi.org/10.3390/ijms23158648

107. Matsuura-Tokita K, Takeuchi M, Ichihara A, Mikuriya K, Nakano A (2006) Live imaging of yeast Golgi cisternal maturation. Nature 441:1007–1010. https://doi.org/10.1038/nature04737
108. McCaughey J, Stevenson NL, Cross S, Stephens DJ (2019) ER-to-Golgi trafficking of procollagen in the absence of large carriers. J Cell Biol 218:929–948. https://doi.org/10.1083/jcb.201806035
109. McGinness AJ, Schoberer J, Pain C, Brandizzi F, Kriechbaumer V (2022) On the nature of the plant ER exit sites. Front Plant Sci 7(13):1010569. https://doi.org/10.3389/fpls.2022.1010569
110. Micaroni M, Perinetti G, Di Giandomenico D, Bianchi K, Spaar A, Mironov AA (2010) Synchronous intra-Golgi transport induces the release of Ca2+ from the Golgi apparatus. Exp Cell Res 13:2071–2086
111. Mironov, A, Luini A, Mironov A (1998) A synthetic model of intra-Golgi traffic The FASEB Journal 12(2) 249–252. https://doi.org/10.1096/fasebj.12.2.249
112. Mironov AA, Arvan P (2008) Chapter 3.11: Origin of the regulated secretory pathway. In: Mironov AA, Pavelka M (eds) Golgi apparatus. Springer, Wien, pp 482–515
113. Mironov AA, Beznoussenko GV (2008) Chapter 3.2. Intra-Golgi transport. In: Mironov AA, Pavelka M (eds) The Golgi apparatus. State of the art 110 years after Camillo Golgi's discovery. Springer-Verlag, Wien, pp 342–357
114. Mironov AA, Beznoussenko GV, Nicoziani P, Martella O, Trucco A, Kweon HS, Di Giandomenico D, Polishchuk RS, Fusella A, Lupetti P, Berger EG, Geerts WJ, Koster AJ, Burger KN, Luini A (2001) Small cargo proteins and large aggregates can traverse the Golgi by a common mechanism without leaving the lumen of cisternae. J Cell Biol 155:1225–1238. https://doi.org/10.1083/jcb.200108073
115. Mironov AA, Beznoussenko GV (2012) The kiss-and-run model of intra-Golgi transport. Int J Mol Sci 13(6):6800–6819. https://doi.org/10.3390/ijms13066800
116. Mironov AA, Beznoussenko GV (2019) Models of intracellular transport: pros and cons. Front Cell Dev Biol 7:146. https://doi.org/10.3389/fcell.2019.00146
117. Mironov AA, Beznoussenko GV (2011) Molecular mechanisms responsible for formation of Golgi ribbon. Histol Histopathol 26(1):117–133. https://doi.org/10.14670/HH-26.117
118. Mironov AA, Beznoussenko GV (2023) The regulated secretion and models of intracellular transport: the goblet cell as an example. Int J Mol Sci 24(11):9560. https://doi.org/10.3390/ijms24119560
119. Mironov AA, Mironov A, Derganc J, Beznoussenko GV (2020) Membrane curvature, trans-Membrane area asymmetry, budding, fission and organelle geometry. Int J Mol Sci 21(20):7594. https://doi.org/10.3390/ijms21207594
120. Mironov AA, Mironov A, Sanavio B, Krol S, Beznoussenko GV (2023) Intracellular membrane transport in vascular endothelial cells. Int J Mol Sci 24(6):5791. https://doi.org/10.3390/ijms24065791
121. Mironov AA, Mironov AA Jr, Beznoussenko GV, Trucco A, Lupetti P, Smith JD, Geerts WJ, Koster AJ, Burger KN, Martone ME, Deerinck TJ, Ellisman MH, Luini A (2003) ER-to-Golgi carriers arise through direct en bloc protrusion and multistage maturation of specialized ER exit domains. Dev Cell 5(4):583–594. https://doi.org/10.1016/s1534-5807(03)00294-6
122. Mironov AA, Sesorova IS, Seliverstova EV, Beznoussenko GV (2017) Different Golgi ultrastructure across species and tissues: implications under functional and pathological conditions, and an attempt at classification. Tissue Cell 49(2 Pt A):186–201. https://doi.org/10.1016/j.tice.2016.12.002
123. Mironov AA, Weidman P, Luini A (1997) Variations on the intracellular transport theme: maturing cisternae and trafficking tubules. J Cell Biol 138:481–484
124. Mogre S, Brown AI, Koslover EF (2020) Getting around the cell: physical transport in the intracellular world. Phys Biol 17(6):061003. https://doi.org/10.1088/1478-3975/aba5e5
125. Morin-Ganet MN, Rambourg A, Deitz SB, Franzusoff A, Képès F (2000) Morphogenesis and dynamics of the yeast Golgi apparatus. Traffic 1:56–68. https://doi.org/10.1034/j.1600-0854.2000.010109.x

126. Morre DJ, Keenan TW, Huang CM (1974) Membrane flow and differentiation: origin of Golgi apparatus membranes from endoplasmic reticulum. Adv Cytopharmacol 2:107–125
127. Mossessova E, Bickford LC, Goldberg J (2003) SNARE selectivity of the COPII coat. Cell 114:483–495. https://doi.org/10.1016/s0092-8674(03)00608-1
128. Nakano A (2022) The Golgi apparatus and its next-door neighbors. Front Cell Dev Biol 10:884360. https://doi.org/10.3389/fcell.2022.884360
129. Nilsson T, Au CE, Bergeron JJ (2009) Sorting out glycosylation enzymes in the Golgi apparatus. FEBS Lett 583:3764–3769. https://doi.org/10.1016/j.febslet.2009.10.064
130. Novikoff PM, Yam A (1978) Sites of lipoprotein particles in normal rat hepatocytes. J Cell Biol 76:1–11. https://doi.org/10.1083/jcb.76.1.1
131. Omari S, Makareeva E, Roberts-Pilgrim A, Mirigian L, Jarnik M, Ott C, Lippincott-Schwartz J, Leikin S (2018) Noncanonical autophagy at ER exit sites regulates procollagen turnover. Proc Natl Acad Sci USA 115(43):E10099–E10108. https://doi.org/10.1073/pnas.1814552115
132. Oprins A, Rabouille C, Posthuma G, Klumperman J, Geuze HJ, Slot JW (2001) The ER to Golgi interface is the major concentration site of secretory proteins in the exocrine pancreatic cell. Traffic 2:831–838. https://doi.org/10.1034/j.1600-0854.2001.21112.x
133. Orci L, Amherdt M, Ravazzola M, Perrelet A, Rothman JE (2000a) Exclusion of golgi residents from transport vesicles budding from Golgi cisternae in intact cells. J Cell Biol 150(6):1263–1270. https://doi.org/10.1083/jcb.150.6.1263
134. Orci L, Perrelet A, Rothman JE (1998) Vesicles on strings: morphological evidence for processive transport within the Golgi stack. Proc Natl Acad Sci USA 95(5):2279–2283. https://doi.org/10.1073/pnas.95.5.2279
135. Orci L, Ravazzola M, Volchuk A, Engel T, Gmachl M, Amherdt M, Perrelet A, Sollner TH, Rothman JE (2000b) Anterograde flow of cargo across the golgi stack potentially mediated via bidirectional "percolating" COPI vesicles. Proc Natl Acad Sci USA 97(19):10400–10405
136. Orci L, Schekman R, Perrelet A (1996) Interleaflet clear space is reduced in the membrane of COP I and COP II-coated buds/vesicles. Proc Natl Acad Sci USA 93(17):8968–8970. https://doi.org/10.1073/pnas.93.17.8968
137. Orzech E, Cohen S, Weiss A, Aroeti B (2000) Interactions between the exocytic and endocytic pathways in polarized Madin-Darby canine kidney cells. J Biol Chem 275(20):15207–15219
138. Pagano RE, Sepanski MA, Martin OC (1989) Molecular trapping of a fluorescent ceramide analogue at the Golgi apparatus of fixed cells: interaction with endogenous lipids provides a trans-Golgi marker for both light and electron microscopy. J Cell Biol 109:2067–2079. https://doi.org/10.1083/jcb.109.5.2067
139. Palade G (1975) Intracellular aspects of the process of protein synthesis. Science 189:347–359. https://doi.org/10.1126/science.1096303
140. Park SY, Yang JS, Schmider AB, Soberman RJ, Hsu VW (2015) Coordinated regulation of bidirectional COPI transport at the Golgi by CDC42. Nature 521(7553):529–532. https://doi.org/10.1038/nature14457
141. Patterson GH, Hirschberg K, Polishchuk RS, Gerlich D, Phair RD, Lippincott-Schwartz J (2008) Transport through the Golgi apparatus by rapid partitioning within a two-phase membrane system. Cell 133:1055–1067. https://doi.org/10.1016/j.cell.2008.04.044
142. Peters A, Palay SL, Webster H (1991) Fine structure of the nervous system: neurons and their supporting cells, 3rd edn. Oxford University Press, Oxford/New York/Toronto, p 528
143. Pfeffer SR (2010) How the Golgi works: a cisternal progenitor model. Proc Natl Acad Sci USA 107:19614–19618. https://doi.org/10.1073/pnas.1011016107
144. Polishchuk RS, Mironov AA (2004) Structural aspects of Golgi function. Cell Mol Life Sci 61:146–158. https://doi.org/10.1007/s00018-003-3353-8
145. Polishchuk RS, Polishchuk EV, Marra P, Alberti S, Buccione R, Luini A, Mironov AA (2000) Correlative light-electron microscopy reveals the tubular-saccular ultrastructure of carriers operating between Golgi apparatus and plasma membrane. J Cell Biol 148(1):45–58. https://doi.org/10.1083/jcb.148.1.45

146. Polishchuk RS, Polishchuk EV, Mironov AA (1999) Coalescence of Golgi fragments in microtubule-deprived living cells. Eur J Cell Biol 78(3):170–185. https://doi.org/10.1016/S0171-9335(99)80096-X
147. Polishchuk EV, Di Pentima A, Luini A, Polishchuk RS (2003) Mechanism of constitutive export from the Golgi: bulk flow via the formation, protrusion, and en bloc cleavage of large trans-Golgi network tubular domains. Mol Biol Cell 14:4470–4485. https://doi.org/10.1091/mbc.e03-01-0033
148. Polishchuk RS, San Pietro E, Di Pentima A, Tete S, Bonifacino JS (2006) Ultrastructure of long-range transport carriers moving from the trans Golgi network to peripheral endosomes. Traffic 7:1092–1103
149. Pothukuchi P, Agliarulo I, Pirozzi M, Rizzo R, Russo D, Turacchio G, Nüchel J, Yang JS, Gehin C, Capolupo L, Hernandez-Corbacho MJ, Biswas A, Vanacore G, Dathan N, Nitta T, Henklein P, Thattai M, Inokuchi JI, Hsu VW, Plomann M, Obeid LM, Hannun YA, Luini A, D'Angelo G, Parashuraman S (2021) GRASP55 regulates intra-Golgi localization of glycosylation enzymes to control glycosphingolipid biosynthesis. EMBO J 40(20):e107766. https://doi.org/10.15252/embj.2021107766
150. Presley JF, Cole NB, Schroer TA, Hirschberg K, Zaal KJ, Lippincott- Schwartz J (1997) ER-to-Golgi transport visualized in living cells. Nature 389:81–85. https://doi.org/10.1038/38001
151. Rambourg A, Clermont Y, Chretien M, Olivier L (1992) Formation of secretory granules in the Golgi apparatus of prolactin cells in the rat pituitary gland: a stereoscopic study. Anat Rec 232:169–179
152. Raote I, Malhotra V (2019) Protein transport by vesicles and tunnels. J Cell Biol 218(3):737–739. https://doi.org/10.1083/jcb.201811073
153. Rexach MF, Latterich M, Schekman RW (1994) Characteristics of endoplasmic reticulum-derived transport vesicles. J Cell Biol 126(5):1133–1148. https://doi.org/10.1083/jcb.126.5.1133
154. Rizzo R, Parashuraman S, Mirabelli P, Puri C, Lucocq J, Luini A (2013) The dynamics of engineered resident proteins in the mammalian Golgi complex relies on cisternal maturation. J Cell Biol 201:1027–1036. https://doi.org/10.1083/jcb.201211147
155. Rothman JE, Bursztyn-Pettegrew H, Fine RE (1980) Transport of the membrane glycoprotein of vesicular stomatitis virus to the cell surface in two stages by clathrin-coated vesicles. J Cell Biol 1980(86):162–171. https://doi.org/10.1083/jcb.86.1.162
156. Rudolf R, Salm T, Rustom A, Gerdes HH (2001) Dynamics of immature secretory granules: role of cytoskeletal elements during transport, cortical restriction, and F-actin-dependent tethering. Mol Biol Cell 12(5):1353–1365
157. Rutter GA, Hill EV (2006) Insulin vesicle release: walk, kiss, pause … then run. Physiology (Bethesda) 21:189–196. https://doi.org/10.1152/physiol.00002.2006
158. Rutter GA, Tsuboi T (2004) Kiss and run exocytosis of dense core secretory vesicles. Neuroreport 15(1):79–81. https://doi.org/10.1097/00001756-200401190-00016
159. Sasaki T, Motegi N, Higashi S (1984) Morphological analysis of the Golgi apparatus in rat amelogenesis as revealed by the Ur-Pb-cu block staining method and freeze fracture replication. J Electron Microsc 33:19–33
160. Scales SJ, Pepperkok R, Kreis TE (1997) Visualization of ER- to-Golgi transport in living cells reveals a sequential mode of action for COPII and COPI. Cell 90:1137–1148. https://doi.org/10.1016/S0092-8674(00)80379-7
161. Scheel J, Pepperkok R, Lowe M, Griffiths G, Kreis TE (1997) Dissociation of coatomer from membranes is required for brefeldin A-induced transfer of Golgi enzymes to the endoplasmic reticulum. J Cell Biol 137(2):319–333. https://doi.org/10.1083/jcb.137.2.319
162. Schekman R, Mellman I (1997) Does COPI go both ways? Cell 90:197–200
163. Sesorova IS, Karelina NR, Kazakova TE, Parashuraman S, Zdorikova MA, Dimov ID, Seliverstova EV, Beznoussenko GV, Mironov AA (2020) 2020. Structure of the enterocyte transcytosis compartments during lipid absorption. Histochem Cell Biol 153(6):413–429. https://doi.org/10.1007/s00418-020-01851-3

164. Sesorova IS, Sesorov VV, Soloviev PB, Lakunin KY, Dimov ID, Mironov AA (2022) Role of endothelial regeneration and overloading of enterocytes with lipids in capturing of lipoproteins by basement Membrane of rat aortic endothelium. Biomedicine 10:2858. https://doi.org/10.3390/biomedicines10112858
165. Sesso A, de Faria FP, Iwamura ESM, Correa H (1994) A three-dimensional reconstruction study of the rough ER-Golgi interface in serial thin sections of the pancreatic acinar cell of the rat. J Cell Sci 107:517–528
166. Siddiqi SA, Gorelick FS, Mahan JT, Mansbach CM 2nd (2003) COPII proteins are required for Golgi fusion but not for endoplasmic reticulum budding of the pre-chylomicron transport vesicle. J Cell Sci 116(Pt 2):415–427
167. Slot JW, Geuze HJ (1983) Immunoelectron microscopic exploration of the Golgi complex. J Histochem Cytochem 31:1049–1056
168. Stalder D, Gershlick DC (2020) Direct trafficking pathways from the Golgi apparatus to the plasma membrane. Semin Cell Dev Biol 107:112–125. https://doi.org/10.1016/j.semcdb.2020.04.001
169. Stephens DJ (2003) De novo formation, fusion and fission of mammalian COPII-coated endoplasmic reticulum exit sites. EMBO Rep 4:210–217
170. Stephens DJ, Lin-Marq N, Pagano A, Pepperkok R, Paccaud JP (2000) COPI-coated ER-to-Golgi transport complexes segregate from COPII in close proximity to ER exit sites. J Cell Sci 113:2177–2185. Available online at: https://jcs.biologists.org/content/joces/113/12/2177.full.pdf?with-ds=yes
171. Stinchcombe JC, Nomoto H, Cutler DF, Hopkins CR (1995) Anterograde and retrograde traffic between the rough endoplasmic reticulum and the Golgi complex. J Cell Biol 131(6 Pt 1):1387–1401
172. Storrie B, Desjardins M (1996) The biogenesis of lysosomes: is it a kiss and run, continuous fusion and fission process? BioEssays 18(11):895–903. https://doi.org/10.1002/bies.950181108
173. Sumya FT, Aragon-Ramirez WS, Lupashin VV (2024) bioRxiv Comprehensive Proteomic Characterization of the Intra-Golgi Trafficking Intermediates. [Preprint]. 2024.10.25.620336. https://doi.org/10.1101/2024.10.25.620336
174. Sumya FT, Pokrovskaya ID, D'Souza Z, Lupashin VV (2023) Acute COG complex inactivation unveiled its immediate impact on Golgi and illuminated the nature of intra-Golgi recycling vesicles. Traffic 24(2):52–75. https://doi.org/10.1111/tra.12876
175. Tanaka K (1992) Cell fine structures observed by scanning electron microscopy. Hum Cell 5:211–217
176. Tanaka K, Mitsushima A, Fukudome H, Kashima Y (1986) Three- dimensional architecture of the Golgi complex observed by high resolution scanning electron microscopy. J Submicrosc Cytol 18:1–9
177. Tie HC, Mahajan D, Lu L (2022) Visualizing intra-Golgi localization and transport by side-averaging Golgi ministacks. J Cell Biol 221(6):e202109114. https://doi.org/10.1083/jcb.202109114
178. Trucco A, Polishchuk RS, Martella O, Di Pentima A, Fusella A, Di Giandomenico D, San Pietro E, Beznoussenko GV, Polishchuk EV, Baldassarre M, Buccione R, Geerts WJ, Koster AJ, Burger KN, Mironov AA, Luini A (2004) Secretory traffic triggers the formation of tubular continuities across Golgi sub-compartments. Nat Cell Biol 6(11):1071–1081. https://doi.org/10.1038/ncb1180
179. Uchiyama Y (1982) The membrane association among the rough- and smooth- surfaced endoplasmic reticulum and Golgi apparatus in rat hepatocytes at certain circadian stages. Tohoku J Exp Med 136:299–302. https://doi.org/10.1620/tjem.136.299
180. van Kempen GT, van der Leest HT, van den Berg RJ et al (2011) Three distinct modes of exocytosis revealed by amperometry in neuroendocrine cells. Biophys J 100:968–977

181. Volchuk A, Amherdt M, Ravazzola M, Brügger B, Rivera VM, Clackson T et al (2000) Megavesicles implicated in the rapid transport of intracisternal aggregates across the Golgi stack. Cell 102:335–348. https://doi.org/10.1016/S0092-8674(00)00039-8
182. Watson P, Townley AK, Koka P, Palmer KJ, Stephens DJ (2006) Sec16 defines endoplasmic reticulum exit sites and is required for secretory cargo export in mammalian cells. Traffic 7(12):1678–1687. https://doi.org/10.1111/j.1600-0854.2006.00493.x
183. Weidman P, Roth R, Heuser J (1993) Golgi membrane dynamics imaged by freeze-etch electron microscopy: views of different membrane coatings involved in tubulation versus vesiculation. Cell 75(1):123-133.80089-80093
184. Weigert R, Silletta MG, Spanò S, Turacchio G, Cericola C, Colanzi A, Mancini R, Polishchuk EV, Salmona M, Facchiano F, Burger KNJ, Mironov A, Luini A, Corda D (1999) CtBP/BARS induces fission of Golgi membranes by acylating lysophosphatidic acid. Nature 402(6760):429–433
185. Williams LM, Lafontane JC (1985) An electron microscope thick section study of endomembrane organization in the Myx-amoebae of Physarum polycephalum. Protoplasma 124:42–51. https://doi.org/10.1007/BF01279722
186. Wong M, Gillingham AK, Munro S (2017) The golgin coiled-coil proteins capture different types of transport carriers via distinct N-terminal motifs. BMC Biol 15(1):3. https://doi.org/10.1186/s12915-016-0345-3
187. Wong M, Munro S (2014) Membrane trafficking. The specificity of vesicle traffic to the Golgi is encoded in the golgin coiled-coil proteins. Science 346(6209):1256898. https://doi.org/10.1126/science.1256898
188. Wu Q, Zhang Q, Liu B, Li Y, Wu X, Kuo S, Zheng L, Wang C, Zhu F, Zhou Z (2019) Dynamin 1 Restrains Vesicular Release to a Subquantal Mode In Mammalian Adrenal Chromaffin Cells. J Neurosci 39(2):199–211. https://doi.org/10.1523/JNEUROSCI.1255-18.2018
189. Xu J, Liang Y, Li N, Dang S, Jiang A, Liu Y, Guo Y, Yang X, Yuan Y, Zhang X, Yang Y, Du Y, Shi A, Liu X, Li D, He K (2024) Clathrin-associated carriers enable recycling through a kiss-and-run mechanism. Nat Cell Biol 26(10):1652–1668. https://doi.org/10.1038/s41556-024-01499-4
190. Yan R, Chen K, Wang B, Xu K (2022) SURF4-induced tubular ERGIC selectively expedites ER-to-Golgi transport. Dev Cell 57(4):512–525.e8. https://doi.org/10.1016/j.devcel.2021.12.018
191. Yang JS, Lee SY, Gao M, Bourgoin S, Randazzo PA, Premont RT, Hsu VW (2002) ARFGAP1 promotes the formation of COPI vesicles, suggesting function as a component of the coat. J Cell Biol 159(1):69–78
192. Yang JS, Lee SY, Spanò S, Gad H, Zhang L, Nie Z, Bonazzi M, Corda D, Luini A, Hsu VW (2005) A role for BARS at the fission step of COPI vesicle formation from Golgi membrane. EMBO J 24(23):4133–4143
193. Zhang Q, Liu B, Wu Q, Liu B, Li Y, Sun S, Wang Y, Wu X, Chai Z, Jiang X, Liu X, Hu M, Wang Y, Yang Y, Wang L, Kang X, Xiong Y, Zhou Y, Chen X, Zheng L, Zhang B, Wang C, Zhu F, Zhou Z (2019) Differential co-release of two neurotransmitters from a vesicle fusion pore in mammalian adrenal chromaffin cells. Neuron 102(1):173–183.e4. https://doi.org/10.1016/j.neuron.2019.01.031

Chapter 9
Biomolecular Condensates in and Around the ER–Golgi Interface

Miguel Ruiz, Savannah M. Bogus, and Andreas M. Ernst

Abstract Biomolecular condensates represent a new paradigm of cytoplasmic organization. In recent years, an increasing amount of evidence suggests that protein condensates can directly impact the morphology and function of membrane compartments. The early secretory pathway, which comprises the ER and Golgi compartments, is emerging as a hotspot for biomolecular condensates. In this chapter, we discuss recent reports suggesting diverse roles for condensates in cargo sorting, regulating membrane traffic, and shaping organelle morphology. This promotes a view in which biomolecular condensates may significantly contribute to defining the anatomy and specific functions of the Golgi and its surrounding membranes.

Keywords Liquid-liquid phase separation · Condensates · Self-organization · Early secretory pathway · ER exit sites (ERESs) · ER · ER-Golgi interface · Golgi matrix

Introduction

Cells are complex living machines that are organized on different scales to function. With regard to their organization on a micron-length scale, the mesoscale, it has been established that cells utilize membrane-enclosed compartments, the endomembrane system [2]. An additional method of macromolecular organization, which has been gaining recognition in recent years, is that of protein condensates as "open" compartments or membraneless organelles [27, 51]. An increasing amount of evidence has emerged to support the idea that protein collectives formed by

M. Ruiz (✉) · S. M. Bogus · A. M. Ernst (✉)
Department of Cell and Developmental Biology, School of Biological Sciences, University of California San Diego, La Jolla, CA, USA
e-mail: mir047@ucsd.edu; aernst@ucsd.edu

A. Nakano, J. Saraste (eds.), *The Golgi Network, Volume I*, Subcellular Biochemistry 110, https://doi.org/10.1007/978-3-032-06936-8_9

liquid–liquid phase separation (LLPS) contribute to the structure and complexity of cellular organization [6, 15, 38, 43, 54, 76, 96, 100, 101, 106, 129, 135, 146]. Condensates have been shown to perform a variety of functions, such as serving as reaction crucibles, sequestering molecules, and forming organizational hubs [7, 14, 18, 19, 21, 36, 41, 42, 50, 60, 63, 66, 75, 85, 89, 91, 116, 122–124, 130, 136, 147]. It is increasingly appreciated that the driving force that leads to the assembly of such molecular collectives is through weak intermolecular interactions between distinct amino acids encoded in intrinsically disordered regions (IDRs) and low complexity domains (LCDs) [5, 8, 20, 33, 42, 45, 47, 57, 65, 73, 74, 88, 90, 92, 98, 109, 128]. How, if at all, do protein condensates, as an open system of cytoplasmic organization, impact membrane compartments such as the Golgi? Here, we will examine the burgeoning literature of biological condensates that are encountered in the secretory pathway of cells.

Condensates at the Endoplasmic Reticulum (ER)

The ER lumen represents the major site of secretory protein manufacture and initiates the early secretory pathway. In recent years, protein condensates were found to impact ER function in several ways, ranging from sorting and concentration of cargo proteins to sequestering mutant proteins for ER-phagy [94, 125] (Fig. 9.1).

Intracisternal Granules Sort ER Residents (ICGs)

Intracisternal granules (ICGs) within pancreatic cells represent one of the earliest examples of protein condensation in the ER that occurs within the ER lumen. ICGs, which were first observed via electron microscopy by George Palade as spherical granules in the ER, sort proteins within the non-acidic environment of the lumen of the rough endoplasmic reticulum (RER) [93]. ICG condensates sequester and thereby restrict over-abundant soluble RER resident proteins, thus acting as a sorting factor [118]. For example, zymogens, which are sets of secreted pancreatic digestive enzymes and proenzymes, will co-condense and be sequestered in the lumen of the RER for autophagy when zymogens have been overproduced, such as in response to stresses like starvation [125].

Proinsulin Akita: Misfolded Protein Condensates Are Cleared Via ER-Phagy

Another condensation mechanism in the ER lumen restricts the size of misfolded proinsulin protein condensates to protect cell viability. When mutant misfolded proinsulin protein *Akita* condenses into discrete foci, Sec24C and ER-phagy receptor

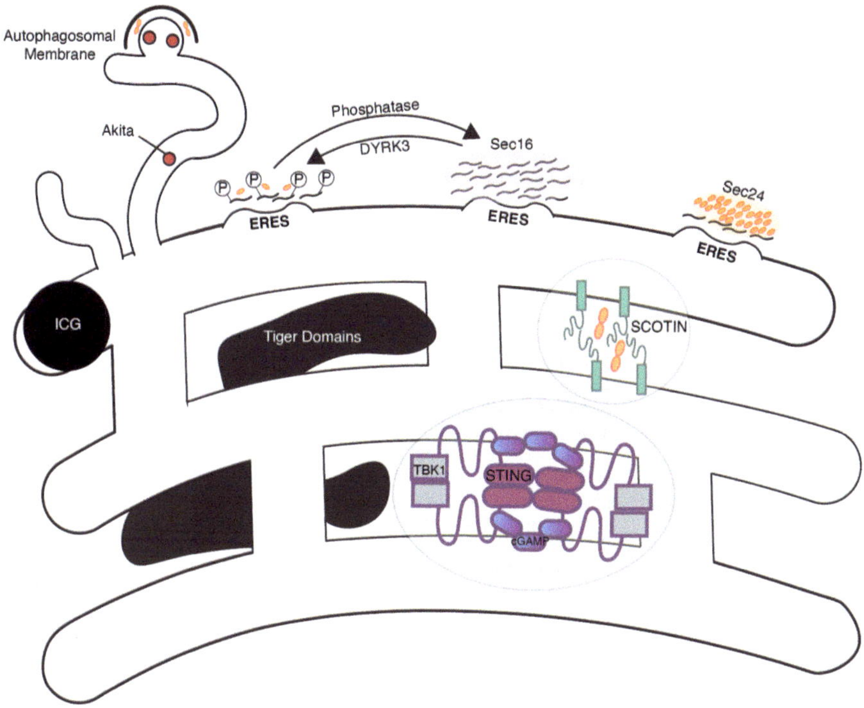

Fig. 9.1 Condensates within the lumen and associated with the surface of the ER. Akita depicted by small punctae in ER tubules and budding into autophagosomes with Sec24C (top left). Intracisternal granules (ICG) depicted by large, black circular punctae (middle left). TIGER domains depicted as large, convoluted black condensate (bottom left). Sec16 condensation is depicted by curved lines in a shaded region at ERES, with its fluidity regulated by the kinase DYRK3 (top middle). Sec24 condensation depicted by small ovals near Sec16 and in isolation (top middle, top right). SCOTIN condensation depicted with boxes attached to IDRs with COPII components depicted by oval-shaped dimers (middle right). Structured STING condensates depicted by larger rod-like shapes, TBK1 depicted by rectangles attached to IDR, and cGAMP depicted by 3 smaller rod-like shapes (lower right)

RTN3 are recruited to transport ER tubules containing the misfolded protein to lysosomes for degradation [94]. When RTN3 or Sec24C was depleted or ER tubules were reduced, *Akita* puncta enlarged. Unlike earlier reports that suggested *Akita* puncta represent immobile aggregates [29], enlarged *Akita* puncta were highly mobile and behaved as dynamic condensates with a diffuse phase surrounding discrete puncta within the ER. The ER-phagy receptor RTN3, together with Sec24C-Sec23A, and LC3B within tubules, play a key role in the condensate elimination mechanism. Co-condensation of *Akita* with RTN3 controls the size of accumulating misfolded proinsulin by promoting targeting of these condensates to lysosomes for ER-phagy. Sequestering and confining cargo proteins into ER tubules thus prevents the formation of enlarged condensates that may become toxic to cells and provides an opportunity to clear accumulated protein via ER-phagy.

The surface of the ER provides a major membrane area allowing interactions with a variety of condensates. While the condensates described here are found outside the ER, they may not directly contribute to the performance of the secretory pathway. However, they demonstrate how versatile condensates are in contributing to ER function, with specific roles including sequestering mRNA for further processing and immune-stimulated regulation of the secretory pathway [59, 70, 139].

TIGER Condensates Recruit Membrane Protein-Encoding mRNAs to the ER

ER-associated granules form an interwoven meshwork throughout the reticular network of the ER to facilitate protein–protein interactions in specialized TIS11B-enriched TIGER (TIS granule-ER) domains [70]. TIGER domains are distinct, membraneless organelles forming a network that juxtaposes the rER. TIGER domains selectively enrich membrane protein-encoding mRNAs with 3′ UTRs that contain long repeating AU-rich elements (AREs), allowing further processing. SET interacts specifically with long UTRs and exhibits reduced mobility upon recruitment to TIGER domains. Highlighting the importance of the biophysical properties of the TIGER domain, the immobilization of cytosolic proteins such as SET may be vital for the processing of membrane protein mRNA containing long UTRs. Intriguingly, the TIS11b assembly mechanism was independent of IDRs, instead relying on particular charge distributions found throughout the protein. Collectively, TIGER domains are an exciting example of a membraneless organelle interfacing with a membrane-enclosed organelle that can orchestrate key protein–protein interactions based on the microenvironment created at the junction between these open and closed compartments.

Condensation of STING Blunts Immune Overactivation

STING, a transmembrane protein that phase separates into an unusual puzzle-like structure at ER membranes [139], is a key regulator of the innate immune response [52, 131]. STING condensation induces spherical membranous biomolecular condensates that engulf STING-TBK1-cGAMP like a "sponge" to blunt immune overactivation in response to DNA viruses. It has been established that condensates can function to sequester macromolecules into the condensate dense phase and that condensation is frequently driven by weak interactions among predominantly cytosolic IDR-containing proteins [6, 15, 88, 129]. In recent years, there has been a growing appreciation for transmembrane proteins to be able to undergo LLPS [7, 106, 143, 146]. Condensation of STING, however, appears to drive the formation of spherical membrane-containing structures. Condensate-like foci form in response to exogenously applied 2′3′-cGAMP or DNA viruses (VACV) that are dynamic and undergo fusion and fission. Surprisingly, STING-enclosed membrane structures exhibit a high mobility of STING within the puzzle-like layer, suggesting that it does not act as a static, coat-like oligomer. STING condensates appear to be flexible, as they were shown to be able to dynamically change their size upon increasing STING expression and upon stimulation with 2′3′cGAMP. Additionally, in vitro data support that STING condensate size depends on concentration, temperature, and the divalent cation Mn^{2+}.

Over time, STING-enclosed condensates transition into a gel-like state that depends on membrane topology and dimerization domains within the transmembrane anchor and the IDR [7, 8, 78, 79, 86]. The puzzle-like morphology of STING condensates sequesters TBK1 but not IRF3, thus acting as a molecular "sponge" for immune response drivers. Importantly, structured condensation of STING effectively reduces interferon (IFN) production even under cGAS overexpression, showcasing that phase separation of proteins can control complex immune responses that can be exacerbated in autoimmune diseases in humans if cytokine production is not limited. Thus, phase separation of STING demonstrates the necessity for a better understanding of molecular condensation mechanisms, as it bears the potential to ameliorate aberrant cytokine release in autoimmune diseases [53, 61, 67].

SCOTIN/SHISA-5 Condensates Halt ER Export during Cellular Stress

ER-to-Golgi transport relies on several cytosolic factors to ensure anterograde transport of biosynthetic cargo [11]. Processes and stimuli, including cell proliferation and stress, can induce changes to the secretory pathway to control the flux of cargo proteins or to prioritize protein machinery elsewhere [3, 35, 68, 115]. Recently, the interferon-inducible protein SCOTIN/SHISA-5 was described to form biomolecular condensates at the ER membrane via a Proline Rich Domain (PRD) that faces the cytosol. The resulting condensates were shown to impede ER-to-Golgi transport by specifically sequestering Sec13/Sec31 heterodimers and thus interfering with COPII carrier biogenesis at ER exit sites (ERESs) [59].

Condensates at Endoplasmic Reticulum Exit Sites (ERESs)

In recent years, there has been an increasing amount of evidence that protein condensates may play a significant direct role in controlling the structure and function of the early secretory pathway [16, 38, 43, 54, 69, 104, 135]. Condensates are specifically enriched at submicron membrane domains of the ER, the ER exit sites (ERESs), where recent reports suggest that they significantly impact both the architecture and function of ERESs (Fig. 9.1, 9.2).

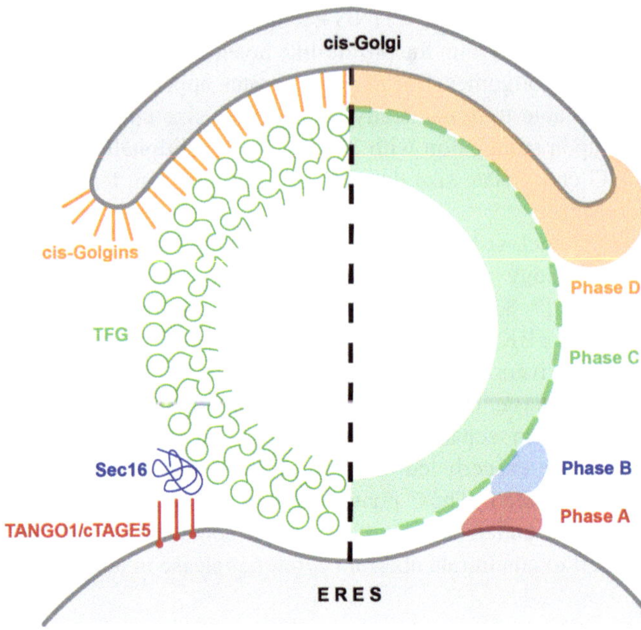

Fig. 9.2 Layering of distinct condensate species at the ER–Golgi interface. Left: Cross-section of phase-separating proteins from ERES to *cis*-Golgi depicting individual molecules of TANGO1/cTAGE5 on the ERES, Sec16 between TANGO1/cTAGE5 and TFG, TFG between Sec16 and *cis*-golgins, and rod-like *cis*-golgin tethers. Right: shaded regions representing layering of distinct condensate species in the early secretory pathway

Sec16 Condensates at ERESs Control the ER Exit of Cargo

The fluidity of Sec16A condensates at ERESs is tightly regulated by the dual specificity kinase DYRK3, which was shown to be an important regulator of the flux of membrane out of ERESs [38]. This parallels early reports suggesting that under cellular stress such as nutrient deprivation, Sec16 can become poly-ADP-ribosylated, resulting in rigid Sec16 "bodies" that halt secretion [1, 24, 127, 142]. DYRKs are termed "dissolvases," as they can directly regulate LLPS of their protein substrate [87, 99]. Upon induction of large Sec16A condensates, DYRK3 was found to specifically colocalize with several markers of the early secretory pathway [38]. The N-terminal portion of Sec16A, which harbors an IDR [97], was sufficient to form dynamic condensates that exhibit hallmarks of a liquid-like state, such as fusion of condensates and the formation of capillary bridges and high protein mobility within the condensate dense phase. Sec16 interacts with the transmembrane ERES protein TANGO1 [71]. Condensates formed by the N-terminal Sec16A construct could recruit early secretory pathway components in cells, including Sec13, TANGO1, cTAGE5, ERGIC-53, and GM130, at concentrations similar to those of full-length endogenous Sec16A [38]. A Tyrosine-Rich Domain (YRD) in Sec16A was crucial for driving condensation, and phosphorylation by DYRK3 and its reversal by phosphatases toggled between more fluid-like or gel-like condensate states. Gel-like condensate states cannot support membrane traffic, which parallels inactivation of Sec16 into "Sec bodies" upon nutrient deprivation [3, 127, 142]. Thus, controlling the material state of condensates at ERESs to enable membrane trafficking is vital, with kinase/phosphatase systems regulating their fluidity.

TANGO1 Rings Outline ERESs to Orchestrate the Export of Bulky Cargo

TANGO1L belongs to the MIA3 gene family and consists of a cytosolic proline-rich domain (PRD), two coiled-coil domains (CC), a TEER domain for recruitment of ERGIC membranes, a transmembrane domain, and a lumenal SH3-like domain that binds Hsp47 for the recruitment of collagen [10, 105, 113, 114]. TANGO1 assembles into 200-nm rings around ER exit sites that recruit COPII coats, thereby putatively catalyzing the formation of carriers that can accommodate bulky cargoes such as collagens, chylomicrons, and antibodies [4, 100, 103, 108]. TANGO1 rings enclose Sec31A-positive foci that are juxtaposed by ERGIC-53, indicating that TANGO1 may contribute to the budding of COPII carriers. TANGO1 rings may thus serve as a boundary, or "fence", to corral COPII components and recruit ERGIC membranes to support the formation of tubules/tunnels that can accommodate bulky cargo [102, 104], as well as separate ER from transitional secretory membranes [104]. TANGO1 phosphorylation by CK1 is required for disassembly of ERESs during mitosis, and dephosphorylation by PP1 allows reassociation via dephosphorylation upon mitotic exit [72]. It has not been demonstrated that TANGO1 can undergo LLPS; however, the close resemblance of its cytosolic moiety with golgin proteins, which contain both coiled-coil domains and IDRs, was shown to form condensates [106, 146], prompting the speculation that phase separation of TANGO1 could contribute to the organization of ERESs and recruitment of the Sec16 condensate layer. This would be supported by the observation that Sec16 condensates can attract both TANGO1 and GM130, a *cis*-Golgi golgin protein [38].

COPII Coats Fine-Tune Anterograde Transport

Activation of Sar1 at ERESs via Sec12 leads to the recruitment of the coat protein heterodimer Sec23/Sec24 through a Sar1 binding site in Sec23, resulting in the formation of the inner layer of the COPII coat [11, 40, 141]. The presence of the Sec23/Sec24 coat protein layer then leads to the recruitment of the outer layer of the COPII coat formed by the heterodimer complex Sec13/Sec31 [17]. The presence of both coat layers leads to membrane deformation and scission of COPII carriers from ERESs [12, 30, 133]. Multivalent interactions resulting in LLPS of COPII coat components were shown to impact cargo transport by promoting the formation of pliable coat networks that can support the secretion of bulky cargo [132]. Specifically, Sec24 phase separation was driven by hydrophobic residues in its N-terminal IDR, and resulted in condensates that enlarge when bulky cargo, such as the lipoprotein

APOB, transits through ERESs. Intriguingly, the outer layer of the COPII coat regulates condensates formed by the inner layer—when Sec23/Sec24 droplets are exposed to Sec13/Sec31, the latter forms a shell around the former, thus limiting the growth of the initial coat protein droplet. Unexpectedly, it was recently demonstrated that secretion continues when Sar1 is absent in cells, challenging the dogma of the requirement of an initiator GTPase to support the formation of COPII carriers [31, 56, 126]. In the absence of Sar1, large COPII condensates engulfed cargo and allowed their export from the ER, albeit with a five-fold reduced efficiency, suggesting that the extent of COPII coat protein condensation could be toggled to accommodate bulky cargo.

TFG Condensates Structure the ER–Golgi Interface to Direct Bidirectional Traffic

The protein Trk-fused gene (TFG) functions in protein secretion and oncogenesis [135]. When TFG is silenced, ER–Golgi interface morphology is distorted, COPII carriers are mislocalized, and cargo export is severely diminished. TFG oligomerizes to form homo octamers via its PB1 domain, resulting in the formation of octameric cups [54]. TFG specifically interacts with Sec16, which in turn interacts with the transmembrane protein TANGO1, localizing it to ERESs [55, 72]. TFG can contribute to COPII carrier uncoating by competing with the outer coat layer for the inner layer [43]. Importantly, depletion of TFG leads to mislocalization of COPII and Sec16 from ERESs. Upon incubation with KOAc, recombinant TFG obtained from bacteria assembled into submicron putative condensates, which matched the submicron dimension of the ER–Golgi interface, and depended on the extensive IDR within TFG [54]. It was thus proposed that TFG can locally concentrate COPII to facilitate protein export from ERESs while serving to link ERESs to the *cis*-Golgi [44, 96]. Based on a recent report, recombinant TFG obtained from human cells phase-transitions to form hollow, anisotropic condensates that again matched the submicron dimensions of the ER–Golgi interface [16]. These cage-like, 300-nm condensates form under physiological conditions in vitro as well as in vivo in cell lines harboring endogenously tagged TFG [16, 56]. TFG "cages" are less fluid than condensates formed by other IDPs. TFG was identified as a phosphoprotein that is dynamically regulated across the cell cycle in a similar fashion to Sec16, controlling its condensation, although the specific kinase for TFG has not been identified [16]. Regulation of higher-order protein assemblies at ERESs via phosphorylation was also reported for TANGO1, which becomes hyperphosphorylated during mitosis [62, 72, 105]. TFG is largely intrinsically disordered (>70%) and harbors regularly spaced hydrophobic residues in its IDR, which were proposed to drive condensation into anisotropic structures [16].

A porous surface was observed in the outer shell of TFG condensates via transmission electron microscopy (TEM), which permitted molecules smaller than 250 kDa to access the condensate core. It was proposed that TFG condensates may act as "molecular sieves" forming a barrier for the large COPI coat complex, while permitting diffusion of the smaller COPII coat layers into the condensate core, thus leading to a spatial separation of COPI vesicle traffic outside TFG condensates [16]. This would further create a diffusion-limited space for anterograde COPII carriers and may contribute to the formation of ERGIC by promoting coalescence of COPII carriers within the condensate core. TFG condensates may therefore act as an extension of the ERESs to support the formation of pleomorphic membrane carriers, including vesicles, tubules, and tunnels. TFG condensates would further provide an explanation to the long-standing question of why ribosomes are excluded from the ER–Golgi interface, as they would be too large to access the condensate core.

Based on the reports outlined above, a view is emerging in which distinct condensate species in the early secretory pathway spontaneously assemble into a layered assembly (Fig. 9.2), much like individual layers of the nucleolus, with their assembly/disassembly dynamically regulated by phosphorylation across the cell cycle [16, 37, 38, 64, 72, 84, 117].

Condensates at the Golgi Apparatus

The surface of the Golgi stack is densely populated by a family of tether proteins, the golgins, resulting in a zone of ribosome exclusion [106, 107, 120, 146]. Golgins are anchored to Golgi membranes either via GTPases or transmembrane domains. Golgins are established to function as vesicle tethers [39], providing specificity for membrane carriers to fuse with specific layers of the stack. Recently, evidence has emerged that suggests an additional role for golgins in dictating the layering and morphology of cisternae via their condensation on Golgi membranes [106] (Fig. 9.3).

GM130 Forms a Membrane-Anchored Condensate Layer at the *Cis*-Golgi

The *cis*-Golgi matrix protein GM130, which is peripherally anchored to the membrane by GRASP65, is the most abundant golgin in the Golgi matrix [82, 83]. Golgins consist predominantly of coiled-coil domains and achieve molecule lengths ranging from 50 to 400 nm [22, 81]. GM130, as a vesicle tether, specifically binds to Rab1-positive carriers that are loaded with p115, leading to its bending toward the membrane [26, 80] and fusion of the carrier with the *cis*-Golgi. GM130 was recently reported to undergo LLPS via its coiled-coil domains, which was unexpected given that the bulk of condensates studied appear to involve interactions predominantly between IDRs or multivalent scaffold domains [106]. The concentration of GM130 in the condensate dense phase was similar to the concentration of GM130 determined at the Golgi, suggesting that endogenous GM130 is present in a phase-separated state at the *cis*-Golgi. Given that the C-terminus of each GM130 molecule is anchored to the membrane by GRASP65, and that the N-terminus binds to p115, proteins within the dense phase would have a polarized orientation,

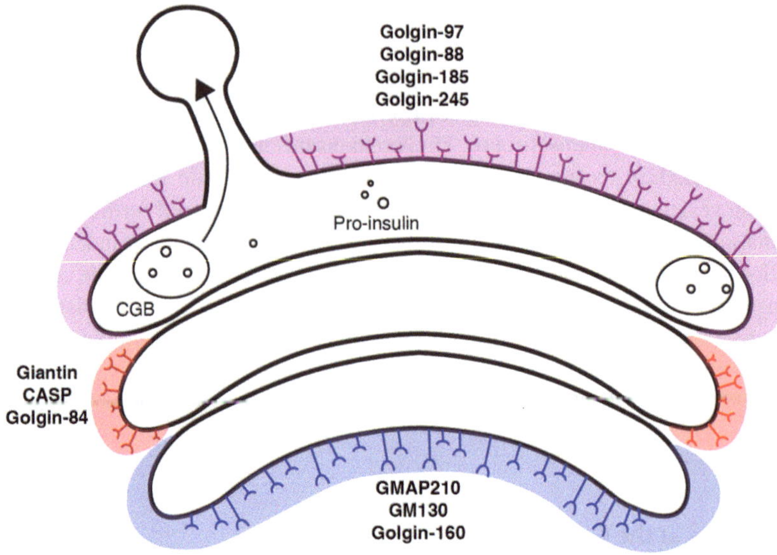

Fig. 9.3 Condensates at the Golgi apparatus. Distinct layers of golgin condensates populate the surface of the Golgi stack. Chromogranin B condensation (CGB) induces its "self-sorting" and budding from the TGN (depicted by large ovals on the left and right with proinsulin represented by small circles in the interior of the *trans*-Golgi). Golgin condensates are represented by shaded regions around Golgi cisternae, with distinct golgins represented by rounded y-shapes extending from cisternae

resulting in a two-dimensional condensate that is one molecule thick—reminiscent of the orientation of molecules within liquid crystals [110].

These observations suggested a model in which golgin condensates dictate the layering of the polarized Golgi stack, with each *cis*-, medial-, and *trans*-Golgi subcompartments spontaneously layering as it is observed for nucleoli that self-organize into three distinct condensate layers [110]. This would imply that the underlying membrane could be deformed by golgin condensates, putatively giving rise to the characteristic cisternal geometry that is important for sorting of enclosed cargo [34]. It further implies that individual *cis*-, medial-, and *trans*-Golgi condensates could form and spontaneously layer even in the absence of membranes, e.g., upon mitotic exit, which suggests that the Golgi matrix directly encodes cisternal identity. Furthermore, it was recently proposed that GM130 co-condenses with RNA via its disordered N-terminus, which may contribute to lateral linking of individual *cis*-Golgi cisternae to support the formation of a Golgi "ribbon" [145]. It remains to be shown whether such mixed condensates formed from GM130 and RNA would functionally resemble TIGER domains at the ER, contributing to local translation of proteins.

Golgins Form Condensates Capable of Unmixing into Distinct Pools

When additional peripherally anchored golgins were studied by overexpression (*cis*-Golgi: golgin-160, GMAP210; *trans*-Golgi: golgin-97, golgin-245, GCC88, GCC185), the resulting condensates exhibited significantly different morphologies [146]. GMAP210 formed spherical condensates, while golgin-160 formed condensates that were extensive branched networks. Overexpression of golgin-97, golgin-245, and GCC185 resulted in spherical condensates, while GCC88 condensates appeared anisotropic and "sponge"-like. Whether these specific condensate morphologies contribute to the overall architecture of the Golgi stack, with disc-shaped cisternae observed toward the *cis*-face, and more convoluted membranes observed at the TGN, remains to be demonstrated. Furthermore, the contribution of the cytosolic domains of transmembrane domain-anchored golgins that localize to medial Golgi cisternae remains to be investigated. Intriguingly, when pairs of *cis*- and *trans*-Golgi-localized golgins are overexpressed, their co-condensation results in structures in which the individual golgin species are juxtaposed, i.e., unmixed into distinct pools that occupy different areas across the dense phase. Although further work is needed, this supports a model in which golgins form homotypic condensates that can adhere to but not mix with other golgin condensates, forming the basis for a golgin condensate-driven layering of the Golgi stack (Fig. 9.3).

"Self-Sorting" of Chromogranin B via Condensation at the TGN

Chromogranins (CG) are key factors that have been shown to contribute to sorting of cargo and formation of secretory granules, which, in regulated secretion, provide a pool of carriers that can rapidly fuse with the plasma membrane upon specific stimuli [25, 28, 49, 58, 95, 111, 138]. Both Chromogranin A and B were reported to form dynamic liquid-like condensates in vitro, which, intriguingly, preferentially occur at a mildly acidic pH as is observed in the TGN lumen [95]. CGB colocalizes with proinsulin at the TGN, where it acts as a "sponge" and is critical for the delivery of cargo to secretory granules. Transport of CGB to secretory granules occurs even upon ectopic expression of constitutively secreted proteins, suggesting that the recruitment of clients to CGB droplets was independent of a conserved sequence and instead based on the concentration of clients in the TGN lumen. Furthermore, the material state impacted client recruitment, as solid-like CGB aggregates failed to recruit cargo, highlighting the importance of the material state of the droplets.

The CG condensation mechanism provides an important example for receptor-independent sorting of cargo at the TGN, and might even contribute directly to budding and scission of carriers loaded with CG-client mixtures.

A Role for Condensates in Controlling Membrane Carrier Biogenesis

A high flux of membrane is critical for the exchange of cargo between compartments of the secretory pathway, and importantly, a recent report suggests that protein condensates may directly induce membrane deformation, specifically elucidating mechanisms for both outward and inward tubule formation. This has direct implications for membrane carrier biogenesis and turnover in the early secretory pathway (Fig. 9.4) and may suggest why distinct condensate pools exist in this space (Fig. 9.2).

Protein Condensates Can Induce Membrane Tubulation

The prevailing view for how membrane curvature can be induced is via the recruitment of proteins that contain highly structured domains, such as BAR domains, which exhibit an intrinsic curvature and thus cause bending upon polymerization [32, 119]. However, this is contrasted by recent reports that suggest highly disordered proteins could directly induce membrane deformations [23, 119]. Recently, it was reported that multiple proteins containing disordered domains can form

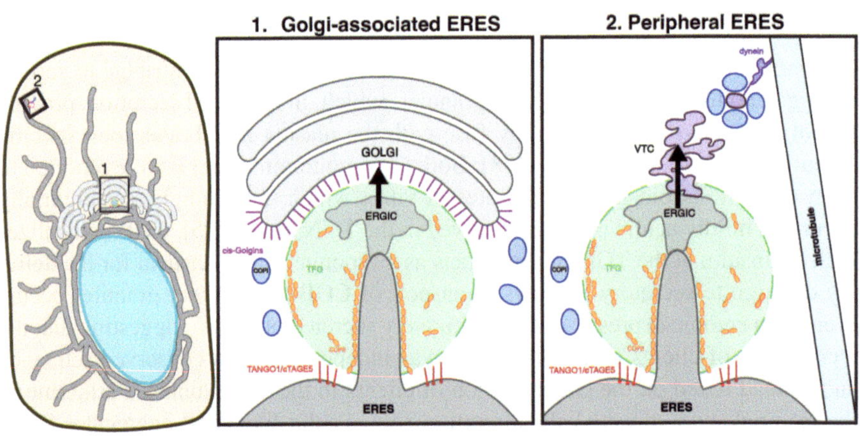

Fig. 9.4 Condensates at ERESs orchestrate the anterograde flux of the membrane. Left: Cartoon of a cell with nucleus, endoplasmic reticulum, and Golgi-associated and peripheral ERESs (boxed structures). Middle: Schematic of Golgi-associated ERESs, which shows exclusion of COPI carriers from the ER–Golgi interface (retrograde moving coat components depicted by large ovals), containment of COPII carriers (anterograde moving coat components depicted by dimers of small ovals), fusion of COPII carriers with ERGIC transport, and direct delivery to the Golgi acceptor membranes. Right: Schematic of peripheral ERES showing an accumulation of ERGIC membranes in the absence of Golgi acceptor membranes, forcing the accumulated membrane to break through the TFG layer and emerge as vesicular-tubular carriers (VTCs) that recruit COPI (large ovals) and migrate via microtubules to the Golgi

networks that have liquid-like properties on membrane surfaces [140]. Specifically, it was demonstrated for the well-characterized intrinsically disordered FUS protein that its phase separation on membranes drives assembly of protein-lined tubules. The tubules appeared to form three distinct classes: undulating tubules, pearled tubules, and a diffraction-limited tubule class that could not be resolved. Inward bending was caused by the reduction in area of the outer protein phase by the disordered proteins being pulled closer together. This likely creates a compressive force and extrudes the relatively rigid membrane inward. Modeling has shown that tubule diameter varies with membrane bending rigidity, driven by lipid composition and protein interaction strength. Intriguingly, simulations also predicted that rigid scaffolding proteins with BAR domains formed different tubules with cylindrical non-undulating morphologies. This suggests that the pearled and undulating morphologies are unique to liquid-like scaffolds. Furthermore, RNPA2-LC, which utilizes a condensation mechanism similar to FUS [112], and the LAF-1 RGG domain, which utilizes a different phase-separating mechanism [33], were both able to form undulating and pearled, inwardly directed membrane tubules. This demonstrates that liquid–liquid phase separation itself is responsible for creating a compressive force that drives membrane deformation. Furthermore, this suggests a role for the condensation of proteins on cytosolic membranes in contributing to the formation of outward membrane protrusions such as filopodia, dendritic spines, viral buds, and cilia. Concomitantly, proteins assembling on the outer surface of cells may contribute to structures that bud into the cell. In this context, it is of note that the mechanism outlined here refers to phase-separating proteins that form compressive forces as opposed to self-avoiding proteins, which may exert a repulsive force among disordered domains and form outward protrusions. In the early secretory pathway, it remains to be explored whether compressive or repulsive forces among membrane-proximal condensate layers directly contribute to membrane deformations.

TANGO1/cTAGE5 Templates Large COPII-Coated Carriers

Relatively small cargos traveling through the secretory pathway are packaged into small (~60–80 nm) COPII-coated carriers [12, 121, 144]. However, human cells must accommodate a variety of bulky cargo, such as chylomicrons, mucins, and procollagens (300–400 nm) that exceed a "standard" COPII vesicle [105]. As discussed previously, TANGO1 forms a ring around ERESs critical for export of procollagen, putatively via tubule-like structures [101]. Importantly, TANGO1/cTAGE5 creates multivalent binding sites that enable the recruitment of an abundance of COPII coats to ERESs, putatively resulting in the formation of a helical lattice at coated buds and in the creation of "mega-carriers" [69].

Employing electron microscopy, such large tubular carriers capable of accommodating procollagen were identified [69]. The tubules harbored a helical array of COPII inner coat components whose orientation was distinct from that observed in vesicles. X-ray crystallography and biochemical analysis support that this helical arrangement is templated by TANGO1/cTAGE5. The inner COPII coat protein

Sec23 interacts with TANGO1/cTAGE5 via a PPP motif in the disordered proline-rich domain (PRD). Intriguingly, these PRDs contain multiple such PPP motifs, which TANGO1/cTAGE5 utilize to concentrate the COPII inner coat at ERESs. Accordingly, PPP motifs in the outer coat protein Sec31 bind to Sec23 as well, leading to a competition with TANGO1/cTAGE5. This implies a mechanism in which TANGO1/cTAGE5 templates COPII coats at the base of a growing tubule, which is further supported by recent findings suggesting that TANGO1 is able to form rings with its coreceptors [101]. Thus, TANGO1 rings may template large tubular carriers that enable transport of large cargos. It remains to be studied whether such tubules would exclusively serve to export bulky cargo, whether they contribute to constitutive export of cargo from ERESs, or if they might be induced in response to cellular queues. Whether TANGO1 forms 2D-condensates, as it is proposed for golgins, that attract both COPII coat condensates and such formed by Sec16 remains to be explored as well. It will also be interesting to identify how the presence of these distinct condensate species can contribute to the formation of different types of carriers at ERESs.

Direct Tubular Connections between the ER and the Golgi

Recently, a remarkable membrane structure within the early secretory pathway that may be impacted by the above-mentioned condensates was described. Rather than a concave membrane dip with clusters of vesicles budding from the ER, ERESs were found to present as intertwined tubular clusters [134]. Importantly, these observations were made using whole-cell focused ion beam scanning electron microscopy (FIB-SEM) [137] with cryo-structured illumination microscopy (cryo-SIM) [46], which has greater Z resolution than conventional TEM serial sectioning approaches for studying ERESs [9, 48, 77, 144]. Using this approach, ERESs appeared to form a distinct membrane compartment from which tubular transport intermediates are extruded to serve as carriers for cargo. Here, a central tubular entity within ERESs had a diameter of 350–400 nm, with tubules protruding from the main tubule bearing a diameter of 40–60 nm. Cargo was shown to be exchanged rapidly between the ER and the ERES "hubs". These "hubs" could also differentiate into cholesterol-rich membrane domains that selectively retained membrane proteins.

Therefore, a view emerges in which ERESs are dynamic membrane structures that can fluctuate in size and morphology during cargo release. Furthermore, the kinetics of cargo entering and exiting ERES structures significantly differ, indicating a gating mechanism into and out of the "hub". Partitioning of cargo into ERES "hubs" was proposed to involve the dynamic recruitment of COPII coats, since inhibition of Sar1 abolished their access to ERES structures. Interestingly, the spatial organization of COPI versus COPII on ERES "hubs" shows that COPII is proximal to the ER, while COPI localizes to its distal moiety. COPI localization to ERESs occurs downstream of Sar1 GTPase activity, implying a sequential role for COPII and COPI activity in these structures. The COPI-positive moiety correlated with transport intermediates emanating from ERES "hubs" and exhibited "beaded" membrane morphology. Such beaded or pearled membrane morphologies are a hallmark of membrane tension [13], which could be generated by microtubule-associated motors tugging on transport intermediates associated with ERES "hubs" [68]. In accordance with this hypothesis, ERES-derived transport intermediates were also shown to be motile structures that could deliver cargo directly from ERESs to the Golgi.

Collectively, these findings propose a novel model for the transport of cargo from the ER to the Golgi in mammalian cells–ERESs act as a sorting hub with independent entry and exit points that produce multiple tubules. These tubules can become associated with microtubules and COPI, but not COPII coats. Rather than budding of COPII vesicles, here, COPII coats serve to stabilize a neck region at the base of ERES "hubs". It can be speculated that the "hub" may be further supported or stabilized by the formation of TANGO1 rings and TFG "cages" at the base of the neck region. This would lead to the recruitment of a flexible layer of phase-separated COPII coat proteins. In this view, COPII would remain on ERES "hubs" and not move forward, extruding membranes from the ER rather than forming distinct coated carriers, providing evidence for the adaptability of ERESs that may be required to facilitate export of bulky cargo from the ER.

A Role for Condensates in Controlling Membrane Flux at the ER–Golgi Interface

Biomolecular condensates can act as liquid "scaffolds", and at ERESs, the golgin-like TANGO1/cTAGE5 was shown to exhibit multivalent interactions to recruit the COPII inner coat proteins Sec23/Sec24, forming large tubular structures that may facilitate transport of bulky cargo [69, 140]. TANGO1 action may thus result in laying the groundwork for an interwoven tubular network that allows for a continuous transport between the ER and Golgi [134]. Compatible with this view, cage-like TFG condensates extending from the ERES may serve to stabilize neck regions by insulating them from COPI coats while concentrating COPII coat proteins [16]. Furthermore, assembly of TANGO1/cTAGE5 into ring-like structures at ERESs may serve as a "fence" that could contribute to membrane deformation, recruitment of bulky cargo, and serve as a diffusion barrier for membrane proteins and lipids that separates ER from post-ER membranes [100–102, 104]. Sec16 is recruited by TANGO1 and would thus anchor the TFG cage to ERES [71]. The "cage-like" TFG condensate, as a cytosolic extension of ERES, would accumulate anterograde membranes, thereby contributing to the formation of ERGIC [16]. In this model, ERGIC membranes either fuse with *cis*-Golgi acceptor membranes (for ERES in close apposition to the Golgi ribbon) or break from the TFG cage, resulting in the recruitment of COPI coats and a microtubule-dependent long-range transport to the Golgi (for peripheral ERES via VTCs) (Fig. 9.4).

Conclusion

This chapter aims to highlight recent reports indicating the presence of multiple distinct protein condensates in the early secretory pathway. Each condensate species is proposed to fulfil specific functions, including sequestration of molecules, providing distinct biophysical and chemical environments, regulating secretion, and dictating the anatomy of membrane compartments in the early secretory pathway. In the future, it is expected that additional examples of biomolecular condensates will emerge that shed light on the question of how self-assembling protein networks can translate information throughout the mesoscale to impact the architecture and function of the early secretory pathway.

Acknowledgments A.M.E. is supported by NIGMS grant R35GM142433 and NIDA grant R21DA061421. M.R. is supported by UCSD Pathways in Biological Sciences Training Program NIGMS T32 GM133351.

Conflict of Interest The authors declare no conflicts of interest.

References

1. Aguilera-Gomez A, van Oorschot MM, Veenendaal T, Rabouille C (2016) In vivo vizualisation of mono-ADP-ribosylation by dPARP16 upon amino-acid starvation. elife 5:e21475. https://doi.org/10.7554/eLife.21475
2. Alberts B, Johnson AD, Lewis J, et al (2020) Molecular biology of the cell (sixth edition) EBook folder
3. Amodio G, Venditti R, De Matteis MA et al (2013) Endoplasmic reticulum stress reduces COPII vesicle formation and modifies Sec23a cycling at ERESs. FEBS Lett 587:3261–3266. https://doi.org/10.1016/j.febslet.2013.08.021
4. Arnolds O, Stoll R (2023) Characterization of a fold in TANGO1 evolved from SH3 domains for the export of bulky cargos. Nat Commun 14. https://doi.org/10.1038/s41467-023-37705-4
5. Banani SF, Rice AM, Peeples WB et al (2016) Compositional control of phase-separated cellular bodies. Cell 166:651–663. https://doi.org/10.1016/j.cell.2016.06.010
6. Banani SF, Lee HO, Hyman AA, Rosen MK (2017) Biomolecular condensates: organizers of cellular biochemistry. Nat Rev Mol Cell Biol 18:285–298. https://doi.org/10.1038/nrm.2017.7
7. Banjade S, Rosen MK (2014) Phase transitions of multivalent proteins canpromote clustering of membrane receptors. eLife 3. https://doi.org/10.7554/eLife.04123
8. Banjade S, Wu Q, Mittal A et al (2015) Conserved interdomain linker promotes phase separation of the multivalent adaptor protein Nck. Proc Natl Acad Sci USA 112:E6426–E6435. https://doi.org/10.1073/pnas.1508778112
9. Bannykh SI, Rowe T, Balch WE (1996) The organization of endoplasmic reticulum export complexes. J Cell Biol 135:19–35. https://doi.org/10.1083/jcb.135.1.19
10. Bard F, Casano L, Mallabiabarrena A et al (2006) Functional genomics reveals genes involved in protein secretion and Golgi organization. Nature 439:604–607. https://doi.org/10.1038/nature04377
11. Barlowe C, Helenius A (2016) Cargo capture and bulk flow in the early secretory pathway. Annu Rev Cell Dev Biol 32:197–222. https://doi.org/10.1146/annurev-cellbio-111315-125016

12. Barlowe C, Orci L, Yeung T et al (1994) COPII: a membrane coat formed by sec proteins that drive vesicle budding from the endoplasmic reticulum. Cell 77:895–907
13. Bar-Ziv R, Moses E (1994) Instability and "pearling" states produced in tubular membranes by competition of curvature and tension. Phys Rev Lett 73:1392–1395. https://doi.org/10.1103/PhysRevLett.73.1392
14. Bergeron-Sandoval L-P, Kumar S, Heris HK et al (2021) Endocytic proteins with prion-like domains form viscoelastic condensates that enable membrane remodeling. Proc Natl Acad Sci USA 118. https://doi.org/10.1073/pnas.2113789118
15. Boeynaems S, Alberti S, Fawzi NL et al (2018) Protein phase separation: a new phase in cell biology. Trends Cell Biol 28:420–435. https://doi.org/10.1016/j.tcb.2018.02.004
16. Bogus SM, Wegeng WR, Ruiz M et al (2025) A hollow TFG condensate spatially compartmentalizes the early secretory pathway. Nat Commun 16:3715. https://doi.org/10.1038/s41467-025-59118-1
17. Bonifacino JS, Glick BS (2004) The mechanisms of vesicle budding and fusion. Cell 116:153–166. https://doi.org/10.1016/s0092-8674(03)01079-1
18. Brangwynne CP, Eckmann CR, Courson DS et al (2009) Germline P granules are liquid droplets that localize by controlled dissolution/condensation. Science 324:1729–1732. https://doi.org/10.1126/science.1172046
19. Brangwynne CP, Mitchison TJ, Hyman AA (2011) Active liquid-like behavior of nucleoli determines their size and shape in Xenopus laevis oocytes. Proc Natl Acad Sci USA 108:4334–4339. https://doi.org/10.1073/pnas.1017150108
20. Brangwynne CP, Tompa P, Pappu RV (2015) Polymer physics of intracellular phase transitions. Nat Phys 11:899–904. https://doi.org/10.1038/nphys3532
21. Bunnell SC, Hong DI, Kardon JR et al (2002) T cell receptor ligation induces the formation of dynamically regulated signaling assemblies. J Cell Biol 158:1263–1275. https://doi.org/10.1083/jcb.200203043
22. Burkhard P, Stetefeld J, Strelkov SV (2001) Coiled coils: a highly versatile protein folding motif. Trends Cell Biol 11:82–88. https://doi.org/10.1016/s0962-8924(00)01898-5
23. Busch DJ, Houser JR, Hayden CC et al (2015) Intrinsically disordered proteins drive membrane curvature. Nat Commun 6:7875. https://doi.org/10.1038/ncomms8875
24. Catara G, Grimaldi G, Schembri L et al (2017) PARP1-produced poly-ADP-ribose causes the PARP12 translocation to stress granules and impairment of Golgi complex functions. Sci Rep 7. https://doi.org/10.1038/s41598-017-14156-8
25. Chanat E, Huttner WB (1991) Milieu-induced, selective aggregation of regulated secretory proteins in the trans-Golgi network. J Cell Biol 115:1505–1519. https://doi.org/10.1083/jcb.115.6.1505
26. Cheung P-YP, Limouse C, Mabuchi H, Pfeffer SR (2015) Protein flexibility is required for vesicle tethering at the Golgi. elife 4. https://doi.org/10.7554/eLife.12790
27. Chong PA, Forman-Kay JD (2016) Liquid-liquid phase separation in cellular signaling systems. Curr Opin Struct Biol 41:180–186. https://doi.org/10.1016/j.sbi.2016.08.001
28. Colomer V, Kicska GA, Rindler MJ (1996) Secretory granule content proteins and the luminal domains of granule membrane proteins aggregate in vitro at mildly acidic pH. J Biol Chem 271:48–55. https://doi.org/10.1074/jbc.271.1.48
29. Cunningham CN, Williams JM, Knupp J et al (2019) Cells deploy a two-pronged strategy to rectify misfolded proinsulin aggregates. Mol Cell 75:442–456.e4. https://doi.org/10.1016/j.molcel.2019.05.011
30. D'Arcangelo JG, Stahmer KR, Miller EA (2013) Vesicle-mediated export from the ER: COPII coat function and regulation. Biochim Biophys Acta 1833:2464–2472. https://doi.org/10.1016/j.bbamcr.2013.02.003
31. d'Enfert C, Barlowe C, Nishikawa S et al (1991) Structural and functional dissection of a membrane glycoprotein required for vesicle budding from the endoplasmic reticulum. Mol Cell Biol 11:5727–5734. https://doi.org/10.1128/mcb.11.11.5727-5734.1991

32. Daumke O, Roux A, Haucke V (2014) BAR domain scaffolds in dynamin-mediated membrane fission. Cell 156:882–892. https://doi.org/10.1016/j.cell.2014.02.017
33. Elbaum-Garfinkle S, Kim Y, Szczepaniak K et al (2015) The disordered P granule protein LAF-1 drives phase separation into droplets with tunable viscosity and dynamics. Proc Natl Acad Sci USA 112:7189–7194. https://doi.org/10.1073/pnas.1504822112
34. Ernst AM, Syed SA, Zaki O et al (2018) S-palmitoylation sorts membrane cargo for anterograde transport in the Golgi. Dev Cell 47:479–493.e7. https://doi.org/10.1016/j.devcel.2018.10.024
35. Farmaki T, Ponnambalam S, Prescott AR et al (1999) Forward and retrograde trafficking in mitotic animal cells. ER-Golgi transport arrest restricts protein export from the ER into COPII-coated structures. J Cell Sci 112 (Pt 5):589–600. https://doi.org/10.1242/jcs.112.5.589
36. Feric M, Vaidya N, Harmon TS et al (2016) Coexisting liquid phases underlie nucleolar subcompartments. Cell 165:1686–1697. https://doi.org/10.1016/j.cell.2016.04.047
37. Frottin F, Schueder F, Tiwary S et al (2019) The nucleolus functions as a phase-separated protein quality control compartment. Science 365:342–347. https://doi.org/10.1126/science.aaw9157
38. Gallo R, Rai AK, McIntyre ABR et al (2023) DYRK3 enables secretory trafficking by maintaining the liquid-like state of ER exit sites. Dev Cell 58:1880–1897.e11. https://doi.org/10.1016/j.devcel.2023.08.005
39. Gillingham AK, Munro S (2016) Finding the Golgi: Golgin coiled-coil proteins show the way. Trends Cell Biol 26:399–408. https://doi.org/10.1016/j.tcb.2016.02.005
40. Gillon AD, Latham CF, Miller EA (2012) Vesicle-mediated ER export of proteins and lipids. Biochim Biophys Acta 1821:1040–1049. https://doi.org/10.1016/j.bbalip.2012.01.005
41. Graham K, Chandrasekaran A, Wang L et al (2024) Liquid-like condensates mediate competition between actin branching and bundling. Proc Natl Acad Sci USA 121:e2309152121. https://doi.org/10.1073/pnas.2309152121
42. Han TW, Kato M, Xie S et al (2012) Cell-free formation of RNA granules: bound RNAs identify features and components of cellular assemblies. Cell 149:768–779. https://doi.org/10.1016/j.cell.2012.04.016
43. Hanna MG IV, Block S, Frankel EB et al (2017) TFG facilitates outer coat disassembly on COPII transport carriers to promote tethering and fusion with ER–Golgi intermediate compartments. Proc Natl Acad Sci USA 114:E7707–E7716. https://doi.org/10.1073/pnas.1709120114
44. Hanna MG, Peotter JL, Frankel EB, Audhya A (2018) Membrane transport at an organelle Interface in the early secretory pathway: take your coat off and stay a while: evolution of the metazoan early secretory pathway. BioEssays 40:e1800004. https://doi.org/10.1002/bies.201800004
45. Harmon TS, Holehouse AS, Rosen MK, Pappu RV (2017) Intrinsically disordered linkers determine the interplay between phase separation and gelation in multivalent proteins. bioRxiv
46. Hoffman DP, Shtengel G, Xu CS et al (2020) Correlative three-dimensional super-resolution and block-face electron microscopy of whole vitreously frozen cells. Science 367:eaaz5357. https://doi.org/10.1126/science.aaz5357
47. Holehouse AS, Kragelund BB (2023) The molecular basis forcellular function of intrinsically disordered protein regions. Nat RevMol Cell Biol 25:187–211
48. Horstmann H, Ng CP, Tang BL, Hong W (2002) Ultrastructural characterization of endoplasmic reticulum--Golgi transport containers (EGTC). J Cell Sci 115:4263–4273. doi:https://doi.org/10.1242/jcs.00115
49. Huh YH, Jeon SH, Yoo SH (2003) Chromogranin B-induced secretory granule biogenesis: comparison with the similar role of chromogranin a. J Biol Chem 278:40581–40589. https://doi.org/10.1074/jbc.M304942200
50. Hyman T, Brangwynne C (2012) In retrospect: the origin of life. Nature 491:524–525. https://doi.org/10.1038/491524a

51. Hyman AA, Weber CA, Jülicher F (2014) Liquid-liquid phase separation in biology. Annu Rev Cell Dev Biol 30:39–58. https://doi.org/10.1146/annurev-cellbio-100913-013325
52. Ishikawa H, Barber GN (2008) STING is an endoplasmic reticulum adaptor that facilitates innate immune signalling. Nature 455:674–678. https://doi.org/10.1038/nature07317
53. Jeremiah N, Neven B, Gentili M et al (2014) Inherited STING-activating mutation underlies a familial inflammatory syndrome with lupus-like manifestations. J Clin Invest 124:5516–5520. https://doi.org/10.1172/JCI79100
54. Johnson A, Bhattacharya N, Hanna M et al (2015) TFG clusters COPII-coated transport carriers and promotes early secretory pathway organization. EMBO J 34:811–827. https://doi.org/10.15252/embj.201489032
55. Kanadome T, Shibata H, Kuwata K et al (2017) The calcium-binding protein ALG-2 promotes endoplasmic reticulum exit site localization and polymerization of Trk-fused gene (TFG) protein. FEBS J 284:56–76. https://doi.org/10.1111/febs.13949
56. Kasberg W, Luong P, Hanna MG et al (2023) The Sar1 GTPase is dispensable for COPII-dependent cargo export from the ER. Cell Rep 42:112635. https://doi.org/10.1016/j.celrep.2023.112635
57. Kato M, Han TW, Xie S et al (2012) Cell-free formation of RNA granules: low complexity sequence domains form dynamic fibers within hydrogels. Cell 149:753–767. https://doi.org/10.1016/j.cell.2012.04.017
58. Kim T, Tao-Cheng JH, Eiden LE, Loh YP (2001) Chromogranin a, an "on/off" switch controlling dense-core secretory granule biogenesis. Cell 106:499–509. https://doi.org/10.1016/s0092-8674(01)00459-7
59. Kim N, Kim T-H, Kim C et al (2023) Intrinsically disordered region-mediated condensation of IFN-inducible SCOTIN/SHISA-5 inhibits ER-to-Golgi vesicle transport. Dev Cell 58:1950–1966.e8. https://doi.org/10.1016/j.devcel.2023.08.030
60. Koga S, Williams DS, Perriman AW, Mann S (2011) Peptide-nucleotide microdroplets as a step towards a membrane-free protocell model. Nat Chem 3:720–724. https://doi.org/10.1038/nchem.1110
61. König N, Fiehn C, Wolf C et al (2017) Familial chilblain lupus due to a gain-of-function mutation in STING. Ann Rheum Dis 76:468–472. https://doi.org/10.1136/annrheumdis-2016-209841
62. Kors S, Schlaitz A-L (2024) Dynamic remodelling of the endoplasmic reticulum for mitosis. J Cell Sci 137. https://doi.org/10.1242/jcs.261444
63. Kusumi A, Tsunoyama TA, Suzuki KGN et al (2024) Transient, nano-scale, liquid-like molecular assemblies coming of age. Curr Opin Cell Biol 89:102394. https://doi.org/10.1016/j.ceb.2024.102394
64. Lafontaine DLJ, Riback JA, Bascetin R, Brangwynne CP (2021) The nucleolus as a multiphase liquid condensate. Nat Rev Mol Cell Biol 22:165–182. https://doi.org/10.1038/s41580-020-0272-6
65. Li P, Banjade S, Cheng H-C et al (2012) Phase transitions in the assembly of multivalent signalling proteins. Nature 483:336–340. https://doi.org/10.1038/nature10879
66. Litschel T, Kelley CF, Cheng X et al (2024) Membrane-induced 2D phase separation of the focal adhesion protein Talin. Nat Commun 15:4986. https://doi.org/10.1038/s41467-024-49222-z
67. Liu Y, Jesus AA, Marrero B et al (2014) Activated STING in a vascular and pulmonary syndrome. N Engl J Med 371:507–518. https://doi.org/10.1056/NEJMoa1312625
68. Liu L, Cai J, Wang H et al (2019) Coupling of COPII vesicle trafficking to nutrient availability by the IRE1α-XBP1s axis. Proc Natl Acad Sci USA 116:11776–11785. https://doi.org/10.1073/pnas.1814480116
69. Ma W, Goldberg J (2016) TANGO1/cTAGE5 receptor as a polyvalent template for assembly of large COPII coats. Proc Natl Acad Sci USA 113:10061–10066. https://doi.org/10.1073/pnas.1605916113

70. Ma W, Mayr C (2018) A membraneless organelle associated with the endoplasmic reticulum enables 3'UTR-mediated protein-protein interactions. Cell 175:1492–1506.e19. https://doi.org/10.1016/j.cell.2018.10.007
71. Maeda M, Katada T, Saito K (2017) TANGO1 recruits Sec16 to coordinately organize ER exit sites for efficient secretion. J Cell Biol 216:1731–1743. https://doi.org/10.1083/jcb.201703084
72. Maeda M, Komatsu Y, Saito K (2020) Mitotic ER exit site disassembly and reassembly are regulated by the phosphorylation status of TANGO1. Dev Cell 55:237–250.e5. https://doi.org/10.1016/j.devcel.2020.07.017
73. Martin EW, Holehouse AS (2020) Intrinsically disordered pro-tein regions and phase separation: sequence determinants of assemblyor lack thereof. Emerg Top Life Sci 4:307–329
74. Martin EW, Holehouse AS, Peran I et al (2020) Valence and patterning of aromatic residues determine the phase behavior of prion-like domains. Science 367:694–699. https://doi.org/10.1126/science.aaw8653
75. Mendes LFS, Oliveira CG, Simões KF et al (2024) Exploring liquid-liquid phase separation in the organisation of Golgi matrix proteins. Biochim Biophys Acta Proteins Proteom 1872:141029. https://doi.org/10.1016/j.bbapap.2024.141029
76. Milovanovic D, Wu Y, Bian X, De Camilli P (2018) A liquid phase of synapsin and lipid vesicles. Science 361:604–607. https://doi.org/10.1126/science.aat5671
77. Mironov AA, Mironov AA Jr, Beznoussenko GV et al (2003) ER-to-Golgi carriers arise through direct en bloc protrusion and multistage maturation of specialized ER exit domains. Dev Cell 5:583–594. https://doi.org/10.1016/s1534-5807(03)00294-6
78. Mohanty P, Kapoor U, Sundaravadivelu Devarajan D et al (2022) Principles governing the phase separation of multidomain proteins. Biochemistry 61:2443–2455. https://doi.org/10.1021/acs.biochem.2c00210
79. Mondal S, Narayan K, Botterbusch S et al (2022) Multivalent interactions between molecular components involved in fast endophilin mediated endocytosis drive protein phase separation. Nat Commun 13:5017. https://doi.org/10.1038/s41467-022-32529-0
80. Moyer BD, Allan BB, Balch WE (2001) Rab1 interaction with a GM130 effector complex regulates COPII vesicle cis–Golgi tethering: Rab1-GM130 complex regulates COPII-Golgi transport. Traffic 2:268–276. doi:https://doi.org/10.1034/j.1600-0854.2001.1o007.x
81. Munro S (2011) The golgin coiled-coil proteins of the Golgi apparatus. Cold Spring Harb Perspect Biol 3. https://doi.org/10.1101/cshperspect.a005256
82. Nakamura N (2010) Emerging new roles of GM130, a cis-Golgi matrix protein, in higher order cell functions. J Pharmacol Sci 112:255–264. https://doi.org/10.1254/JPHS.09R03CR
83. Nakamura N, Rabouille C, Watson R et al (1995) Characterization of a cis-Golgi matrix protein, GM130. J Cell Biol 131:1715–1726. https://doi.org/10.1083/jcb.131.6.1715
84. Nakamura N, Lowe M, Levine TP et al (1997) The vesicle docking protein p115 binds GM130, a cis-Golgi matrix protein, in a mitotically regulated manner. Cell 89:445–455. https://doi.org/10.1016/s0092-8674(00)80225-1
85. Namitz KEW, Showalter SA, Cosgrove MS (2023) Phase separation promotes a highly active oligomeric scaffold of the MLL1 core complex for regulation of histone H3K4 methylation. J Biol Chem 299:105204. https://doi.org/10.1016/j.jbc.2023.105204
86. Nandi SK, Österle D, Heidenreich M et al (2022) Affinity and valence impact the extent and symmetry of phase separation of multivalent proteins. Phys Rev Lett 129:128102. https://doi.org/10.1103/PhysRevLett.129.128102
87. Nosella ML, Forman-Kay JD (2021) Phosphorylation-dependent regulation of messenger RNA transcription, processing and translation within biomolecular condensates. Curr Opin Cell Biol 69:30–40. https://doi.org/10.1016/j.ceb.2020.12.007
88. Nott TJ, Petsalaki E, Farber P et al (2015) Phase transition of a disordered nuage protein generates environmentally responsive membraneless organelles. Mol Cell 57:936–947. https://doi.org/10.1016/j.molcel.2015.01.013

89. Nott TJ, Craggs TD, Baldwin AJ (2016) Membraneless organelles can melt nucleic acid duplexes and act as biomolecular filters. Nat Chem 8:569–575. https://doi.org/10.1038/nchem.2519
90. Obradovic Z, Peng K, Vucetic S et al (2003) Predicting intrinsic disorder from amino acid sequence. Proteins 53(Suppl 6):566–572. https://doi.org/10.1002/prot.10532
91. Osborne CS, Chakalova L, Brown KE et al (2004) Active genes dynamically colocalize to shared sites of ongoing transcription. Nat Genet 36:1065–1071. https://doi.org/10.1038/ng1423
92. Pak CW, Kosno M, Holehouse AS et al (2016) Sequence determinants of intracellular phase separation by complex Coacervation of a disordered protein. Mol Cell 63:72–85. https://doi.org/10.1016/j.molcel.2016.05.042
93. Palade GE (1956) Intracisternal granules in the exocrine cells of the pancreas. J Biophys Biochem Cytol 2:417–422. https://doi.org/10.1083/jcb.2.4.417
94. Parashar S, Chidambaram R, Chen S et al (2021) Endoplasmic reticulum tubules limit the size of misfolded protein condensates. elife 10:e71642. https://doi.org/10.7554/eLife.71642
95. Parchure A, Tian M, Stalder D et al (2022) Liquid-liquid phase separation facilitates the biogenesis of secretory storage granules. J Cell Biol 221. https://doi.org/10.1083/jcb.202206132
96. Peotter J, Kasberg W, Pustova I, Audhya A (2019) COPII-mediated trafficking at the ER/ERGIC interface. Traffic 20:491–503. https://doi.org/10.1111/tra.12654
97. Pietrosemoli N, Pancsa R, Tompa P (2013) Structural disorder provides increased adaptability for vesicle trafficking pathways. PLoS Comput Biol 9:e1003144. https://doi.org/10.1371/journal.pcbi.1003144
98. Quiroz FG, Chilkoti A (2015) Sequence heuristics to encode phase behaviour in intrinsically disordered protein polymers. Nat Mater 14:1164–1171. https://doi.org/10.1038/nmat4418
99. Rai A, Chen J-X, Selbach M, Pelkmans L (2018) Kinase-controlled phase transition of membraneless organelles in mitosis. Nature 559:211–216. https://doi.org/10.1038/s41586-018-0279-8
100. Raote I, Malhotra V (2021) Tunnels for protein export from the endoplasmic reticulum. Annu Rev Biochem 90:605–630. https://doi.org/10.1146/annurev-biochem-080120-022017
101. Raote I, Ortega Bellido M, Pirozzi M et al (2017) TANGO1 assembles into rings around COPII coats at ER exit sites. J Cell Biol 216:901–909. https://doi.org/10.1083/jcb.201608080
102. Raote I, Ortega-Bellido M, Santos AJ et al (2018) TANGO1 builds a machine for collagen export by recruiting and spatially organizing COPII, tethers and membranes. elife 7. https://doi.org/10.7554/eLife.32723
103. Raote I, Chabanon M, Walani N et al (2020a) A physical mechanism of TANGO1-mediated bulky cargo export. elife 9. https://doi.org/10.7554/eLife.59426
104. Raote I, Ernst AM, Campelo F et al (2020b) TANGO1 membrane helices create a lipid diffusion barrier at curved membranes. elife 9. https://doi.org/10.7554/eLife.57822
105. Raote I, Saxena S, Campelo F, Malhotra V (2021) TANGO1 marshals the early secretory pathway for cargo export. Biochim Biophys Acta Biomembr 1863:183700. https://doi.org/10.1016/j.bbamem.2021.183700
106. Rebane AA, Ziltener P, LaMonica LC et al (2020) Liquid-liquid phase separation of the Golgi matrix protein GM130. FEBS Lett 594:1132–1144. https://doi.org/10.1002/1873-3468.13715
107. Reinert J, Ursprung H, Baxter R (1971) Origin andContinuity of cell organelles. Springer, Berlin, NY
108. Ríos-Barrera LD, Sigurbjörnsdóttir S, Baer M, Leptin M (2017) Dual function for Tango1 in secretion of bulky cargo and in ER-Golgi morphology. Proc Natl Acad Sci USA 114:E10389–E10398. https://doi.org/10.1073/pnas.1711408114
109. Romero P, Obradovic Z, Li X et al (2001) Sequence complexity of disordered protein. Proteins 42:38–48. https://doi.org/10.1002/1097-0134(20010101)42:1<38::aid-prot50>3.0.co;2-3
110. Rothman JE (2019) Jim's view: is the Golgi stack a phase-separated liquid crystal? FEBS Lett 593:2701–2705. https://doi.org/10.1002/1873-3468.13609

111. Rustom A, Bajohrs M, Kaether C et al (2002) Selective delivery of secretory cargo in Golgi-derived carriers of nonepithelial cells: selective transport of secretory cargo. Traffic 3:279–288. https://doi.org/10.1034/j.1600-0854.2002.030405.x
112. Ryan VH, Dignon GL, Zerze GH et al (2018) Mechanistic view of hnRNPA2 low-complexity domain structure, interactions, and phase separation altered by mutation and arginine methylation. Mol Cell 69:465–479.e7. https://doi.org/10.1016/j.molcel.2017.12.022
113. Saito K, Chen M, Bard F et al (2009) TANGO1 facilitates cargo loading at endoplasmic reticulum exit sites. Cell 136:891–902. https://doi.org/10.1016/j.cell.2008.12.025
114. Santos AJM, Raote I, Scarpa M et al (2015) TANGO1 recruits ERGIC membranes to the endoplasmic reticulum for procollagen export. elife 4:e10982. https://doi.org/10.7554/eLife.10982
115. Sargeant J, Seiler DK, Costain T et al (2021) ALG-2 and peflin regulate COPII targeting and secretion in response to calcium signaling. J Biol Chem 297:101393. https://doi.org/10.1016/j.jbc.2021.101393
116. Shin Y, Brangwynne CP (2017) Liquid phase condensation in cell physiology and disease. Science 357. https://doi.org/10.1126/science.aaf4382
117. Shorter J, Warren G (2002) Golgi architecture and inheritance. Annu Rev Cell Dev Biol 18:379–420. https://doi.org/10.1146/annurev.cellbio.18.030602.133733
118. Smith RE, Farquhar MG (1966) Lysosome function in the regulation of the secretory process in cells of the anterior pituitary gland. J Cell Biol 31:319–347. https://doi.org/10.1083/jcb.31.2.319
119. Snead WT, Zeno WF, Kago G et al (2019) BAR scaffolds drive membrane fission by crowding disordered domains. J Cell Biol 218:664–682. https://doi.org/10.1083/jcb.201807119
120. Staehelin LA, Kang BH (2008) Nanoscalearchitecture of endoplasmic reticulum export sites andof Golgi membranes as determined by electron tomography
121. Stagg SM, Gürkan C, Fowler DM et al (2006) Structure of the Sec13/31 COPII coat cage. Nature 439:234–238. https://doi.org/10.1038/nature04339
122. Strulson CA, Molden RC, Keating CD, Bevilacqua PC (2012) RNA catalysis through compartmentalization. Nat Chem 4:941–946. https://doi.org/10.1038/nchem.1466
123. Su X, Ditlev JA, Hui E et al (2016) Phase separation of signaling molecules promotes T cell receptor signal transduction. Science 352:595–599. https://doi.org/10.1126/science.aad9964
124. Takahara T, Maeda T (2012) Transient sequestration of TORC1 into stress granules during heat stress. Mol Cell 47:242–252. https://doi.org/10.1016/j.molcel.2012.05.019
125. Tooze J, Kern HF, Fuller SD, Howell KE (1989) Condensation-sorting events in the rough endoplasmic reticulum of exocrine pancreatic cells. J Cell Biol 109:35–50. https://doi.org/10.1083/jcb.109.1.35
126. Van der Verren SE, Zanetti G (2023) The small GTPase Sar1, control Centre of COPII trafficking. FEBS Lett 597:865–882. https://doi.org/10.1002/1873-3468.14595
127. van Leeuwen W, Nguyen DTM, Grond R et al (2022) Stress-induced phase separation of ERES components into sec bodies precedes ER exit inhibition in mammalian cells. J Cell Sci 135. https://doi.org/10.1242/jcs.260294
128. Vucetic S, Brown CJ, Dunker AK, Obradovic Z (2003) Flavors of protein disorder. Proteins 52:573–584. https://doi.org/10.1002/prot.10437
129. Wang Z, Zhang H (2019) Phase separation, transition, and autophagic degradation of proteins in development and pathogenesis. Trends Cell Biol 29:417–427. https://doi.org/10.1016/j.tcb.2019.01.008
130. Wang Q, Sawyer IA, Sung M-H et al (2016) Cajal bodies are linked to genome conformation. Nat Commun 7:10966. https://doi.org/10.1038/ncomms10966
131. Wang C, Guan Y, Lv M et al (2018) Manganese increases the sensitivity of the cGAS-STING pathway for double-stranded DNA and is required for the host defense against DNA viruses. Immunity 48:675–687.e7. https://doi.org/10.1016/j.immuni.2018.03.017
132. Wang X, Huang R, Wang Y et al (2023) Manganese regulation of COPII condensation controls circulating lipid homeostasis. Nat Cell Biol 25:1650–1663. https://doi.org/10.1038/s41556-023-01260-3

133. Watson P, Stephens DJ (2005) ER-to-Golgi transport: form and formation of vesicular and tubular carriers. Biochim Biophys Acta 1744:304–315. https://doi.org/10.1016/j.bbamcr.2005.03.003
134. Weigel AV, Chang C-L, Shtengel G et al (2021) ER-to-Golgi protein delivery through an interwoven, tubular network extending from ER. Cell 184:2412–2429.e16. https://doi.org/10.1016/j.cell.2021.03.035
135. Witte K, Schuh AL, Hegermann J et al (2011) TFG-1 function in protein secretion and oncogenesis. Nat Cell Biol 13:550–558. https://doi.org/10.1038/ncb2225
136. Woodruff JB, Ferreira Gomes B, Widlund PO et al (2017) The centrosome is a selective condensate that nucleates microtubules by concentrating tubulin. Cell 169:1066–1077.e10. https://doi.org/10.1016/j.cell.2017.05.028
137. Xu CS, Hayworth KJ, Lu Z et al (2017) Enhanced FIB-SEM systems for large-volume 3D imaging. elife 6. https://doi.org/10.7554/eLife.25916
138. Yoo SH (1995) pH- and ca(2+)-induced conformational change and aggregation of chromogranin B. Comparison with chromogranin a and implication in secretory vesicle biogenesis. J Biol Chem 270:12578–12583. https://doi.org/10.1074/jbc.270.21.12578
139. Yu X, Zhang L, Shen J et al (2021) The STING phase-separator suppresses innate immune signalling. Nat Cell Biol 23:330–340. https://doi.org/10.1038/s41556-021-00659-0
140. Yuan F, Alimohamadi H, Bakka B et al (2021) Membrane bending by protein phase separation. Proc Natl Acad Sci USA 118:e2017435118. https://doi.org/10.1073/pnas.2017435118
141. Zanetti G, Pahuja KB, Studer S et al (2011) COPII and the regulation of protein sorting in mammals. Nat Cell Biol 14:20–28. https://doi.org/10.1038/ncb2390
142. Zappa F, Wilson C, Di Tullio G et al (2019) The TRAPP complex mediates secretion arrest induced by stress granule assembly. EMBO J 38:e101704. https://doi.org/10.15252/embj.2019101704
143. Zeng M (2018) Reconstituted postsynaptic density as a molecular platform forunderstanding synapse formation and plasticity. Cell 174:1172–1187
144. Zeuschner D, Geerts WJC, van Donselaar E et al (2006) Immuno-electron tomography of ER exit sites reveals the existence of free COPII-coated transport carriers. Nat Cell Biol 8:377–383. https://doi.org/10.1038/ncb1371
145. Zhang Y, Seemann J (2024) RNA scaffolds the Golgi ribbon by forming condensates with GM130. Nat Cell Biol 26:1139–1153. https://doi.org/10.1038/s41556-024-01447-2
146. Ziltener P, Rebane AA, Graham M et al (2020) The golgin family exhibits a propensity to form condensates in living cells. FEBS Lett 594:3086–3094. https://doi.org/10.1002/1873-3468.13884
147. Zwicker D, Seyboldt R, Weber CA et al (2017) Growth and division of active droplets provides a model for protocells. Nat Phys 13:408–413. https://doi.org/10.1038/nphys3984

Chapter 10
Plant Golgi and GECCO: The Universality of ER–Golgi Interface

Yoko Ito and Tomohiro Uemura

Abstract In spite of its highly conserved molecular constituents, the Golgi apparatus shows diverse morphologies depending on the organism and cell type. In particular, in plant cells, the distribution and dynamics of the Golgi differ considerably from that seen in their mammalian or yeast counterparts. The architecture of the ER–Golgi interface, the first transfer station of the whole secretory traffic, has also been thought to differ between plants and the better-known mammalian system that possesses the ER–Golgi intermediate compartment (ERGIC). However, detailed observations with plant cells gave rise to the identification of GECCO, the ER–Golgi intermediate that is likely to correspond to the mammalian ERGIC. The existence of compartments that are seemingly different, but still share common features across multiple eukaryotic kingdoms, highlights the universality of the ER–Golgi interface. In this chapter, after summarizing the characteristics of the plant Golgi, we will introduce the findings about GECCO by comparing plants with mammals and yeast. In addition, we will discuss the universal identity of the ER–Golgi intermediate compartments in order to transcend the morphological diversity for a deeper understanding of the ER–Golgi interface.

Keywords Plant · Golgi · GECCO · ERGIC · Membrane traffic

Introduction

Similarly to the other eukaryotes, plant cells possess the Golgi apparatus, the organelle that is essential for the cell's life. The fundamental roles of the Golgi are shared between plants and other eukaryotes. The Golgi is positioned at the center of the

Y. Ito (✉)
Institute for Human Life Science, Ochanomizu University, Tokyo, Japan
e-mail: ito.yoko@ocha.ac.jp

T. Uemura
Graduate School of Humanities and Sciences, Ochanomizu University, Tokyo, Japan

A. Nakano, J. Saraste (eds.), *The Golgi Network, Volume I*, Subcellular Biochemistry 110, https://doi.org/10.1007/978-3-032-06936-8_10

intracellular transport network and decorates the cargo molecules, such as proteins and lipids, with sugar chains. The plant Golgi apparatus is also the place for the production of non-cellulosic cell wall polysaccharides, such as hemicelluloses and pectins [45], which also shares similarities with the function of mammalian Golgi in producing extensively glycosylated extracellular substances such as mucins [41]. Most of the components of the ER–Golgi and intra-Golgi traffic systems, including Sar/Arf and Rab GTPases, COPI and COPII coat proteins, tethering and membrane fusion factors like COGs and SNAREs, are extremely well conserved across the eukaryotic kingdom [15, 52, 56, 85, 92]. Disruption of their functions in the early secretory pathway frequently results in gametophytic lethality in the model plant *Arabidopsis thaliana*, presumably reflecting their critical roles in cell maintenance at the very basic level [22, 28, 64, 89, 117, 118, 130]. Since the shared molecular mechanisms operating at the ER–Golgi boundary will be summarized elsewhere in this book, we are not going to focus on them in this chapter. Instead, we describe the structural and dynamic features of plant Golgi, and their relation to the organization of ER–Golgi interface by comparing with the knowledge in mammals and yeast.

Shape and Distribution of Plant Golgi

Plant Golgi Exists as Discrete Stacks

Considering the universality of the function of the Golgi apparatus and the molecular machinery supporting it, the Golgi shows striking diversity in terms of its appearance. Camillo Golgi described the structure he found in the Purkinje cells as "*apparato reticolare interno*" (internal reticular apparatus) in his first report in 1898 [34]. Later on, he also wrote that the reticular structure was positioned at the perinuclear region embracing the nucleus [11]. His observations were correct and also reproduced by subsequent studies using cutting-edge microscopic techniques (which will be described in other sections of this book). However, they led to controversy because the intracellular structures that were stained with Golgi's method in other species looked completely different. In insects and mollusks, they were not reticular networks but discrete granules [8], and similar structures were also found in many kinds of plant cells [14]. Before electron microscopy revealed the fine morphology of these structures, their homology with the mammalian Golgi was subject to debate for more than half a century after Golgi's discovery [7, 9].

This discrepancy was solely due to the differences in the spatial arrangement of the Golgi apparatus across species, especially between the cells of vertebrates and most other eukaryotes. The minimum Golgi unit is typically a stack of several flat cisternae resembling piled-up pancakes. There are also certain exceptional organisms, such as the budding yeast *Saccharomyces cerevisiae* with unstacked Golgi (Fig. 10.1a), or some unicellular parasites that do not have an obvious Golgi structure. They were previously thought to display an ancient shape of the Golgi;

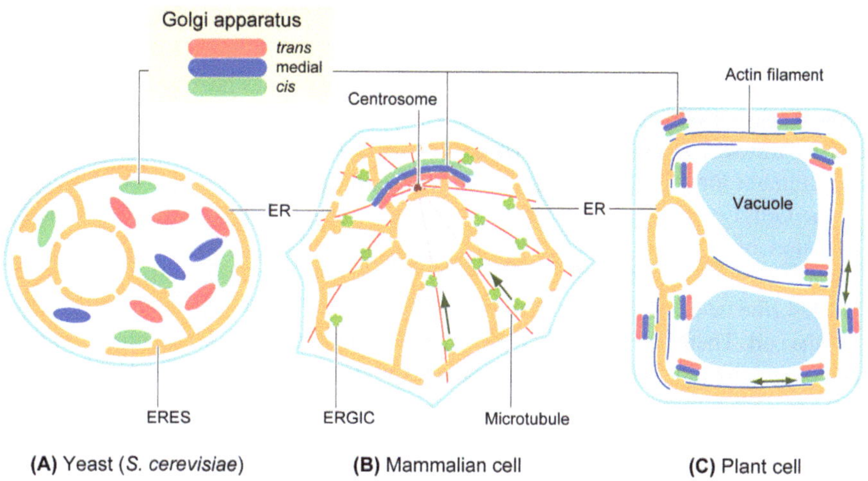

Fig. 10.1 Comparison of Golgi organization in different eukaryotes. (**a**) In budding yeast *S. cerevisiae*, the Golgi cisternae rarely form stacks, and individual cisternae are dispersed in the cytoplasm. (**b**) In most mammalian cells, the stacked Golgi are gathered adjacent to the centrosome by microtubules and interconnected to form the Golgi ribbon. There are intermediate compartments called ERGIC between the ER and the Golgi. (**c**) Plant cells have discrete Golgi stacks spread in the cytoplasm. The stacks are associated with the ERES, with *cis*-side orienting to the ERES, and their movement is mostly dependent on the actomyosin system. (Adapted from [49] with permission from Elsevier)

however, phylogenetic analyses have revealed that they evolved from organisms with stacked Golgi, indicating cisternal stacking as the common ancestral feature of the Golgi apparatus [69]. In most vertebrate cells, the stacks are collected close to the centrosome by a radial array of microtubules, and laterally interconnected to form an extensive network called the "Golgi ribbon" that was initially described by Golgi in the nineteenth century (Fig. 10.1b; [95]). However, this is not the case for the rest of the eukaryotes, since most of them—including plants—have the discrete Golgi stacks dispersed in the cytoplasm (Fig. 10.1c; [124]).

Land plants such as tobacco and *Arabidopsis*, which have frequently been the subject of electron microscopic studies, have Golgi stacks consisting of 5–7 cisternae, with a diameter of approximately 700–800 nm [25, 73, 84, 110]. Many hundreds of such stacks can be observed in a single cell, but usually they are not packed like the Golgi ribbon (Fig. 10.1c; [43]). Formation of the Golgi ribbon in vertebrate cells is thought to contribute to polarized secretion, since the Golgi ribbon, together with the centrosome, re-orients to the direction of movement upon cell migration. Moreover, its dispersion or mislocalization perturbs directional cell movement [17, 47, 125]. Plant cells also demonstrate very efficient polar secretion, especially those of tip-growing root hairs and pollen tubes. In these cells, the Golgi stacks localize to the region that is behind the tip area, while the region within 2–3 μm from the tip is largely filled with vesicles, and the region that lies behind the Golgi is occupied by the nucleus and the vacuole [19]. Even during the tip growth, the plant Golgi stacks are not interconnected like the Golgi ribbon, and their accumulation in the growing region is not absolute [26]. Additionally, Golgi distribution during cell plate formation, which requires massive polar secretion, has been analyzed by electron microscopy. In contrast to mammalian Golgi, which is fragmented into vesicles during mitosis, plant Golgi remains intact and functional in order to fulfill the requirement for rapid polar secretion to the division plane [26, 104]. The orientation of the Golgi stacks was not biased toward the secretory target; however, the accumulation of the Golgi stacks around the phragmoplast during cell division was observed in tobacco BY-2 cells and *Arabidopsis* shoot meristem cells [74, 102]. As we describe later in this chapter, the plant Golgi stacks show great mobility throughout the cell. Therefore, the plant Golgi might accomplish polar secretion by dynamically regulating the position of each stack, rather than forming a single Golgi ribbon. Also in certain vertebrate cells, such as myoblasts, uroepithelial cells, and gastric parietal cells that during differentiation undergo Golgi ribbon fragmentation into individual dispersed stacks—or neurons that contain both the ribbon in the soma and individual stacks (Golgi outposts) along the dendrites—it is suggested that the Golgi organization is based on a balance between the concentrated and dispersed states [53]. The spatial organization of the Golgi should thus be considered a dynamic process.

Despite the discrete nature of the plant Golgi stacks, utilizing live-cell imaging one can frequently observe Golgi stacks that are interconnected by tubules, and sometimes even two stacks can share a single cisterna (Fig. 10.2). Such Golgi stacks that seem to be under fission (or fusion) have been described by electron microscopy already in the 1960s [110]. These observations are sometimes considered as supporting evidence for Golgi biogenesis via division of preexisting stacks, although the dynamics and mechanism of this process remain unknown. In analogy to mitochondria, which are thought to keep their integrity by exchanging DNA through continuous fusion and fission [3], one may speculate that the discrete Golgi stacks also keep uniformity by mixing contents with their neighbors. Whether resident Golgi components or transported cargos are exchanged between the plant Golgi stacks has not been well addressed. It has been recently reported that XYLT2 (a glycosylation enzyme for glycosaminoglycan synthesis) in mammalian cells moves rapidly between the interconnected Golgi stacks (referred to as "ministacks" or "Golgi units" in this paper) that constitute the Golgi ribbon. By contrast, GALANT6 (an *O*-glycosylation enzyme) shows a mobility that predominantly takes place within each unit, suggesting that inter-Golgi stack protein exchange is subject to regulation [42].

In this context, it is also of interest to ask the question whether the Golgi stacks within a plant cell are all identical. In *Drosophila* oocytes, in which discrete Golgi stacks are evenly distributed throughout the cell, only the stacks in the dorsal/anterior corner transport a specific protein necessary for body axis establishment [44]. It has also been reported in *Drosophila* that in a single imaginal disk cell different

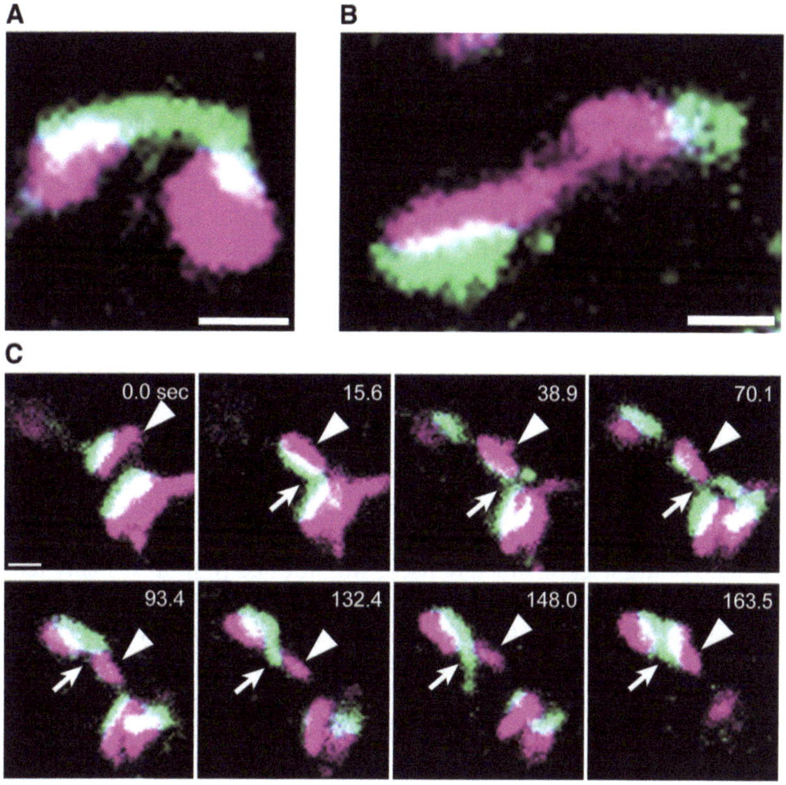

Fig. 10.2 Plant Golgi stacks are occasionally connected. Confocal images of the Golgi stacks in tobacco BY-2 cells expressing GFP-SYP31 (*cis*/GECCO marker, green) and ST-mRFP (*trans*-Golgi marker, magenta). (**a**) Two Golgi stacks that seem to share the *cis*-cisterna. (**b**) Two Golgi stacks with the *trans*-cisternae interconnected by a tubule. (**c**) Time-lapse imaging. A Golgi stack (arrowhead) first interacts with another stack at its *cis*-side, and subsequently associates with yet another stack. Arrows indicate possible interconnections between the Golgi stacks. Scale bars = 1 μm

glycosylation-related proteins localize to different subsets of Golgi stacks [127]. Even in mammalian cells with the interconnected Golgi ribbon, human sialyltransferase isoforms, all of which are thought to localize to the *trans*-Golgi cisternae, show nuanced variations in localization within a ribbon [126]. The rapid movements of the plant Golgi stacks in 3D make it difficult to track them individually throughout the cell. As a result, the degree of differentiation of the plant Golgi is not known. The advancement of image-analyzing techniques such as machine learning might help to resolve this question.

What Tethers the Golgi Cisternae into a Stack?

The unique pancake-like stacked structure of the Golgi apparatus, which is shared by almost all eukaryotes [69], is thought to be maintained by a group of proteins collectively referred to as the Golgi matrix. They correspond to the proteins that can be isolated from the Golgi after membrane solubilization and Golgi enzyme removal [107]. Electron microscopy of both animal and plant cells has revealed that cytoplasmic ribosomes are excluded from the area surrounding the Golgi apparatus [51, 63, 111], and this area is suggested to be occupied by the Golgi matrix proteins. Among them, GRASPs (or GORASPs) became of great interest because of their presumed function in Golgi cisternal stacking. One of the two GRASPs of metazoans, GRASP65 (GORASP1), was first identified as a protein necessary for post-mitotic reassembly of the fragmented mammalian Golgi membranes into stacks [6]. Somewhat later, its homolog GRASP55 (GORASP2) was also shown to be required for the stacking of Golgi cisternae in vitro [105]. However, subsequent in vivo knockdown/knockout experiments with one or both of the proteins gave controversial results, as some of them showed that only the linking of the stacks into a Golgi ribbon, but not the stacked structure itself, was affected [66]. It was only recently clearly demonstrated that the function of mammalian GRASPs is not related to Golgi stacking but involves the lateral connection of the stacks [36, 129]. Most recent studies have also suggested that GRASPs are involved in the sorting of Golgi glycosylation enzymes within the Golgi or unconventional secretion that bypasses the Golgi stacks [66, 80, 81]. In good agreement, the non-vertebrate GRASP homologs, such as the ones of *Drosophila* and *Pichia pastoris*, are not significantly involved in Golgi stacking, and *S. cerevisiae*, an organism with unstacked Golgi, still possesses a GRASP homolog Grh1 [10]. On top of these facts, green plants are thought to have lost the GRASPs during evolution without losing the ability for Golgi stacking [52]. Obviously, the mechanism operating in the formation of the stacked Golgi structure still remains unknown.

Evidently, we have to return to the starting point in our search for the "Golgi stacking factor." The most potential candidates so far are the Golgins, the major components of the Golgi matrix. They represent a family of proteins with long coiled-coil domains on the cytoplasmic surface of Golgi membranes, with each member displaying a specific intra-Golgi localization [33, 62]. Whether there exists a specific "stack-forming Golgin" remains unknown; however, the shared molecular structure and diverse localizations of the Golgin family have attracted the attention of Golgi researchers [84, 87]. Numerous mammalian Golgins are already known to function in the tethering of transport vesicles before their fusion with the Golgi membranes [33, 62], and one of them—giantin—was recently reported to be important for lateral connection of the stacks [42, 97]. Some conserved Golgin homologs are also found in plants. Although some of them have been shown to function in ER–Golgi trafficking or ER–Golgi membrane tethering, their involvement in Golgi stacking—similar to their mammalian counterparts—remains unknown [75, 77, 90, 114, 123]. Regarding plants, electron microscopists have long suggested that there are intercisternal filaments that seem to cross-link the adjacent cisternae [84]. Interestingly, except for p115/MAG4, which localizes to the *cis* side of the Golgi, and GRIP, which is suggested to localize to the *trans*/TGN side [51, 57, 114], many of the plant Golgins show a doughnut-like localization [58, 123]. This indicates that the Golgins in question mainly reside at the cisternal rims and are absent from the cisternal surfaces facing the neighboring cisternae. Since the components of the intercisternal filaments have not been identified, there might still be completely unknown factors working in Golgi stacking.

Plant Golgi Dynamics

Another characteristic feature of the plant Golgi is its mobility. Contrasting with the static images of plant cells, the intracellular world of plant cells is highly dynamic. Similarly to many other plant organelles, Golgi stacks move along actin filaments by the myosin motor proteins [103, 108]. This is in sharp contrast to the vertebrate Golgi, whose dynamics and distribution are mainly dependent on microtubules [27]. The plant Golgi movements are known to be largely inhibited by actin-depolymerizing drugs or a myosin ATPase inhibitor [13, 73], while microtubule-disrupting drugs do not have significant effects [73]. Studies of tobacco BY-2 cells have shown that as the Golgi stacks travel they change their leading side by rolling, with a slight preference for placing the *cis* side at the front [65]. Their speed is several microns per second, reaching up to 7 μm/s in a highly vacuolated *Arabidopsis* root [1].

Ever since it became possible to observe the Golgi stacks in living cells, it has been known that their movements are discontinuous. They repeatedly pause for seconds to minutes between rapid directional movements, hence the designation "stop-and-go" motion [73]. This motion was first presumed to be connected with anterograde ER–Golgi trafficking events, with cargo loading to the Golgi taking place during the immobile periods [51, 72, 73]. However, a fluorescence recovery after photobleaching (FRAP) analysis suggested that ER-to-Golgi protein transport occurs while the Golgi stacks are moving [24]. Although this conclusion has been criticized based on the argument that it was not clear whether the "transported" cargo really reaches the Golgi or remains concentrated in the COPII vesicles located at the *cis*-side of the Golgi [51], it is still technically difficult to distinguish the two. On the other hand, the Golgi-to-ER retrograde vesicles have been suggested to fuse with the ER when the Golgi stacks pause at the "ER import sites" [21, 59]; however, due to technical issues, cargo transport has never been directly observed. The stop-and-go movement is also considered in a more general context because not only the Golgi apparatus, but also other plant organelles show similar behavior. Although the pausing sites overlap with cortical microtubules, the pauses are not affected by microtubule depolymerization [38]. Stable spots of cortical microtubules have been reported to contribute as "anchoring points" as the ER forms multiway junctions, but their relation to Golgi movement is unknown [39]. Additionally, cellulose synthase complexes are suggested to be inserted into the plasma membrane as the Golgi stacks pause close to the cortical microtubules, indicating a possible connection between Golgi motility and post-Golgi transport events [23, 37].

ER–Golgi Interface

Close Association of the ER and Golgi in Plant Cells

The intracellular location that is mostly affected by the variations in Golgi morphology and dynamics is the ER–Golgi interface. At this location, cargo molecules are loaded into COPII vesicles that bud from the specialized domains of the ER called ER exit sites (ERES) and then transported to the Golgi. Some proteins are sorted into COPI vesicles and retrieved back to the ER [16]. Although the contribution of free vesicles in transport is still under debate, it has been well established that the rapid cycle between the ER and Golgi is important for the maintenance of the Golgi structure and function [86]. In order to achieve this efficient bidirectional trafficking, it is reasonable to think that the best way is to place the ER and Golgi close together. Indeed, an electron tomographic study of *Arabidopsis* root meristem cells reported that the majority of the Golgi stacks are adjacent to ERES within 300 nm distance, with the *cis*-side facing the ERES [51]. Since the size of the budding COPII vesicles is approximately 60 nm, the distance of transport between the ER and Golgi is considered to be very short. There is a discrepancy in live-imaging or immunostaining data regarding the relationship between plant Golgi and ERES, since some have reported that they are always linked together as ERES-Golgi units, while others claim that there are also non-Golgi-associated free ERES. Nevertheless, the general idea is that ER-to-Golgi transport occurs from the Golgi-associated ERES in either model [15, 21]. The most recent study visualizing the real-time dynamics of Golgi and ERES has revealed that the free ERES are captured by the Golgi stacks for association, and the associated ones are released presumably to facilitate recycling [113]. It remains to be investigated how this capture-and-release dynamics is involved in the ER–Golgi transport, but Takagi and coworkers suggest that cargo transfer takes place as the two compartments associate. It is worth noting that, even in mammalian cells that have the Golgi ribbon, the Golgi ministacks are located close to the ERES when microtubules are disrupted [40]. The close positioning of the Golgi to the ERES could be the fundamental feature of this organelle.

In addition, it has been shown that plant Golgi stacks are physically tethered with the ER membrane because ER tubules follow the Golgi stacks that are grabbed and displaced by optical tweezers [109]. CASP, one of the proteins of the Golgin family, has been suggested to partially contribute to this ER–Golgi tethering [77]. The Golgi matrix is also thought to function in capturing vesicles at the tethered plant ER–Golgi interface [51]. Because of the rapid Golgi movements in plants, such physical tethering might be necessary.

GECCO as the ER–Golgi Intermediate in Plant Cells

In contrast to the plant Golgi, mammalian Golgi is typically organized as a single giant Golgi ribbon adjacent to the centrosome, as described above. By contrast, the ER network spreads all over the cell, and about half of the ERES are located at a considerable distance from the centralized Golgi [112]. Therefore, secretory cargos have to travel a long distance from the peripheral ERES to the Golgi ribbon. Mammalian cells achieve this long-way traffic by using the intermediate compartment between the ER and Golgi. It is referred to either as ERGIC (ER–Golgi intermediate compartment) or simply IC (intermediate compartment), and also regarded as the same compartment that was named VTC (vesicular tubular cluster), based on its morphological appearance in the electron microscope [2, 94]. ERGIC contains both anterograde and retrograde cargos that are transported between the ER and Golgi, and may interact with both COPI and COPII coats [5, 78, 98, 106]. Therefore, ERGIC is thought to function as the sorting station at the ER–Golgi interface. Instead of the main Golgi body, ERGIC is frequently observed next to the ERES, supporting the idea that it works as the acceptor compartment to receive ERES-derived transport vesicles [5, 46, 93, 94].

Because of the close association of the ER and Golgi in plants, an ERGIC-like compartment that is distinct from the Golgi stacks has not been unambiguously identified. An electron microscopic study reported that small compartments that would be the very first cisternae with the size of 3 to 5 COPII vesicles are present at the *cis*-face of the plant Golgi stacks [25]. The cisternae at the *cis*-most side of the plant Golgi, including such small compartments, contain the ER chaperone BiP, which is expected to be transported back to the ER from those cisternae. In addition, α-1,2 mannosidase I (ManI), the enzyme that mediates the first *N*-glycosylation reaction at the Golgi, does not localize to the *cis* cisternae, but mainly localizes to medial-Golgi [25]. Based on these results, the *cis*-most cisternae of the plant Golgi have been suggested to represent the site of cargo sorting or stack assembly, but be devoid of glycosylation activity. Importantly, the most recent study using lifetime tau-STimulated Emission Depletion (τ-STED) and Expansion Microscopy (ExM) super-resolution techniques reported the existence in *Arabidopsis* cells of pre-Golgi tubular structures that contain early Golgi SNARE MEMB12 [29]. Therefore, more research regarding this compartment is required.

One tool that has provided interesting information about the ER–Golgi interface is the drug Brefeldin A (BFA). BFA was first found to inhibit the ER export of secretory proteins in mammalian cells [68, 115], and microscopic observations revealed the reversible disappearance of the Golgi apparatus upon drug treatment due to relocalization of Golgi resident enzymes to the ER [30, 60]. The target of BFA is the GEF (guanine nucleotide exchange factor) for Arf GTPases, including Arf1, which is known to regulate COPI vesicle formation for Golgi-to-ER retrograde trafficking [88]. A similar effect of BFA on the Golgi apparatus can also be observed in plant cells, as long as the target Arf GEF is sensitive to the drug. There are both BFA-sensitive and -insensitive Arf GEFs depending on the structure of the BFA binding site, which can be engineered by one amino acid substitution [32, 82]. For example, similarly to mammals, the Golgi stacks disappear in tobacco cells subjected to BFA treatment, but not in *Arabidopsis* root cells because the Arf GEF GNL1 (GNOM-LIKE 1) expressed in the latter is BFA-insensitive [55, 83, 120].

Strikingly, BFA-mediated Golgi disassembly in tobacco cells revealed that specific Golgi proteins do not relocalize to the ER like the other Golgi components. SYP31 (plant homolog of yeast Sed5 and mammalian Syntaxin5) and RER1 (Golgi-to-ER retrieval receptor for membrane proteins) were the first Golgi components that were discovered to behave in this way. These proteins localized to unknown punctate structures in close vicinity to ERES, instead of being redistributed to the ER (Fig. 10.3; [48]). Subsequently, those proteins were shown to localize to the *cis*-most side of the stacks in untreated cells, while the proteins that relocalized to the ER after BFA addition were found to reside in the medial or *trans* cisternae [50]. ManI, which marks the starting point of Golgi glycosylation as described above, was one of those that relocalize to the ER with BFA, indicating that the proteins of the earliest *cis* cisternae, corresponding to a putative "sorting station," show this special behavior [50]. After BFA removal, ER export is reinitiated, and the Golgi stacks undergo reassembly [54]. At the onset of this Golgi reformation, 4D super-resolution live-imaging demonstrated that the punctate structures containing the *cis*-Golgi proteins receive the redistributed Golgi components from the ER and act as the scaffold for rebuilding of the stacks [50]. These results encouraged us to propose the name Golgi Entry Core Compartment (GECCO) for the punctate drug-resistant structures and—in a wider context—for the *cis*-most cisternae of the unperturbed plant Golgi [50]. Recent studies have further reported that MNS3, an ER-type

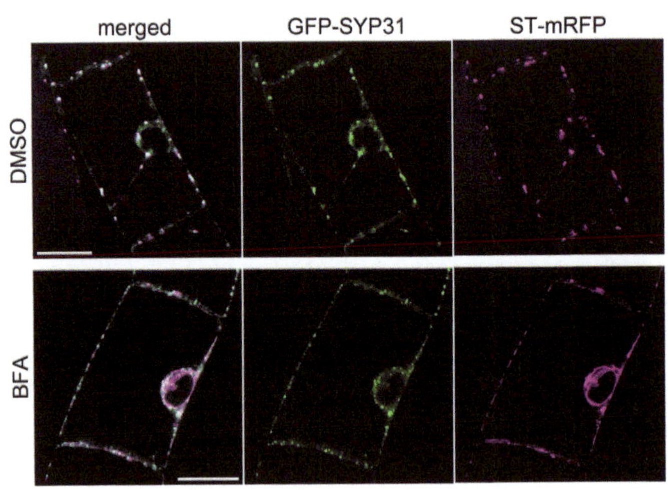

Fig. 10.3 BFA treatment identifies GECCO. Confocal images of tobacco BY-2 cells expressing GFP-SYP31 (*cis*/GECCO marker, green) and ST-mRFP (*trans*-Golgi marker, magenta) following a 90 min BFA (50 μM) treatment. GFP-SYP31 remains punctate upon BFA treatment while ST-mRFP relocalizes to the ER. Scale bars = 20 μm

α-mannosidase that acts before ManI, and Golgin-84a—one of the Golgin family proteins—show similar behavior as SYP31 and RER1B; that is, they originally localize to the *cis*-side of the Golgi and remain in punctate structures upon BFA treatment [100, 123], which we refer to as GECCO-type behavior in this chapter. Similarly, MEMB12, which is a component of the pre-Golgi tubular structures mentioned above [29], also localizes to the punctate structures that persist after BFA treatment (Ito et al., unpublished).

Considering its position and the role at the ER–Golgi interface, GECCO can be thought of as the counterpart of mammalian ERGIC, even though the major part of it is not separated from the main Golgi stack. Actually, proteins that localize to mammalian ERGIC are known to remain in punctate structures upon Golgi disassembly by BFA treatment, similarly to the GECCO components [31, 61, 96, 119]. In *S. cerevisiae*, it is impossible to define the ERGIC as a compartment distinct from the Golgi stacks because the majority of the cisternae are already separated from each other. However, the most recent study clearly demonstrated that proteins localizing to the earliest cisternae remain in the cytoplasmic puncta upon BFA treatment, unlike proteins of later Golgi cisternae that are absorbed into the ER [122]. These findings indicate the existence of underlying commonality of the compartments operating at the ER–Golgi interface, regardless of the variability in their dynamics and morphological appearance (Fig. 10.4).

Although BFA has been used for a long time as a convenient tool to block the ER export and its targets—the Arf GEFs—have been identified, the mechanisms of drug-induced Golgi disassembly and absorption into the ER remain to be elucidated. In particular, the relocalization of medial- and *trans*-Golgi proteins to the ER under conditions when retrograde Golgi-to-ER transport is affected due to Arf inhibition appears quite contradictory. Because blocking of anterograde COPII traffic by the expression of dominant mutants of the regulator GTPase SAR1 largely has the same effect as BFA treatment [67, 76, 116, 128], there has been reason to think that the imbalance between anterograde and retrograde transport at the ER–Golgi interface makes the whole Golgi unstable. In BFA-treated tobacco cells, stacked ER, which looks like a hybrid structure derived from both Golgi and ER, has been observed by electron microscopy [83]. Therefore, it has been hypothesized that the disruption of Golgi-to-ER transport results in an over-accumulation at the Golgi of proteins, which normally function in the fusion of Golgi-derived vesicles to the ER, leading as a result to the direct fusion of the Golgi stacks with the ER. However, experimental support for this hypothesis is still missing. Since the detailed effects of BFA on Golgi organization are still not clear, why the GECCO/ERGIC proteins show distinct behavior also remains to be established. Elucidation of this problem would be important for understanding the ER–Golgi system operating in the different eukaryotes.

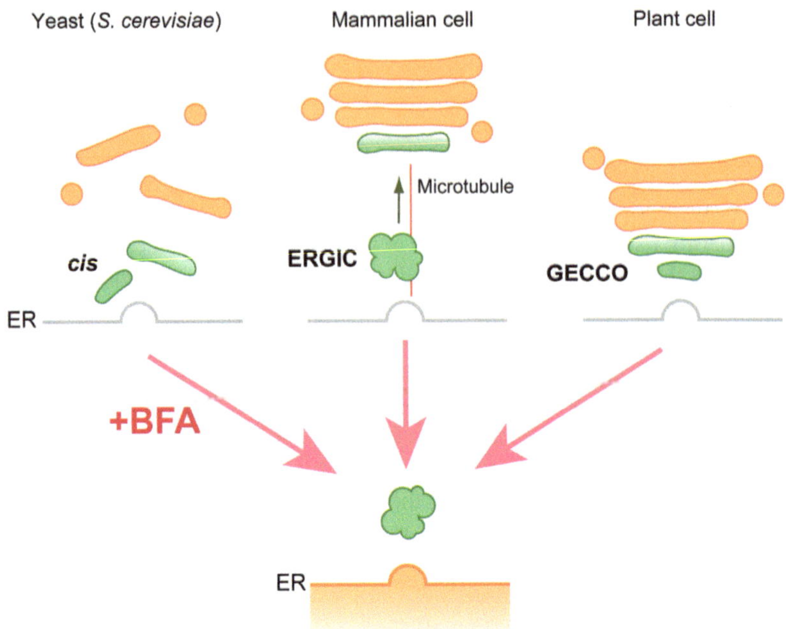

Fig. 10.4 The effect of BFA on the ER–Golgi interface. A model of BFA effects on Golgi and ER–Golgi compartments in different organisms. Despite their different morphological appearances, a portion of the early post-ER compartments (indicated in green) generally remain punctate upon Golgi disassembly by BFA treatment

What Should We Call "Intermediate"?

When discussing the ER–Golgi intermediate, it is not easy to decide to what extent we can use the term "intermediate." Although mammalian ERGIC can sometimes be identified based on its tubulo-vesicular morphology and the dynamics that differ from both the ER and Golgi, these criteria cannot be easily applied to the other eukaryotes. This is the reason why it has been previously thought that there are no intermediate compartments in plants, yeasts, or invertebrates. Mammalian ERGIC can also be defined as a compartment where the marker protein ERGIC-53 is localized [101]; however, the ERGIC-53 homolog has not been identified in plants.

One possibility we propose is to classify as "intermediate" the *cis* (or earlier) compartments that precede the compartment where Golgi-type glycosylation begins. *N*-glycans on glycoproteins are converted from the ER-type (high-mannose) to Golgi-type glycans as the proteins are transported across the ER–Golgi interface. In particular, the process of mannose trimming from the middle branch of $Man_9GlcNAc_2$ into $Man_8GlcNAc_2$ is the last step that is conserved among yeast, mammals, and plants. Furthermore, ER-α-mannosidase I, the enzyme that is responsible for this process, is known to be important for ER-mediated protein quality control [20, 100]. In plants, the resulting $Man_8GlcNAc_2$ subsequently undergoes further trimming by Golgi-α-mannosidase I within the Golgi apparatus [70, 99, 100]. As mentioned above, *Arabidopsis* ER-α-mannosidase I MNS3 shows GECCO-type behavior, while Golgi-α-mannosidase I, ManI (or MNS1), localizes to the medial-Golgi cisternae that are disassembled by BFA [25, 50, 100]. This indicates that the boundary between BFA-defined GECCO and the Golgi proper coincides with the boundary between ER- and Golgi-*N*-glycosylation. Yeast and mammalian homologs of ER-α-mannosidase I, Mns1p and ERManI (MAN1B1), respectively, were first reported to be the ER-resident proteins [18, 35]. However, immunolocalization of endogenous mammalian ERManI using monoclonal antibodies later demonstrated that the protein localizes exclusively to the Golgi apparatus [79, 91]. Moreover, a recent study that systematically mapped the detailed localization of various mammalian Golgi proteins also reported that MAN1B1-Myc localizes at the main body of the Golgi [121]. Although there are reports suggesting that ERManI localizes to additional compartments called quality control compartments or QC vesicles [4, 12], its Golgi localization might indicate that not only ERGIC but also some compartments that seem to be a part of the main Golgi body could be classified as "intermediate" also in mammalian cells. Supporting this idea, GM130, a widely used mammalian *cis*-Golgi marker, remains in punctate structures upon BFA treatment, colocalizing with ERGIC-53 [71]. More discussion would be necessary, but establishing a criterion to identify the "ER–Golgi intermediate" that is generally applicable to eukaryotic cells with different types of Golgi would contribute to a deeper understanding of the early secretory pathway.

Conclusion and Future Perspectives

The universality and diversity of the Golgi apparatus raise a lot of questions. Why are the morphology and distribution of the Golgi so different across species despite its conserved functions? What is the physiological significance of the stacked structure? What is the remarkably rapid movement of plant Golgi required for? Although not addressed in detail in this chapter, the process of Golgi biogenesis is still mysterious, as well. The plant Golgi stacks remain intact throughout the cell cycle in contrast to the mammalian Golgi that breaks down upon mitosis [26, 74]. Nevertheless, how the new Golgi stacks are generated in plant cells remains unknown.

The identification of plant GECCO, the ER–Golgi intermediate that was previously indistinguishable from the other Golgi cisternae, could be one of the milestones for comparing different types of Golgi in a straightforward fashion. Discoveries of this kind will continue thanks to the new techniques that allow visualization of trafficking in various kinds of living cells. We believe that the new aspects of the Golgi apparatus will be revealed by taking advantage of the universality and diversity of this organelle, and the plant Golgi will surely be a key contributor in these studies.

Compliance with Ethical Standards The authors declare that the research was conducted in the absence of any commercial or financial relationships that could be construed as a potential conflict of interest. This work was supported by Grants-in-Aid for Scientific Research from the Ministry of Education, Culture, Sports, Science and Technology of Japan (grant numbers: 22 K06260 and 23H00382 to YI, 22H02643 and 22 K19327 to TU) and JST (grant number: JPMJCR20E5 to TU).

References

1. Akkerman M, Overdijk EJR, Schel JHN, Emons AMC, Ketelaar T (2011) Golgi body motility in the plant cell cortex correlates with actin cytoskeleton organization. Plant Cell Physiol 52(10):1844–1855. https://doi.org/10.1093/pcp/pcr122
2. Appenzeller-Herzog C, Hauri H-P (2006) The ER-Golgi intermediate compartment (ERGIC): in search of its identity and function. J Cell Sci 119(11):2173–2183. https://doi.org/10.1242/jcs.03019
3. Arimura S, Yamamoto J, Aida GP, Nakazono M, Tsutsumi N (2004) Frequent fusion and fission of plant mitochondria with unequal nucleoid distribution. Proc Natl Acad Sci 101(20):7805–7808. https://doi.org/10.1073/pnas.0401077101
4. Avezov E, Frenkel Z, Ehrlich M, Herscovics A, Lederkremer GZ (2008) Endoplasmic reticulum (ER) mannosidase I is compartmentalized and required for N-glycan trimming to Man5–6GlcNAc2 in glycoprotein ER-associated degradation. Mol Biol Cell 19(1):216–225. https://doi.org/10.1091/mbc.e07-05-0505
5. Bannykh SI, Rowe T, Balch WE (1996) The organization of endoplasmic reticulum export complexes. J Cell Biol 135(1):19–35. https://doi.org/10.1083/jcb.135.1.19
6. Barr FA, Puype M, Vandekerckhove J, Warren G (1997) GRASP65, a protein involved in the stacking of Golgi cisternae. Cell 91(2):253–262. https://doi.org/10.1016/s0092-8674(00)80407-9
7. Beams HW, Kessel RG (1968) The Golgi apparatus: structure and function. In: Danielli JF, Jeon KW (eds) Bourne GH. Academic Press, New York and London, pp 209–276
8. Beams HW, King RL (1932) Cytoplasmic structures in the ganglion cells of certain Orthoptera, with special reference to the Golgi bodies, mitochondria, 'vacuome,' intracellular trabeculae (trophospongium), and neurofibrillae. J Morphol 53(1):59–95. doi:https://doi.org/10.1002/jmor.1050530104
9. Beams HW, Tahmisian TN, Devine RL, Anderson E (1956) Electron microscope studies on the DICTYOSOMES and ACROBLASTS in the male germ cells of the cricket. J Cell Biol 2(4):123–128. https://doi.org/10.1083/jcb.2.4.123
10. Behnia R, Barr FA, Flanagan JJ, Barlowe C, Munro S (2007) The yeast orthologue of GRASP65 forms a complex with a coiled-coil protein that contributes to ER to Golgi traffic. J Cell Biol 176(3):255–261. https://doi.org/10.1083/jcb.200607151
11. Bentivoglio M (1998) 1898: The Golgi apparatus emerges from nerve cells. Trends Neurosci 21(5):195–200. https://doi.org/10.1016/s0166-2236(98)01229-6
12. Benyair R, Ogen-Shtern N, Mazkereth N, Shai B, Ehrlich M, Lederkremer GZ (2015) Mammalian ER mannosidase I resides in quality control vesicles, where it encounters its glycoprotein substrates. Mol Biol Cell 26(2):172–184. https://doi.org/10.1091/mbc.e14-06-1152
13. Boevink P, Oparka K, Cruz S, Martin B, Betteridge A, Hawes C (1998) Stacks on tracks: the plant Golgi apparatus traffics on an actin/ER network. Plant J 15(3). https://doi.org/10.1046/j.1365-313x.1998.00208.x
14. Bowen RH (1928) Studies on the structure of plant protoplasm. Z Zellforsch Mikrosk Anat 6(5):689–725. https://doi.org/10.1007/bf02583419
15. Brandizzi F (2018) Transport from the endoplasmic reticulum to the Golgi in plants: where are we now? Semin Cell Dev Biol 80:94–105. https://doi.org/10.1016/j.semcdb.2017.06.024

16. Brandizzi F, Barlowe C (2013) Organization of the ER–Golgi interface for membrane traffic control. Nat Rev Mol Cell Bio 14(6):382–392. https://doi.org/10.1038/nrm3588
17. Brownhill K, Wood L, Allan V (2009) Molecular motors and the Golgi complex: staying put and moving through. Semin Cell Dev Biol 20(7):784–792. https://doi.org/10.1016/j.semcdb.2009.03.019
18. Burke J, Lipari F, Igdoura S, Herscovics A (1996) The Saccharomyces cerevisiae processing alpha 1,2-mannosidase is localized in the endoplasmic reticulum, independently of known retrieval motifs. Eur J Cell Biol 70(4):298–305
19. Campanoni P, Blatt MR (2006) Membrane trafficking and polar growth in root hairs and pollen tubes. J Exp Bot 58(1):65–74. https://doi.org/10.1093/jxb/erl059
20. Caramelo JJ, Parodi AJ (2008) Getting in and out from Calnexin/Calreticulin cycles. J Biol Chem 283(16):10221–10225. https://doi.org/10.1074/jbc.r700048200
21. Cheung AY, Cosgrove DJ, Hara-Nishimura I, Jürgens G, Lloyd C, Robinson DG, Staehelin LA, Weijers D (2021) A rich and bountiful harvest: key discoveries in plant cell biology. Plant Cell 34(1):53–71. https://doi.org/10.1093/plcell/koab234
22. Conger R, Chen Y, Fornaciari S, Faso C, Held MA, Renna L, Brandizzi F (2011) Evidence for the involvement of the Arabidopsis SEC24A in male transmission. J Exp Bot 62(14):4917–4926. https://doi.org/10.1093/jxb/err174
23. Crowell EF, Bischoff V, Desprez T, Rolland A, Stierhof Y-D, Schumacher K, Gonneau M, Höfte H, Vernhettes S (2009) Pausing of Golgi bodies on microtubules regulates secretion of cellulose synthase complexes in Arabidopsis. Plant Cell 21(4):1141–1154. https://doi.org/10.1105/tpc.108.065334
24. daSilva LLP, Snapp EL, Denecke J, Lippincott-Schwartz J, Hawes C, Brandizzi F (2004) Endoplasmic reticulum export sites and Golgi bodies behave as single Mobile secretory units in plant cells. Plant Cell 16(7):1753–1771. https://doi.org/10.1105/tpc.022673
25. Donohoe BS, Kang B-H, Gerl MJ, Gergely ZR, McMichael CM, Bednarek SY, Staehelin LA (2013) Cis -Golgi Cisternal assembly and biosynthetic activation occur sequentially in plants and algae. Traffic 14(5):551–567. https://doi.org/10.1111/tra.12052
26. Dupree P, Sherrier DJ (1998) The plant Golgi apparatus. Biochimica Et Biophysica Acta Bba - Mol Cell Res 1404(1–2):259–270. https://doi.org/10.1016/s0167-4889(98)00061-5
27. Egea G, Ríos RM (2008) The Golgi apparatus, state of the art 110 years after Camillo Golgi's discovery. In: Mironov AA, Pavelka M (eds) The Golgi apparatus, state of the art 110 years after Camillo Golgi's discovery. Springer, Vienna, pp 270–300
28. El-Kasmi F, Pacher T, Strompen G, Stierhof Y-D, Müller LM, Koncz C, Mayer U, Jürgens G (2011) Arabidopsis SNARE protein SEC22 is essential for gametophyte development and maintenance of Golgi-stack integrity. Plant J Cell Mol Biology 66(2):268–279. https://doi.org/10.1111/j.1365-313x.2011.04487.x
29. Fougère L, Grison M, Laquel P, Montrazi M, Cordelières F, Fernández-Monreal M, Poujol C, Uemura T, Nakano A, Ito Y, Boutté Y (2025) ER-to-Golgi trafficking through a dynamic intermediate *cis*-Golgi tubular network in Arabidopsis. Nat Cell Biol (in press)
30. Fujiwara T, Oda K, Yokota S, Takatsuki A, Ikehara Y (1988) Brefeldin a causes disassembly of the Golgi complex and accumulation of secretory proteins in the endoplasmic reticulum. J Biol Chem 263(34):18545–18552
31. Füllekrug J, Sönnichsen B, Schäfer U, Van PN, Söling H-D, Mieskes G (1997) Characterization of brefeldin a induced vesicular structures containing cycling proteins of the intermediate compartment/cis-Golgi network. FEBS Lett 404(1):75–81. https://doi.org/10.1016/s0014-5793(97)00097-5
32. Geldner N, Anders N, Wolters H, Keicher J, Kornberger W, Muller P, Delbarre A, Ueda T, Nakano A, Jürgens G (2003) The Arabidopsis GNOM ARF-GEF mediates endosomal recycling, auxin transport, and auxin-dependent plant growth. Cell 112(2):219–230. https://doi.org/10.1016/s0092-8674(03)00003-5
33. Gillingham AK, Munro S (2016) Finding the Golgi: Golgin coiled-coil proteins show the way. Trends Cell Biol 26(6):399–408. https://doi.org/10.1016/j.tcb.2016.02.005

34. Golgi C (1898) Sur la structure des cellules nerveuses. Arch Ital Biol 30:60–71
35. Gonzalez DS, Karaveg K, Vandersall-Nairn AS, Lal A, Moremen KW (1999) Identification, expression, and characterization of a cDNA encoding human endoplasmic reticulum mannosidase I, the enzyme that catalyzes the first mannose trimming step in mammalian Asn-linked oligosaccharide biosynthesis. J Biol Chem 274(30):21375–21386. https://doi.org/10.1074/jbc.274.30.21375
36. Grond R, Veenendaal T, Duran JM, Raote I, Es JH van, Corstjens S, Delfgou L, Haddouti BE, Malhotra V, Rabouille C (2020) The function of GORASPs in Golgi apparatus organization in vivo. J Cell Biol 219(9):e202004191. doi:https://doi.org/10.1083/jcb.202004191
37. Gutierrez R, Lindeboom JJ, Paredez AR, Emons AMC, Ehrhardt DW (2009) Arabidopsis cortical microtubules position cellulose synthase delivery to the plasma membrane and interact with cellulose synthase trafficking compartments. Nat Cell Biol 11(7):797–806. https://doi.org/10.1038/ncb1886
38. Hamada T, Tominaga M, Fukaya T, Nakamura M, Nakano A, Watanabe Y, Hashimoto T, Baskin TI (2012) RNA processing bodies, peroxisomes, Golgi bodies, mitochondria, and endoplasmic reticulum tubule junctions frequently pause at cortical microtubules. Plant Cell Physiol 53(4):699–708. https://doi.org/10.1093/pcp/pcs025
39. Hamada T, Ueda H, Kawase T, Hara-Nishimura I (2014) Microtubules contribute to tubule elongation and anchoring of endoplasmic reticulum, resulting in high network complexity in Arabidopsis. Plant Physiol 166(4):1869–1876. https://doi.org/10.1104/pp.114.252320
40. Hammond AT, Glick BS (2000) Dynamics of transitional endoplasmic reticulum sites in vertebrate cells. Mol Biol Cell 11(9):3013–3030. https://doi.org/10.1091/mbc.11.9.3013
41. Hang HC, Bertozzi CR (2005) The chemistry and biology of mucin-type O-linked glycosylation. Bioorg Med Chem 13(17):5021–5034. https://doi.org/10.1016/j.bmc.2005.04.085
42. Harada A, Kunii M, Kurokawa K, Sumi T, Kanda S, Zhang Y, Nadanaka S, Hirosawa KM, Tokunaga K, Tojima T, Taniguchi M, Moriwaki K, Yoshimura S, Yamamoto-Hino M, Goto S, Katagiri T, Kume S, Hayashi-Nishino M, Nakano M, Miyoshi E, Suzuki KGN, Kitagawa H, Nakano A (2024) Dynamic movement of the Golgi unit and its glycosylation enzyme zones. Nat Commun 15(1):4514. https://doi.org/10.1038/s41467-024-48901-1
43. Hawes C, Osterrieder A, Sparkes I (2008) The Golgi apparatus, state of the art 110 years after Camillo Golgi's discovery. In: Mironov AA, Pavelka M (eds) The Golgi apparatus, state of the art 110 years after Camillo Golgi's discovery. Springer, Vienna, pp 611–622
44. Herpers B, Rabouille C (2004) mRNA localization and ER-based protein sorting mechanisms dictate the use of transitional endoplasmic reticulum-Golgi units involved in Gurken transport in drosophila oocytes. Mol Biol Cell 15(12):5306–5317. https://doi.org/10.1091/mbc.e04-05-0398
45. Hoffmann N, King S, Samuels AL, McFarlane HE (2021) Subcellular coordination of plant cell wall synthesis. Dev Cell 56(7):933–948. https://doi.org/10.1016/j.devcel.2021.03.004
46. Hughes H, Budnik A, Schmidt K, Palmer KJ, Mantell J, Noakes C, Johnson A, Carter DA, Verkade P, Watson P, Stephens DJ (2009) Organisation of human ER-exit sites: requirements for the localisation of Sec16 to transitional ER. J Cell Sci 122(16):2924–2934. https://doi.org/10.1242/jcs.044032
47. Hurtado L, Caballero C, Gavilan MP, Cardenas J, Bornens M, Rios RM (2011) Disconnecting the Golgi ribbon from the centrosome prevents directional cell migration and ciliogenesis. J Cell Biol 193(5):917–933. https://doi.org/10.1083/jcb.201011014
48. Ito Y, Uemura T, Shoda K, Fujimoto M, Ueda T, Nakano A (2012) Cis-Golgi proteins accumulate near the ER exit sites and act as the scaffold for Golgi regeneration after brefeldin a treatment in tobacco BY-2 cells. Mol Biol Cell 23(16):3203–3214. https://doi.org/10.1091/mbc.e12-01-0034
49. Ito Y, Uemura T, Nakano A (2014) Formation and maintenance of the Golgi apparatus in plant cells. In: Kwang JW (ed) International review of cell and molecular biology. Academic Press, Burlington, pp 221–287. https://doi.org/10.1016/B978-0-12-800180-6.00006-2

50. Ito Y, Uemura T, Nakano A (2018) The Golgi entry core compartment functions as a COPII-independent scaffold for ER-to-Golgi transport in plant cells. J Cell Sci 131(2):jcs203893. https://doi.org/10.1242/jcs.203893
51. Kang B-H, Staehelin LA (2008) ER-to-Golgi transport by COPII vesicles in Arabidopsis involves a ribosome-excluding scaffold that is transferred with the vesicles to the Golgi matrix. Protoplasma 234(1–4):51–64. https://doi.org/10.1007/s00709-008-0015-6
52. Klute MJ, Melancon P, Dacks JB (2011) Evolution and diversity of the Golgi. Csh Perspect Biol 3(8):a007849–a007849. https://doi.org/10.1101/cshperspect.a007849
53. Kulkarni-Gosavi P, Makhoul C, Gleeson PA (2019) Form and function of the Golgi apparatus: scaffolds, cytoskeleton and signalling. FEBS Lett 593(17):2289–2305. https://doi.org/10.1002/1873-3468.13567
54. Langhans M, Hawes C, Hillmer S, Hummel E, Robinson DG (2007) Golgi regeneration after Brefeldin a treatment in BY-2 cells entails stack enlargement and Cisternal growth followed by division. Plant Physiol 145(2):527–538. https://doi.org/10.1104/pp.107.104919
55. Langhans M, Förster S, Helmchen G, Robinson DG (2011) Differential effects of the brefeldin a analogue (6R)-hydroxy-BFA in tobacco and Arabidopsis. J Exp Bot 62(8):2949–2957. https://doi.org/10.1093/jxb/err007
56. Latijnhouwers M, Hawes C, Carvalho C (2005a) Holding it all together? Candidate proteins for the plant Golgi matrix. Curr Opin Plant Biol 8(6):632–639. https://doi.org/10.1016/j.pbi.2005.09.014
57. Latijnhouwers M, Hawes C, Carvalho C, Oparka K, Gillingham AK, Boevink P (2005b) An Arabidopsis GRIP domain protein locates to the trans-Golgi and binds the small GTPase ARL1. Plant J 44(3):459–470. https://doi.org/10.1111/j.1365-313x.2005.02542.x
58. Latijnhouwers M, Gillespie T, Boevink P, Kriechbaumer V, Hawes C, Carvalho CM (2007) Localization and domain characterization of Arabidopsis golgin candidates. J Exp Bot 58(15–16):4373–4386. https://doi.org/10.1093/jxb/erm304
59. Lerich A, Hillmer S, Langhans M, Scheuring D, van Bentum P, Robinson DG (2012) ER import sites and their relationship to ER exit sites: a new model for bidirectional ER-Golgi transport in higher plants. Front Plant Sci 3:143. https://doi.org/10.3389/fpls.2012.00143
60. Lippincott-Schwartz J, Yuan LC, Bonifacino JS, Klausner RD (1989) Rapid redistribution of Golgi proteins into the ER in cells treated with brefeldin a: evidence for membrane cycling from Golgi to ER. Cell 56(5):801–813. https://doi.org/10.1016/0092-8674(89)90685-5
61. Lippincott-Schwartz J, Donaldson JG, Schweizer A, Berger EG, Hauri H-P, Yuan LC, Klausner RD (1990) Microtubule-dependent retrograde transport of proteins into the ER in the presence of brefeldin a suggests an ER recycling pathway. Cell 60(5):821–836. https://doi.org/10.1016/0092-8674(90)90096-w
62. Lowe M (2019) The physiological functions of the Golgin vesicle tethering proteins. Front Cell Dev Biol 7:94. https://doi.org/10.3389/fcell.2019.00094
63. Lucocq J, Pryde J, Berger E, Warren G (1987) A mitotic form of the Golgi apparatus in HeLa cells. J Cell Biol 104(4):865–874. https://doi.org/10.1083/jcb.104.4.865
64. Ma T, Li E, Li L, Li S, Zhang Y (2021) The Arabidopsis R-SNARE protein YKT61 is essential for gametophyte development. J Integr Plant Biol 63(4):676–694. https://doi.org/10.1111/jipb.13017
65. Madison SL, Nebenführ A (2011) Live-cell imaging of dual-labeled Golgi stacks in tobacco BY-2 cells reveals similar behaviors for different cisternae during movement and brefeldin a treatment. Mol Plant 4(5):896–908. https://doi.org/10.1093/mp/ssr067
66. Mendes LFS, Fontana NA, Reddy ST, Uversky VN, Costa-Filho AJ (2020) The exquisite structural biophysics of the Golgi reassembly and stacking proteins. Int J Biol Macromol 164:3632–3644. https://doi.org/10.1016/j.ijbiomac.2020.08.203
67. Miles S, McManus H, Forsten KE, Storrie B (2001) Evidence that the entire Golgi apparatus cycles in interphase HeLa cells. J Cell Biol 155(4):543–556. https://doi.org/10.1083/jcb.200103104

68. Misumi Y, Misumi Y, Miki K, Takatsuki A, Tamura G, Ikehara Y (1986) Novel blockade by brefeldin a of intracellular transport of secretory proteins in cultured rat hepatocytes. J Biol Chem 261(24):11398–11403
69. Mowbrey K, Dacks JB (2009) Evolution and diversity of the Golgi body. FEBS Lett 583(23):3738–3745. https://doi.org/10.1016/j.febslet.2009.10.025
70. Nagashima Y, von Schaewen A, Koiwa H (2018) Function of N-glycosylation in plants. Plant Sci 274:70–79. https://doi.org/10.1016/j.plantsci.2018.05.007
71. Nakamura N, Rabouille C, Watson R, Nilsson T, Hui N, Slusarewicz P, Kreis TE, Warren G (1995) Characterization of a *cis*-Golgi matrix protein, GM130. J Cell Biol 131(6):1715–1726. https://doi.org/10.1083/jcb.131.6.1715
72. Nebenführ A, Staehelin LA (2001) Mobile factories: Golgi dynamics in plant cells. Trends Plant Sci 6(4):160–167. https://doi.org/10.1016/s1360-1385(01)01891-x
73. Nebenführ A, Gallagher LA, Dunahay TG, Frohlick JA, Mazurkiewicz AM, Meehl JB, Staehelin LA (1999) Stop-and-go movements of plant Golgi stacks are mediated by the Acto-myosin System1. Plant Physiol 121(4):1127–1141. https://doi.org/10.1104/pp.121.4.1127
74. Nebenführ A, Frohlick JA, Staehelin LA (2000) Redistribution of Golgi stacks and other organelles during mitosis and cytokinesis in plant cells. Plant Physiol 124(1):135–152. https://doi.org/10.1104/pp.124.1.135
75. Osterrieder A (2012) Tales of tethers and tentacles: golgins in plants. J Microsc 247(1):68–77. https://doi.org/10.1111/j.1365-2818.2012.03620.x
76. Osterrieder A, Hummel E, Carvalho CM, Hawes C (2010) Golgi membrane dynamics after induction of a dominant-negative mutant Sar1 GTPase in tobacco. J Exp Bot 61(2):405–422. https://doi.org/10.1093/jxb/erp315
77. Osterrieder A, Sparkes IA, Botchway SW, Ward A, Ketelaar T, de Ruijter N, Hawes C (2017) Stacks off tracks: a role for the golgin AtCASP in plant endoplasmic reticulum-Golgi apparatus tethering. J Exp Bot 68(13):3339–3350. https://doi.org/10.1093/jxb/erx167
78. Palokangas H, Ying M, Väänänen K, Saraste J (1998) Retrograde transport from the pre-Golgi intermediate compartment and the Golgi complex is affected by the vacuolar H+-ATPase inhibitor Bafilomycin A1. Mol Biol Cell 9(12):3561–3578. https://doi.org/10.1091/mbc.9.12.3561
79. Pan S, Wang S, Utama B, Huang L, Blok N, Estes MK, Moremen KW, Sifers RN (2011) Golgi localization of ERManI defines spatial separation of the mammalian glycoprotein quality control system. Mol Biol Cell 22(16):2810–2822. https://doi.org/10.1091/mbc.e11-02-0118
80. Pothukuchi P, Agliarulo I, Pirozzi M, Rizzo R, Russo D, Turacchio G, Nüchel J, Yang J, Gehin C, Capolupo L, Hernandez-Corbacho MJ, Biswas A, Vanacore G, Dathan N, Nitta T, Henklein P, Thattai M, Inokuchi J, Hsu VW, Plomann M, Obeid LM, Hannun YA, Luini A, D'Angelo G, Parashuraman S (2021) GRASP55 regulates intra-Golgi localization of glycosylation enzymes to control glycosphingolipid biosynthesis. EMBO J 40(20):e107766. https://doi.org/10.15252/embj.2021107766
81. Rabouille C (2017) Pathways of unconventional protein secretion. Trends Cell Biol 27(3):230–240. https://doi.org/10.1016/j.tcb.2016.11.007
82. Richter S, Geldner N, Schrader J, Wolters H, Stierhof Y-D, Rios G, Koncz C, Robinson DG, Jürgens G (2007) Functional diversification of closely related ARF-GEFs in protein secretion and recycling. Nature 448(7152):488–492. https://doi.org/10.1038/nature05967
83. Ritzenthaler C, Nebenführ A, Movafeghi A, Stussi-Garaud C, Behnia L, Pimpl P, Staehelin LA, Robinson DG (2002) Reevaluation of the effects of Brefeldin a on plant cells using tobacco bright yellow 2 cells expressing Golgi-targeted green fluorescent protein and COPI antisera. Plant Cell 14(1):237–261. https://doi.org/10.1105/tpc.010237
84. Robinson DG (2020) Plant Golgi ultrastructure. J Microsc-oxford 280(2):111–121. https://doi.org/10.1111/jmi.12899
85. Robinson DG, Herranz M-C, Bubeck J, Pepperkok R, Ritzenthaler C (2007) Membrane dynamics in the early secretory pathway. Crit Rev Plant Sci 26(4):199–225. https://doi.org/10.1080/07352680701495820

86. Robinson DG, Brandizzi F, Hawes C, Nakano A (2015) Vesicles versus tubes: is endoplasmic reticulum-Golgi transport in plants fundamentally different from other eukaryotes? Plant Physiol 168(2):393–406. https://doi.org/10.1104/pp.15.00124
87. Rothman JE (2019) Jim's view: is the Golgi stack a phase-separated liquid crystal? FEBS Lett 593(19):2701–2705. https://doi.org/10.1002/1873-3468.13609
88. Rothman JE, Wieland FT (1996) Protein sorting by transport vesicles. Science 272(5259):227–234. https://doi.org/10.1126/science.272.5259.227
89. Rui Q, Tan X, Liu F, Li Y, Liu X, Li B, Wang J, Yang H, Qiao L, Li T, Fang S, Gao R, Wang W, Bednarek SY, Bao Y (2021) Syntaxin of plants31 (SYP31) and SYP32 is essential for Golgi morphology maintenance and pollen development. Plant Physiol 186(1):330–343. https://doi.org/10.1093/plphys/kiab049
90. Rui Q, Tan X, Liu F, Bao Y (2022) An update on the key factors required for plant Golgi structure maintenance. Front Plant Sci 13:933283. https://doi.org/10.3389/fpls.2022.933283
91. Rymen D, Peanne R, Millón MB, Race V, Sturiale L, Garozzo D, Mills P, Clayton P, Asteggiano CG, Quelhas D, Cansu A, Martins E, Nassogne M-C, Gonçalves-Rocha M, Topaloglu H, Jaeken J, Foulquier F, Matthijs G (2013) MAN1B1 deficiency: an unexpected CDG-II. PLoS Genet 9(12):e1003989. https://doi.org/10.1371/journal.pgen.1003989
92. Sanderfoot A (2007) Increases in the number of SNARE genes parallels the rise of multicellularity among the green plants. Plant Physiol 144(1):6–17. https://doi.org/10.1104/pp.106.092973
93. Sannerud R, Marie M, Nizak C, Dale HA, Pernet-Gallay K, Perez F, Goud B, Saraste J (2006) Rab1 defines a novel pathway connecting the pre-Golgi intermediate compartment with the cell periphery. Mol Biol Cell 17(4):1514–1526. https://doi.org/10.1091/mbc.e05-08-0792
94. Saraste J, Marie M (2018) Intermediate compartment (IC): from pre-Golgi vacuoles to a semi-autonomous membrane system. Histochem Cell Biol 150(5):407–430. https://doi.org/10.1007/s00418-018-1717-2
95. Saraste J, Prydz K (2019) A new look at the functional organization of the Golgi ribbon. Front Cell Dev Biology 7:171. https://doi.org/10.3389/fcell.2019.00171
96. Saraste J, Svensson K (1991) Distribution of the intermediate elements operating in ER to Golgi transport. J Cell Sci 100(3):415–430. https://doi.org/10.1242/jcs.100.3.415
97. Satoh A, Hayashi-Nishino M, Shakuno T, Masuda J, Koreishi M, Murakami R, Nakamura Y, Nakamura T, Abe-Kanoh N, Honjo Y, Malsam J, Yu S, Nishino K (2019) The Golgin protein Giantin regulates interconnections between Golgi stacks. Front Cell Dev Biol 7:160. https://doi.org/10.3389/fcell.2019.00160
98. Scales SJ, Pepperkok R, Kreis TE (1997) Visualization of ER-to-Golgi transport in living cells reveals a sequential mode of action for COPII and COPI. Cell 90(6):1137–1148. https://doi.org/10.1016/s0092-8674(00)80379-7
99. Schoberer J, Strasser R (2011) Sub-compartmental Organization of Golgi-Resident N-glycan processing enzymes in plants. Mol Plant 4(2):220–228. https://doi.org/10.1093/mp/ssq082
100. Schoberer J, König J, Veit C, Vavra U, Liebminger E, Botchway SW, Altmann F, Kriechbaumer V, Hawes C, Strasser R (2019) A signal motif retains Arabidopsis ER-α-mannosidase I in the cis-Golgi and prevents enhanced glycoprotein ERAD. Nat Commun 10(1):3701. https://doi.org/10.1038/s41467-019-11686-9
101. Schweizer A, Fransen JA, Bächi T, Ginsel L, Hauri HP (1988) Identification, by a monoclonal antibody, of a 53-kD protein associated with a tubulo-vesicular compartment at the cis-side of the Golgi apparatus. J Cell Biol 107(5):1643–1653. https://doi.org/10.1083/jcb.107.5.1643
102. Seguí-Simarro JM, Staehelin LA (2006) Cell cycle-dependent changes in Golgi stacks, vacuoles, clathrin-coated vesicles and multivesicular bodies in meristematic cells of *Arabidopsis thaliana*: a quantitative and spatial analysis. Planta 223(2):223–236. https://doi.org/10.1007/s00425-005-0082-2
103. Shimmen T, Yokota E (2004) Cytoplasmic streaming in plants. Curr Opin Cell Biol 16(1):68–72. https://doi.org/10.1016/j.ceb.2003.11.009

104. Shorter J, Warren G (2002) Golgi architecture and inheritance. Annu Rev Cell Dev Biol 18(1):379–420. https://doi.org/10.1146/annurev.cellbio.18.030602.133733
105. Shorter J, Watson R, Giannakou M-E, Clarke M, Warren G, Barr FA (1999) GRASP55, a second mammalian GRASP protein involved in the stacking of Golgi cisternae in a cell-free system. 18(18). https://doi.org/10.1093/emboj/18.18.4949
106. Simpson JC, Nilsson T, Pepperkok R (2006) Biogenesis of tubular ER-to-Golgi transport intermediates. Mol Biol Cell 17(2):723–737. https://doi.org/10.1091/mbc.e05-06-0580
107. Slusarewicz P, Nilsson T, Hui N, Watson R, Warren G (1994) Isolation of a matrix that binds medial Golgi enzymes. J Cell Biol 124(4):405–413. https://doi.org/10.1083/jcb.124.4.405
108. Sparkes I (2011) Recent advances in understanding plant myosin function: life in the fast lane. Mol Plant 4(5):805–812. https://doi.org/10.1093/mp/ssr063
109. Sparkes IA, Ketelaar T, de Ruijter NCA, Hawes C (2009) Grab a Golgi: laser trapping of Golgi bodies reveals in vivo interactions with the endoplasmic reticulum. Traffic 10(5):567–571. https://doi.org/10.1111/j.1600-0854.2009.00891.x
110. Staehelin LA, Kang B-H (2008) Nanoscale architecture of endoplasmic reticulum export sites and of Golgi membranes as determined by electron tomography. Plant Physiol 147(4):1454–1468. https://doi.org/10.1104/pp.108.120618
111. Staehelin LA, Giddings TH, Kiss JZ, Sack FD (1990) Macromolecular differentiation of Golgi stacks in root tips of *Arabidopsis* and *Nicotiana* seedlings as visualized in high pressure frozen and freeze-substituted samples. Protoplasma 157(1–3):75–91. https://doi.org/10.1007/bf01322640
112. Stephens DJ (2003) De novo formation, fusion and fission of mammalian COPII-coated endoplasmic reticulum exit sites. EMBO Rep 4(2):210–217. https://doi.org/10.1038/sj.embor.embor736
113. Takagi J, Kimori Y, Shimada T, Hara-Nishimura I (2020) Dynamic capture-and-release of endoplasmic reticulum exit sites by Golgi stacks in Arabidopsis. iScience 23(7):101265. https://doi.org/10.1016/j.isci.2020.101265
114. Takahashi H, Tamura K, Takagi J, Koumoto Y, Hara-Nishimura I, Shimada T (2010) MAG4/Atp115 is a golgi-localized tethering factor that mediates efficient anterograde transport in Arabidopsis. Plant Cell Physiol 51(10):1777–1787. https://doi.org/10.1093/pcp/pcq137
115. Takatsuki A, Tamura G (1985) Brefeldin a, a specific inhibitor of intracellular translocation of vesicular stomatitis virus G protein: intracellular accumulation of high-mannose type G protein and inhibition of its cell surface expression. Agr Biol Chem Tokyo 49(3):899–902. https://doi.org/10.1271/bbb1961.49.899
116. Takeuchi M, Ueda T, Sato K, Abe H, Nagata T, Nakano A (2000) A dominant negative mutant of Sar1 GTPase inhibits protein transport from the endoplasmic reticulum to the Golgi apparatus in tobacco and Arabidopsis cultured cells. Plant J 23(4):517–525. https://doi.org/10.1046/j.1365-313x.2000.00823.x
117. Tan X, Cao K, Liu F, Li Y, Li P, Gao C, Ding Y, Lan Z, Shi Z, Rui Q, Feng Y, Liu Y, Zhao Y, Wu C, Zhang Q, Li Y, Jiang L, Bao Y (2016) Arabidopsis COG complex subunits COG3 and COG8 modulate Golgi morphology, vesicle trafficking homeostasis and are essential for pollen tube growth. PLoS Genet 12(7):e1006140. https://doi.org/10.1371/journal.pgen.1006140
118. Tanaka Y, Nishimura K, Kawamukai M, Oshima A, Nakagawa T (2013) Redundant function of two Arabidopsis COPII components, AtSec24B and AtSec24C, is essential for male and female gametogenesis. Planta 238(3):561–575. https://doi.org/10.1007/s00425-013-1913-1
119. Tang BL, Wong SH, Qi XL, Low SH, Hong W (1993) Molecular cloning, characterization, subcellular localization and dynamics of p23, the mammalian KDEL receptor. J Cell Biol 120(2):325–338. https://doi.org/10.1083/jcb.120.2.325
120. Teh O, Moore I (2007) An ARF-GEF acting at the Golgi and in selective endocytosis in polarized plant cells. Nature 448(7152):493–496. https://doi.org/10.1038/nature06023
121. Tie HC, Ludwig A, Sandin S, Lu L (2018) The spatial separation of processing and transport functions to the interior and periphery of the Golgi stack. elife 7:e41301. https://doi.org/10.7554/elife.41301

122. Tojima T, Suda Y, Jin N, Kurokawa K, Nakano A (2024) Spatiotemporal dissection of the Golgi apparatus and the ER-Golgi intermediate compartment in budding yeast. elife 13:e92900. https://doi.org/10.7554/elife.92900
123. Vieira V, Pain C, Wojcik S, Rossi TS, Denecke J, Osterrieder A, Hawes C, Kriechbaumer V (2020) Living on the edge: the role of Atgolgin-84A at the plant ER–Golgi interface. J Microsc 280(2):158–173. https://doi.org/10.1111/jmi.12946
124. Wei J-H, Seemann J (2010) Unraveling the Golgi ribbon. Traffic 11(11):1391–1400. https://doi.org/10.1111/j.1600-0854.2010.01114.x
125. Yadav S, Linstedt AD (2011) Golgi positioning. Cold Spring Harb Perspect Biol 3(5):a005322. https://doi.org/10.1101/cshperspect.a005322
126. Yagi H, Tateo S, Saito T, Ohta Y, Nishi E, Obitsu S, Suzuki T, Seetaha S, Hellec C, Nakano A, Tojima T, Kato K (2024) Deciphering the sub-Golgi localization of glycosyltransferases via 3D super-resolution imaging. Cell Struct Funct:24008. https://doi.org/10.1247/csf.24008
127. Yano H, Yamamoto-Hino M, Abe M, Kuwahara R, Haraguchi S, Kusaka I, Awano W, Kinoshita-Toyoda A, Toyoda H, Goto S (2005) Distinct functional units of the Golgi complex in *drosophila* cells. Proc Natl Acad Sci 102(38):13467–13472. https://doi.org/10.1073/pnas.0506681102
128. Zaal KJM, Smith CL, Polishchuk RS, Altan N, Cole NB, Ellenberg J, Hirschberg K, Presley JF, Roberts TH, Siggia E, Phair RD, Lippincott-Schwartz J (1999) Golgi membranes are absorbed into and reemerge from the ER during mitosis. Cell 99(6):589–601. https://doi.org/10.1016/s0092-8674(00)81548-2
129. Zhang Y, Seemann J (2020) Rapid degradation of GRASP55 and GRASP65 reveals their immediate impact on the Golgi structure. J Cell Biol 220(1):e202007052. https://doi.org/10.1083/jcb.202007052
130. Zhao B, Shi H, Wang W, Liu X, Gao H, Wang X, Zhang Y, Yang M, Li R, Guo Y (2016) Secretory COPII protein SEC31B is required for Pollen Wall development. Plant Physiol 172(3):1625–1642. https://doi.org/10.1104/pp.16.00967

Chapter 11
Post-Golgi Trafficking in Plant Cells

Elizabeth Berryman, Ariadna González Solís, Ethan Weiner, and Marisa S. Otegui

Abstract Post-Golgi trafficking in plants regulates transport to and from the cell surface, vacuolar trafficking, and recycling pathways within the endomembrane system. Endosomes serve as central hubs in these pathways, managing the composition of lipids and proteins in the plasma membrane and vacuole in response to developmental signals or environmental changes. Once internalized via endocytosis, plasma membrane proteins are directed to the *trans*-Golgi Network (TGN), which acts as an early endosome. From the TGN, proteins can either be sent back to the plasma membrane or trafficked to multivesicular endosomes (MVEs) for further sorting and ultimate delivery to the vacuole for degradation. Key molecular assemblies such as the retromer, the ESCRT (Endosomal Sorting Complex Required for Transport) machinery, small GTPases, adaptor proteins, and SNAREs associate with distinct domains of endosomal membranes to facilitate protein sorting and membrane remodeling. This review focuses on the roles of endosomes in post-Golgi trafficking, mechanisms of cargo sorting, and membrane remodeling.

Keywords Endosomes · *trans*-Golgi Network · ESCRT · Retromer · Endocytosis · Exocytosis · Retrograde · Anterograde

Abbreviations

AP	Adaptor protein complexes
ARF	ADP-ribosylation factors
ESCRT	Endosomal Sorting Complex Required for Transport
GAP	GTPase-activating proteins

E. Berryman · A. G. Solís · E. Weiner · M. S. Otegui (✉)
Department of Botany and Center for Quantitative Cell Imaging, University of Wisconsin-Madison, Madison, WI, USA
e-mail: otegui@wisc.edu

A. Nakano, J. Saraste (eds.), *The Golgi Network, Volume I*, Subcellular Biochemistry 110, https://doi.org/10.1007/978-3-032-06936-8_11

GEF	Guanine nucleotide Exchange Factor
MVE	Multivesicular endosome
RAB	Ras-related in brain
SNARE	Soluble N-ethylmaleimide sensitive factor attachment protein receptor
SNX	Sorting nexin
SYP	Syntaxin of Plants
TGN	*trans*-Golgi network
TRAPP	Transport Protein Particle
VAMP	Vesicle-Associated Membrane Proteins
VPS	Vacuolar protein sorting

Introduction

The endomembrane system of plant cells transports molecules essential for plant development, cell signaling, and responses to biotic and abiotic stresses. Many developmental and environmental signals are decoded through the endomembrane system, allowing plants to quickly adjust the trafficking of proteins, lipids, polysaccharides, and other macromolecules. Trafficking within the endomembrane system is highly dynamic and operates in more than one direction. For example, anterograde trafficking transports cargo molecules from their site of synthesis (e.g., endoplasmic reticulum or ER, Golgi) to their site of function (extracellular space, cell wall, plasma membrane, vacuole). Retrograde trafficking refers to the mechanisms that recycle cargo and receptors back to their donor compartments [97]. Within the endomembrane system, endosomes serve as sorting hubs for both anterograde and retrograde trafficking pathways.

Plant cells contain two functionally interconnected types of endosomes: the *trans*-Golgi Network (TGN) and multivesicular endosomes (MVE), also known as the multivesicular bodies or prevacuolar compartments [81]. Both lipids and cargo move through the endomembrane system as ER-derived vesicles fuse to form new cisternae on the *cis*-side of the Golgi apparatus. Through the process of cisternal maturation, the *trans*-most cisterna matures into a new TGN. The TGN regulates post-Golgi trafficking, directing biosynthetic cargo to either the plasma membrane and cell wall through exocytosis/secretion or multivesicular endosomes and vacuoles [90]. However, the TGN also receives cargo internalized from the plasma membrane through endocytosis [20, 59, 116] and either recycles it back to the plasma membrane or redirects it to MVEs for degradation inside vacuoles. The spatial and temporal segregation of cargo proteins and sorting machineries at endosomes is crucial to maintain proper cell function.

In this chapter, we discuss the main features of plant endosomes and their associated molecular machineries as part of their functions in post-Golgi and endocytic trafficking. However, it is important to note that endosomes comprise a continuum of compartments across a maturation gradient, and thus, subpopulations of TGNs and MVEs exist in cells.

The Endosomal System and Its Trafficking Machinery

Endosomal trafficking pathways include exocytosis/secretion at the plasma membrane, transport to the vacuole of both newly synthesized and endocytosed molecules, and retrograde (recycling). Recognition and sorting of cargo through these pathways is tightly regulated by post-translational modifications and targeting signals in the cargo molecules, such as amino acid motifs in proteins [2], and specific trafficking machineries that can recognize these sorting motifs.

Maturation of endosomes and transport of cargo is driven mainly through the formation of transport vesicles. This process requires extensive changes in membrane morphology and composition, and it is regulated by membrane-remodeling proteins. Like yeast and animals, plant cells form several types of vesicles that differ in their biogenesis and the target compartment with which they fuse. The formation of a vesicle requires the selection and sequestration of specific cargo molecules (e.g., transmembrane or soluble proteins) and the recruitment of (a) specific coat proteins (e.g., clathrin) and their protein adaptors (e.g., AP complexes), (b) proteins mediating membrane deformation and scission to release vesicles from their donor membranes (e.g., epsins, dynamins, ESCRT, retromer), (c) small GTPases (e.g., RABs, ARFs) that provide energy for key steps in membrane remodeling and fusion, **d)** motor proteins for transport (e.g., myosins), and (e) tethering complexes and SNAREs (Soluble N-ethylmaleimide sensitive factor Attachment protein Receptor) that mediate vesicle recognition, tethering, and fusion with the target membrane (Fig. 11.1).

There are five types of adaptor protein complexes (AP1-AP5) that share a heterotetrameric organization with two large (β and $\alpha/\gamma/\delta/\varepsilon/\zeta$), one medium ($\mu$), and one small subunit (σ) [40]. These complexes have distinct subcellular localizations and mediate the trafficking of cargo proteins by recognizing and binding their sorting motifs. In addition, AP complexes also interact with clathrin to promote membrane deformation and vesicle formation. Both AP1 and AP4 localize to subdomains of the TGN [100]. APs recognize the tyrosine-based sorting signal (YXXØ; Y is tyrosine; X is any amino acid; Ø is an amino acid with a bulky hydrophobic group) and a dileucine sorting signal ([D/E]XXXL[L/I]; D is aspartic acid; E is glutamic acid; L is leucine) in cargo proteins [111].

Many steps in membrane trafficking require energy input. Small GTPases use the energy released from GTP hydrolysis to mediate the association and dissociation of effector proteins and support membrane deformation. The two major types of plant endosomal small GTPases are RABs (Ras-related in brain) and ARFs (ADP-Ribosylation Factors). These recruit effector proteins to confined membrane areas to regulate protein sorting, membrane remodeling, vesicle transport, and/or tethering [77]. They cycle between the GTP-bound (active) and the GDP-bound (inactive) state and function as molecular switches interacting with effector proteins, like molecular tethers, fusion regulators, and sorting adaptors. The activity of these small GTPases is regulated by specific GEFs (guanine exchange factors) that stimulate GDP/GTP exchange and GAPs (GTPase-activating proteins) that promote

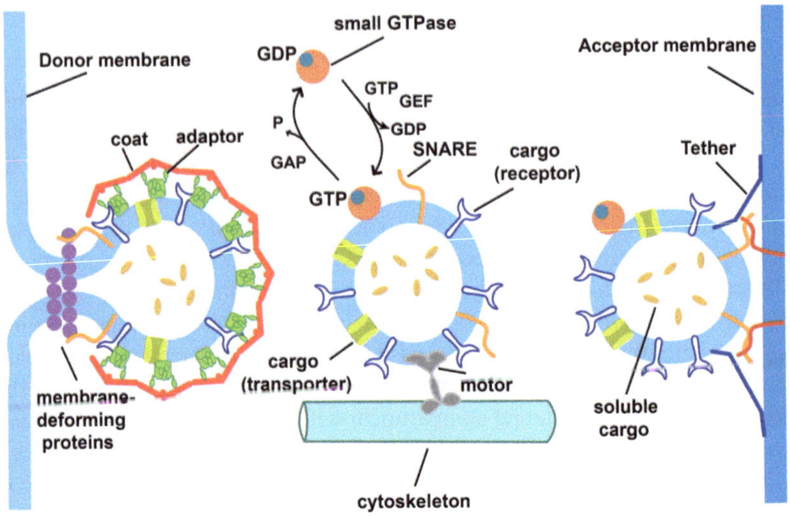

Fig. 11.1 The major components involved in transport vesicle formation, trafficking, and fusion. Transmembrane cargos (e.g., receptors and transporters) and soluble protein cargos are recognized by adaptor proteins, which recruit proteins to form a coat around the vesicle, and the vesicle is released by membrane-deforming proteins. Motor proteins carry the free vesicle along the cytoskeleton toward the acceptor membrane. Small GTPase interaction with the vesicle is regulated by GAPs and GEFs. Tether complexes and SNAREs mediate vesicle docking and fusion with the acceptor membrane

hydrolysis of GTP. In *Arabidopsis thaliana*, there are 57 RABs classified into eight families, A to H [77], and 12 ARF GTPases. Members of the RABA family localize to distinct TGN subdomains, whereas the RABF GTPases localize to subpopulations of MVEs [22, 32, 46, 47, 112]. The ARFs are involved in diverse trafficking functions, including endocytosis, endosomal recycling, and vacuolar trafficking [102].

Tethering factors are proteins or protein complexes that usually reside in the target membrane and bind specific transport vesicles to mediate initial membrane contact, whereas SNAREs facilitate docking and fusion of the vesicle to the target membrane. Recruitment of tethers to specific membranes is mediated by RAB GTPases. Adding to the high specificity of membrane fusion events, complementary SNAREs are differentially distributed in vesicles and target membranes. There are 54 SNAREs encoded in the *Arabidopsis* genome, including 24 SYPs (SYNTAXIN OF PLANTS) and 14 VAMPs (VESICLE-ASSOCIATED MEMBRANE PROTEINS) [113]. Based on amino acid sequence similarity, the plant SYPs can be divided into 8 subfamilies (SYP1- SYP8) [94], with SYP4 and SYP6 SNAREs localized to TGNs and some SYP2 and SYP5 members to MVEs [113]. Among the 14 VAMPs, isoforms of VAMP7 have been detected in endosomes [113]. Four members of another group of SNAREs, the VTI1 (VPS TEN INTERACTING1)-type, localize to both TGN and MVEs [113].

Other protein complexes such as ESCRT (Endosomal Sorting Complex Required for Transport), retromer, and retriever are able to both bind and select cargo proteins and deform endosomal membranes [15, 79].

Another important function at endosomes is luminal pH and ion homeostasis, as pH is critical for allowing specific enzymatic reactions, molecular interactions, and cargo trafficking in the different populations of endosomes [86]. The average luminal pH has been estimated to be 6.1–6.2 and 6.6 for TGN and MVE, respectively [68, 86], which is more acidic than the cytoplasm. The acidification of the TGN is largely dependent on a V-ATPase complex that contains the subunit VHA-a1. The electrochemical gradient generated enables the activity of many transporters, including H+ antiporters and other transporters located at the endosomal membranes that, in turn, increase the lumenal pH to maintain ion homeostasis in the compartments [7, 96, 117]. The TGN V-ATPase is crucial for trafficking, which has been experimentally demonstrated for plasma membrane-localized proteins such as brassinosteroid receptor BRI1 (BRASSINOSTEROID INSENSITIVE 1) and cellulose synthase complexes [13, 20, 68].

The TGN

During the process of cisternal maturation, the *trans*-most cisterna of a Golgi stack loses 30–35% of its membrane and becomes a TGN, which is highly dynamic and characterized by tubular and vesicle budding profiles [51]. While the TGN remains attached to the Golgi, it is known as a Golgi-associated TGN (GA-TGN) until eventually it detaches and becomes a separate organelle, the Golgi-independent TGN (GI-TGN). The GI-TGN undergoes vesiculation and membrane fission, generating vesicles and MVEs [51, 110, 114, 115] (Fig. 11.2). Based on three-dimensional microscopy reconstructions, it has been estimated that 35 Golgi stacks and GA-TGNs exist in an *Arabidopsis* shoot apical meristematic cell at interphase [95].

The GA- and GI-TGNs play distinct trafficking functions. For example, the GA-TGN but not GI-TGNs are labeled with the endocytic tracer FM4–64 [115] while the SNAREs known to facilitate protein secretion to the plasma membrane, VAMP721 (Vesicle-Associated Membrane Protein 721) and VAMP722, are found in the GI-TGNs, suggesting that endocytic-related functions are mainly carried out by the GA-TGN whereas the GI-TGNs are primarily involved in exocytosis [45, 88, 115]. In addition, other studies have shown that there are functional subpopulations of GI-TGNs, with one being responsible for trafficking cargo to the plasma membrane, and the other responsible for generating MVEs [19]. To make this functional sub-specialization even more complex, the GA-TGN is organized into distinct sorting subdomains, each one defined by specific trafficking machineries, cargo molecules, and lipids [37, 38, 100, 124]. In the following sections, we will discuss the molecular machinery and trafficking pathways associated with the TGN.

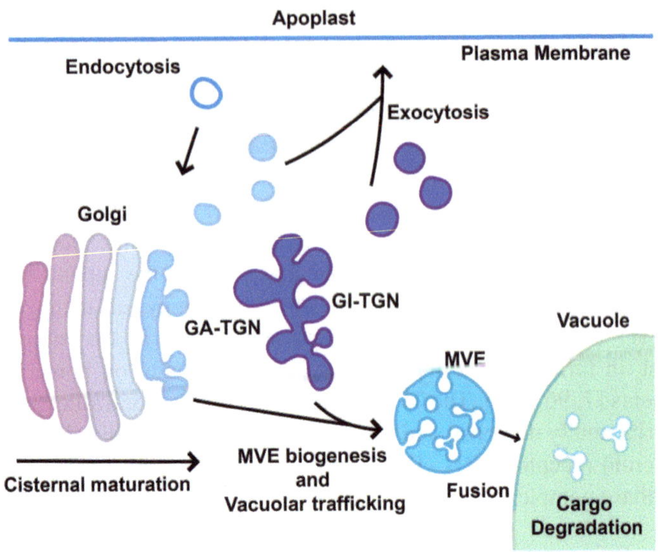

Fig. 11.2 TGN trafficking pathways. A Golgi stack undergoes cisternal maturation and forms a GA-TGN (Golgi-associated TGN), which matures into a GI-TGN (Golgi-independent TGN). Both the GA and GI-TGNs send cargo to the plasma membrane/apoplast via exocytosis and form MVEs for the vacuolar trafficking. Endocytic vesicles carry plasma membrane cargo to the TGN. The MVE forms ILVs with cargo destined for degradation. The MVE fuses with the vacuole, delivering its internal contents for degradation

TGN Trafficking Pathways

The TGN is considered an early endosome because it is the first organelle that receives plasma membrane proteins internalized by endocytosis [20, 59, 116]. Endocytosis is one of the most important mechanisms that cells employ to control the composition of their plasma membrane and modulate signals triggered at the cell surface by receptors. These can be developmental or environmental signals recognized by plasma membrane receptors. During the initiation of endocytosis, adaptor complexes (e.g., AP-2, TPLATE) are recruited to the plasma membrane, where they recognize both specific linear motifs in proteins as well as post-translational modifications such as ubiquitination [31]. They also mediate the recruitment and assembly of the clathrin cage. Clathrin-coated vesicles are released from the plasma membrane and, upon losing their coats, fuse with the TGN. Endocytosed membrane proteins that lose their ubiquitin moieties are recycled back to the plasma membrane. If they remain ubiquitinated, they are trafficked to the membrane for further sorting and vacuolar degradation. Examples of plant plasma membrane proteins that become endocytosed upon ubiquitination include the metal transporter IRON-REGULATED TRANSPORTER1 (IRT1) [6, 101], brassinosteroid receptor BRASSINOSTEROID INSENSITIVE1 (BRI1), auxin efflux carrier PIN FORMED2 (PIN2) [57, 60], boron transporter REQUIRES HIGH BORON1

(BOR1) [54, 132], immunity membrane-associated receptor BOTRYTIS-INDUCED KINASE1 (BIK1)[67, 134], and immunity receptor FLAGELLIN-SENSITIVE2 (FLS2) [66].

As part of the anterograde pathway, newly synthesized proteins, lipids, and polysaccharides (e.g., pectins, hemicelluloses) that move through the Golgi are also sorted at the TGN for delivery to the plasma membrane/cell wall (exocytosis) or to MVEs for vacuolar delivery (membrane or lumen) [81].

As the TGN matures from the *trans*-most Golgi cisterna, it becomes enriched in specific trafficking machinery, including Rab GTPases RAB-A2a and RAB-A4b, phosphatidylinositol 4-kinase PI4Kb1, vacuolar V-ATPase subunit VHA1a, and the SNAREs SYP61 (Syntaxin of Plants 61), SYP43, VAMP721, VAMP722, and VAMP727 [17, 20, 52, 136]. Some of these proteins are commonly used as fluorescent markers to image or biochemically isolate TGN compartments.

Within the GA-TGN, there are at least two sorting domains: the secretory (exocytic) and the vacuolar-trafficking domains. The secretory domain generates vesicles destined for the plasma membrane and is enriched in SNARE VAMP721, adaptor complex AP1, accessory protein EPSIN1, and clathrin. RABA2a and RABA3 colocalize with clathrin at the TGN and mediate transport to the cell plate in a specialized type of exocytosis [16]. The vacuolar trafficking domain generates vesicles targeted to MVEs and ultimately, to the vacuole, and is enriched in VAMP727, adaptor complex AP-4, and accessory protein MODIFIED TRANSPORT TO THE VACUOLE1 (MTV1) [38, 100]. AP4 is important for trafficking soluble proteins to the vacuole through interactions with a vacuolar sorting receptor VSR1 [24, 29].

Although these TGN sorting domains show concentration of specific trafficking factors and functions, they are not completely independent. For example, although AP1 localizes to the exocytic domain and facilitates exocytosis [65, 82, 109, 130], it is also involved in vacuolar sorting [82]. AP1 binds the dileucine motif of the transporter VIT1 (VACUOLAR IRON TRANSPORTER 1) for its sorting to the vacuolar membrane [120]. Consistent with the broad functions of AP-1, *Arabidopsis* mutants for AP-1 subunits show severe developmental and growth defects [23, 35, 82, 99, 109, 118, 122, 123, 131]. Additionally, the SNARE SYP61 mediates exocytosis of the aquaporin PIP2;7 (PLANT PLASMA MEMBRANE INTRINSIC PROTEINS 2;7) [33] and cell wall polysaccharides (e.g., pectins and hemicelluloses) [87] as well as vacuolar trafficking [139]. Interestingly, SYP61 interacts with and is ubiquitinated by the ubiquitin ligase ATL31 (ARABIDOPSIS TOXICOS EN LEVADURA 31), which triggers its endosomal sorting for degradation. The ubiquitination level of SYP61 is regulated in response to carbon/nitrogen status [36].

Several exocytic pathways for plasma membrane proteins and Golgi-synthesized polysaccharides are active at the plant TGN. For example, 7TM (seven transmembrane domain-containing) regulates exocytosis of cellulose synthases, but not other plasma membrane proteins [72]. The TGN-localized protein ECHIDNA is required for the secretion of cell wall polysaccharides (e.g., pectins, xyloglucans) [27, 28, 70] and auxin influx carrier AUX1 [11], but not for the auxin efflux carriers PIN2 (PIN-FORMED 2), PIN3 [11], or the wax transporters ABCG11 (ATP-BINDING

CASSETTE G 11) and ABCG12 [71]. Further supporting the versatility of the endosomal trafficking machinery, ECHIDNA also regulates vacuolar trafficking of storage proteins in embryos [42]. One of the ECHIDNA-independent trafficking pathways is mediated by the Transport Protein Particle II (TRAPPII) complex [85], which in yeast and metazoans acts as a GEF for Rab GTPases [128]. In addition to its role as GEFs, the TRAPP complexes serve as tethering factors that bring together membranes to facilitate their fusion [93]. Whereas ECHIDNA is required for general cell growth, TRAPPII mediates exocytosis of polarized proteins at the PM and trafficking to the cell plate during cytokinesis [84, 91]. In addition, a recent study has shown that activity of the TRAPP complex in plants is regulated by phosphorylation in response to stress conditions and for mediating adaptive growth [126].

MVEs

Plant MVEs are spherical and approximately 200–500 nm in diameter. They are derived from TGN membranes and fuse with the vacuole upon reaching maturity. Meristematic cells of *Arabidopsis* contain 17–20 MVEs, usually in close proximity to the GA-TGN [95]. Their distinctive morphological feature is the presence of 35–45 nm intraluminal vesicles that sequester membrane proteins for degradation upon the fusion of MVEs with vacuoles. MVEs receive newly synthesized proteins targeted to the vacuolar membrane or the vacuolar lumen, as well as membrane proteins that have been ubiquitinated at the plasma membrane, internalized into endocytic vesicles, and delivered to TGNs through endocytosis. Those plasma membrane proteins that remain ubiquitinated are recognized by the ubiquitin-binding domains of several ESCRT proteins and sorted into intralumenal vesicles (Fig. 11.3). Proteins that are deubiquitinated may be destined for recycling and are sorted into membrane domains that bud into the cytosol through the participation of the retromer and retriever complexes (Fig. 11.3).

MVE Trafficking Pathways

As mentioned above, the MVE limiting membrane is capable of budding into either the cytosol for endosomal recycling functions or into the lumen for degradation in the vacuole. Recycling from MVEs is thought to be primarily mediated by the retromer and retriever complexes. The retromer complex has been reported to mediate retrograde recycling, transport of biosynthetic cargo receptors back to the TGN/Golgi, and delivery of endocytic cargo back to the plasma membrane. Originally characterized in yeast, the retromer complex consists of two subcomplexes: the core vacuolar protein sorting (VPS) heterotrimer that consists of Vps35, Vps36, and Vps29 and the sorting nexin (SNX) subcomplex, which includes Vps5 and Vps7. The mammalian homologs of Vps5 and Vps7 are SNX1/SNX2 and SNX5/SNX6,

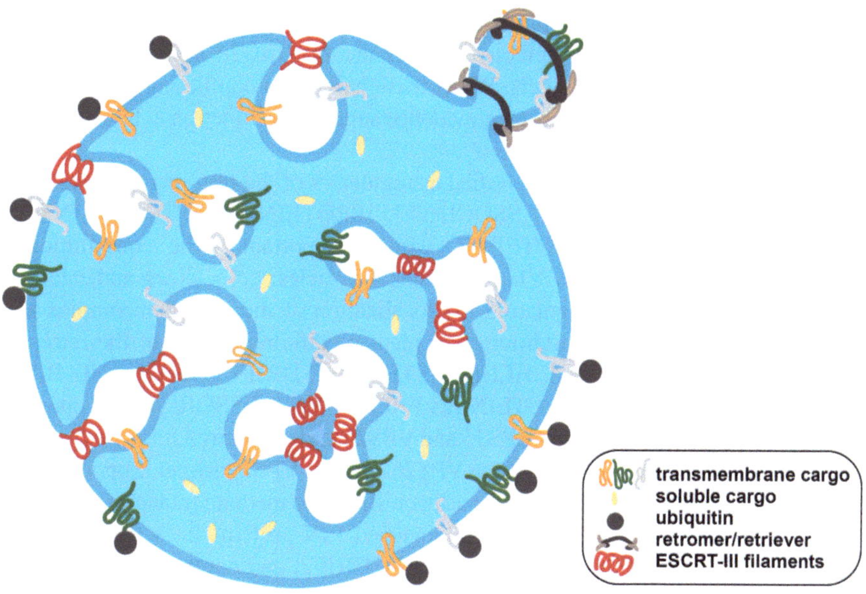

Fig. 11.3 MVE trafficking pathways. Protein cargo sequestration in intralumenal vesicles (ILVs) is controlled by ESCRT proteins, which are recruited from the cytosol and bind ubiquitin to cluster transmembrane cargo. The late-acting ESCRT-III components polymerize into spiral filaments to constrict the ILV bud necks and remain associated with the membrane as budding continues. Recycling of cargo is controlled by the retromer and retriever complexes to form vesicles in the opposite topology of the ESCRT components

respectively. Sorting nexins bind to phosphoinositide 3-phosphate and induce tubulation of endosomal membranes into the cytosol through their arch-shaped BAR domains [76], whereas both the core retromer and SNX proteins are capable of binding specific cargo proteins. Different from yeast, in other eukaryotes including plants, the core retromer and the SNX proteins seem to mediate different pathways [18].

In *Arabidopsis*, there are three VPS35 isoforms, two VPS26 isoforms, and one VPS29 isoform [48], along with six SNX genes encoding SNX1, SNX2a, SNX2b, SNX3 (producing three splicing variants), SNX4, and SNX5 [39]. The plant core retromer plays a crucial role in recycling the vacuolar receptors VSR1 back to the Golgi [53, 80], and mutations in *vps26, vps9*, and *vps35* lead to mis-sorting of vacuolar seed storage proteins, abnormal endosomal recycling of PIN proteins back to the plasma membrane, and severe developmental defects [48, 56, 98, 129, 135]. SNX1 and SNX2 are not required for the recruitment of the core retromer to plant endosomes. Instead, RABG3f, which is enriched in MVEs, interacts with VPS35 to recruit the core retromer complex to endosomes, a process that is also important for the HOPS (Homotypic Fusion and Protein Sorting) tethering complex and MVE fusion with the vacuole. Which SNX proteins work with the core retromer in plants remains unclear. Supporting the idea that the core retromer and SNX proteins may have distinct roles in endosomal recycling, mutations in core retromer subunits cause severe developmental defects, whereas the *snx1 snx2a snx2b* triple mutant exhibits only mild growth defects. While there have been conflicting reports about the localization of the plant retromer, evidence suggests its function at both the TGN and MVEs [48, 78].

Several proteins have been identified as regulators of the retromer. For example, the ESCRT protein ALIX (ASSOCIATED PROTEIN APOPTOSIS-LINKED GENE-2 INTERACTING PROTEIN X), which is important for the sequestration of ubiquitinated proteins into MVE intralumenal vesicles, colocalizes and interacts with VPS26 and VPS29, regulating the localization of the retromer complex and vacuolar transport [41]. In addition, two plant-specific regulators of the retromer have been recently described, BLI (BLISTER) and DRIF1 (DEAH AND RING DOMAIN-CONTAINING PROTEIN AS FREE1 SUPPRESSOR 1). BLI interacts with the retromer core subunits VPS35 and VPS29 and is required for proper assembly and endosomal recruitment of the VPS26-VPS35-VPS29 trimer. Supporting its function in retromer regulation, BLI has been shown to mediate transport of soluble vacuolar proteins and recycling of endocytosed PIN proteins back to the plasma membrane [63]. DRIF1 has a close homolog, DRIF2, and combined mutations in both genes lead to early embryonal lethality, enlarged MVEs with more vesicles, and abnormal PIN2 trafficking [138]. DRIF1 interacts with SNX1 and regulates its partition between TGN and MVEs [138].

The retriever complex was identified more recently in mammals and other eukaryotes, including plants, but not in yeast [73]. Retriever shares subunits with the retromer complex. For example, the mammalian retriever consists of three subunits: DSCR3/VPS26C, C16orf62 (renamed as VPS35-like), and VPS29, and associates with a SNX protein, SNX17. In *Arabidopsis*, a *vps26c* mutant shows abnormal root hair development, which is rescued by expression of mammalian DSCR3/VPS26C, suggesting that a retriever function may also exist in plants [49]. Considering the presence of common subunits between the two endosomal recycling complexes, it will be important to re-evaluate whether mutant phenotypes attributed to abnormal retromer function could instead be due to defects in retriever-dependent pathways [62].

The formation of the distinctive intralumenal vesicles inside MVEs is mediated by ESCRT proteins. The ESCRT machinery consists of several complexes that are recruited sequentially from the cytoplasm onto the endosomal membranes. These complexes are ESCRT-0, -I, -II, -III, and the AAA ATPase SKD1 (also called Vps4 in yeast and metazoans) with its cofactor LIP5. The early ESCRT components (e.g., ESCRT-0, -I, -II) bind ubiquitin and concentrate cargo proteins in domains of the membrane, whereas the late components (ESCRT-III, and SKD1-LIP5) deform the membrane away from the cytosol to form an intralumenal vesicle bud, then constrict the neck, and finally releasing the vesicle. Plants contain putative orthologs for most of the ESCRT proteins originally identified in metazoans and fungi with the exception of the canonical ESCRT-0 [61, 127] and several plant-specific ESCRT components. However, a group of proteins called TOL (TOML1-LIKE) with the capacity of binding ubiquitinated cargo proteins is likely to play the role of

ESCRT-0 in plants [57, 74]. There are nine TOL genes in *Arabidopsis* with redundant functions critical for plant development [57]. The ubiquitin-binding domains in TOL/ESCRT-0, -I, and -II interact with the ubiquitinated cargo proteins at endosomal membranes. By forming protein assemblies, these ESCRT complexes contribute to the clustering of ubiquitinated cargo on the endosomal membrane.

Arabidopsis mutants for the VPS23A/ELC ESCRT-I subunit show abnormal trichomes and defective cytokinesis [104]. Plants also have an ESCRT-I-related protein named FYVE DOMAIN PROTEIN REQUIRED FOR ENDOSOMAL SORTING 1 (FREE1), which aids in early cargo recruitment by binding PI3P and ubiquitinated proteins [25]. Transcript knock-out mutants for *FREE1* are seedling-lethal with aberrant MVEs [25].

ESCRT-III and other proteins like plant ALIX facilitate the recruitment of deubiquitinating enzymes to remove ubiquitin from cargo proteins prior to sequestration into intralumenal vesicles [1, 44, 50, 55]. The ESCRT-III complex that associates with endosomal membranes consists of seven subunits: VPS2, VPS20, VPS24, SNF7, CHMP1, IST1, and VPS60 [4, 92]. All these subunits are found in plants and in most cases, represented by several isoforms [13, 127]. Mutations in ESCRT-III subunits, such as CHMP1 [105] and VPS2 [55], result in severe developmental defects and embryonal or seedling lethality; the overexpression of a dominant version of SKD1, unable to hydrolyze ATP, results in cell death [32], whereas mutations in LIP5 lead to hypersensitive stress responses [119, 121] and abnormal root gravitropic responses [13]. FYVE4, a plant-specific ESCRT-III-related protein that interacts with SNF7 and modulates SKD1 association with endosomal membranes, has been shown to act synergistically with LIP5 [64] and to be required for normal plant development.

How exactly intralumenal vesicle formation occurs in vivo is unclear; the most accepted model postulates that sequential replacement of ESCRT-III subunits mediated by the ATPase activity of SKD1/Vps4 results in tighter and tighter spirals on the limiting membrane, constricting the neck of the forming vesicle [83].

In fungi and metazoans, MVE vesicles are assumed to form in single budding events followed but the recycling of ESCRT-III subunits back to the cytoplasm [3]. In plants, the endosomal limiting membrane undergoes multiple cycles of budding without membrane scission, leading to concatenated networks of vesicles [14]. ESCRT proteins remain inside the constricted membrane of the bud necks and are not recycled. This feature has recently been shown to be widely spread in all plants, although to different extents [125].

MVEs in plants associate with members of the RABF/RAB5 GTPase subfamily. In *Arabidopsis*, there are two canonical RABF2/RAB5 GTPases, RHA1/RabF2a and ARA7/RabF2b, which are involved in endocytic and vacuolar transport. Additionally, there is a plant-specific RAB GTPase, ARA6/AtRabF1, that mediates transport between MVEs and the plasma membrane [10, 21, 58, 103]. All three RABF GTPases in *Arabidopsis* share the same guanine nucleotide exchange factor (GEF), VPS9a [30], which is required for plant development and pathogen responses [43]. Interestingly, though, the activation of ARA6/F1 and the canonical RABF2s requires different binding domains in VPS9a [107]. Moreover, PUF2

(PLANT-UNIQUE RAB5 EFFECTOR 2) interacts with the active GTP-bound form of ARA1/RABF1, the inactive/GDP-bound forms of the two canonical RABF2s, and with VPS9a. PUF2 could bring inactive RABF2s close to VPS9a for activation, promoting vacuolar trafficking, whereas ARA6 negatively regulates the RABF2s by sequestering PUF2 [47].

As MVEs mature, they gradually dissociate from the RABF/RAB5 and associate with RABG/RAB7 GTPases. In yeast and animals, the tethering complexes CORVET (class C core vacuole/endosome tethering) and HOPS are recruited by RAB5 and RAB7 GTPases. These complexes share four of their core subunits (VPS11, VPS16, VPS18, and VPS33), which associate with VPS3 and VPS8 in CORVET and with VPS39 and VPS41 in HOPS [5]. *Arabidopsis* mutants for these subunits show severe developmental defects and abnormal pollen tube growth through the stigma [12, 34, 89, 108]. CORVET, together with a VAMP727-containing SNARE complex, mediates MVE to vacuole fusion, while HOPS and a VAMP71-containing SNARE complex facilitate homotypic vacuolar membrane fusion [12, 108]. MVEs in metazoans can fuse with autophagosomes to generate organelles known as amphisomes [9]. A similar phenomenon has been observed to occur in plants [137].

Although it is reasonable to think that most, if not all, plasma membrane proteins are sorted for degradation at MVEs, only few have been experimentally demonstrated to be MVE cargo in plants, for example, proteins involved in auxin transport such as PINs and AUX1 [106], the brassinosteroid receptor BRI1 [69], and the abscisic acid transporter ABCG25 [75], among others. Interestingly, some PYR/PYL/RCAR (PYRABACTIN RESISTANCE1/PYR1-LIKE/REGULATORY COMPONENTS OF ABA RECEPTORS ABA receptors) proteins, which are soluble abscisic acid receptors that associate with membranes to mediate stress responses, are recognized by ESCRT proteins and sorted for endosomal degradation [8, 26, 75, 133].

Conclusions and Future Directions

Endosomes coordinate several trafficking pathways within the endomembrane systems with remarkable examples of sub-functionalization, coordination, and integration. Thus, several exocytotic routes exist for different cargoes and both TGN and MVEs are organized in subdomains with distinct and sometimes opposite functions. The continuous discovery of plant-specific trafficking factors, as well as the increased resolution provided by imaging and molecular methods, is quickly revealing new unanticipated mechanisms and filling gaps in our understanding of the complex endosomal system of plant cells.

Acknowledgements: Research on endosomal trafficking in the Otegui laboratory is supported by grant NSF MCB 2114603.

References

1. Amerik A, Sindhi N, Hochstrasser M (2006) A conserved late endosome-targeting signal required for Doa4 deubiquitylating enzyme function. J Cell Biol 175:825–835
2. Arora D, Damme DV (2021) Motif-based endomembrane trafficking. Plant Physiol 186:221–238
3. Azmi I, Davies B, Dimaano C, Payne J, Eckert D, Babst M, Katzmann DJ (2006) Recycling of ESCRTS by the AAA-ATPase Vps4 is regulated by a conserved VSL region in Vta1. J Cell Biol 172:705–717
4. Babst M, Katzmann DJ, Estepa-Sabal EJ, Meerloo T, Emr SD (2002) ESCRT-III: an endosome-associated heterooligomeric protein complex required for MVB sorting. Dev Cell 3:271–282
5. Balderhaar HJK, Ungermann C (2013) Corvet and Hops tethering complexes—coordinators of endosome and lysosome fusion. J Cell Sci 126:1307–1316
6. Barberon M, Zelazny E, Robert S, ConéJéro G, Curie C, Friml J, Vert G (2011) Monoubiquitin-dependent endocytosis of the iron-regulated transporter 1 (Irt1) transporter controls iron uptake in plants. Proc Natl Acad Sci USA 108:E450–E458
7. Bassil E, Ohto MA, Esumi T, Tajima H, Zhu Z, Cagnac O, Belmonte M, Peleg Z, Yamaguchi T, Blumwald E (2011) The Arabidopsis intracellular Na+/H+ antiporters NHX5 and NHX6 are endosome associated and necessary for plant growth and development. Plant Cell 23:224–239
8. Belda-Palazon B, Rodriguez L, Fernandez MA, Castillo MC, Anderson EA, Gao C, Gonzalez-Guzman M, Peirats-Llobet M, Zhao Q, De Winne N, Gevaert K, De Jaeger G, Jiang L, Leon J, Mullen RT, Rodriguez PL (2016) FYVE1/FREE1 interacts with the PYl4 ABA receptor and mediates its delivery to the vacuolar degradation pathway. Plant Cell 28:2291–2311
9. Berg TO, Fengsrud M, Strømhaug PE, Berg T, Seglen PO (1998) Isolation and characterization of rat liver amphisomes. Evidence for fusion of autophagosomes with both early and late endosomes. J Biol Chem 273:21883–21892
10. Bottanelli F, Gershlick DC, Denecke J (2012) Evidence for sequential action of Rab5 and Rab7 GTPases in prevacuolar organelle partitioning. Traffic 13:338–354
11. Boutte Y, Jonsson K, Mcfarlane HE, Johnson E, Gendre D, Swarup R, Friml J, Samuels L, Robert S, Bhalerao RP (2013) ECHIDNA-mediated post-Golgi trafficking of auxin carriers for differential cell elongation. Proc Natl Acad Sci USA 110:16259–16264
12. Brillada C, Zheng J, Krüger F, Rovira-Diaz E, Askani JC, Schumacher K, Rojas-Pierce M (2018) Phosphoinositides control the localization of HOPS subunit VPS41, which together with Vps33 mediates vacuole fusion in plants. Proc Natl Acad Sci USA 115:E8305–e8314
13. Brüx A, T.-Y. Liu M, Krebs Y.-D, Stierhof JU, Lohmann O, Miersch C, Wasternack K, Schumacher (2008)Reduced V-ATPase activity in the trans-Golgi network causes oxylipin-dependent hypocotyl growth inhibition inArabidopsis. Plant Cell 20(4): 1088-1100
14. Buono RA, Paez-Valencia J, Miller ND, Goodman K, Spitzer C, Spalding EP, Otegui MS (2016) Role of SKD1 regulators LIP5 and IST1-LIKE1 in endosomal sorting and plant development. Plant Physiol 171:251–264
15. Buono RA, Leier A, Paez-Valencia J, Pennington J, Goodman K, Miller N, Ahlquist P, Marquez-Lago T, Otegui MS (2017) ESCRT-mediated vesicle concatenation in plant endosomes. J Cell Biol 216:2167–2177
16. Chen K-E, Healy MD, Collins BM (2019) Towards a molecular understanding of endosomal trafficking by Retromer and retriever. Traffic 20:465–478
17. Chow C-M, Neto H, Foucart C, Moore I (2008a) Rab-A2 and Rab-A3 GTPases define a trans-Golgi endosomal membrane domain in *Arabidopsis* that contributes substantially to the cell plate. Plant Cell 20:101–123
18. Chow CM, Neto H, Foucart C, Moore I (2008b) Rab-A2 and Rab-A3 GTPases define a trans-golgi endosomal membrane domain in Arabidopsis that contributes substantially to the cell plate. Plant Cell 20:101–123

19. Cui Y, Carosi JM, Yang Z, Ariotti N, Kerr MC, Parton RG, Sargeant TJ, Teasdale RD (2018) Retromer has a selective function in cargo sorting via endosome transport carriers. J Cell Biol 218:615–631
20. De Caroli M, Manno E, Perrotta C, De Lorenzo G, Di Sansebastiano GP, Piro G (2020) CesA6 and PGIP2 endocytosis involves different subpopulations of TGN-related endosomes. Front Plant Sci 11:350
21. Dettmer J, Hong-Hermesdorf A, Stierhof YD, Schumacher K (2006) Vacuolar H^+-ATPase activity is required for endocytic and secretory trafficking in *Arabidopsis*. Plant Cell 18:715–730
22. Ebine K, Fujimoto M, Okatani Y, Nishiyama T, Goh T, Ito E, Dainobu T, Nishitani A, Uemura T, Sato MH, Thordal-Christensen H, Tsutsumi N, Nakano A, Ueda T (2011) A membrane trafficking pathway regulated by the plant-specific RAB GTPase ARA6. Nat Cell Biol 13:853–859
23. Elliott L, Moore I, Kirchhelle C (2020) Spatio-temporal control of post-Golgi exocytic trafficking in plants. J Cell Sci 133
24. Feng C, Wang JG, Liu HH, Li S, Zhang Y (2017) Arabidopsis adaptor protein 1G is critical for pollen development. J Integr Plant Biol 59:594–599
25. Fuji K, Shirakawa M, Shimono Y, Kunieda T, Fukao Y, Koumoto Y, Takahashi H, Hara-Nishimura I, Shimada T (2016) The adaptor complex AP-4 regulates vacuolar protein sorting at the trans-Golgi network by interacting with vacuolar sorting Receptor1. Plant Physiol 170:211–219
26. Gao C, Luo M, Zhao Q, Yang R, Cui Y, Zeng Y, Xia J, Jiang L (2014) A unique plant ESCRT component, FREE1, regulates multivesicular body protein sorting and plant growth. Curr Biol 24:2556–2563
27. García-LeóN M, Cuyas L, El-Moneim DA, Rodriguez L, Belda-PalazóN B, Sanchez-Quant E, Fernández Y, Roux B, ZamarreñO Á, M., García-Mina, J. M., Nussaume, L., Rodriguez, P. L., Paz-Ares, J., Leonhardt, N. & Rubio, V. (2019) Arabidopsis ALIX regulates stomatal aperture and turnover of abscisic acid receptors. Plant Cell 31:2411–2429
28. Gendre D, Oh J, Boutté Y, Best JG, Samuels L, Nilsson R, Uemura T, Marchant A, Bennett MJ, Grebe M, Bhalerao RP (2011) Conserved *Arabidopsis* ECHIDNA protein mediates trans–Golgi-network trafficking and cell elongation. Proc Natl Acad Sci USA 108:8048–8053
29. Gendre D, Mcfarlane HE, Johnson E, Mouille G, Sjödin A, Oh J, Levesque-Tremblay G, Watanabe Y, Samuels L, Bhalerao RP (2013) Trans-Golgi network localized ECHIDNA/Ypt interacting protein complex is required for the secretion of cell wall polysaccharides in Arabidopsis. Plant Cell 25:2633–2646
30. Gershlick DC, Lousa Cde M, Foresti O, Lee AJ, Pereira EA, Dasilva LL, Bottanelli F, Denecke J (2014) Golgi-dependent transport of vacuolar sorting receptors is regulated by COPII, AP1, and AP4 protein complexes in tobacco. Plant Cell 26:1308–1329
31. Goh T, Uchida W, Arakawa S, Ito E, Dainobu T, Ebine K, Takeuchi M, Sato K, Ueda T, Nakano A (2007) VPS9a, the common activator for two distinct types of Rab5 GTPases, is essential for the development of Arabidopsis thaliana. Plant Cell 19:3504–3515
32. Grones P, De Meyer A, Pleskot R, Mylle E, Kraus M, Vandorpe M, Yperman K, Eeckhout D, Dragwidge JM, Jiang Q, Nolf J, Pavie B, De Jaeger G, De Rybel B, Van Damme D (2022) The endocytic TPLATE complex internalizes ubiquitinated plasma membrane cargo. Nat Plants 8:1467–1483
33. Haas TJ, Sliwinski MK, D.E., M., Preuss, M., Ebine, K., Ueda, T., Nielsen, E., Odorizzi, G. & Otegui, M. S. (2007) The *Arabidopsis* AAA ATPase SKD1 is involved in multivesicular endosome function and interacts with its positive regulator Lyst-interacting protein 5. Plant Cell 19:1295–1312
34. Hachez C, Laloux T, Reinhardt H, Cavez D, Degand H, Grefen C, De Rycke R, Inzé D, Blatt MR, Russinova E, Chaumont F (2014) Arabidopsis Snares SYP61 and SYP121 coordinate the trafficking of plasma membrane aquaporin PIP2;7 to modulate the cell membrane water permeability. Plant Cell 26:3132–3147

35. Hao L, Liu J, Zhong S, Gu H, Qu L-J (2016) AtVPS41-mediated endocytic pathway is essential for pollen tube–stigma interaction in Arabidopsis. Proc Natl Acad Sci USA 113:6307–6312
36. Happel N, Honing S, Neuhaus J-M, Paris N, Robinson DG, Holstein SEH (2004) *Arabidopsis* mu A-adaptin interacts with the tyrosine motif of the vacuolar sorting receptor VSR-PS1. Plant J 37:678–693
37. Hasegawa Y, Huarancca Reyes T, Uemura T, Baral A, Fujimaki A, Luo Y, Morita Y, Saeki Y, Maekawa S, Yasuda S, Mukuta K, Fukao Y, Tanaka K, Nakano A, Takagi J, Bhalerao RP, Yamaguchi J, Sato T (2022) The Tgn/EE SNARE protein Syp61 and the ubiquitin ligase ATL31 cooperatively regulate plant responses to carbon/nitrogen conditions in Arabidopsis. Plant Cell 34:1354–1374
38. Heinze L, Freimuth N, Rößling A-K, Hahnke R, Riebschläger S, Fröhlich A, Sampathkumar A, Mcfarlane HE, Sauer M (2020a) EPSIN1 and MTV1 define functionally overlapping but molecularly distinct trans-Golgi network subdomains in Arabidopsis. Proc Natl Acad Sci USA 117:25880–25889
39. Heinze L, Freimuth N, Rossling AK, Hahnke R, Riebschlager S, Frohlich A, Sampathkumar A, Mcfarlane HE, Sauer M (2020b) EPSIN1 and MTV1 define functionally overlapping but molecularly distinct trans-Golgi network subdomains in Arabidopsis. Proc Natl Acad Sci USA 117:25880–25889
40. Heucken N, Ivanov R (2018) The retromer, sorting nexins and the plant endomembrane protein trafficking. J Cell Sci 131:jcs203695
41. Hirst J, Barlow LD, Francisco GC, Sahlender DA, Seaman MN, Dacks JB, Robinson MS (2011) The fifth adaptor protein complex. PLoS Biol 9:e1001170
42. Hu S, Li B, Wu F, Zhu D, Zouhar J, Gao C, Shimada T, Rojo E, Hara-Nishimura I, Jiang L, Shen J (2022) Plant ESCRT protein ALIX coordinates with retromer complex in regulating receptor-mediated sorting of soluble vacuolar proteins. Proc Natl Acad Sci USA 119:e2200492119
43. Ichino T, Maeda K, Hara-Nishimura I, Shimada T (2020) Arabidopsis ECHIDNA protein is involved in seed coloration, protein trafficking to vacuoles, and vacuolar biogenesis. J Exp Bot 71:3999–4009
44. Inoue T, Kondo Y, Naramoto S, Nakano A, Ueda T (2013) RAB5 activation is required for multiple steps in *Arabidopsis thaliana* root development. Plant Cell Physiol 54:1648–1659
45. Isono E, Katsiarimpa A, Müller IK, Anzenberger F, Stierhof Y-D, Geldner N, Chory J, Schwechheimer C (2010) The deubiquitinating enzyme AMSH3 is required for intracellular trafficking and vacuole biogenesis in *Arabidopsis thaliana*. Plant Cell 22:1826–1837
46. Ito Y, Boutte Y (2020) Differentiation of trafficking pathways at Golgi entry Core compartments and post-Golgi subdomains. Front Plant Sci 11:609516
47. Ito E, Uemura T (2022) RAB GTPases and SNAREs at the trans-Golgi network in plants. J Plant Res 135:389–403
48. Ito E, Ebine K, Choi S-W, Ichinose S, Uemura T, Nakano A, Ueda T (2018) Integration of two RAB5 groups during endosomal transport in plants. elife 7:e34064
49. Jaillais Y, Santambrogio M, Rozier F, Fobis-Loisy I, Miege C, Gaude T (2007) The retromer protein VPS29 links cell polarity and organ initiation in plants. Cell 130:1057–1070
50. Jha SG, Larson ER, Humble J, Domozych DS, Barrington DS, Tierney ML (2018) Vacuolar protein sorting 26C encodes an evolutionarily conserved large retromer subunit in eukaryotes that is important for root hair growth in Arabidopsis thaliana. Plant J 94:595–611
51. Kalinowska K, Nagel MK, Goodman K, Cuyas L, Anzenberger F, Alkhofer A, Paz-Ares J, Braun P, Rubio V, Otegui MS, Isono E (2015) Arabidopsis ALIX is required for the endosomal localization of the deubiquitinating enzyme AMSH3. Proc Natl Acad Sci USA 112:E5543–E5551
52. Kang B-H, Nielsen E, Preuss ML, Mastronarde D, Staehelin LA (2011a) Electron tomography of RabA4b- and PI-4Kβ1-labeled trans Golgi network compartments in Arabidopsis. Traffic 12:313–329
53. Kang B-H, Nielsen E, Preuss ML, Mastronarde D, Staehelin LA (2011b) Electron tomography of RabA4b- and PI-4Kβ1-labeled trans Golgi network compartments in Arabidopsis. Traffic (Copenhagen, Denmark) 12:313–329

54. Kang H, Kim SY, Song K, Sohn EJ, Lee Y, Lee DW, Hara-Nishimura I, Hwang I (2012) Trafficking of vacuolar proteins: the crucial role of Arabidopsis vacuolar protein sorting 29 in recycling vacuolar sorting receptor. Plant Cell 24:5058–5073
55. Kasai K, Takano J, Miwa K, Toyoda A, Fujiwara T (2011) High boron-induced ubiquitination regulates vacuolar sorting of the BOR1 borate transporter in *Arabidopsis thaliana.* J Biol Chem 286:6175–6183
56. Katsiarimpa A, Anzenberger F, Schlager N, Neubert S, Hauser M-T, Schwechheimer C, Isono E (2011) The Arabidopsis deubiquitinating enzyme AMSH3 interacts with ESCRT-III subunits and regulates their localization. Plant Cell 23:3026–3040
57. Kleine-Vehn J, Leitner J, Zwiewka M, Sauer M, Abas L, Luschnig C, Friml J (2008) Differential degradation of PIN2 auxin efflux carrier by retromer-dependent vacuolar targeting. Proc Natl Acad Sci USA 105:17812–17817
58. Korbei B, Moulinier-Anzola J, De-Araujo L, Lucyshyn D, Retzer K, Khan MA, Luschnig C (2013) Arabidopsis TOL proteins act as gatekeepers for vacuolar sorting of PIN2 plasma membrane protein. Curr Biol 23:2500–2505
59. Kotzer AM, Brandizzi F, Neumann U, Paris N, Moore I, Hawes C (2004) AtRabF2b (Ara7) acts on the vacuolar trafficking pathway in tobacco leaf epidermal cells. J Cell Sci 117:6377–6389
60. Lam SK, Siu CL, Hillmer S, Jang S, An G, Robinson DG, Jiang L (2007) Rice Scamp1 defines clathrin-coated, trans-Golgi-located tubular-vesicular structures as an early endosome in tobacco BY-2 cells. Plant Cell 19:296–319
61. Leitner J, Petrášek J, Tomanov K, Retzer K, Pařezová M, Korbei B, Bachmair A, Zažímalová E, Luschnig C (2012) Lysine63-linked ubiquitylation of PIN2 auxin carrier protein governs hormonally controlled adaptation of Arabidopsis root growth. Proc Natl Acad Sci USA 109:8322–8327
62. Leung KF, Dacks JB, Field MC (2008) Evolution of the multivesicular body ESCRT machinery; retention across the eukaryotic lineage. Traffic 9:1698–1716
63. Lewis CD, Tierney ML (2024) Contrasting retromer with a newly described retriever in *Arabidopsis thaliana. Plants (Basel)*:13
64. Li H, Huang R, Liao Y, Yang S, Feng B, Qin H, Zhou J, Zeng Y, Shen J, Zhuang X, Jiang L, Otegui MS, Zhang S, Gao C (2023) A plant-unique protein BLISTER coordinates with core retromer to modulate endosomal sorting of plasma membrane and vacuolar proteins. Proc Natl Acad Sci USA 120:e2211258120
65. Liu C, Zeng Y, Li H, Yang C, Shen W, Xu M, Xiao Z, Chen T, Li B, Cao W, Jiang L, Otegui MS, Gao C (2021) A plant unique ESCRT component, FYVE4, regulates multivesicular endosome biogenesis and plant growth. New Phytol 231:193–209
66. Liu C, Li Z, Tian D, Xu M, Pan J, Wu H, Wang C, Otegui MS (2022) AP1/2β-mediated exocytosis of tapetum-specific transporters is required for pollen development in Arabidopsis thaliana. Plant Cell 34:3961–3982
67. Lu D, Lin W, Gao X, Wu S, Cheng C, Avila J, Heese A, Devarenne TP, He P, Shan L (2011) Direct ubiquitination of pattern recognition receptor FLS2 attenuates plant innate immunity. Science 332:1439–1442
68. Luo Y, Scholl S, Doering A, Zhang Y, Irani NG, Rubbo SD, Neumetzler L, Krishnamoorthy P, Van Houtte I, MylleE, Bischoff V, Vernhettes S, Winne J, Friml J, Stierhof YD, Schumacher K, Persson S,Russinova E (2015) V-ATPase activity in the TGN/EE is required for exocytosis and recycling in Arabidopsis. NatPlants 1: 15094
69. Ma X, Claus LAN, Leslie ME, Tao K, Wu Z, Liu J, Yu X, Li B, Zhou J, Savatin DV, Peng J, Tyler BM, Heese A, Russinova E, He P, Shan L (2020) Ligand-induced monoubiquitination of Bik1 regulates plant immunity. Nature 581:199–203
70. Martiniere A, Bassil E, Jublanc E, Alcon C, Reguera M, Sentenac H, Blumwald E, Paris N (2013) In vivo intracellular pH measurements in tobacco and Arabidopsis reveal an unexpected pH gradient in the endomembrane system. Plant Cell 25:4028–4043
71. Martins S, Dohmann EM, Cayrel A, Johnson A, Fischer W, Pojer F, Satiat-Jeunemaitre B, Jaillais Y, Chory J, Geldner N, Vert G (2015) Internalization and vacuolar targeting of the brassinosteroid hormone receptor BRI1 are regulated by ubiquitination. Nat Commun 6:6151

72. Mcfarlane HE, Watanabe Y, Gendre D, Carruthers K, Levesque-Tremblay G, Haughn GW, Bhalerao RP, Samuels L (2013) Cell wall polysaccharides are mislocalized to the vacuole in echidna mutants. Plant Cell Physiol 54:1867–1880
73. Mcfarlane HE, Watanabe Y, Yang W, Huang Y, Ohlrogge J, Samuels AL (2014) Golgi- and trans-Golgi network-mediated vesicle trafficking is required for wax secretion from epidermal cells. Plant Physiol 164:1250–1260
74. Mcfarlane HE, Mutwil-Anderwald D, Verbancic J, Picard KL, Gookin TE, Froehlich A, Chakravorty D, Trindade LM, Alonso JM, Assmann SM, Persson S (2021) A G protein-coupled receptor-like module regulates cellulose synthase secretion from the endomembrane system in Arabidopsis. Dev Cell 56:1484–1497 e7
75. Mcnally KE, Faulkner R, Steinberg F, Gallon M, Ghai R, Pim D, Langton P, Pearson N, Danson CM, Nägele H, Morris LL, Singla A, Overlee BL, Heesom KJ, Sessions R, Banks L, Collins BM, Berger I, Billadeau DD, Burstein E, Cullen PJ (2017) Retriever is a multiprotein complex for retromer-independent endosomal cargo recycling. Nat Cell Biol 19:1214–1225
76. Moulinier-Anzola J, Schwihla M, De-Araújo L, Artner C, Jörg L, Konstantinova N, Luschnig C, Korbei B (2020) Tols function as ubiquitin receptors in the early steps of the ESCRT pathway in higher plants. Mol Plant 13:717–731
77. Moulinier-Anzola J, Schwihla M, Lugsteiner R, Leibrock N, Feraru MI, Tkachenko I, Luschnig C, Arcalis E, Feraru E, Lozano-Juste J, Korbei B (2024) Modulation of abscisic acid signaling via endosomal TOL proteins. New Phytol 243:1065–1081
78. Mukadam AS, Seaman MNJ (2015) Retromer-mediated endosomal protein sorting: the role of unstructured domains. FEBS Lett 589:2620–2626
79. Nielsen E (2020) The small GTPase superfamily in plants: a conserved regulatory module with novel functions. Ann Rev Plant Biol 71:247–272
80. Niemes S, Langhans M, Viotti C, Scheuring D, San Wan Yan M, Jiang L, Hillmer S, Robinson DG, Pimpl P (2010) Retromer recycles vacuolar sorting receptors from the trans-Golgi network. Plant J 61:107–121
81. Norris A, Grant BD (2020) Endosomal microdomains: formation and function. Curr Opin Cell Biol 65:86–95
82. Oliviusson P, Heinzerling O, Hillmer S, Hinz G, Tse YC, Jiang L, Robinson DG (2006) Plant retromer, localized to the prevacuolar compartment and microvesicles in *Arabidopsis*, may interact with vacuolar sorting receptors. Plant Cell 18:1239–1252
83. Paez Valencia J, Goodman K, Otegui MS (2016) Endocytosis and endosomal trafficking in plants. Ann Rev Plant Biol 67:309–335
84. Park M, Song K, Reichardt I, Kim H, Mayer U, Stierhof YD, Hwang I, Jürgens G (2013) Arabidopsis μ-adaptin subunit AP1M of adaptor protein complex 1 mediates late secretory and vacuolar traffic and is required for growth. Proc Natl Acad Sci USA 110:10318–10323
85. Pfitzner A-K, Mercier V, Jiang X, Moser Von Filseck J, Baum B, Šarić A, Roux A (2020) An ESCRT-III polymerization sequence drives membrane deformation and fission. Cell 182:1140–1155.e18
86. Qi X, Kaneda M, Chen J, Geitmann A, Zheng H (2011) A specific role for Arabidopsis Trappii in post-Golgi trafficking that is crucial for cytokinesis and cell polarity. Plant J 68:234–248
87. Ravikumar R, Kalbfuß N, Gendre D, Steiner A, Altmann M, Altmann S, Rybak K, Edelmann H, Stephan F, Lampe M, Facher E, Wanner G, Falter-Braun P, Bhalerao RP, Assaad FF (2018) Independent yet overlapping pathways ensure the robustness and responsiveness of trans-Golgi network functions in Arabidopsis. Development 145
88. Reguera M, Bassil E, Tajima H, Wimmer M, Chanoca A, Otegui MS, Paris N, Blumwald E (2015) pH regulation by NHX-type antiporters is required for receptor-mediated protein trafficking to the vacuole in Arabidopsis. Plant Cell 27:1200–1217
89. Ren G, Rosquete MR, Peralta AG, Pattathil S, Hahn MG, Wilkop T, Drakakaki G (2020) Isolation and Glycomic analysis of trans-Golgi network vesicles in plants. Methods Mol Biol 2177:153–167

90. Renna L, Stefano G, Slabaugh E, Wormsbaecher C, Sulpizio A, Zienkiewicz K, Brandizzi F (2018) TGNap1 is required for microtubule-dependent homeostasis of a subpopulation of the plant trans-Golgi network. Nat Commun 9
91. Rojo E, Gillmor CS, Kovaleva V, Somerville CR, Raikhel NV (2001) Vacuoleless1 is an essential gene required for vacuole formation and morphogenesis in *Arabidopsis*. Dev Cell 1:303–310
92. Rosquete MR, Davis DJ, Drakakaki G (2018) The plant trans-Golgi network: not just a matter of distinction. Plant Phyisiol 176:187–198
93. Rosquete MR, Worden N, Ren G, Sinclair RM, Pfleger S, Salemi M, Phinney BS, Domozych D, Wilkop T, Drakakaki G (2019) AtTRAPPC11/ROG2: a role for TRAPPs in maintenance of the plant trans-Golgi network/early endosome organization and function. Plant Cell 31:1879–1898
94. Rue SM, Mattei S, Saksena S, Emr SD (2008) Novel ist1-did2 complex functions at a late step in multivesicular body sorting. Mol Biol Cell 19:475–484
95. Sacher M, Barrowman J, Wang W, Horecka J, Zhang Y, Pypaert M, Ferro Novick S (2001) TRAPP I implicated in the specificity of tethering in ER-to-Golgi transport. Mol Cell 7:433–442
96. Sanderfoot AA, Assaad FF, Raikhel NV (2000) The Arabidopsis genome. An abundance of soluble N-ethylmaleimide-sensitive factor adaptor protein receptors. Plant Physiol 124:1558–1569
97. Segui-Simarro JM, Staehelin LA (2006) Cell cycle-dependent changes in Golgi stacks, vacuoles, clathrin-coated vesicles and multivesicular bodies in meristematic cells of Arabidopsis thaliana: a quantitative and spatial analysis. Planta 223:223–236
98. Shang J, Mu G, Qi Y, Zhang X, Shen W, Xie Y, Ge M, He Y, Qiao F, Qiu QS (2024) NHX5/NHX6/SPY22 complex regulates BRI1 and brassinosteroid signaling in Arabidopsis. J Plant Physiol 302:154318
99. Shao X, Xu H, Pimpl P (2023) Nanobody-based VSR7 tracing shows clathrin-dependent TGN to Golgi recycling. Nat Commun 14:6926
100. Shimada T, Koumoto Y, Li L, Yamazaki M, Kondo M, Nishimura M, Hara-Nishimura I (2006) AtVPS29, a putative component of a retromer complex, is required for the efficient sorting of seed storage proteins. Plant Cell Physiol 47:1187–1194
101. Shimada T, Kunieda T, Sumi S, Koumoto Y, Tamura K, Hatano K, Ueda H, Hara-Nishimura I (2018) The AP-1 complex is required for proper mucilage formation in *Arabidopsis* seeds. Plant Cell Physiol 59:2331–2338
102. Shimizu Y, Takagi J, Ito E, Ito Y, Ebine K, Komatsu Y, Goto Y, Sato M, Toyooka K, Ueda T, Kurokawa K, Uemura T, Nakano A (2021) Cargo sorting zones in the trans-Golgi network visualized by super-resolution confocal live imaging microscopy in plants. Nat Commun 12:1901–1901
103. Shin LJ, Lo JC, Chen GH, Callis J, Fu H, Yeh KC (2013) IRT1 degradation factor1, a ring E3 ubiquitin ligase, regulates the degradation of iron-regulated transporter1 in Arabidopsis. Plant Cell 25:3039–3051
104. Singh MK, Jürgens G (2018) Specificity of plant membrane trafficking – ARFs, regulators and coat proteins. Semin Cell Dev Biol 80:85–93
105. Sohn EJ, Kim ES, Zhao M, Kim SJ, Kim H, Kim Y-W, Lee YJ, Hillmer S, Sohn U, Jiang L, Hwang I (2003) Rha1, an Arabidopsis Rab5 homolog, plays a critical role in the vacuolar trafficking of soluble cargo proteins. Plant Cell 15:1057–1070
106. Spitzer C, Schellmann S, Sabovljevic A, Shahriari M, Keshavaiah C, Bechtold N, Herzog M, Muller S, Hanisch FG, Hulskamp M (2006) The *Arabidopsis elch* mutant reveals functions of an ESCRT component in cytokinesis. Development 133:4679–4689
107. Spitzer C, Reyes FC, Buono R, Sliwinski MK, Haas TJ, Otegui MS (2009) The ESCRT-related CHMP1A and B proteins mediate multivesicular body sorting of auxin carriers in *Arabidopsis* and are required for plant development. Plant Cell 21:749–766

108. Spitzer C, Li F, Buono R, Roschzttardtz H, Chung T, Zhang M, Osteryoung KW, Vierstra RD, Otegui MS (2015) The endosomal protein charged multivesicular body Protein1 regulates the autophagic turnover of plastids in Arabidopsis. Plant Cell 27:391–402
109. Sunada M, Goh T, Ueda T, Nakano A (2016) Functional analyses of the plant-specific C-terminal region of VPS9a: the activating factor for RAB5 in Arabidopsis thaliana. J Plant Res 129:93–102
110. Takemoto K, Ebine K, Askani JC, Kruger F, Gonzalez ZA, Ito E, Goh T, Schumacher K, Nakano A, Ueda T (2018) Distinct sets of tethering complexes, SNARE complexes, and Rab GTPases mediate membrane fusion at the vacuole in Arabidopsis. Proc Natl Acad Sci USA 115:E2457–e2466
111. Teh OK, Shimono Y, Shirakawa M, Fukao Y, Tamura K, Shimada T, Hara-Nishimura I (2013) The AP-1 μ adaptin is required for KNOLLE localization at the cell plate to mediate cytokinesis in Arabidopsis. Plant Cell Physiol 54:838–847
112. Toyooka K, Goto Y, Asatsuma S, Koizumi M, Mitsui T, Matsuoka K (2009) A mobile secretory vesicle cluster involved in mass transport from the Golgi to the plant cell exterior. Plant Cell 21:1212–1229
113. Traub LM, Bonifacino JS (2013) Cargo recognition in clathrin-mediated endocytosis. Cold Spring Harb Perspect Biol 5:a016790
114. Ueda T, Yamaguchi M, Uchimiya H, Nakano A (2001) Ara6, a plant unique novel type Rab Gtpase, functions in the endocytic pathway of *Arabidopsis thaliana*. EMBO J 20:4730–4741
115. Uemura T, Ueda T, Ohniwa RL, Nakano A, Takeyasu K, Sato MH (2004) Systematic analysis of SNARE molecules in *Arabidopsis*: dissection of the post-Golgi network in plant cells. Cell Struct Funct 29:49–65
116. Uemura T, Suda Y, Ueda T, Nakano A (2014) Dynamic behavior of the trans-Golgi network in root tissues of Arabidopsis revealed by super-resolution live imaging. Plant Cell Physiol 55:694–703
117. Uemura T, Nakano RT, Takagi J, Wang Y, Kramer K, Finkemeier I, Nakagami H, Tsuda K, Ueda T, Schulze-Lefert P, Nakano A (2019) A Golgi-released subpopulation of the trans-Golgi network mediates protein secretion in Arabidopsis. Plant Physiol 179:519–532
118. Viotti C, Bubeck J, Stierhof YD, Krebs M, Langhans M, Van Den Berg W, Van Dongen W, Richter S, Geldner N, Takano J, Jurgens G, De Vries SC, Robinson DG, Schumacher K (2010) Endocytic and secretory traffic in *Arabidopsis* merge in the trans-Golgi network/early endosome, an independent and highly dynamic organelle. Plant Cell 22:1344–1357
119. Von Der Fecht-Bartenbach J, Bogner M, Krebs M, Stierhof YD, Schumacher K, Ludewig U (2007) Function of the anion transporter AtCLC-d in the trans-Golgi network. Plant J 50:466–474
120. Wang J-G, Li S, Zhao X-Y, Zhou L-Z, Huang G-Q, Feng C, Zhang Y (2013) HAPLESS13, the *Arabidopsis* μ1 adaptin, is essential for protein sorting at the trans-Golgi network/early endosome. Plant Physiol 162:1897–1910
121. Wang F, Shang Y, Fan B, Yu JQ, Chen Z (2014a) Arabidopsis LIP5, a positive regulator of multivesicular body biogenesis, is a critical target of pathogen-responsive MAPK cascade in plant basal defense. PLoS Pathog 10:e1004243
122. Wang X, Cai Y, Wang H, Zeng Y, Zhuang X, Li B, Jiang L (2014b) Trans-Golgi network-located AP1 gamma adaptins mediate dileucine motif-directed vacuolar targeting in Arabidopsis. Plant Cell 26:4102–4118
123. Wang F, Yang Y, Wang Z, Zhou J, Fan B, Chen Z (2015) A critical role of LIP5, a positive regulator of multivesicular body biogenesis, in plant responses to heat and salt stresses. Plant Physiol 169:497–511
124. Wang JG, Feng C, Liu HH, Ge FR, Li S, Li HJ, Zhang Y (2016) HAPLESS13-mediated trafficking of STRUBBELIG is critical for ovule development in *Arabidopsis*. PLoS Genet 12:e1006269
125. Wang JG, Feng C, Liu HH, Feng QN, Li S, Zhang Y (2017) AP1G mediates vacuolar acidification during synergid-controlled pollen tube reception. Proc Natl Acad Sci USA 114:E4877–e4883

126. Wattelet-Boyer V, Brocard L, Jonsson K, Esnay N, Joubes J, Domergue F, Mongrand S, Raikhel N, Bhalerao RP, Moreau P, Boutte Y (2016) Enrichment of hydroxylated C24- and C26-acyl-chain sphingolipids mediates PIN2 apical sorting at trans-Golgi network subdomains. Nat Commun 7:12788
127. Weiner E, Berryman E, Frey F, González SolíS A, Leier A, Marquez-Lago T, Šarić A, Otegui MS (2024) Endosomal membrane budding patterns in plants. Proc Natl Acad Sci USA, in press 121
128. Wiese C, Abele M, Al B, Altmann M, Steiner A, Kalbfuß N, Strohmayr A, Ravikumar R, Park CH, Brunschweiger B, Meng C, Facher E, Ehrhardt DW, Falter-Braun P, Wang Z-Y, Ludwig C, Assaad FF (2024) Regulation of adaptive growth decisions via phosphorylation of the TRAPPII complex in Arabidopsis. J Cell Biol 223
129. Winter V, Hauser M-T (2006) Exploring the Escrting machinery in eukaryotes. Trends Plant Sci 11:115–123
130. Yamasaki A, Menon S, Yu S, Barrowman J, Meerloo T, Oorschot V, Klumperman J, Satoh A, Ferro-Novick S (2009) mTrs130 is a component of a mammalian TRAPPII complex, a Rab1 GEF that binds to COPI-coated vesicles. Mol Biol Cell 20:4205–4215
131. Yamazaki M, Shimada T, Takahashi H, Tamura K, Kondo M, Nishimura M, Hara-Nishimura I (2008) *Arabidopsis* VPS35, a retromer component, is required for vacuolar protein sorting and involved in plant growth and leaf senescence. Plant Cell Physiol 49:142–156
132. Yan X, Wang Y, Xu M, Dahhan DA, Liu C, Zhang Y, Lin J, Bednarek SY, Pan J (2021) Cross-talk between clathrin-dependent post-Golgi trafficking and clathrin-mediated endocytosis in Arabidopsis root cells. Plant Cell 33:3057–3075
133. Yin GM, Fang YR, Wang JG, Liu Y, Xiang X, Li S, Zhang Y (2024) Arabidopsis HAPLESS13/AP-1μ is critical for pollen sac formation and tapetal function. Plant Sci 341:111998
134. Yoshinari A, Hosokawa T, Beier MP, Oshima K, Ogino Y, Hori C, Takasuka TE, Fukao Y, Fujiwara T, Takano J (2020) Transport-coupled ubiquitination of the borate transporter BOR1 for its boron-dependent degradation. Plant Cell 33:420–438
135. Yu F, Lou L, Tian M, Li Q, Ding Y, Cao X, Wu Y, Belda-Palazon B, Rodriguez PL, Yang S, Xie Q (2016) Escrt-I component Vps23A affects aba signaling by recognizing ABA receptors for endosomal degradation. Mol Plant 9:1570–1582
136. Yu G, Derkacheva M, Rufian JS, Brillada C, Kowarschik K, Jiang S, Derbyshire P, Ma M, Defalco TA, Morcillo RJL, Stransfeld L, Wei Y, Zhou JM, Menke FLH, Trujillo M, Zipfel C, Macho AP (2022) The Arabidopsis E3 ubiquitin ligase PUB4 regulates BIK1 and is targeted by a bacterial type-III effector. EMBO J 41:e107257
137. Zelazny E, Santambrogio M, Pourcher M, Chambrier P, Berne-Dedieu A, Fobis-Loisy I, Miege C, Jaillais Y, Gaude T (2013) Mechanisms governing the endosomal membrane recruitment of the core retromer in Arabidopsis. J Biol Chem 288:8815–8825
138. Zhang L, Zhang H, Liu P, Hao H, Jin JB, Lin J (2011) Arabidopsis R-SNARE proteins Vamp721 and Vamp722 are required for cell plate formation. PLoS One 6:e26129
139. Zhao J, Bui MT, Ma J, Künzl F, Picchianti L, De La Concepcion JC, Chen Y, Petsangouraki S, Mohseni A, GarcíA-Leon M, Gomez MS, Giannini C, Gwennogan D, Kobylinska R, Clavel M, Schellmann S, Jaillais Y, Friml J, Kang BH, Dagdas Y (2022) Plant autophagosomes mature into amphisomes prior to their delivery to the central vacuole. J Cell Biol 221
140. Zhu Y, Zhao Q, Cao W, Huang S, Ji C, Zhang W, Trujillo M, Shen J, Jiang L (2023) The plant-unique protein DRIF1 coordinates with sorting nexin 1 to regulate membrane protein homeostasis. Plant Cell 35:4217–4237
141. Zouhar J, Rojo E, Bassham DC (2009) AtVPS45 is a positive regulator of the SYP41/SYP61/VTI12 snare complex involved in trafficking of vacuolar cargo. Plant Physiol 149:1668–1678

Chapter 12
Cisternal Maturation of the Golgi Apparatus in Budding Yeast

Takuro Tojima and Akihiko Nakano

Abstract The Golgi apparatus in the budding yeast *Saccharomyces cerevisiae*, unlike those in mammalian and plant cells, does not form stacks but exists as multiple separate cisternae scattered throughout the cytoplasm. This unique morphological feature has allowed us to observe the spatiotemporal dynamics of individual Golgi cisternae by live cell imaging. Nearly two decades ago, we developed super-resolution confocal live imaging microscopy (SCLIM) and visualized that the yeast Golgi truly matures, i.e., a single cisterna gradually changes its nature from *cis* to *trans* cisternae over time. Since then to date, many researchers have observed a wide variety of proteins residing in and around the Golgi, as well as cargoes, in live yeast cells, unveiling the detailed step-by-step processes of Golgi cisternal maturation. Notably, our recent studies identified the ER–Golgi intermediate compartment (ERGIC) in budding yeast, which matures into the *cis* Golgi. After *cis*-to-*trans* maturation, the *trans* Golgi further matures into the *trans*-Golgi network (TGN), which can be divided into at least two sub-stages, the early and late TGN. Altogether, accumulating evidence revealed that the Golgi apparatus is not a static, but rather a highly dynamic organelle that appears transiently with a short lifetime along with cargo traffic.

Keywords *Saccharomyces cerevisiae* · Golgi apparatus · ER-Golgi intermediate compartment (ERGIC) · *trans*-Golgi network (TGN) · Cisternal maturation · Live cell imaging

T. Tojima (✉) · A. Nakano
RIKEN Center for Advanced Photonics, Wako, Saitama, Japan
e-mail: takuro.tojima@riken.jp

A. Nakano, J. Saraste (eds.), *The Golgi Network, Volume I*, Subcellular Biochemistry 110, https://doi.org/10.1007/978-3-032-06936-8_12

Introduction

Eukaryotic cells have membranous organelles such as the endoplasmic reticulum (ER), Golgi apparatus, endosomes, and lysosomes/vacuoles. The Golgi apparatus consists of multiple membrane sacs called cisternae, which can be classified as *cis*, medial, and *trans* cisternae. The Golgi is a major site of post-translational modifications and a hub for multiple membrane traffic pathways [1–4]. Newly synthesized secretory and membrane proteins that exit from the ER enter the Golgi from the *cis* cisterna, move to the medial and then to *trans* cisterna, where they undergo various modifications, including glycosylation. Subsequently, the cargo proteins are transferred to the *trans*-Golgi network (TGN), where the cargoes are sorted and exported toward individual intracellular compartments and extracellular spaces.

In plant and insect cells, a set of Golgi cisternae (*cis*, medial, *trans*, and TGN) are piled up together to form a Golgi stack, and many Golgi stacks are distributed throughout the cytoplasm. In mammalian cells, almost all the Golgi stacks are concentrated and joined together in the vicinity of the nucleus, forming a single giant "Golgi ribbon" structure. In contrast, the Golgi in the budding yeast *Saccharomyces cerevisiae* does not form stacks but exists as separate cisternae moving around the cytoplasm [5–7] (Fig. 12.1). This unique structural feature of the yeast Golgi gave us a great advantage to investigate the spatiotemporal dynamics of single Golgi cisternae by live cell imaging.

The visualization of Golgi dynamics has been facilitated by recent advances in fluorescence microscopy technology. Researchers need to achieve high-speed, high-resolution, three-dimensional (3D), and simultaneous multi-color imaging all at once, because single Golgi cisternae in budding yeast are small (approximately 1 μm in diameter) and move at high speed within the 3D cytoplasmic space. To meet these requirements, various types of confocal laser scanning microscopy have been

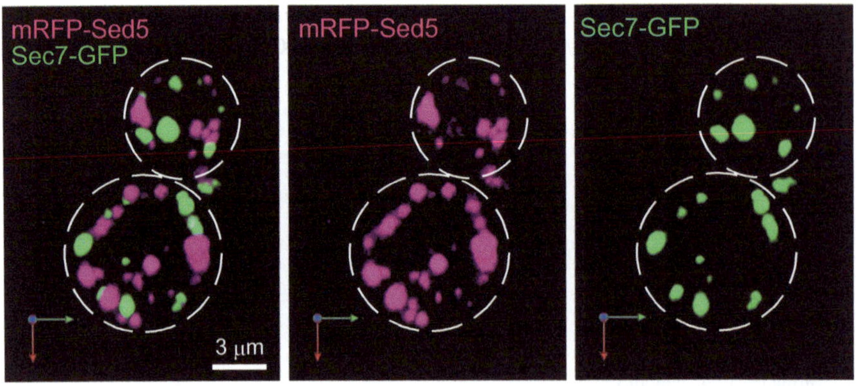

Fig. 12.1 Golgi apparatus in budding yeast. Dual-color three-dimensional (3D) SCLIM imaging of budding yeast cells expressing mRFP-Sed5 (early Golgi marker, magenta) and Sec7-GFP (late Golgi marker, green). Both markers exhibit multiple punctate signals that do not form stacks. The white broken lines indicate the edge of the cells

employed. In particular, our group originally developed the super-resolution confocal live imaging microscopy (SCLIM), which consists of a spinning-disk confocal scanner, a custom-built fast piezo actuator, three excitation lasers, an emission splitter unit for simultaneous three-color acquisition, and three EM-CCD/CMOS cameras attached to custom-built cooled image intensifiers [8, 9].

In 2006, we and Glick's group successfully visualized the Golgi cisternal maturation in budding yeast [10, 11]. In these studies, two different cisternal marker proteins were simultaneously labeled by green and red fluorescent protein tags in a live cell. A typical example of our experiments is that Sed5 (early Golgi marker) and Sec7 (late Golgi marker) were tagged with mRFP and GFP, respectively. If the classic vesicular transport model is correct, each cisterna should be stable and static, and its fluorescence colors (red or green) should remain unchanged for a long time. However, simultaneous dual-color SCLIM imaging of a single cisterna showed a gradual color change from red to green within a few minutes. This indicates that the cisterna, which initially harbored Sed5, gradually released it and also recruited Sec7 complementarily, providing crucial evidence that the Golgi cisterna really matures. From this initial discovery to the present, many research groups including us have extensively investigated the dynamics of a large number (more than 50) of functional proteins that reside in and around the Golgi by live imaging [12–33] (Table 12.1; Figs. 12.2, 12.3, and 12.4), contributing to make a detailed temporal roadmap of Golgi cisternal maturation, in which different functional molecules appear and disappear one after the other throughout the entire lifetime of a single Golgi cisterna. The functional molecules mapped until now include small GTPases with their guanine nucleotide exchange factors (GEFs) and GTPase-activating proteins (GAPs), coat and adaptor proteins, soluble N-ethylmaleimide-sensitive factor attachment protein receptors (SNAREs), glycosylation-related enzymes, cargo receptors, Golgi matrix proteins, and so on (Table 12.1; Figs. 12.3 and 12.4). As a result of this mapping, we recently identified in budding yeast the existence of the ER–Golgi intermediate compartment (ERGIC), which was originally reported in mammalian cells [32]. The yeast ERGIC matures into the *cis* Golgi, and after *cis*-to-*trans* maturation, the *trans* Golgi further matures into the *trans*-Golgi network (TGN) [21, 32]. We also found that the TGN can be divided into at least two functionally distinct sub-stages, the early TGN and the late TGN [31].

Another notable achievement of live imaging was the visualization of cargo proteins passing through the Golgi and the TGN. In 2019, we and Glick's group succeeded, again in parallel, in visualizing secretory cargo proteins during Golgi maturation by triple-color 3D live imaging [14, 21], and found that the cargoes persistently stay inside the cisterna undergoing maturation, as predicted by the cisternal maturation model. In addition, we visualized the entry of cargo from the ER into the Golgi [34].

In this chapter, we will summarize what we have learned about Golgi cisternal maturation from live imaging of the budding yeast *S. cerevisiae*. By integrating the accumulating evidence, we propose to update the major functional criteria of individual maturation stages as follows: (1) the "ERGIC" stage, which mediates ER–Golgi traffic; (2) the "*cis* and medial Golgi" stage, which mediates carbohydrate

Table 12.1 List of proteins that have been examined by live cell imaging and mapped on Golgi maturation stages

Yeast protein	Human homolog[a]	Residing maturation stage(s)[b]	Description	References
Grh1	GRASP55/65	ERGIC	single GRASP homolog of *S. cerevisiae*	[32]
Emp46	ERGIC-53	ERGIC	cargo receptor for the transport of glycoproteins from ER to Golgi	[32]
Ypt1	RAB1A/B	ERGIC and *trans* Golgi–early TGN (appeared twice)	RAB GTPase involved in ER-Golgi and endosome-Golgi traffic	[23, 27, 28, 32], and more
Rer1	RER1	ERGIC	retrieval receptor for Golgi-ER retrograde traffic	[11, 32]
Erd2	KDELR	ERGIC	K/HDEL-dependent retrieval receptor for Golgi-ER retrograde traffic	[32]
Sed5	Syntaxin5	ERGIC–*cis* Golgi	Qa-SNARE involved in ER-Golgi and intra Golgi traffic	[11, 19, 21, 32]
Mnn9	-	*cis* Golgi	Golgi-resident mannosyltransferase	[19, 22, 25, 32], and more
Gea1	GBF1	*cis* Golgi	Golgi-resident ARF GEF	[32]
Glo3	ARFGAP2/3	*cis* Golgi	Golgi-resident ARF GAP	[22]
Vrg4	TMEM241	*cis* Golgi	Golgi-resident GDP-mannose transporter	[10, 14, 24, 32], and more
Anp1 (Mnn8)	–	*cis* Golgi	Golgi-resident mannosyltransferase	[15]
Erd1	–	*cis* Golgi	involved in recycling of early glycosyltransferases from late to early Golgi compartments	[25]
Sec21	COPG1/2 (gamma-COP)	*cis* Golgi–medial Golgi	gamma subunit of the COPI complex that mediates intra-Golgi and Golgi-ER retrograde traffic	[31, 32]
Ret1 (Cop1)	COPA (alpha-COP)	*cis* Golgi–medial Golgi	alpha subunit of the COPI complex that mediates intra-Golgi and Golgi-ER retrograde traffic	[20, 25]
Mnn2	-	*cis* Golgi–medial Golgi	Golgi-resident mannosyltransferase	[32]
Gnt1	-	medial Golgi	Golgi-resident glucose N-acetyltransferase	[19, 31, 32]
Gos1	GOSR1 (GOS28)	medial Golgi	Qb-SNARE protein involved in Golgi transport	[11]
Sys1	SYS1	*trans* Golgi	Golgi membrane protein that recruits Imh1 via Arl1 and Arl3 GTPases	[19, 21, 32]

(continued)

Table 12.1 (continued)

Yeast protein	Human homolog[a]	Residing maturation stage(s)[b]	Description	References
Ypt6	RAB6A/B/C	*trans* Golgi	RAB GTPase that recruits the GARP complex to regulate the fusion of endosome-derived vesicles with late Golgi	[26, 30, 31]
Imh1	-	*trans* Golgi	yeast GRIP-domain Golgin that mediates endosome-Golgi traffic	[32]
Apl6	AP3B1/2	*trans* Golgi	beta subunit of the AP-3 complex involved in Golgi-vacuole traffic	[32]
Apl5	AP3D1	*trans* Golgi	delta subunit of the AP-3 complex involved in Golgi-vacuole traffic	[18]
Aur1	-	*trans* Golgi–early TGN	Phosphatidylinositol:ceramide phosphoinositol transferase	[15]
Rbd2	RHBDD2	*trans* Golgi–early TGN	Golgi rhomboid protease	[15]
Gea2	GBF1	*trans* Golgi–early TGN	Golgi-resident ARF-GEF	[18, 25, 32]
Gcs1	ARFGAP1	*trans* Golgi–early TGN	Golgi-resident ARF-GAP	[22]
Vps74	GOLPH3	*trans* Golgi–early TGN	Golgi PI4K effector that interacts with glycosyltransferases and, in the PI4P-bound state, mediates retention of these enzymes in the Golgi	[25]
Trs130	TRAPPC10	early TGN	component of TRAPPII complex, a multimeric GEF for Ypt31/32	[27]
Gyp1	TBC1D22A/B	early TGN	RAB GAP for Ypt1	[28]
Gyp6	TBC1D5	early TGN	RAB GAP for Ypt6	[30]
Vps10 (Pep1)	SORCS1/2/3	early TGN	receptor for sorting and transport of carboxypeptidase Y to the vacuole	[15]
Nhx1	SLC9A6/7/9	early TGN	Na^+/H^+ and K^+/H^+ exchanger	[15]
Trs85	TRAPPC8	early TGN	component of TRAPPIII complex, a multimeric GEF for Ypt1	[28]
Tlg2	syntaxin 16	early TGN	Qa-SNARE that forms a complex with Tlg1 and Vti1 and mediates fusion of endosome-derived vesicles with the late Golgi	[31–33]
Tlg1	syntaxin 6/10	early TGN	Qc-SNARE that forms a complex with Tlg2 and Vti1 and mediates fusion of endosome-derived vesicles with the late Golgi	[17, 15]

(continued)

Table 12.1 (continued)

Yeast protein	Human homolog[a]	Residing maturation stage(s)[b]	Description	References
Kex2	FURIN	early TGN	serine protease involved in the activation of proproteins of secretory pathway	[24, 15]
Drs2	ATP8A1/2	early TGN	aminophospholipid flippase that maintains membrane lipid asymmetry in post-Golgi secretory pathway	[15]
Ste13	DPP10/6	early TGN	dipeptidyl aminopeptidase	[15]
Stv1	ATP6V0A1/2/4	early TGN	subunit of vacuolar-ATPase that is targeted to the vacuole via AP 3 pathway	[15]
Sec7	BIG1/2	early TGN–late TGN	Golgi-resident ARF GEF mediating carrier formation from the TGN	[10, 11, 16, 20, 23, 25], and more.
Age2	SMAP1/2	early TGN–late TGN	Golgi-resident ARF GAP	[23]
Chs5	-	late TGN	component of the exomer complex	[12, 31]
Clc1	clathrin light chain	late TGN	clathrin light chain	[31]
Chc1	clathrin heavy chain	late TGN	clathrin heavy chain	[16, 20]
Gga1	GGA1	late TGN	GGA protein that acts as a clathrin adaptor	[12, 32]
Gga2	GGA2	late TGN	GGA protein that acts as a clathrin adaptor	[13, 16, 18, 31, 33]
Pik1	PI4KB	late TGN	phosphatidylinositol 4-kinase	[16, 18]
Ent3	ENTHD1/ epsin-2B	late TGN	epsin-related clathrin adaptor	[16]
Apl2	AP1B1	late TGN	beta subunit of the AP-1 clathrin adaptor complex	[14, 16–18, 31], and more
Aps1	AP1S1/2/3	late TGN	sigma subunit of the AP-1 clathrin adaptor complex	[12]
Ent5	–	late TGN	epsin-related clathrin adaptor	[16, 15]
Ypt31	RAB11A/B	late TGN	RAB GTPase implicated in cargo export from the TGN	[18, 23, 33]
Ypt32	RAB11A/B	late TGN	RAB GTPase implicated in cargo export from the TGN	[26, 29]

[a]searched by ALLIENCE of GENOME RESOURCES (viewed 27 June 2024, https://www.alliancegenome.org)

[b]Here we apply the new criteria we propose in this article, which often differ from the original paper(s)

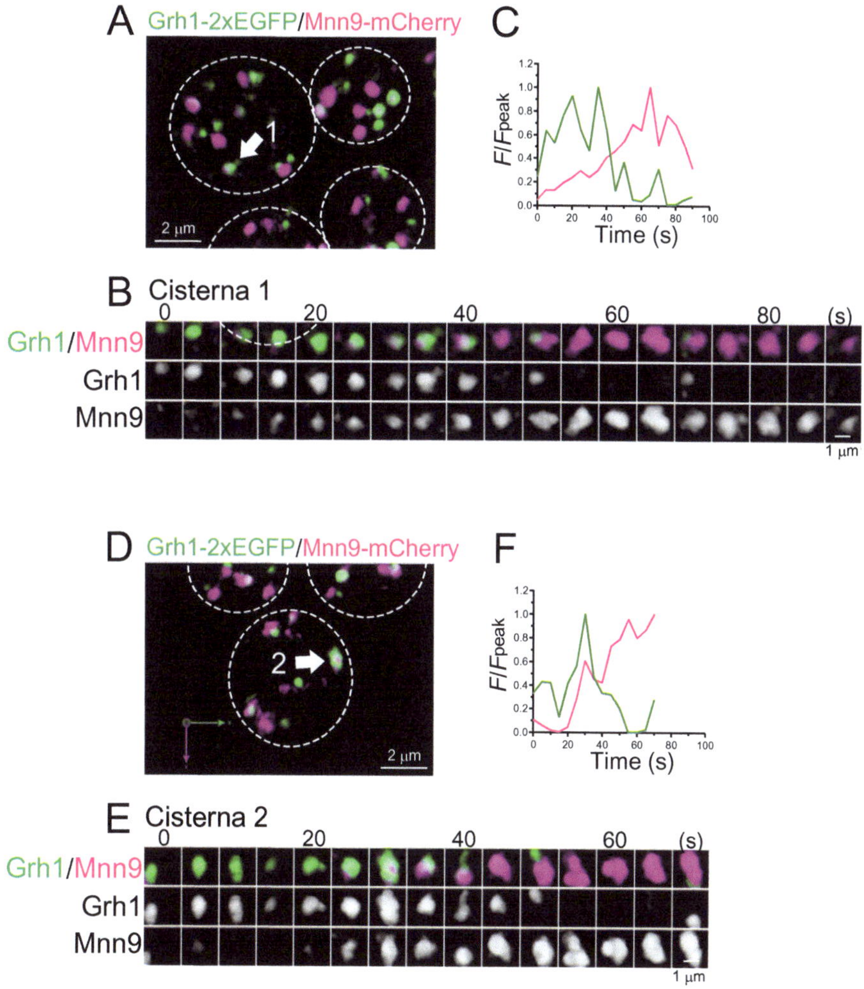

Fig. 12.2 Visualization of Golgi cisternal maturation in budding yeast. Dual-color four-dimensional (4D) SCLIM imaging of single cisternae in budding yeast cells expressing Grh1-2xEGFP (ERGIC marker, green) and Mnn9-mCherry (*cis* Golgi marker, magenta). Two representative cisternae are shown. (A, D) Low-magnification images of the cells. The white broken lines indicate the edge of the cells. (B, E) Time-lapse images of the single cisternae (white arrows in A and B, respectively) in the cells. (C, F) Time course changes in relative fluorescence intensities (F/F_{peak}) of green and magenta channels in A and D, respectively. The data indicate that the ERGIC gradually matures into the *cis* Golgi

synthesis; (3) the "*trans* Golgi" stage, which serves as a scaffold for TGN assembly; (4) the "early TGN" stage, which mediates cargo import into the TGN; and (5) the "late TGN" stage, which mediates cargo export from the TGN [32]. These stages appear in sequence along with maturation but can coexist within a single cisterna during their transition periods. Although it is not strictly possible to draw clear

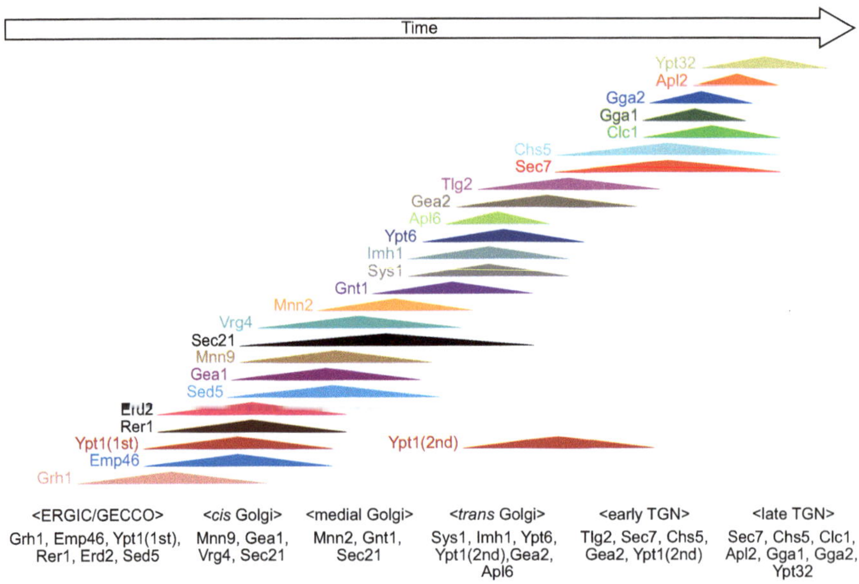

Fig. 12.3 Temporal mapping of 25 functional proteins residing in and around the Golgi in budding yeast, as revealed by the author's group. The appearance periods (oblong triangles) of 25 yeast ERGIC/Golgi/TGN-resident molecules are plotted on the time axis of the Golgi maturation. Time flows from left to right. This figure shows results from our group only [19, 26, 31, 32]

boundaries between these stages, it is important to assign the major functional criteria to each Golgi stage. In the following sections, we describe in detail the current knowledge of these maturation stages.

Yeast ERGIC and Its Maturation to *cis* Golgi

The ERGIC was originally identified in mammalian cells as a mobile carrier that mediates long-distance anterograde cargo traffic from the ER exit sites (ERES) at the cell periphery to the Golgi ribbon near the nucleus [35–40]. The ERGIC is also involved in retrograde cargo traffic from the Golgi to the ER; i.e., ER-resident proteins are retrieved from the ERGIC before reaching the Golgi [41]. In plant and insect cells, the ERES and the Golgi stack are adjacent [42, 43], so long-distance ER–Golgi carriers, as in mammals, do not seem to be necessary. Even in mammalian cells, while a large number of ERGIC are located away from the Golgi ribbon, a subpopulation of ERGIC, as well as ERES, is adjacent to the *cis* side of the Golgi ribbon [32, 44, 45]. We have previously identified the Golgi entry core compartment (GECCO) in plant cells, a specialized membrane compartment that plays a role similar to the mammalian ERGIC [3, 46–48]. The GECCO is located at *cis*-most of the Golgi stack (i.e., the interface between the Golgi and the ERES) and receives

Maturation stages	ERGIC	⇨	*cis* Golgi	⇨	medial Golgi	⇨	*trans* Golgi	⇨	early TGN	⇨	late TGN
ArfGEFs			Gea1					Gea2		Sec7	
ArfGAPs			Glo3					Gcs1		Age2	
Coat proteins					Sec21 Ret1						Clc1 Chc1
Adaptors								Apl5 Apl6		Gga1 Gga2 Ent3 Chs5	Aps1 Apl2 Ent5
Rabs	Ypt1 (1st)						Ypt6		Ypt1 (2nd)		Ypt31 Ypt32
RabGAPs										Gyp1 Gyp6	
TRAPPs									Trs85	Trs130	
SNAREs		Sed5			Gos1				Tlg1 Tlg2		
GRASP, Golgin, and its receptor	Grh1						Imh1 Sys1				
Lipid metabolism							Aur1		Drs2	Pik1 PI4P	
Glycosylation			Mnn9 Anp1 Vrg4	Mnn2	Gnt1						
Glycosylation enzyme recycling			Erd1					Vps74			
Proteases							Rbd2		Ste13 Kex2		
Cargo receptors	Emp46	Erd2 Rer1							Vps10		
Proton pump and exchanger									Nhx1 Stv1		

Fig. 12.4 Temporal mapping of more than 50 functional proteins residing in and around the Golgi in budding yeast, as revealed by many groups. The peak appearance positions of more than 50 yeast ERGIC/Golgi/TGN-resident molecules are plotted on the time axis of the Golgi maturation. Time flows from left to right. This figure integrates results from several research groups [10–33]. Note that exact comparisons of data may be inappropriate due to differences in experimental conditions between groups

ER-derived cargoes for Golgi stack regeneration. However, until recently, it remained unclear whether such ER–Golgi intermediates (mammalian ERGIC or plant GECCO) also exist in the budding yeast *S. cerevisiae*.

In 2024, to examine whether the ERGIC/GECCO exists in budding yeast, we investigated the spatiotemporal dynamics of several key pre- or early Golgi proteins by SCLIM [32]. The proteins we focused on were Emp46 (yeast counterpart of mammalian ERGIC-53, an ERGIC marker), Ypt1 (yeast counterpart of mammalian Rab1, another ERGIC marker), Rer1 and Erd2 (retrieval receptors), and Grh1 (the yeast homolog of mammalian Golgi reassembly-stacking protein of 65 kDa

[GRASP65]). When we visualized these proteins by fluorescent protein tags, we found many mobile fluorescent punctate signals appearing and disappearing with a lifetime of a few minutes. We then performed dual-color SCLIM imaging of each of these proteins versus a typical *cis* Golgi reference marker, Mnn9 (glycosyltransferase). Representative data (Fig. 12.2) shows that the punctate signal of Grh1-2xEGFP appeared first, and then the Mnn9-mCherry signal came up on the pre-existing Grh1-positive compartment. Then, as the Mnn9 signal gradually increased, the Grh1 signal decreased and eventually became invisible. Similar transition patterns were observed when we visualized Emp46, Ypt1, Rer1, and Erd2 versus Mnn9, indicating that these pre/early Golgi proteins assemble prior to the *cis* Golgi stage. Notably, we found that, during the transition periods, the two signals are often spatially segregated within the cisterna. We further investigated the order of appearance of these proteins and found that Grh1 is the earliest protein to appear among those we examined [32]. Considering the molecular functions of Ypt1, Rer1, Erd2, and Grh1, this newly discovered pre-*cis* Golgi compartment should have specialized roles for ER–Golgi traffic, corresponding to mammalian ERGIC and plant GECCO. Together, we identified the presence of the ERGIC for the first time in budding yeast and demonstrated that it matures into the *cis* Golgi.

Visualization of Cargo Traffic from ER to ERGIC/*cis* Golgi

Newly synthesized cargo proteins at the ER are exported through specific ER membrane domains, called the ERES. In this process, it has long been thought that the cargo proteins are incorporated into small coat protein complex II (COPII) vesicles with a diameter of approximately 60 nm that bud from the ER membrane, and that the earliest Golgi compartment is generated by fusion of multiple COPII vesicles. However, in 2014, our live SCLIM observation of the Golgi and the ERES in budding yeast has shown that the pre-existing *cis* Golgi compartments (labeled by Mnn9 or Sed5) approach toward and contact with the ERES (labeled by Sec13, a COPII component), and a cargo protein (Axl2) initially located in the ER is handed over to the *cis* Golgi during this transient contact [34]. We have named this new mode of cargo transfer "hug-and-kiss" behavior. Later, in 2024, we found that the ERGIC in budding yeast (labeled by Grh1) also exhibits the hug-and-kiss [32]. The ERGIC had a higher probability of localizing near the ERES than *cis* Golgi, suggesting that ERGIC exhibits the hug-and-kiss behavior more frequently than *cis* Golgi.

Such a contact-mediated system may contribute to efficient and precise cargo transfer to the recipient compartment, but its molecular mechanism remains obscure. It is possible that, following the classic vesicular transport model, free COPII vesicles are released and immediately captured by the *cis* Golgi/ERGIC, but one can suggest an alternative mechanism independent of small vesicles. In mammalian cells, a recent study using correlative light and electron microscopy has shown that the cell peripheral ERES extends long tubules containing cargo along microtubules toward a perinuclear Golgi ribbon, and that the COPII components remain associated at the neck of the tubules, presumably through a continuous assembly/disassembly cycle controlled by Sar1 GTPase [49]. It is also proposed that the cargo export from the ERES, particularly for large-sized cargoes such as procollagens, is accomplished through direct conduits between the ER and ERGIC [50].

In plant and insect cells, the ERES and the Golgi stack are in close proximity [42, 43]. In mammalian cells, microtubule disruption by nocodazole treatment leads to the disassembly of the Golgi ribbon and the formation of many Golgi ministacks in the vicinity of the ERES [51, 52], resembling the plant and insect Golgi. Even when the microtubule network is intact, substantial amounts of ERES are observed at the *cis* face of the perinuclear Golgi ribbon [32, 44, 45]. Based on these facts, the proximity distribution of ERES and Golgi appears to be a fundamental morphological feature across species, and budding yeast would be an exception: the yeast Golgi/ERGIC moves rapidly through the cytoplasm, and the ERES-Golgi contact is transient.

Distinction Between ERGIC and *cis* Golgi

Since the yeast ERGIC gradually matures into the Golgi, drawing a clear boundary between the two compartments is difficult. We employed the Golgi-disrupting agent Brefeldin A (BFA) to dissect them pharmacologically [32]. BFA is an Arf-GEF inhibitor that causes Golgi disassembly and relocalization of Golgi-resident enzymes into the ER [53–55]. As mentioned above, we have previously shown that plant cells have the GECCO, a specialized compartment acting like mammalian ERGIC [3, 46–48]. The GECCO was originally identified in tobacco BY-2 cells by using BFA. Upon treatment with BFA, several Golgi-resident proteins, ST (rat sialyltransferase, *trans*-Golgi marker), ManI (alpha-1,2-mannosidase I, medial-Golgi marker), and ATCASP (putative Golgi matrix protein), are absorbed into the ER, whereas some pre/early Golgi proteins, SYP31 (homolog of the yeast SNARE Sed5) and RER1B (homolog of yeast Rer1), formed small punctate compartments near the ERES. After the BFA washout, the ER-absorbed Golgi proteins first entered in, and then passed through, the small puncta to regenerate the Golgi stack. These results strongly suggest that the small puncta that appeared upon BFA treatment act like a scaffold for Golgi stack reassembly, and we termed it as Golgi-entry core compartment (GECCO). In mammalian cells, BFA treatment also causes the absorption of many Golgi-resident proteins into the ER [56, 57], but several pre/early Golgi proteins such as the ERGIC marker ERGIC-53 (Emp46 in yeast), GRASP65 (Grh1 in yeast), syntaxin 5 (Sed5 in yeast; SYP31 in plant), and RER1 form similar punctate structures called "Golgi remnants" [58–63]. Therefore, in both plant and mammalian cells, pre-Golgi compartments such as GECCO and ERGIC can be isolated by BFA.

In budding yeast, we also found that, upon BFA treatment, pre/early-Golgi proteins, Grh1, Rer1, and Sed5, accumulate in punctate structures, which we named "yeast GECCO" [32]. In contrast, similar to plant and mammalian cells, Golgi-resident proteins, Mnn2 (glycosyltransferase) and Vrg4 (mannose transporter), were absorbed into the ER upon BFA treatment. Since the molecular components and BFA sensitivity of these punctate structures (plant GECCO, mammalian ERGIC, and yeast GECCO) are quite similar, we propose that they are homologous compartments. Consistent with the spatial distribution of the plant GECCO proteins at the *cis*-most Golgi stack in the absence of BFA [46, 47], Grh1, Rer1, and Sed5 in budding yeast were recruited prior to *cis* Golgi proteins (e.g., Mnn9 and Vrg4) during cisternal maturation [32] as described above. In combination with the BFA experiments, we propose to define the ERGIC-Golgi boundary in budding yeast between Sed5 and Mnn9. By SCLIM imaging, Sed5 appeared slightly earlier than Mnn9, although their periods of appearance largely overlap [32]. However, we occasionally observed different spatial distributions of Sed5 and Mnn9 within a single cisterna, suggesting that the ERGIC and *cis* Golgi form spatially distinct zones, although they may partially overlap.

cis-to-medial Golgi Stage as a Glycosylation Factory

A major function of the Golgi is carbohydrate synthesis. Dynamics of several Golgi-resident glycosylation-related proteins, including Vrg4 (mannose transporter), Mnn9, Mnn2 (mannosyltransferases), and Gnt1 (glucose N-acetyltransferase), was observed in live yeast cells by SCLIM [31, 32]. Vrg4 and Mnn9 are widely used as typical markers for *cis* Golgi in budding yeast [10, 19, 34]. These *cis* Golgi marker proteins assembled at nearly the same timing, but Vrg4 appeared slightly later than Mnn9 [32]. Glick's group has shown that Anp1 (another mannosyltransferase) is recruited at a similar timing as Mnn9 and Vrg4 [15]. We also found that Mnn2 assembled after these proteins, consistent with the fact that Mnn2 acts later than Mnn9 and Anp1 in the N-linked glycosylation pathway [64]. Gnt1, another glycosyltransferase whose physiological role in vivo is unclear [65], appeared much later than Mnn9 and Mnn2, which makes it reasonable to define it as a medial Golgi marker [19, 31, 32].

According to the cisternal maturation model, the glycosylation enzymes must be packaged into coat protein complex I (COPI) vesicles and sent back to earlier cisterna for reuse. In this recycling process, Vps74 (human Golgi phosphoprotein 3 [GOLPH3]) binds both to the COPI complex and to the cytosolic tails of the glycosylation enzymes and plays a central role in their retrograde transport [66]. Live imaging revealed that Vps74 mainly assembles around the *trans* Golgi stage [25] and its deletion in budding yeast causes the prolonged retention of Mnn9 in the Golgi cisterna [66]. Emr's group recently showed that Erd1, a Golgi integral membrane protein, facilitates the stable interaction of glycosylation enzymes with Vps74 to execute successful Golgi enzyme recycling [25]. Erd1 mainly assembles at the *cis* Golgi stage, suggesting a sequential association of Erd1 and Vps74 with their client Golgi enzymes.

COPI Vesicle Formation at the Golgi Stage

The COPI complex mediates the Golgi–ER traffic and intra-Golgi traffic [67]. Our SCLIM imaging showed that Sec21, a COPI subunit, initially appears at almost the same timing as Mnn9, but the Sec21 signal slowly increases and peaks during the decay phase of Mnn9 [32]. This is consistent with the notion that the COPI vesicles are produced throughout the *cis*-to-*trans* Golgi stages. Another COPI subunit, Ret1 (Cop1), also shows similar dynamics to Sec21 [22]. Acute inactivation of COPI subunits by several methods (temperature-sensitive mutants, auxin-inducible protein degradation, and anchor-away system) results in the impairment of Golgi cisternal maturation, indicating the essential role of COPI in this process [19, 24].

COPI vesicle formation at the Golgi is regulated by Arf GTPases [68]. In contrast to mammals, which have at least four Arf isotypes on the Golgi (Arf1, 3, 4, and 5) [69], budding yeast has two presumably redundant Arfs, Arf1 and 2, with Arf1 being the dominant one [70]. Several GEFs (Gea1, Gea2, and Sec7) and GAPs (Glo3, Gcs1, and Age2) for Arf have been identified in budding yeast [71, 72]. By SCLIM imaging, we found that Gea1 appears synchronously with the *cis* Golgi marker Mnn9 [32], and Fromme's group showed that Glo3 also appears synchronously with Mnn9 [22]. These results are consistent with the idea that Gea1 and Glo3 act cooperatively with Arf1 to form COPI vesicles at the *cis* Golgi. We previously found that the Glo3 function requires the Glo3 motif (ISSxxxFG) at the C-terminus, which is not present in Gcs1 [73]. Gcs1 and Gea2 assemble at the *trans* Golgi stage [22, 32], suggesting their contribution to COPI formation, and presumably to different events occurring at the *trans* Golgi stage, such as the recruitment of the Golgin Imh1 and the lipid flippase Drs2 (discussed later). At the TGN stage, where the COPI components are absent, Sec7 and Age2 are recruited [22, 31, 32] to regulate clathrin- and exomer-dependent carrier formation (also discussed later). Thus, Arf1 is thought to be present throughout overall Golgi/TGN maturation stages and is involved in multiple Golgi functions in collaboration with various GEFs and GAPs [74].

trans Golgi as AP-3-Dependent Cargo Export Stage

The adaptor protein complex 3 (AP-3) is known to be involved in cargo transport from the Golgi to the vacuole, bypassing the prevacuolar compartment (PVC), in budding yeast [75, 76]. It has long been believed that the export of cargo proteins occurs mainly at the TGN, but we unexpectedly found that Apl6, a subunit of the AP-3 complex, assembles at the Sys1-positive *trans* Golgi stage, while only a small fraction of Apl6 is located at the Sec7-positive TGN stage [31, 32]. Fromme's group also showed that Apl5, another subunit of AP-3, is recruited much earlier than the clathrin-dependent adaptor proteins, Gga2 (Golgi-associated, gamma adaptin ear containing, ARF binding protein 2) and Apl2 (AP-1 subunit), which are located at the TGN [18]. These results suggest that, unlike other adaptor proteins, including AP-1, GGA, and exomer (discussed later), AP-3-mediated cargo export occurs at the *trans* Golgi stage, where the clathrin coat is absent. This is consistent with the notion that AP-3 functions independently of clathrin [76]. Intriguingly, our SCLIM imaging showed that the Apl6 signal covers a partial area of the *trans* Golgi cisterna even at its peak time point, suggesting the formation of a functional zone for cargo export to the vacuole [32].

Distinction Between Golgi and TGN

The functions of the Golgi and the TGN are distinct: the Golgi is a major site of glycosylation, whereas the TGN mainly performs cargo sorting and export [77–79]. An exception is the AP-3-mediated cargo export at the *trans* Golgi stage, as mentioned above. Another important role of the TGN is the incorporation of retrograde cargoes from the endocytic pathway [4, 80–82]. Despite such differences in functions, the Golgi and the TGN exist in very close proximity. In mammalian and plant cells, the main compartment of the TGN is attached onto the *trans* face of the Golgi stack. In budding yeast, the TGN is progressively generated from the *trans* Golgi via cisternal maturation. Therefore, it is difficult to draw a clear boundary between them, as well as the ERGIC-*cis* Golgi boundary. However, our SCLIM imaging demonstrated that the TGN compartment is often segregated from the Golgi compartment within a single cisterna undergoing maturation [31, 32]. As a typical example, when we observed the transition from Sec21 (Golgi marker) to Sec7 (TGN marker) by simultaneous dual-color SCLIM, the two signals are distributed with complementary mosaic-like patterns [31]. Furthermore, our correlative light and electron microscopy analysis showed that Golgi and TGN markers are segregated within a continuous membrane compartment [21], providing strong evidence that the Golgi and the TGN can transiently coexist as spatially distinct zones within a single cisterna.

As mentioned above, BFA can dissect the Golgi and ERGIC/GECCO in plant and budding yeast [32, 46, 47]: BFA treatment causes the Golgi to disappear, but the ERGIC/GECCO remains. BFA is also able to dissect the TGN and the Golgi in plant cells [83]. Upon BFA treatment, the plant TGN is disrupted and dispersed as numerous small vesicles throughout the cytoplasm. After the BFA washout, the dispersed TGN components initially reassemble independently of the GECCO and Golgi stacks and eventually attach onto the *trans* side of the Golgi stacks. The effect of BFA on the TGN in budding yeast should be examined in future research.

Golgi–TGN Interface as a Scaffold for TGN Assembly

Based on the concept of Golgi cisternal maturation, the *trans* Golgi cisterna should serve as a scaffold for TGN assembly. We propose to focus on this function as a new criterion of the *trans* Golgi stage, because it is located at the border of drastic differences in function and origin between the *trans* Golgi and the TGN [3]. As mentioned above, the TGN receives not only newly synthesized cargoes but also endocytic cargoes [33, 84]. Sys1, an integral membrane protein that localizes to the *trans* Golgi [19, 21], binds to Arl3 GTPase, and Arl3 then recruits Arl1, which in turn recruits Imh1, the GRIP domain-containing yeast Golgin [85–87]. Imh1 acts as a tether for capturing endocytic compartments to facilitate their fusion with the TGN membrane [85, 88, 89]. The Arf-GEF Gea2 forms a complex with Arl1 and the lipid flippase Drs2 to recruit Imh1 [89, 90]. Imh1 also interacts with Ypt6, the yeast counterpart of Rab6 GTPase that mediates endocytic traffic to the TGN [91–93]. Ypt6 executes this process via the recruitment of the tethering factor Golgi-associated retrograde-protein (GARP) complex [94, 95]. The Ypt6 activity is suppressed by its GAP Gyp6, which also acts as an effector for Ypt32 in the Rab-GAP cascade during the Golgi–TGN cisternal maturation [26]. Live imaging studies from different groups [15, 26, 30, 32] have shown that these molecular players that serve as a scaffold for the cargo entry, including Imh1, Gea2, Drs2, Ypt6, and Gyp6, assemble at the transition period from the Sys1-labeled *trans* Golgi stage to the Sec7-positive TGN stage, although their precise order of appearance remains to be examined.

Ypt1, the yeast counterpart of Rab1 GTPase, has been implicated in not only ER–Golgi traffic but also endosome–Golgi traffic and vesicle formation at the late Golgi [23, 96]. We found that, after the initial appearance at the ERGIC stage, Ypt1 again assembles between the Sys1-positive *trans* Golgi stage and the Sec7-positive TGN stage [32]. Fromme's group also showed that Ypt1 assembles at the late Golgi stage [28]. Ypt1 recruits the Arf-GEFs Gea1 and Gea2 [97], consistent with our data that the timings of the first and second Ypt1 recruitments are close to those of Gea1 and Gea2, respectively. Ypt1 activity is regulated by its GEF, the transport protein particle III (TRAPPIII) complex [28]. Trs85, a TRAPPIII component, is recruited slightly prior to Ypt1, and the Ypt1 GAP Gyp1 is assembled during the decay phase of Ypt1 signals [28].

Cargo Traffic Through Golgi and TGN

Based on the cisternal maturation model, cargo proteins should stay continuously in a single cisterna undergoing maturation. In 2019, we and Glick's group have succeeded in visualizing secretory cargo proteins passing through the Golgi and TGN in live yeast cells [14, 21]. We fluorescently labeled *cis* and *trans* Golgi with mCherry-tagged Mnn9 and iRFP-tagged Sys1, respectively, and observed them together with GFP-labeled Axl2 as a secretory cargo by simultaneous three-color SCLIM imaging [21]. During the period of the *cis*-to-*trans* transition, the cargo signals are sustained within the cisterna over time. This is also the case when we observed Sed5 (ERGIC/*cis* Golgi marker) and Sec7 (TGN marker) together with Axl2, and Glick's group observed Vrg4 (*cis* Golgi marker) and Sec7 with a designed secretory cargo. These results provide firm evidence for the cisternal maturation model.

Early TGN as a Cargo Entry Stage

As described above, the *trans* Golgi cisterna harbors Imh1 that can tether transport vesicles to serve as a site for TGN assembly. After being tethered, the vesicles undergo membrane fusion with the Golgi, which is mediated by a set of SNARE proteins including Tlg1 and Tlg2 [98, 99]. We compared the dynamics of Tlg2 versus Sys1 (*trans* Golgi marker) and found that Tlg2 appears later than Sys1, indicating that the Tlg2-mediated cargo entry occurs after the *trans* Golgi stage [31]. Intriguingly, the Sys1-Tlg2 transition proceeded by a gradual decrease in the Sys1-positive area and a complementary increase in the Tlg2-positive area within a single cisterna, and these two areas were often spatially segregated. Glick's group showed that, similar to Tlg2 dynamics, Tlg1 appears prior to Sec7 [17]. Furthermore, we found that the Tlg2 signal peaks during the rising phase of Sec7 (Arf GEF for carrier formation; TGN marker) and Clc1 (clathrin light chain) signals [31], suggesting that the cargo reception occurs prior to the clathrin-mediated cargo export. In the cisterna undergoing Tlg2-Clc1 transition, we also noticed that the two areas are often segregated. Based on these observations, we have proposed that the TGN stage can be divided into two spatiotemporally distinct sub-stages: the early TGN, which mediates cargo entry, and the late TGN, which mediates cargo export.

The dynamics of endocytic cargoes have been visualized. The lipophilic fluorescent dye FM4–64 rapidly incorporated into the yeast TGN [17], consistent with the previous findings in the plant TGN [100–102]. More recently, Toshima's group, in collaboration with us, has shown that fluorescently labeled alpha-factor is incorporated into the Tlg2-positive early TGN compartment, but not into the Sec7-positive compartment, immediately after endocytosis [33]. These results indicate that the early TGN in budding yeast, as well as plant TGN, behaves in the same manner as early endosomes in mammalian cells.

Early TGN as a Proteolytic Processing Site

Kex2, the yeast homolog of mammalian furin, is a membrane-associated protease that catalyzes the cleavage of precursor proteins to produce their mature forms. Kex2 and furin are reported to localize at the TGN [103, 104]. Glick's and Fromme's groups showed in budding yeast that the Kex2 signal peaks slightly earlier than the Sec7 signal in maturing cisterna [23, 24]. This Kex2 temporal dynamics resembles those of Tlg1 and Tlg2. Ste13, another peptidase that acts downstream of Kex2 [105], also exhibited similar dynamics as Kex2 [15]. These results suggest that the proteolytic processing of the glycosylated cargo proteins occurs at the early TGN stage prior to their export at the late TGN stage.

Late TGN as a Cargo Export Stage

We propose to define the late TGN as a cargo export stage in budding yeast. Sec7, an Arf-GEF for cargo export from the TGN, peaks later than Gea2 and Tlg2, and its lifetime spans that of clathrin (Clc1 and Chc1) and various clathrin adaptor proteins, including Gga1, Gga2, Apl2, Aps1, Ent3, and Ent5 [12, 13, 16, 31]. The GGA monomeric clathrin adaptors (Gga1 and Gga2) mediate the cargo export from the TGN to the prevacuolar compartment (PVC) in budding yeast [13]. Here we define the PVC as the Vps21 (Rab5 in mammals)-positive and intraluminal vesicle-containing compartment, also called the prevacuolar endosome (PVE), multivesicular body (MVB), late endosome (LE), etc. [3]. On the other hand, the adaptor protein-1 (AP-1) complex (Apl2 and Aps1) has recently been shown to mediate the recycling of Golgi-resident proteins from the TGN back to the Golgi in budding yeast [15]. It is also suggested in mammalian cells that AP-1 mediates retrograde cargo trafficking in the late secretory pathway [106]. During the clathrin-residing period in the late TGN stage, Gga1 and Gga2 assemble first, and then Apl2 and Aps1 follow [16, 31]. The epsin-like proteins Ent3 and Ent5 are also recruited sequentially to the TGN, overlapping the Gga1/2- and AP-1-residing periods, respectively [15, 16], supporting the notion that Ent3 and Ent5 interact with and act in combination with GGAs and AP-1, respectively [16, 107–109]. The AP-1 recruitment to the TGN is mediated by phosphatidylinositol 4-phosphate (PI4P) [110], and the production of PI4P at the TGN is catalyzed by the yeast PI4-kinase Pik1. The temporal dynamics of Pik1 and PI4P, as visualized by the PI4P-binding probes GFP-P4C and GFP-PH^{OSH1}, are almost the same as Sec7 [16, 18]. Sec7 is also known to regulate the formation of exomer, another coat protein complex that is involved in the direct transport of a subset of secretory cargo proteins in budding yeast. Chs5, an exomer component, exhibited almost the same assembly dynamics as Sec7 [12, 31]. Such differential appearance patterns of various coat/adaptor proteins suggest the presence of multiple spatiotemporally distinct zones within the late TGN stage for efficient packaging and export of cargoes destined for different subcellular compartments.

Glick's group visualized the temporal dynamics of various proteins that depart from the TGN in a GGA-dependent manner (Vps10 and Nhx1) and in an AP-1/Ent5-dependent manner (Aur1, Rbs2, Ste13, Stv1, etc.). They classified their temporal appearance patterns at the TGN into several groups [15]. They also showed that a designed PVC-targeted cargo exits from the TGN before the exit of the secretory cargo [13]. This is consistent with the fact that GGAs, which mediate the TGN-PVC pathway, are assembled prior to Ypt31/32, which mediate the secretory pathway (discussed below).

TGN and Recycling Endosomes

While Rab11 GTPase is identified as a typical marker of recycling endosomes in mammalian cells [111], Ypt31/32, the yeast counterparts of Rab11, is implicated in cargo export from the TGN to the plasma membrane [112]. By our SCLIM imaging, Ypt32 is found to be the last protein to assemble at the late TGN among all the proteins that we have examined [26, 32]. Fromme's group showed that Ypt31 assembles later than Gga2 and Apl2 [18]. In addition, Trs130, a component of the TRAPPII complex, is recruited slightly earlier than Ypt31, consistent with its function as a GEF for Ypt31/32 [27].

In mammalian cells, the recycling endosomes were initially identified as organelles responsible for the slow recycling of endocytosed cargoes. Later, it was shown that they also relay a subset of secretory cargoes from the Golgi to the plasma membrane [113]. Satoh's group demonstrated that, in nocodazole-treated mammalian cells and Drosophila cells, a subpopulation of recycling endosomes is attached onto the *trans* face of the Golgi stacks, and a cargo (EGFP-labeled glycosylphosphatidylinositol) visualized by the retention using the selective hook (RUSH) system is incorporated into the Golgi-associated recycling endosomes after passing through the Golgi [114]. These facts strongly suggest that the late TGN and the recycling endosomes are homologous compartments across species [3].

Conclusion and Future Perspectives

As reviewed in this chapter, recent advances in live cell imaging techniques have demonstrated that Golgi cisternae in budding yeast do indeed mature. A wide variety of Golgi-resident proteins move in and out of a single cisterna in a spatiotemporally controlled manner, while cargo proteins remain inside throughout this process. It is now clear that multiple functional zones exist within a maturing cisterna, although some of these zones overlap. In addition, the entries and exits of various membrane traffic pathways between the Golgi and other organelles are becoming more clearly defined on the maturation process (Fig. 12.5). In the future, many more functional proteins residing on and around the Golgi will be added to this maturation roadmap.

To fully understand the membrane traffic systems in and around the Golgi, it would be necessary to visualize the dynamics of small transport vesicles (less than 100 nm in diameter). Recently, we have succeeded in developing a next-generation SCLIM, named SCLIM2M, with greatly improved spatiotemporal resolution [115].

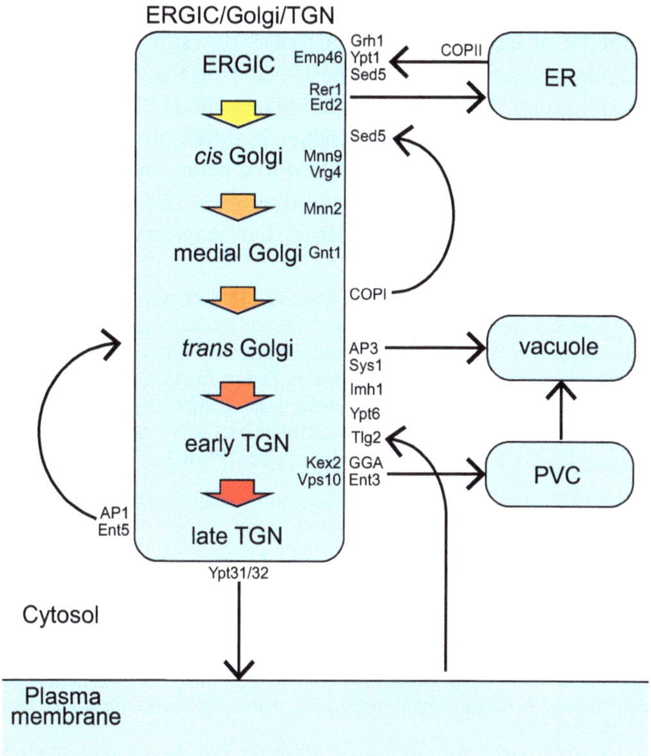

Fig. 12.5 Schematic presentation of the Golgi maturation process and membrane traffic pathways around the Golgi in budding yeast. The colored downward arrows (yellow, orange, and red) represent the gradual transition (maturation) of Golgi stages (ERGIC, Golgi, and TGN). Thin black arrows represent membrane traffic pathways within the Golgi and between the Golgi and other organelles (ER, PVC, and vacuole) and the plasma membrane. Also shown are several key Golgi-resident proteins involved in the membrane traffic pathways and post-translational modification to the cargoes

SCLIM2M is equipped by extremely high-speed CMOS cameras (1000 frames/s) with image intensifiers to detect individual single-photon signals emitted from fluorescent molecules separately, and a novel image reconstruction algorithm is performed to achieve a spatial resolution of less than 100 nm. Also, by using a fast-moving piezo actuator, sequential 3D images can be acquired at a higher speed (20 volumes/s) without time lag. With SCLIM2M, we are now able to capture the movement of small clathrin-coated vesicles budded from the TGN in budding yeast [115].

The identification of the ERGIC and the TGN sub-stages in budding yeast suggests the universality of the Golgi functions and organizations across different eukaryotes, including yeast, mammalian, insect, and plant cells, despite their different morphological features and subcellular distributions. Indeed, the temporal order of yeast Golgi proteins (Figs. 12.3-12.5) is well comparable to the spatial order of localization of mammalian Golgi proteins within the Golgi stack [116]. However, the current criteria of individual Golgi cisternae may still differ somewhat across species; for example, carbohydrate synthesis occurs in the medial and *trans* Golgi, but not in the *cis* Golgi, in mammalian and plant cells [117]. Further comparative studies of the Golgi and its related organelles in different species using advanced technologies, including SCLIM2M, will lead to a better understanding of the universal mechanisms of Golgi functions and dynamics, and eventually allow us to unify the definitions of Golgi cisternae across diverse eucaryotes.

Acknowledgments This article is dedicated to Kazuo Kurokawa, who passed away in April 2024. We thank all the past and present members of the Nakano laboratory for valuable discussions. We apologize to the researchers whose work we could not cite in this article due to space limitations. Recent work by the authors' group was supported by Grants-in-Aid for Scientific Research from the Ministry of Education, Culture, Sports, Science, and Technology (MEXT) of Japan (grant numbers 22 K06213 to T.T.; 18H05275 and 23H00382 to A.N.) and the Japan Science and Technology Agency (JST) CREST program (grant number JPMJCR21E3 to T.T.).

Compliance with Ethical Standards The authors declare that they have no conflicts of interest.

References

1. Glick BS, Nakano A (2009) Membrane traffic within the Golgi apparatus. Annu Rev Cell Dev Biol 25:113–132
2. Nakano A, Luini A (2010) Passage through the Golgi. Curr Opin Cell Biol 22:471–478
3. Nakano A (2022) The Golgi apparatus and its next-door neighbors. Front Cell Dev Biol 10:884360
4. Pantazopoulou A, Glick BS (2019) A kinetic view of membrane traffic pathways can transcend the classical view of Golgi compartments. Front Cell Dev Biol 7:153
5. Beznoussenko GV, Ragnini-Wilson A, Wilson C, Mironov AA (2016) Three-dimensional and immune electron microscopic analysis of the secretory pathway in *Saccharomyces cerevisiae*. Histochem Cell Biol 146:515–527
6. Preuss D, Mulholland J, Franzusoff A, Segev N, Botstein D (1992) Characterization of the *saccharomyces* Golgi complex through the cell cycle by immunoelectron microscopy. Mol Biol Cell 3:789–803
7. Rambourg A, Jackson CL, Clermont Y (2001) Three dimensional configuration of the secretory pathway and segregation of secretion granules in the yeast *Saccharomyces cerevisiae*. J Cell Sci 114:2231–2239
8. Kurokawa K, Nakano A (2020) Live-cell imaging by super-resolution confocal live imaging microscopy (SCLIM): simultaneous three-color and four-dimensional live cell imaging with high space and time resolution. BioProtoc 10:e3732
9. Tojima T, Miyashiro D, Kosugi Y, Nakano A (2023) Super-resolution live imaging of cargo traffic through the Golgi apparatus in mammalian cells. Methods Mol Biol 2557:127–140
10. Losev E, Reinke CA, Jellen J, Strongin DE, Bevis BJ, Glick BS (2006) Golgi maturation visualized in living yeast. Nature 441:1002–1006
11. Matsuura-Tokita K, Takeuchi M, Ichihara A, Mikuriya K, Nakano A (2006) Live imaging of yeast Golgi cisternal maturation. Nature 441:1007–1010

12. Anton-Plagaro C, Sanchez N, Valle R, Mulet JM, Duncan MC, Roncero C (2021) Exomer complex regulates protein traffic at the TGN through differential interactions with cargos and clathrin adaptor complexes. FASEB J 35:e21615
13. Casler JC, Glick BS (2020) A microscopy-based kinetic analysis of yeast vacuolar protein sorting. eLife 9:e56844
14. Casler JC, Papanikou E, Barrero JJ, Glick BS (2019) Maturation-driven transport and AP-1-dependent recycling of a secretory cargo in the Golgi. J Cell Biol 218:1582–1601
15. Casler JC, Johnson N, Krahn AH, Pantazopoulou A, Day KJ, Glick BS (2022) Clathrin adaptors mediate two sequential pathways of intra-Golgi recycling. J Cell Biol 221:e202103199
16. Daboussi L, Costaguta G, Payne GS (2012) Phosphoinositide-mediated clathrin adaptor progression at the *trans*-Golgi network. Nat Cell Biol 14:239–248
17. Day KJ, Casler JC, Glick BS (2018) Budding yeast has a minimal endomembrane system. Dev Cell 44:56–72
18. Highland CM, Fromme JC (2021) Arf1 directly recruits the Pik1-Frq1 PI4K complex to regulate the final stages of Golgi maturation. Mol Biol Cell 32:1064–1080
19. Ishii M, Suda Y, Kurokawa K, Nakano A (2016) COPI is essential for Golgi cisternal maturation and dynamics. J Cell Sci 129:3251–3261
20. Kim JJ, Lipatova Z, Majumdar U, Segev N (2016) Regulation of Golgi cisternal progression by Ypt/Rab GTPases. Dev Cell 36:440–452
21. Kurokawa K, Osakada H, Kojidani T, Waga M, Suda Y, Asakawa H, Haraguchi T, Nakano A (2019) Visualization of secretory cargo transport within the Golgi apparatus. J Cell Biol 218:1602–1618
22. Manzer KM, Fromme JC (2023) The Arf-GAP Age2 localizes to the late-Golgi via a conserved amphipathic helix. Mol Biol Cell 34:ar119
23. McDonold CM, Fromme JC (2014) Four GTPases differentially regulate the Sec7 Arf-GEF to direct traffic at the *trans*-golgi network. Dev Cell 30:759–767
24. Papanikou E, Day KJ, Austin J, Glick BS (2015) COPI selectively drives maturation of the early Golgi. eLife 4:e13232
25. Sardana R, Highland CM, Straight BE, Chavez CF, Fromme JC, Emr SD (2021) Golgi membrane protein Erd1 is essential for recycling a subset of Golgi glycosyltransferases. eLife 10:e70774
26. Suda Y, Kurokawa K, Hirata R, Nakano A (2013) Rab GAP cascade regulates dynamics of Ypt6 in the Golgi traffic. Proc Natl Acad Sci USA 110:18976–18981
27. Thomas LL, Fromme JC (2016) GTPase cross talk regulates TRAPPII activation of Rab11 homologues during vesicle biogenesis. J Cell Biol 215:499–513
28. Thomas LL, Joiner AMN, Fromme JC (2018) The TRAPPIII complex activates the GTPase Ypt1 (Rab1) in the secretory pathway. J Cell Biol 217:283–298
29. Thomas LL, van der Vegt SA, Fromme JC (2019) A steric gating mechanism dictates the substrate specificity of a Rab-GEF. Dev Cell 48:100–114
30. Thomas LL, Highland CM, Fromme JC (2021) Arf1 orchestrates Rab GTPase conversion at the *trans*-Golgi network. Mol Biol Cell 32:1104–1120
31. Tojima T, Suda Y, Ishii M, Kurokawa K, Nakano A (2019) Spatiotemporal dissection of the *trans*-Golgi network in budding yeast. J Cell Sci 132:jcs231159
32. Tojima T, Suda Y, Jin N, Kurokawa K, Nakano A (2024) Spatiotemporal dissection of the Golgi apparatus and the ER-Golgi intermediate compartment in budding yeast. eLife 13:e92900
33. Toshima JY, Tsukahara A, Nagano M, Tojima T, Siekhaus DE, Nakano A, Toshima J (2023) The yeast endocytic early/sorting compartment exists as an independent sub-compartment within the *trans*-Golgi network. eLife 12:e84850
34. Kurokawa K, Okamoto M, Nakano A (2014) Contact of *cis*-Golgi with ER exit sites executes cargo capture and delivery from the ER. Nat Commun 5:3653
35. Presley JF, Cole NB, Schroer TA, Hirschberg K, Zaal KJ, Lippincott-Schwartz J (1997) ER-to-Golgi transport visualized in living cells. Nature 389:81–85

36. Saraste J, Svensson K (1991) Distribution of the intermediate elements operating in ER to Golgi transport. J Cell Sci 100:415–430
37. Saraste J, Palade GE, Farquhar MG (1987) Antibodies to rat pancreas Golgi subfractions: identification of a 58-kD *cis*-Golgi protein. J Cell Biol 105:2021–2029
38. Scales SJ, Pepperkok R, Kreis TE (1997) Visualization of ER-to-Golgi transport in living cells reveals a sequential mode of action for COPII and COPI. Cell 90:1137–1148
39. Schweizer A, Fransen JA, Bächi T, Ginsel L, Hauri HP (1988) Identification, by a monoclonal antibody, of a 53-kD protein associated with a tubulo-vesicular compartment at the *cis*-side of the Golgi apparatus. J Cell Biol 107:1643–1653
40. Schweizer A, Fransen JA, Matter K, Kreis TE, Ginsel L, Hauri HP (1990) Identification of an intermediate compartment involved in protein transport from endoplasmic reticulum to Golgi apparatus. Eur J Cell Biol 53:185–196
41. Stornaiuolo M, Lotti LV, Borgese N, Torrisi MR, Mottola G, Martire G, Bonatti S (2003) KDEL and KKXX retrieval signals appended to the same reporter protein determine different trafficking between endoplasmic reticulum, intermediate compartment, and Golgi complex. Mol Biol Cell 14:889–902
42. Takagi J, Kimori Y, Shimada T, Hara-Nishimura I (2020) Dynamic capture and release of endoplasmic reticulum exit sites by Golgi stacks in Arabidopsis iScience 23:101265
43. Yang K, Liu M, Feng Z, Rojas M, Zhou L, Ke H, Pastor-Pareja JC (2021) ER exit sites in drosophila display abundant ER-Golgi vesicles and pearled tubes but no megacarriers. Cell Rep 36:109707
44. Ladinsky MS, Mastronarde DN, McIntosh JR, Howell KE, Staehelin LA (1999) Golgi structure in three dimensions: functional insights from the normal rat kidney cell. J Cell Biol 144:1135–1149
45. Mellman I, Simons K (1992) The Golgi complex: in vitro veritas? Cell 68:829–840
46. Ito Y, Uemura T, Shoda K, Fujimoto M, Ueda T, Nakano A (2012) *Cis*-Golgi proteins accumulate near the ER exit sites and act as the scaffold for Golgi regeneration after brefeldin a treatment in tobacco BY-2 cells. Mol Biol Cell 23:3203–3214
47. Ito Y, Uemura T, Nakano A (2018) The Golgi entry core compartment functions as a COPII-independent scaffold for ER-to-Golgi transport in plant cells. J Cell Sci 131:jcs203893
48. Ito Y, Boutté Y (2020) Differentiation of trafficking pathways at Golgi entry core compartments and post-Golgi subdomains. Front Plant Sci 11:609516
49. Weigel AV, Chang CL, Shtengel G, Xu CS, Hoffman DP, Freeman M, Iyer N, Aaron J, Khuon S, Bogovic J, Qiu W, Hess HF, Lippincott-Schwartz J (2021) ER-to-Golgi protein delivery through an interwoven, tubular network extending from ER. Cell 184:2412–2429
50. Raote I, Malhotra V (2021) Tunnels for protein export from the endoplasmic reticulum. Annu Rev Biochem 90:605–630
51. Cole NB, Sciaky N, Marotta A, Song J, Lippincott-Schwartz J (1996) Golgi dispersal during microtubule disruption: regeneration of Golgi stacks at peripheral endoplasmic reticulum exit sites. Mol Biol Cell 7:631–650
52. Hammond AT, Glick BS (2000) Dynamics of transitional endoplasmic reticulum sites in vertebrate cells. Mol Biol Cell 11:3013–3030
53. Chardin P, McCormick F (1999) Brefeldin A: the advantage of being uncompetitive. Cell 97:153–155
54. Fujiwara T, Oda K, Yokota S, Takatsuki A, Ikehara Y (1988) Brefeldin A causes disassembly of the Golgi complex and accumulation of secretory proteins in the endoplasmic reticulum. J Biol Chem 263:18545–18552
55. McCloud TG, Burns MP, Majadly FD, Muschik GM, Miller DA, Poole KK, Roach JM, Ross JT, Lebherz WB 3rd (1995) Production of brefeldin-A. J Ind Microbiol 15:5–9
56. Barzilay E, Ben-Califa N, Hirschberg K, Neumann D (2005) Uncoupling of brefeldin A-mediated coatomer protein complex-I dissociation from Golgi redistribution. Traffic 6:794–802

57. Scheel J, Pepperkok R, Lowe M, Griffiths G, Kreis TE (1997) Dissociation of coatomer from membranes is required for brefeldin A-induced transfer of Golgi enzymes to the endoplasmic reticulum. J Cell Biol 137:319–333
58. Füllekrug J, Sönnichsen B, Schäfer U, Nguyen Van P, Söling HD, Mieskes G (1997) Characterization of brefeldin A induced vesicular structures containing cycling proteins of the intermediate compartment/*cis*-Golgi network. FEBS Lett 404:75–81
59. Lippincott-Schwartz J, Donaldson JG, Schweizer A, Berger EG, Hauri HP, Yuan LC, Klausner RD (1990) Microtubule-dependent retrograde transport of proteins into the ER in the presence of brefeldin A suggests an ER recycling pathway. Cell 60:821–836
60. Nakamura N, Rabouille C, Watson R, Nilsson T, Hui N, Slusarewicz P, Kreis TE, Warren G (1995) Characterization of a *cis*-Golgi matrix protein, GM130. J Cell Biol 131:1715–1726
61. Seemann J, Jokitalo E, Pypaert M, Warren G (2000) Matrix proteins can generate the higher order architecture of the Golgi apparatus. Nature 407:1022–1026
62. Tang BL, Wong SH, Qi XL, Low SH, Hong W (1993) Molecular cloning, characterization, subcellular localization and dynamics of p23, the mammalian KDEL receptor. J Cell Biol 120:325–338
63. Ward TH, Polishchuk RS, Caplan S, Hirschberg K, Lippincott-Schwartz J (2001) Maintenance of Golgi structure and function depends on the integrity of ER export. J Cell Biol 155:557–570
64. Orlean P (2012) Architecture and biosynthesis of the *Saccharomyces cerevisiae* cell wall. Genetics 192:775–818
65. Yoko-o T, Wiggins CA, Stolz J, Peak-Chew SY, Munro S (2003) An N-acetylglucosaminyltransferase of the Golgi apparatus of the yeast *Saccharomyces cerevisiae* that can modify N-linked glycans. Glycobiology 13:581–589
66. Rizzo R, Russo D, Kurokawa K, Sahu P, Lombardi B, Supino D, Zhukovsky MA, Vocat A, Pothukuchi P, Kunnathully V, Capolupo L, Boncompain G, Vitagliano C, Zito Marino F, Aquino G, Montariello D, Henklein P, Mandrich L, Botti G, Clausen H, Mandel U, Yamaji T, Hanada K, Budillon A, Perez F, Parashuraman S, Hannun YA, Nakano A, Corda D, D'Angelo G, Luini A (2021) Golgi maturation-dependent glycoenzyme recycling controls glycosphingolipid biosynthesis and cell growth via GOLPH3. EMBO J 40:e107238
67. Popoff V, Adolf F, Brügger B, Wieland F (2011) COPI budding within the Golgi stack. Cold Spring Harb Perspect Biol 3:a005231
68. Jackson CL, Bouvet S (2014) Arfs at a glance. J Cell Sci 127:4103–4109
69. Wong-Dilworth L, Rodilla-Ramirez C, Fox E, Restel SD, Stockhammer A, Adarska P, Bottanelli F (2023) STED imaging of endogenously tagged ARF GTPases reveals their distinct nanoscale localizations. J Cell Biol 222:e202205107
70. Yahara N, Ueda T, Sato K, Nakano A (2001) Multiple roles of Arf1 GTPase in the yeast exocytic and endocytic pathways. Mol Biol Cell 12:221–238
71. Casanova JE (2007) Regulation of Arf activation: the Sec7 family of guanine nucleotide exchange factors. Traffic 8:1476–1485
72. Spang A, Shiba Y, Randazzo PA (2010) Arf GAPs: gatekeepers of vesicle generation. FEBS Lett 584:2646–2651
73. Yahara N, Sato K, Nakano A (2006) The Arf1p GTPase-activating protein Glo3p executes its regulatory function through a conserved repeat motif at its C-terminus. J Cell Sci 119:2604–2612
74. Bhave M, Papanikou E, Iyer P, Pandya K, Jain BK, Ganguly A, Sharma C, Pawar K, Austin J 2nd, Day KJ, Rossanese OW, Glick BS, Bhattacharyya D (2014) Golgi enlargement in Arf-depleted yeast cells is due to altered dynamics of cisternal maturation. J Cell Sci 127:250–257
75. Cowles CR, Odorizzi G, Payne GS, Emr SD (1997) The AP-3 adaptor complex is essential for cargo-selective transport to the yeast vacuole. Cell 91:109–118
76. Vowels JJ, Payne GS (1998) A dileucine-like sorting signal directs transport into an AP-3-dependent, clathrin-independent pathway to the yeast vacuole. EMBO J 17:2482–2493
77. De Matteis MA, Luini A (2008) Exiting the Golgi complex. Nat Rev Mol Cell Biol 9:273–284

78. Guo Y, Sirkis DW, Schekman R (2014) Protein sorting at the *trans*-Golgi network. Annu Rev Cell Dev Biol 30:169–206
79. Paczkowski JE, Richardson BC, Fromme JC (2015) Cargo adaptors: structures illuminate mechanisms regulating vesicle biogenesis. Trends Cell Biol 25:408–416
80. Pavelka M, Ellinger A, Debbage P, Loewe C, Vetterlein M, Roth J (1998) Endocytic routes to the Golgi apparatus. Histochem Cell Biol 109:555–570
81. Malsam J, Söllner TH (2011) Organization of SNAREs within the Golgi stack. Cold Spring Harb Perspect Biol 3:a005249
82. Bonifacino JS, Rojas R (2006) Retrograde transport from endosomes to the *trans*-Golgi network. Nat Rev Mol Cell Biol 7:568–579
83. Ito Y, Toyooka K, Fujimoto M, Ueda T, Uemura T, Nakano A (2017) The *trans*-Golgi network and the Golgi stacks behave independently during regeneration after Brefeldin a treatment in tobacco BY-2 cells. Plant Cell Physiol 58:811–821
84. Papanikou E, Glick BS (2014) Golgi compartmentation and identity. Curr Opin Cell Biol 29:74–81
85. Behnia R, Panic B, Whyte JR, Munro S (2004) Targeting of the Arf-like GTPase Arl3p to the Golgi requires N-terminal acetylation and the membrane protein Sys1p. Nat Cell Biol 6:405–413
86. Graham TR (2004) Flippases and vesicle-mediated protein transport. Trends Cell Biol 14:670–677
87. Ishida M, Bonifacino JS (2019) ARFRP1 functions upstream of ARL1 and ARL5 to coordinate recruitment of distinct tethering factors to the *trans*-Golgi network. J Cell Biol 218:3681–3696
88. Setty SR, Strochlic TI, Tong AH, Boone C, Burd CG (2004) Golgi targeting of ARF-like GTPase Arl3p requires its Nalpha-acetylation and the integral membrane protein Sys1p. Nat Cell Biol 6:414–419
89. Yu CJ, Lee FJ (2017) Multiple activities of Arl1 GTPase in the *trans*-Golgi network. J Cell Sci 130:1691–1699
90. Tsai PC, Hsu JW, Liu YW, Chen KY, Lee FJ (2013) Arl1p regulates spatial membrane organization at the *trans*-Golgi network through interaction with Arf-GEF Gea2p and flippase Drs2p. Proc Natl Acad Sci USA 110:E668–E677
91. Bensen ES, Yeung BG, Payne GS (2001) Ric1p and the Ypt6p GTPase function in a common pathway required for localization of *trans*-Golgi network membrane proteins. Mol Biol Cell 12:13–26
92. Siniossoglou S, Peak-Chew SY, Pelham HR (2000) Ric1p and Rgp1p form a complex that catalyses nucleotide exchange on Ypt6p. EMBO J 19:4885–4894
93. Tsukada M, Gallwitz D (1996) Isolation and characterization of SYS genes from yeast, multicopy suppressors of the functional loss of the transport GTPase Ypt6p. J Cell Sci 109:2471–2481
94. Conibear E, Cleck JN, Stevens TH (2003) Vps51p mediates the association of the GARP (Vps52/53/54) complex with the late Golgi t-SNARE Tlg1p. Mol Biol Cell 14:1610–1623
95. Siniossoglou S, Pelham HR (2002) Vps51p links the VFT complex to the SNARE Tlg1p. J Biol Chem 277:48318–48324
96. Sclafani A, Chen S, Rivera-Molina F, Reinisch K, Novick P, Ferro-Novick S (2010) Establishing a role for the GTPase Ypt1p at the late Golgi. Traffic 11:520–532
97. Gustafson MA, Fromme JC (2017) Regulation of Arf activation occurs via distinct mechanisms at early and late Golgi compartments. Mol Biol Cell 28:3660–3671
98. Abeliovich H, Grote E, Novick P, Ferro-Novick S (1998) Tlg2p, a yeast syntaxin homolog that resides on the Golgi and endocytic structures. J Biol Chem 273:11719–11727
99. Chen Y, Gan BQ, Tang BL (2010) Syntaxin 16: unraveling cellular physiology through a ubiquitous SNARE molecule. J Cell Physiol 225:326–332

100. Chow CM, Neto H, Foucart C, Moore I (2008) Rab-A2 and Rab-A3 GTPases define a *trans*-golgi endosomal membrane domain in Arabidopsis that contributes substantially to the cell plate. Plant Cell 20:101–123
101. Dettmer J, Hong-Hermesdorf A, Stierhof YD, Schumacher K (2006) Vacuolar H^+-ATPase activity is required for endocytic and secretory trafficking in Arabidopsis. Plant Cell 18:715–730
102. Lam SK, Tse YC, Robinson DG, Jiang L (2007) Tracking down the elusive early endosome. Trends Plant Sci 12:497–505
103. Redding K, Seeger M, Payne GS, Fuller RS (1996) The effects of clathrin inactivation on localization of Kex2 protease are independent of the TGN localization signal in the cytosolic tail of Kex2p. Mol Biol Cell 7:1667–1677
104. Takahashi S, Nakagawa T, Banno T, Watanabe T, Murakami K, Nakayama K (1995) Localization of furin to the *trans*-Golgi network and recycling from the cell surface involves Ser and Tyr residues within the cytoplasmic domain. J Biol Chem 270:28397–28401
105. Fuller RS, Sterne RE, Thorner J (1988) Enzymes required for yeast prohormone processing. Annu Rev Physiol 50:345–362
106. Robinson MS, Antrobus R, Sanger A, Davies AK, Gershlick DC (2024) The role of the AP-1 adaptor complex in outgoing and incoming membrane traffic. J Cell Biol 223:e202310071
107. Copic A, Starr TL, Schekman R (2007) Ent3p and Ent5p exhibit cargo-specific functions in trafficking proteins between the *trans*-Golgi network and the endosomes in yeast. Mol Biol Cell 18:1803–1815
108. Costaguta G, Duncan MC, Fernández GE, Huang GH, Payne GS (2006) Distinct roles for TGN/endosome epsin-like adaptors Ent3p and Ent5p. Mol Biol Cell 17:3907–3920
109. Hung CW, Duncan MC (2016) Clathrin binding by the adaptor Ent5 promotes late stages of clathrin coat maturation. Mol Biol Cell 27:1143–1153
110. Wang YJ, Wang J, Sun HQ, Martinez M, Sun YX, Macia E, Kirchhausen T, Albanesi JP, Roth MG, Yin HL (2003) Phosphatidylinositol 4 phosphate regulates targeting of clathrin adaptor AP-1 complexes to the Golgi. Cell 114:299–310
111. Welz T, Wellbourne-Wood J, Kerkhoff E (2014) Orchestration of cell surface proteins by Rab11. Trends Cell Biol 24:407–415
112. Jedd G, Mulholland J, Segev N (1997) Two new Ypt GTPases are required for exit from the yeast *trans*-Golgi compartment. J Cell Biol 137:563–580
113. Taguchi T (2013) Emerging roles of recycling endosomes. J Biochem 153:505–510
114. Fujii S, Kurokawa K, Inaba R, Hiramatsu N, Tago T, Nakamura Y, Nakano A, Satoh T, Satoh AK (2020) Recycling endosomes attach to the *trans*-side of Golgi stacks in Drosophila and mammalian cells. J Cell Sci 133:jcs236935
115. Miyashiro D, Tojima T, Nakano A (2024) Extremely high spatiotemporal resolution microscopy for live cell imaging by single photon counting, noise elimination, and a novel restoration algorithm based on probability calculation. Front Cell Dev Biol 12:1324906
116. Tie HC, Ludwig A, Sandin S, Lu L (2018) The spatial separation of processing and transport functions to the interior and periphery of the Golgi stack. eLife 7:e41301
117. Day KJ, Staehelin LA, Glick BS (2013) A three-stage model of Golgi structure and function. Histochem Cell Biol 140:239–249

Chapter 13
Rethinking the Yeast Endomembrane System

Benjamin S. Glick

Abstract The budding yeast *Saccharomyces cerevisiae* is tremendously useful for revealing basic principles of cell biology. An example is the characterization of the endomembrane system, which consists of secretory, endocytic, and autophagic compartments that exchange material by membrane traffic pathways and undergo dynamic transformations. Yeast studies of this topic continue to uncover components and molecular mechanisms. However, questions persist about high-level features of the yeast endomembrane system, including the identities and life cycles of endocytic compartments and the traffic pathways followed by resident proteins of the Golgi apparatus. Time-lapse imaging of endosome and Golgi compartments points to a new synthesis that revises several long-standing assumptions. This updated picture has implications for understanding conserved aspects of the endomembrane system.

Keywords Yeast · Secretory pathway · Endocytic pathway · Golgi · Cisternal maturation · Endosomes · Endosome maturation · AP-1

Introduction

Yeast research has deeply influenced our view of the endomembrane system. Among the many insights obtained from yeast are identification of core components of the ER translocation machinery [37], discovery of the COPII vesicles that mediate ER export [8], identification of receptors for ER export and for retrieval of escaped ER proteins [13, 94, 107, 111], discovery of pathways and components involved in

B. S. Glick (✉)
Department of Molecular Genetics and Cell Biology, University of Chicago, Chicago, IL, USA
e-mail: bsglick@uchicago.edu

A. Nakano, J. Saraste (eds.), *The Golgi Network, Volume I*, Subcellular Biochemistry 110, https://doi.org/10.1007/978-3-032-06936-8_13

autophagy [78], discovery of the role of small GTPases in regulating membrane traffic [106, 110], discovery of the involvement of phosphoinositides in protein traffic [108], identification of components involved in Golgi-to-endosome traffic [6, 104], and identification of the ESCRT machinery that generates intraluminal vesicles in endosomes [3, 4, 62].

Looking ahead, *S. cerevisiae* is poised to yield a rich understanding of the Golgi apparatus. Golgi cisternae in *S. cerevisiae* are not organized into stacks but instead are dispersed throughout the cytoplasm [45, 91, 97]. Nevertheless, components of the Golgi traffic machinery are conserved between yeast and other eukaryotes, and yeast exhibits a polarized distribution of resident Golgi proteins resembling that seen in mammalian cells [40, 91]. Two considerations suggest that the core operating principles of the Golgi are the same in *S. cerevisiae* as in organisms with stacked cisternae. First, Golgi stacking has been lost in multiple lineages during eukaryotic evolution, indicating that this feature is dispensable [80]. Second, the budding yeast *Pichia pastoris* has a stacked Golgi but otherwise resembles *S. cerevisiae* [79, 91], and Golgi function in these two closely related organisms is likely similar. In support of a unified view of Golgi operation, studies of algal and mammalian cells provided indirect evidence that Golgi cisternae progressively mature [11, 16], and then time-lapse microscopy of individual cisternae in *S. cerevisiae* provided direct evidence for this phenomenon [72, 75]. To visualize cisternal maturation in yeast, two resident Golgi proteins are fused to different fluorescent tags, and a cisterna is tracked to quantify the relative arrival and departure times of the tagged proteins (Fig. 13.1). Each resident Golgi protein has a characteristic kinetic signature during the lifetime of a maturing cisterna. Based on this finding, Golgi polarity has a natural explanation: the kinetic signature of a given resident Golgi protein determines how that protein is distributed across the Golgi [89].

Despite many advances, there are gaps in our knowledge about how proteins travel between endomembrane system compartments and about how those compartments maintain their identities. Once again, yeast can reveal conserved principles. The simplicity of the yeast endomembrane system confers some advantages. For example, although mammalian lysosomes and late endosomes are hard to distinguish [109], the yeast lysosome-like vacuole is clearly distinct from the late-endosome-like prevacuolar endosome (PVE) compartments that associate with the vacuolar surface [95]. A less straightforward aspect of yeast is the relationship between the Golgi and the endocytic pathway. One source of ambiguity is that the dispersed nature of the yeast Golgi limits the use of morphological landmarks. Notably, position alone cannot indicate whether a compartment is a Golgi cisterna or an endosome. Another source of ambiguity is that Golgi dynamics have implications that were not historically appreciated. Cisternal maturation renews the entire yeast Golgi within five minutes, so experimentally observed changes that require tens of minutes may not reflect the major pathways of Golgi traffic. Moreover, static images of the Golgi are misleading because a given cisterna changes composition. Figure 13.2 illustrates that if two resident Golgi proteins have partially overlapping kinetic signatures, then at a fixed time point, a cisterna may contain either the earlier protein or the later protein or both proteins. When this factor is overlooked,

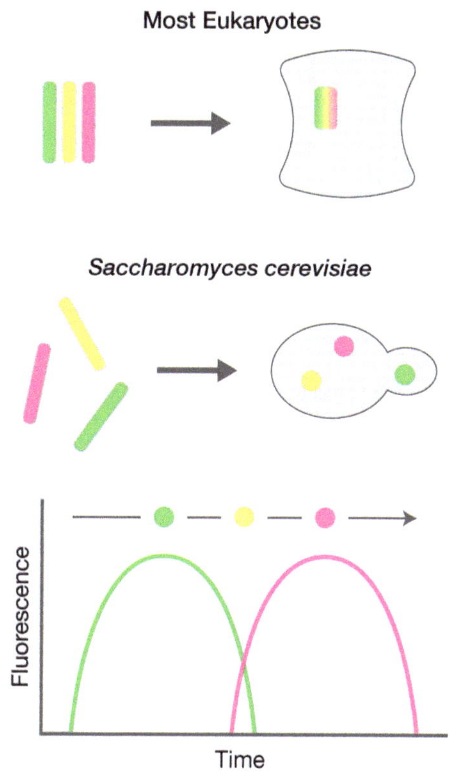

Fig. 13.1 Visualizing cisternal maturation in *S. cerevisiae*. In most eukaryotes (top), Golgi cisternae are stacked and cannot be optically resolved, but in *S. cerevisiae* (middle), Golgi cisternae are not stacked and can be resolved as individual fluorescent spots. If an early Golgi protein is fused to a green fluorescent tag while a late Golgi protein is fused to a red fluorescent tag, then one set of spots will glow green, a second set of spots will glow red, and a third set of spots will glow yellow because both red and green fluorescence are present. When an individual spot is imaged over time to track cisternal maturation (bottom), the green fluorescence will rise and fall and then the red fluorescence will rise and fall as each tagged protein arrives and then departs

colocalization analysis of Golgi proteins can create an illusion of multiple independent compartments. Such experimental barriers must be overcome to harness the full potential of the yeast system. We need a clear picture of which compartments exist, and of how and when those compartments exchange membranes.

The purpose here is to take a fresh look at the secretory and endocytic pathways of yeast. Autophagy will not be covered, except to state that this subject presents additional fascinating puzzles [55]. Some established interpretations about the yeast endomembrane system are questionable due not only to technical issues, but also to a conceptual framework derived from mammalian cell studies that are themselves imperfect. The diagram of yeast cell compartments and pathways shown in Fig. 13.3 represents a view that was a general consensus until recently and that persists today. Figure 13.3 includes findings accepted by most yeast researchers, namely, the role

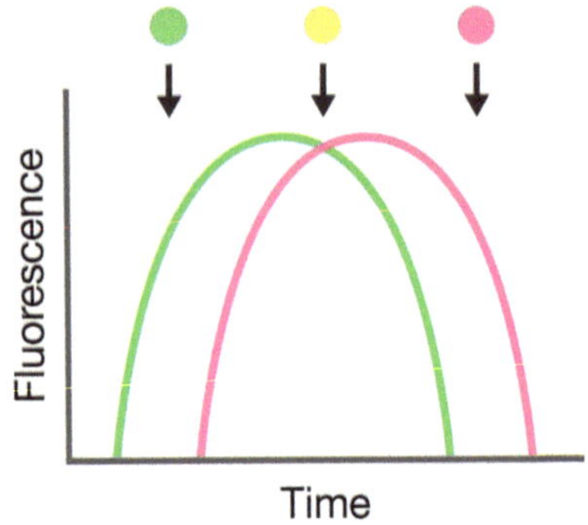

Fig. 13.2 Interpretation of static colocalization data for maturing Golgi cisternae labeled with two resident proteins that show extensive but incomplete kinetic overlap. The fluorescent labeling scheme is as in Fig. 13.1. When a snapshot is taken by fluorescence microscopy, an individual cisterna may contain only the tagged earlier Golgi protein (green circle), or only the tagged later Golgi protein (red circle), or both tagged proteins (yellow circle). Those labeling patterns represent different maturation stages rather than independent compartments

of COPII vesicles in transporting proteins from the endoplasmic reticulum to the Golgi, the role of COPI vesicles in recycling proteins from the Golgi to the ER, and the role of COPI vesicles in recycling proteins within the Golgi during cisternal maturation [9, 60, 65, 90]. However, other aspects of Fig. 13.3 are problematic. Discussed below are arguments for considering the alternative diagram shown in Fig. 13.4. Because the evidence is still incomplete, the intent here is not to resolve every debate, but rather to examine long-standing ideas and to evaluate the broader implications.

Timing of Departure from the Golgi to PVE Compartments and the Vacuole

It has long been assumed that during passage through the Golgi, proteins destined for the plasma membrane travel in the company of proteins destined for endosomes and lysosomes, with separation occurring in a final sorting stage at the *trans*-Golgi network (TGN) [36, 50] (Fig. 13.3). Although the TGN is not precisely defined, this part of the Golgi is generally understood to be marked by clathrin and its adaptors [101]. In mammalian cells, clathrin associates only with the last cisterna of a Golgi stack [66], but in yeast, clathrin arrives about halfway through the maturation process [34, 63, 116]. The relatively long TGN phase in yeast allows for a nuanced analysis of when proteins depart from the Golgi for PVE compartments or the vacuole. There are two relevant pathways: Golgi-to-PVE traffic dependent on the GGA and Ent3 clathrin adaptors, and Golgi-to-vacuole traffic dependent on the AP-3 complex [81].

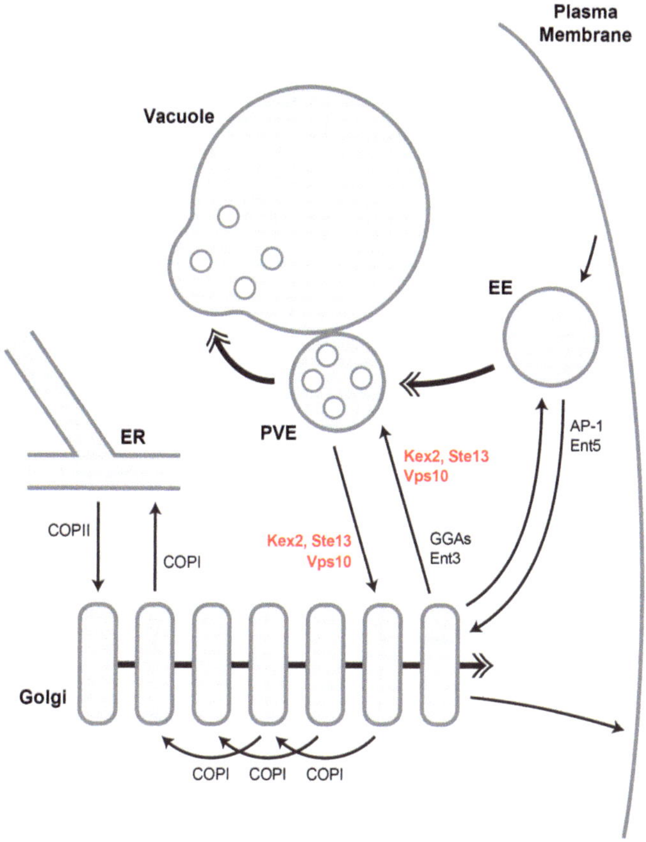

Fig. 13.3 Conventional view of the yeast secretory and endocytic pathways. Single-headed arrows indicate pathways mediated by vesicles or other transport carriers, and double-headed arrows indicate pathways that involve transformation of a compartment. The following concepts are depicted. COPII vesicles bud from the endoplasmic reticulum (ER) and deliver biosynthetic cargo proteins to a nascent Golgi cisterna, which progressively matures. During cisternal maturation, COPI recycles certain components to the ER, and COPI also recycles a subset of the resident Golgi transmembrane proteins in a series of retrograde intra-Golgi pathways. At the terminal stage of Golgi maturation, biosynthetic cargo proteins and resident TGN transmembrane proteins are sorted into one of three pathways: transport to prevacuolar endosome (PVE) compartments with the aid of GGAs and Ent3, transport to early endosomes (EE), and transport to the plasma membrane in secretory vesicles. Not shown is AP-3-dependent delivery of proteins from the TGN to the vacuole. Early endosomes are represented as existing compartments that receive traffic from the plasma membrane and TGN, although they could also arise from homotypic fusion of endocytic vesicles or from maturation of the TGN [92]. The postulated traffic pathway from the TGN to early endosomes has not been characterized, but it has been proposed to involve AP-1 [71]. AP-1 and Ent5 mediate retrograde traffic from early endosomes to the TGN. Early endosomes mature into PVE compartments, which then fuse with the vacuole to deliver their contents, including intraluminal vesicles. Kex2 and Ste13 travel together with Vps10 from the TGN to PVE compartments, and then all three proteins return to the Golgi. Retrieval from PVE compartments involves sorting nexins [10, 73]

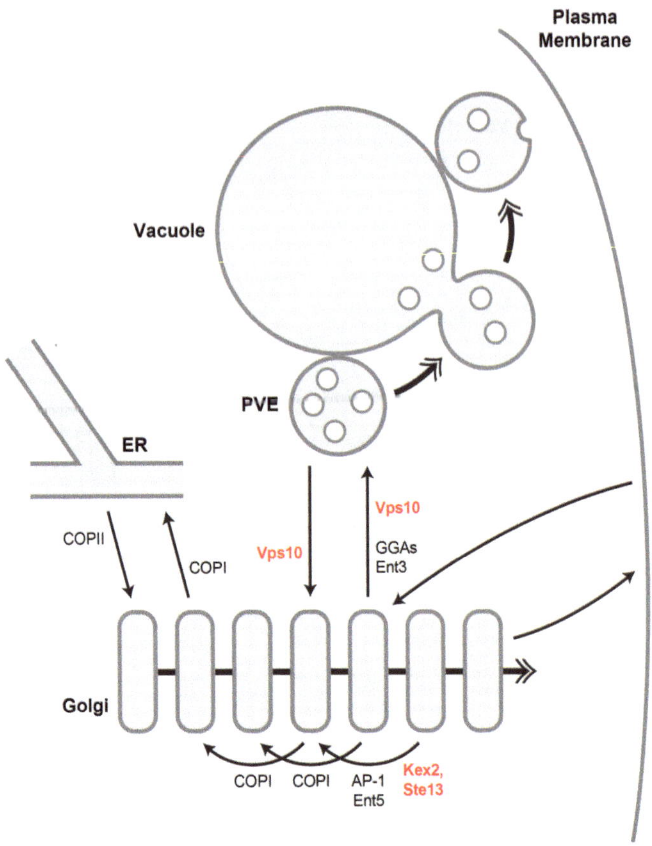

Fig. 13.4 Updated view of the yeast secretory and endocytic pathways. Compared to Fig. 13.3, the symbols and abbreviations are the same, but five concepts are different. (1) Transport from the Golgi to PVE compartments, and AP-3-dependent transport from the Golgi to the vacuole (not shown), begin about halfway through cisternal maturation, well before biosynthetic cargo proteins are sorted into secretory vesicles. (2) Endocytic vesicles fuse with maturing Golgi cisternae, which function as early endosomes. This fusion of endocytic vesicles occurs about halfway through cisternal maturation. Internalized protein cargoes then move rapidly from the Golgi to PVE compartments. (3) PVE compartments are long-lived organelles. They undergo transient kiss-and-run fusion events with the vacuole, delivering a portion of their contents each time. (4) AP-1 and Ent5 act at the late Golgi, where they mediate recycling of resident TGN transmembrane proteins. (5) Unlike Vps10, which cycles between the Golgi and PVE compartments, Kex2 and Ste13 recycle within the Golgi in the AP-1/Ent5 pathway

The GGA adaptors in yeast are Gga1 and Gga2, with Gga2 being the major isoform [15, 32, 132]. GGAs transport the vacuolar hydrolase receptor Vps10 and its bound ligands from the Golgi to PVE compartments [43]. After Vps10 releases its ligands, this receptor is retrieved from PVE compartments to Golgi cisternae, where it is available for another round of binding to vacuolar hydrolases [17, 23, 24]. The timing of Vps10 departure from the Golgi should depend on the timing of GGA arrival. Gga2 arrives at a cisterna about halfway through maturation, around the same time as the Arf guanine nucleotide exchange factor Sec7 [23, 34]. As expected, when Gga2 arrives, Vps10 and its ligands begin to depart from a maturing cisterna [23, 24]. For example, an artificial fluorescent vacuolar cargo containing a Vps10 recognition signal begins to depart from a maturing cisterna about halfway through maturation, whereas an artificial fluorescent secretory cargo destined for the extracellular space persists in a cisterna until the very end of maturation [23, 25]. GGAs operate in conjunction with the Ent3 adaptor, which arrives at a maturing cisterna at the same time as Gga2 [34, 81]. Ent3 carries specific cargoes such as the SNARE Vti1 to PVE compartments [28]. Presumably, the kinetics of Ent3-dependent transport are similar to those of GGA-dependent transport. The net result is that PVE-targeted cargoes depart from a maturing cisterna at the beginning of the TGN phase whereas plasma membrane-targeted cargoes depart at the end of the TGN phase (Fig. 13.4).

Although mammalian cell data are more limited, a similar temporal separation may exist, with sorting to endosomes taking place before sorting to the plasma membrane. There is evidence that a mammalian GGA adaptor is active early in the TGN phase [33]. Moreover, at early time points during passage through the mammalian Golgi, proteins destined for endosomes segregate from proteins destined for the plasma membrane [27].

AP-3 transports certain transmembrane proteins directly from the Golgi to the yeast vacuole or the mammalian lysosome [87]. AP-3 is related to the AP-1 and AP-2 clathrin adaptors, but AP-3 seems to operate independently of clathrin [126]. Yeast AP-3 arrives at a Golgi cisterna shortly before Sec7 [35, 117]. Thus, the departure of AP-3-dependent cargoes probably occurs about halfway through yeast Golgi maturation.

What might be the benefits of a relatively early departure from the Golgi of proteins destined for endosomes or vacuoles/lysosomes? A hint comes from the finding that in mutants defective for sorting of yeast vacuolar hydrolases or mammalian lysosomal hydrolases, those hydrolases are secreted [6, 64, 104]. The same type of missorting could presumably occur in wild-type cells if hydrolases were present in the TGN during secretory vesicle biogenesis. By carrying out GGA-dependent sorting before the terminal stage of cisternal maturation, the cell may reduce the likelihood that hydrolases will escape to the plasma membrane. A more speculative idea is that additional fidelity might be provided by the AP-1 clathrin adaptor (see below), which acts downstream of the GGA and Ent3 adaptors [23, 24, 33, 34]. If a newly synthesized hydrolase initially escapes GGA-dependent sorting and remains in a maturing cisterna, then AP-1-dependent recycling to an earlier TGN stage [25] would potentially give the hydrolase another chance to be captured by the GGA machinery. These considerations suggest that the TGN phase of Golgi maturation can be divided into an early stage specialized for sorting to endosomes or vacuoles/lysosomes, followed by a late stage specialized for sorting to the plasma membrane.

Relationship Between Early Endosomes and the Golgi

The endocytic organelles of mammalian cells have been intensively studied [59, 109]. These organelles can be classified into three types: early endosomes, late endosomes, and recycling endosomes. Endocytic cargoes arrive at early endosomes, which are thought to mature into late endosomes, which in turn are thought to fuse with lysosomes. Some endocytic cargo molecules travel from early endosomes to recycling endosomes before returning to the cell surface. Yeast cells were assumed to have a similar set of endosome types [74, 92]. For example, internalization of the dye FM 4–64 initially labels punctate compartments that were designated early endosomes [57, 122]. After longer chase times, FM 4–64 labels vacuole-associated PVE compartments and then labels the vacuole [122]. Internalized FM 4–64 can also recycle to the plasma membrane, so some of the dye molecules reach a compartment equivalent to a recycling endosome [128]. Thus, in functional terms, the yeast endosomal system has the expected properties.

Yet the endosomal system in yeast seems to be simpler than in mammalian cells. Yeast early endosomes were described as sharing properties with the TGN [48, 61, 121], and eventually it was shown that early endosomes marked by newly internalized FM 4–64 also label with the TGN marker Sec7, suggesting that yeast early endosomes are identical to the TGN [14]. A proposed marker for yeast early endosomes was the Qc-SNARE Tlg1 [69, 96, 129], but kinetic analysis indicated that Tlg1 resides exclusively in the Golgi, where it arrives and departs shortly before Sec7 [35]. Other potential markers for different classes of yeast endosomes were shown to label either TGN cisternae or PVE compartments (or both in the case of Vps10) [35]. The implication is that the TGN serves as an early endosome in yeast, as it does in plant cells [35, 38, 124]. A direct test of this hypothesis was to image FM 4–64-labeled endocytic vesicles, which were seen to fuse directly with TGN cisternae but not with PVE compartments [35]. Fusion of endocytic vesicles with a maturing cisterna occurs at about the same time that Sec7 is arriving. Therefore, a yeast Golgi cisterna acquires early endosome characteristics about halfway through maturation.

As an alternative to the bulk membrane dye FM 4–64, transmembrane proteins at the yeast cell surface can be used to track endocytosis. For example, the Ste2 receptor is internalized after it binds fluorescently tagged derivatives of the pheromone α-factor [2, 119]. FM 4–64 and α-factor-bound Ste2 behave differently after endocytosis. Internalized FM 4–64 gradually disappears from the TGN and appears in PVE compartments, achieving maximal labeling of PVE compartments after about 10 min [35]. By contrast, internalized α-factor-bound Ste2 becomes detectable in PVE compartments within 1 min, achieving maximal labeling of PVE compartments after about 5 min. At early time points after internalization, α-factor-bound Ste2 shows transient labeling of the TGN as marked by Sec7 [35]. Detection of this TGN intermediate is enhanced by marking the TGN instead with Tlg1 or Tlg2, which are endocytic Q-SNAREs that arrive at a maturing cisterna shortly before Sec7 [35, 116, 120]. The residence of α-factor-bound Ste2 in the TGN can be prolonged by deletion of Gga1 and Gga2 [35, 120]. Therefore, two independent studies provided evidence that the TGN serves as an early endosome for α-factor-bound Ste2 [35, 120]. The combined data favor the following scenario. Endocytic vesicles arrive at a maturing cisterna at the beginning of the TGN phase. If the endocytic vesicles carry FM 4–64, then this dye partitions into three pathways: GGA/Ent3-dependent transport to PVE compartments, recycling to the plasma membrane in secretory vesicles, and intra-Golgi recycling by the AP-1/Ent5 pathway described below. Intra-Golgi recycling causes FM 4–64 to persist in the TGN for many minutes [24], but an increasing fraction of the dye molecules move to PVE compartments or recycle to the cell surface. If the endocytic vesicles contain α-factor-bound Ste2, then this receptor undergoes signal-mediated and GGA-dependent transport from the TGN to PVE compartments. As a result, α-factor-bound Ste2 exits the TGN rapidly, and it becomes concentrated in PVE compartments much sooner than FM 4–64. In addition to α-factor-bound Ste2, other internalized plasma membrane proteins probably also transit through the TGN on their way to PVE compartments [12, 22].

A challenge to this view comes from Mup1, a yeast plasma membrane permease that responds to methionine addition by undergoing endocytosis and eventual degradation in the vacuole [70, 77]. When Mup1 is tracked after endocytosis, it appears quickly in PVE compartments and shows minimal colocalization with Sec7, even after deletion of Gga1 and Gga2 [35, 67]. One possibility is that endocytic vesicles containing Mup1 fuse directly with a class of PVE compartments that function as early endosomes [67]. If so, then Mup1-containing vesicles must be a small subset of the endocytic vesicles, because endocytic vesicles should all carry FM 4–64 and yet little FM 4–64 is delivered to PVE compartments soon after internalization [35]. A more likely possibility is that endocytic vesicles carrying Mup1 fuse with TGN cisternae, and then Mup1 departs rapidly for PVE compartments in a GGA-independent manner, perhaps with the aid of Ent3. Further investigation will be needed to settle this issue.

Based on the existence of endocytic cargoes such as Mup1 and α-factor-bound Ste2, the timing of endocytic vesicle arrival during cisternal maturation makes sense according to the following logic. PVE compartments are an intermediate stop en route to degradation of endocytic cargoes in the vacuole [17, 123]. Because the GGA/Ent3-dependent Golgi-to-PVE shuttle operates early in the TGN phase, endocytic vesicles need to fuse with a maturing cisterna by the beginning of the TGN phase so that their cargoes can catch the Golgi-to-PVE shuttle. Thus, endocytic cargoes destined for degradation arrive at a Golgi cisterna and then move almost immediately to PVE compartments (Fig. 13.4).

This revised view of the yeast endocytic pathway is not universally accepted, and some researchers continue to believe that yeast early endosomes are distinct from the Golgi [47, 67, 93]. But as described in the next section, additional results have challenged long-standing ideas about yeast endosomes.

Origin and Fate of PVE Compartments

PVE compartments are multivesicular endosomes [86, 95], and they were thought to be generated by a process analogous to the mammalian conversion of early endosomes to multivesicular late endosomes [2, 59, 99]. Yeast cells contain a Mon1-Ccz1 complex related to the mammalian complex that drives maturation of early endosomes labeled with the Rab5 GTPase to late endosomes labeled with the Rab7 GTPase [68]. The expectation was that a yeast early endosome would mature into a PVE compartment, which would be consumed by fusion with the vacuole [5] (Fig. 13.3). But if yeast cells lack distinct early endosomes, then where do PVE compartments come from?

This question can be addressed by time-lapse imaging of yeast endosomal compartments. For example, the endocytic cargo α-factor was tracked relative to various endosomal markers [2]. Those markers all showed strong overlap, and even though the results were interpreted in the framework of endosomal maturation, no maturation was detected. This issue was revisited by performing 4D microscopy of PVE compartments [35], which were labeled by tagging the Vps8 subunit of the CORVET tether [2]. Unlike Golgi cisternae, which are readily observed to mature, PVE compartments seem to be indefinitely long-lived. The results were taken to mean that PVE compartments are not in fact transient intermediates in a maturation pathway, but instead are stable organelles (Fig. 13.4). PVE compartments undergo fission and homotypic fusion at similar rates, and the proliferation of PVE compartments in growing cells can be understood by assuming that fission slightly outpaces homotypic fusion [35].

Further evidence that PVE compartments do not arise from early endosomes came from protein localization analysis [35]. Sec7 was used as the TGN marker, and Vps8 was used as the PVE marker. Various endosomal proteins label either TGN cisternae or PVE compartments as described above, and none of the tested proteins label separate early endosomes. Particularly revealing are yeast orthologs of Rab5, the canonical early endosome marker in mammalian cells [131]. Unstressed yeast cells contain the Rab5 orthologs Vps21 and Ypt52 [83, 112]. Both proteins colocalize with Vps8 at PVE compartments, suggesting that PVE compartments combine properties of mammalian early and late endosomes [35, 118].

Stable PVE compartments must have a way to transfer luminal proteins and intraluminal vesicles to the vacuole [5, 17]. Because full fusion with the vacuole apparently does not happen, a likely alternative is transient "kiss-and-run" fusion [35]. Analogous kiss-and-run events are seen during the fusion of mammalian late endosomes with lysosomes [19, 20, 105]. The kiss-and-run hypothesis was tested by tracking a soluble vacuolar cargo as it moved from PVE compartments to the vacuole [23]. Delivery of the cargo to the vacuole occurred in multiple steps rather than all at once, consistent with transient partial fusion events. This interpretation is supported by a striking electron tomogram from the Odorizzi lab, accessible at https://elifesciences.org/articles/56844/figures-fig 8s2, showing apparent partial fusion of PVE compartments with the vacuole [76].

To preserve the concept of yeast endosomal maturation, one could postulate that yeast TGN cisternae function like mammalian early endosomes by maturing into PVE compartments. So far, there is no evidence for such a process. However, PVE compartments do receive material from the TGN. Cargoes such as newly synthesized vacuolar hydrolases and internalized plasma membrane proteins travel from the TGN to PVE compartments. Moreover, an intriguing study reported that Vps9, the guanine nucleotide exchange factor for Vps21, is first recruited to the TGN and then delivered to PVE compartments [82]. This result suggests that the TGN has an incompletely understood role in regulating PVE function.

Additional questions remain. It has been proposed that PVE compartments can be divided into classes with distinct properties and partially overlapping protein compositions [47]. The relationship between these different classes of PVE compartments is still obscure. Despite these uncertainties, the data indicate that Vps8-containing PVE compartments are much more long-lived than originally expected, and that the concept of a linear pathway in which transient early endosomes mature into transient PVE compartments needs to be amended.

Roles of the AP-1 and Ent5 Clathrin Adaptors

The heterotetrameric AP-1 clathrin adaptor was identified decades ago, but its function has been mysterious [41, 102]. Mammalian AP-1 localizes to the TGN and to recycling endosomes [51], which are TGN-related compartments [46, 49]. Similarly, yeast AP-1 is present at the final stage of Golgi maturation [24, 25, 34, 35]. Mammalian AP-1 was originally thought to mediate the transport of lysosomal hydrolases from the TGN to endosomes. However, the story became more complex with the discovery that the GGA clathrin adaptors mediate the transport of hydrolases from the TGN to endosomes in both mammalian and yeast cells [15, 32, 54, 98, 132]. In the case of mammalian cells, AP-1 and GGAs were proposed to function together in TGN-to-endosome traffic [39], but in the case of yeast, AP-1 was shown to be dispensable for TGN-to-PVE traffic [15, 23, 32, 56, 130]. These findings raise the question: which cargoes are transported by AP-1 vesicles and to which destination?

Insights came from experiments implicating yeast AP-1 in the Golgi retention of certain transmembrane proteins, including the chitin synthase Chs3 [121], the cell fusion protein Fus1 [7], and the prion-like domain-containing protein Pin2 [100]. The interpretation was that AP-1 is needed to recycle proteins from early endosomes to the TGN. Thus, AP-1 vesicles were proposed to mediate retrograde endosome-to-TGN traffic rather than anterograde TGN-to-endosome traffic [121] (Fig. 13.3). But the endosome-to-TGN model was challenged by evidence that yeast cells lack distinct early endosomes and that yeast AP-1 resides exclusively at the Golgi [35]. In retrospect, the earlier data can be reinterpreted by postulating that yeast AP-1 mediates retrograde intra-Golgi recycling from older to younger cisternae (Fig. 13.4).

The transport of Chs3, Fus1, and Pin2 to the plasma membrane requires a fungal-specific protein complex known as exomer [88, 113]. Loss of exomer causes Chs3, Fus1, and Pin2 to be retained in the TGN, but transport to the plasma membrane can be restored by inactivating AP-1 [7, 100, 114]. In Chs3, the exomer-binding signal overlaps with the AP-1-binding signal [114, 127], and it was suggested that exomer competes with AP-1 for binding to Chs3 [1]. Thus, for certain proteins, exomer apparently serves as a switch that suppresses AP-1-dependent intra-Golgi recycling and enables delivery to the plasma membrane in secretory vesicles.

A complicating factor is that the yeast Ent5 clathrin adaptor arrives together with AP-1 and cooperates functionally with AP-1 [24, 30, 31, 34, 42]. The precise nature of this cooperation is unclear, but AP-1 and Ent5 seem to have partially redundant roles in driving intra-Golgi recycling. Double deletion of AP-1 and Ent5 inhibits the recycling of bulk membrane within the late Golgi as determined by tracking internalized FM 4–64 [24]. Moreover, the AP-1/Ent5 double deletion perturbs the traffic of late Golgi transmembrane proteins, including the processing protease Kex2 [24]. Mutant cells lacking AP-1 and Ent5 show reduced levels of Kex2 in the Golgi, and when the mutant cells are treated with an inhibitor of endocytosis, Kex2 becomes trapped outside the Golgi. Those findings led to two inferences. First, Kex2 normally recycles within the late Golgi in an AP-1/Ent5-dependent pathway (Fig. 13.4). Second, when the AP-1/Ent5 pathway fails to operate, Kex2 travels to the plasma membrane and then returns to the late Golgi in endocytic vesicles.

Ent5 appears to be particularly important for recycling SNARE proteins at the late Golgi. Even when AP-1 is present, deletion of Ent5 perturbs traffic of the Qc-SNARE Tlg1 [24, 58]. The story may be similar for the Qa-SNARE Tlg2. It was reported that Ent5 mediates recycling of Tlg2 from early endosomes to the TGN [93], but those findings are equally consistent with Ent5-dependent intra-Golgi recycling of Tlg2.

Other late Golgi proteins also recycle with the aid of AP-1 and/or Ent5 [24, 71]. The best characterized example is the dipeptidyl aminopeptidase Ste13 [44, 85]. Kex2 and Ste13 arrive at a maturing cisterna simultaneously, and then they begin to depart from the cisterna when AP-1 and Ent5 become detectable [24]. Presumably, Kex2 and Ste13 recycle in the same vesicles. This prediction was confirmed by an in vivo vesicle capture assay [65]. A septin at the yeast bud neck was tagged with the FK506-binding protein FKBP, and Kex2 was tagged with the FKBP-rapamycin binding domain FRB. Rapamycin addition resulted in the capture of Kex2-containing Golgi-derived vesicles at the bud neck. When Ste13 was tagged with GFP, rapamycin addition caused GFP fluorescence to appear at the bud neck, indicating that tagged Ste13 was in the same vesicles as Kex2. Still unknown is whether individual vesicles contain both AP-1 and Ent5 or whether separate populations of vesicles contain either AP-1 or Ent5. However, the combined data indicate that AP-1 and Ent5 mediate the intra-Golgi recycling of a set of resident Golgi proteins at a late stage of maturation. This interpretation fits with a recent study of AP-1 function in mammalian cells [103].

AP-1-driven retrograde intra-Golgi traffic might have functions beyond the retention of transmembrane proteins in the late Golgi. A soluble artificial secretory cargo was found to recycle within the yeast Golgi in an AP-1-dependent manner [25]. The physiological significance of this observation is unclear, but AP-1 vesicles seem to have some capacity for bulk transport.

Traffic Pathways of Late Golgi Proteins

The conclusions in the previous section about intra-Golgi recycling of Kex2 and Ste13 are at odds with the conventional view shown in Fig. 13.3. It has long been thought that Kex2 and Ste13 cycle together with Vps10 between the TGN and PVE compartments [10, 17, 125]. The key evidence is that inactivation of PVE-localized sorting nexins causes Vps10, Kex2, and Ste13 to accumulate in PVE compartments and the vacuole, implying that like Vps10, Kex2 and Ste13 normally undergo retrieval from PVE compartments to the TGN [26, 29, 84, 125]. Indeed, PVE-to-TGN retrieval signals have been identified in the cytoplasmic tails of Kex2 and Ste13 [18, 21]. But unlike Vps10, Kex2 and Ste13 have no known role in PVE compartments. Moreover, unlike Vps10, Kex2 is normally present only at very low levels in PVE compartments [35]. The traffic of Kex2 and Ste13 differs from that of Vps10 in three additional ways: (1) Kex2 and Ste13 arrive at a maturing cisterna slightly later than Vps10 [24]. (2) Deletion of the synaptojanin-like protein Inp53, or deletion of AP-1 plus Ent5, perturbs the traffic of Kex2 and Ste13 but not of Vps10 [24, 52]. (3) Kex2 and Ste13 move from the TGN to PVE compartments slowly, with half-times of tens of minutes, whereas Vps10 makes this journey rapidly, with a half-time of minutes [21, 44, 52, 53]. How can these various observations be reconciled?

The likely answer is that Vps10 continually travels to PVE compartments during its normal itinerary, whereas Kex2 and Ste13 undergo AP-1/Ent5-dependent intra-Golgi recycling but also sometimes visit PVE compartments. Consistent with this model, the cytoplasmic tail of Ste13 contains a TGN localization signal that interacts with AP-1 [44, 85]. The occasional delivery of Kex2 and Ste13 to PVE compartments can be understood by noting that protein sorting in membrane traffic tends to be leaky. This phenomenon is well documented for ER proteins, which occasionally escape to post-ER compartments and then are retrieved by salvage pathways [9]. Something similar probably happens with Golgi proteins. For example, Kex2 and Ste13 recycle within the Golgi approximately once per minute [24, 65], so even occasional missorting would be disastrous in the absence of retrieval mechanisms. Consider the possibilities for missorting of Kex2. If a Kex2 molecule fails to recycle within the Golgi in the AP-1/Ent5 pathway, then the molecule might experience one of two fates (Fig. 13.5). The first possible fate for a missorted Kex2 molecule is delivery to the plasma membrane in secretory vesicles. At that point, an endocytosis signal in the Kex2 cytoplasmic tail [115] will direct recycling of the Kex2 molecule to the Golgi. If AP-1 and Ent5 are inactivated, then escape of Kex2 to the plasma membrane will be enhanced, and endocytosis will presumably switch from being a Kex2 salvage pathway to being the major pathway for keeping Kex2 in

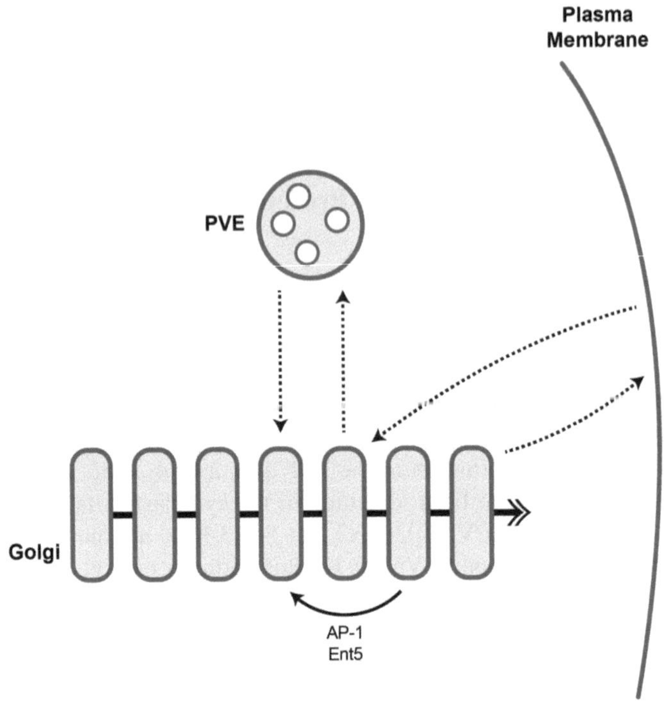

Fig. 13.5 Proposed traffic pathways for Kex2. The major pathway, indicated with a solid arrow, is AP-1/Ent5-dependent intra-Golgi recycling. Two minor pathways, indicated with dashed arrows, involve missorting followed by salvage mechanisms for retrieval. Kex2 probably travels occasionally to the plasma membrane, after which it recycles to the Golgi in endocytic vesicles. In addition, Kex2 probably travels occasionally to prevacuolar endosome (PVE) compartments, after which it recycles to the Golgi with the aid of sorting nexins

the Golgi [24]. The second possible fate for a missorted Kex2 molecule is delivery to a PVE compartment. At that point, sorting nexin-dependent retrieval will return the Kex2 molecule to the Golgi. If a crucial sorting nexin is inactivated, then Kex2 will accumulate in PVE compartments. Thus, observations about Kex2 and Ste13 retrieval from PVE compartments [10, 125] can be explained by assuming that those studies highlighted a PVE-to-TGN salvage pathway rather than the primary intra-Golgi recycling pathway.

The emerging picture is that a resident Golgi protein such as Kex2 or Ste13 utilizes a major sorting pathway—in this case, AP-1/Ent5-dependent intra-Golgi recycling—but also undergoes occasional missorting followed by retrieval in salvage pathways (Fig. 13.5). This view helps to explain why the cytoplasmic tails of Kex2 and Ste13 contain multiple sorting signals [18, 21, 115]. More generally, the likely existence of salvage pathways that preserve Golgi homeostasis is a promising topic for further research.

Conclusion and Future Perspectives

To take full advantage of yeast, we need to know the intracellular compartments and membrane traffic pathways in this organism. As summarized here, analysis of endosome and Golgi dynamics has inspired a reassessment of fundamental ideas about the yeast endomembrane system. A range of findings can be accommodated by the models presented in Figs. 13.4 and 13.5. These concepts may also prompt a new look at aspects of the mammalian endomembrane system, such as the identities of endosomal compartments and the role of AP-1 in localizing proteins to the TGN.

Acknowledgments This work was supported by NIH grant R35 GM144050. Thanks to the Glick lab, Greg Odorizzi, and Chris Fromme for helpful comments on the manuscript.

Compliance with Ethical Standards The author discloses no conflict of interest.

References

1. Anton-Plagaro C, Sanchez N, Valle R, Miguel Mulet J, Duncan MC, Roncero C (2021) Exomer complex regulates protein traffic at the TGN through differential interactions with cargos and clathrin adaptor complexes. FASEB J 35:e21615
2. Arlt H, Auffarth K, Kurre R, Lisse D, Piehler J, Ungermann C (2015) Spatiotemporal dynamics of membrane remodeling and fusion proteins during endocytic transport. Mol Biol Cell 26:1357–1370
3. Babst M, Katzmann DJ, Estepa-Sabal EJ, Meerloo T, Emr SD (2002a) ESCRT-III: an endosome-associated heterooligomeric protein complex required for MVB sorting. Dev Cell 3:271–282
4. Babst M, Katzmann DJ, Snyder WB, Wendland B, Emr SD (2002b) Endosome-associated complex, ESCRT-II, recruits transport machinery for protein sorting at the multivesicular body. Dev Cell 3:283–289
5. Balderhaar HJ, Ungermann C (2013) CORVET and HOPS tethering copmplexes—coordinators of endosome and lysosome fusion. J Cell Sci 126:1307–1316
6. Bankaitis VA, Johnson LM, Emr SD (1986) Isolation of yeast mutants defective in protein targeting to the vacuole. Proc Natl Acad Sci USA 83:9075–9079
7. Barfield RM, Fromme JC, Schekman R (2009) The exomer coat complex transports Fus1p to the plasma membrane via a novel plasma membrane sorting signal in yeast. Mol Biol Cell 20:4985–4996
8. Barlowe C, Orci L, Yeung T, Hosobuchi M, Hamamoto S, Salama N, Rexach MF, Ravazzola M, Amherdt M, Schekman R (1994) COPII: a membrane coat formed by sec proteins that drive vesicle budding from the endoplasmic reticulum. Cell 77:895–907
9. Barlowe CK, Miller EA (2013) Secretory protein biogenesis and traffic in the early secretory pathway. Genetics 193:383–410
10. Bean BDM, Davey M, Conibear E (2017) Cargo selectivity of yeast sorting nexins. Traffic 18:110–122
11. Becker B, Melkonian M (1996) The secretory pathway of protists: spatial and functional organization and evolution. Microbiol Rev 60:697–721
12. Becuwe M, Léon S (2014) Integrated control of transporter endocytosis and recycling by the arrestin-related protein Rod1 and the ubiquitin ligase Rsp5. *eLife* 3:e03307

13. Belden WJ, Barlowe C (2001) Distinct roles for the cytoplasmic tail sequences of Emp24p and Erv25p in transport between the endoplasmic reticulum and Golgi complex. J Biol Chem 276:43040–43048
14. Bhave M, Papanikou E, Iyer P, Pandya K, Jain BK, Ganguly A, Sharma C, Pawar K, Austin JI, Day KJ, Rossanese OW, Glick BS, Bhattacharyya D (2014) Golgi enlargement in Arf-depleted yeast cells is due to altered dynamics of cisternal maturation. J Cell Sci 127:250–257
15. Black MW, Pelham HR (2000) A selective transport route from Golgi to late endosomes that requires the yeast GGA proteins. J Cell Biol 151:587–600
16. Bonfanti L, Mironov AA Jr, Martínez-Menárguez J, Martella O, Fusella A, Baldassarre M, Buccione R, Geuze HJ, Mironov AA, Luini A (1998) Procollagen traverses the Golgi stack without leaving the lumen of cisternae: evidence for cisternal maturation. Cell 95:993–1003
17. Bowers K, Stevens TH (2005) Protein transport from the late Golgi to the vacuole in the yeast *Saccharomyces cerevisiae*. Biochim Biophys Acta 1744:438–454
18. Brickner JH, Fuller RS (1997) *SOI1* encodes a novel, conserved protein that promotes TGN-endosomal cycling of Kex2p and other membrane proteins by modulating the function of two TGN localization signals. J Cell Biol 139:23–36
19. Bright NA, Davis LJ, Luzio JP (2016) Endolysosomes are the principal intracellular sites of acid hydrolase activity. Curr Biol 26:2233–2245
20. Bright NA, Gratian MJ, Luzio JP (2005) Endocytic delivery to lysosomes mediated by concurrent fusion and kissing events in living cells. Curr Biol 15:360–365
21. Bryant NJ, Stevens TH (1997) Two separate signals act independently to localize a yeast late Golgi membrane protein through a combination of retrieval and retention. J Cell Biol 136:287–297
22. Buelto D, Hung CW, Aoh QL, Lahiri S, Duncan MC (2020) Plasma membrane to vacuole traffic induced by glucose starvation requires Gga2-dependent sorting at the *trans*-Golgi network. Biol Cell 112:349–367
23. Casler JC, Glick BS (2020) A microscopy-based kinetic analysis of yeast vacuolar protein sorting. elife 9:e56844
24. Casler JC, Johnson N, Krahn AH, Pantazopoulou A, Day KJ, Glick BS (2022) Clathrin adaptors mediate two sequential pathways of intra-Golgi recycling. J Cell Biol 221:e202103199
25. Casler JC, Papanikou E, Barrero JJ, Glick BS (2019) Maturation-driven transport and AP-1-dependent recycling of a secretory cargo in the Golgi. J Cell Biol 218:1582–1601
26. Cereghino JL, Marcusson EG, Emr SD (1995) The cytoplasmic tail domain of the vacuolar protein sorting receptor Vps10p and a subset of *VPS* gene products regulate receptor stability, function, and localization. Mol Biol Cell 6:1089–1102
27. Chen Y, Gershlick DC, Park SY, Bonifacino JS (2017) Segregation in the Golgi complex precedes export of endolysosomal proteins in distinct transport carriers. J Cell Biol 216:4141–4151
28. Chidambaram S, Müllers N, Wiederhold K, Haucke V, Fischer von Mollard G (2004) Specific interaction between SNAREs and epsin N-terminal homology (ENTH) domains of epsin-related proteins in *trans*-Golgi network to endosome transport. J Biol Chem 279:4175–4179
29. Cooper AA, Stevens TH (1996) Vps10p cycles between the late-Golgi and prevacuolar compartments in its function as the sorting receptor for multiple yeast vacuolar hydrolases. J Cell Biol 133:529–541
30. Čopič A, Starr TL, Schekman R (2007) Ent3p and Ent5p exhibit cargo-specific functions in trafficking proteins between the *trans*-Golgi network and the endosomes in yeast. Mol Biol Cell 18:1803–1815
31. Costaguta G, Duncan MC, Fernández GE, Huang GH, Payne GS (2006) Distinct roles for TGN/endosome epsin-like adaptors Ent3p and Ent5p. Mol Biol Cell 17:3907–3920
32. Costaguta G, Stefan CJ, Bensen ES, Emr SD, Payne GS (2001) Yeast Gga coat proteins function with clathrin in Golgi to endosome transport. Mol Biol Cell 12:1885–1896

33. Daboussi L, Costaguta G, Ghukasyan R, Payne GS (2017) Conserved role for Gga proteins in phosphatidylinositol 4-kinase localization to the *trans*-Golgi network. Proc Natl Acad Sci USA 114:3433–3438
34. Daboussi L, Costaguta G, Payne GS (2012) Phosphoinositide-mediated clathrin adaptor progression at the *trans*-Golgi network. Nat Cell Biol 14:239–248
35. Day KJ, Casler JC, Glick BS (2018) Budding yeast has a minimal endomembrane system. Dev Cell 44:56–72
36. De Matteis MA, Luini A (2008) Exiting the Golgi complex. Nat Rev Mol Cell Biol 9:273–284
37. Deshaies R, Sanders SL, Feldheim DA, Schekman R (1991) Assembly of yeast sec proteins involved in translocation into the endoplasmic reticulum into a membrane-bound multisubunit complex. Nature 349:806–808
38. Dettmer J, Hong-Hermesdorf A, Stierhof YD, Schumacher K (2006) Vacuolar H^+-ATPase activity is required for endocytic and secretory trafficking in *Arabidopsis*. Plant Cell 18:715–730
39. Doray B, Ghosh P, Griffith J, Geuze HJ, Kornfeld S (2002) Cooperation of GGAs and AP-1 in packaging MPRs at the trans-Golgi network. Science 297:1700–1703
40. Duden R, Schekman R (1997) Insights into Golgi function through mutants in yeast and animal cells. In: Berger EG, Roth J (eds) The Golgi apparatus. Birkhäuser Verlag, Basel, pp 219–246
41. Duncan MC (2022) New directions for the clathrin adaptor AP-1 in cell biology and human disease. Curr Opin Cell Biol 76:102079
42. Duncan MC, Costaguta G, Payne GS (2003) Yeast epsin-related proteins required for Golgi-endosome traffic define a gamma-adaptin ear-binding motif. Nat Cell Biol 5:77–81
43. Eising S, Esch B, Wälte M, Vargas Duarte P, Walter S, Ungermann C, Bohnert M, Fröhlich F (2022) A lysosomal biogenesis map reveals the cargo spectrum of yeast vacuolar protein targeting pathways. J Cell Biol 221:e202107148
44. Foote C, Nothwehr SF (2006) The clathrin adaptor complex 1 directly binds to a sorting signal in Ste13p to reduce the rate of its trafficking to the late endosome of yeast. J Cell Biol 173:615–626
45. Franzusoff A, Redding K, Crosby J, Fuller RS, Schekman R (1991) Localization of components involved in protein transport and processing through the yeast Golgi apparatus. J Cell Biol 112:27–37
46. Fujii S, Kurokawa K, Inaba R, Hiramatsu N, Tago T, Nakamura Y, Nakano A, Satoh T, Satoh AK (2020) Recycling endosomes attach to the trans-side of Golgi stacks in *drosophila* and mammalian cells. J Cell Sci 133:jcs236935
47. Gao J, Nicastro R, Péli-Gulli MP, Grziwa S, Chen Z, Kurre R, Piehler J, De Virgilio C, Fröhlich F, Ungermann C (2022) The HOPS tethering complex is required to maintain signaling endosome identity and TORC1 activity. J Cell Biol 221:e202109084
48. Gaynor EC, Chen CY, Emr SD, Graham TR (1998) ARF is required for maintenance of yeast Golgi and endosome structure and function. Mol Biol Cell 9:653–670
49. Grant BD, Donaldson JG (2009) Pathways and mechanisms of endocytic recycling. Nat Rev Mol Cell Biol 10:597–608
50. Griffiths G, Simons K (1986) The *trans* Golgi network: sorting at the exit site of the Golgi complex. Science 234:438–443
51. Guo X, Mattera R, Ren X, Chen Y, Retamal C, González A, Bonifacino JS (2013) The adaptor protein-1 μ1B subunit expands the repertoire of basolateral sorting signal recognition in epithelial cells. Dev Cell 27:353–366
52. Ha SA, Brunch JT, Hama H, DeWald DB, Nothwehr SF (2001) A novel mechanism for localizing membrane proteins to yeast *trans*-Golgi network requires function of synaptojanin-like protein. Mol Biol Cell 12:3175–3190
53. Ha SA, Torabinejad J, DeWald DB, Wenk MR, Lucast L, De Camilli P, Newitt RA, Aebersold R, Nothwehr SF (2003) The synaptojanin-like protein Inp53/Sjl3 functions with clathrin in

a yeast TGN-to-endosome pathway distinct from the GGA protein-dependent pathway. Mol Biol Cell 14:1319–1333
54. Hirst J, Lui WW, Bright NA, Totty N, Seaman MN, Robinson MS (2000) A family of proteins with γ-adaptin and VHS domains that facilitate trafficking between the trans-Golgi network and the vacuole/lysosome. J Cell Biol 149:67–80
55. Hollenstein DM, Kraft C (2020) Autophagosomes are formed at a distinct cellular structure. Curr Opin Cell Biol 65:50–57
56. Huang KM, D'Hondt K, Riezman H, Lemmon SK (1999) Clathrin functions in the absence of heterotetrameric adaptors and AP180-related proteins in yeast. EMBO J 18:3897–3908
57. Huckaba TM, Gay AC, Pantalena LF, Yang HC, Pon LA (2004) Live cell imaging of the assembly, disassembly, and actin cable-dependent movement of endosomes and actin patches in the budding yeast, Saccharomyces cerevisiae. J Cell Biol 167:519–530
58. Hung CW, Duncan MC (2016) Clathrin binding by the adaptor Ent5 promotes late stages of clathrin coat maturation. Mol Biol Cell 27:1143–1153
59. Huotari J, Helenius A (2011) Endosome maturation. EMBO J 30:3481–3500
60. Ishii M, Suda Y, Kurokawa H, Nakano A (2016) COPI is essential for Golgi cisternal maturation and dynamics. J Cell Sci 129:3251–3261
61. Johnston HD, Foote C, Santeford A, Nothwehr SF (2005) Golgi-to-late endosome trafficking of the yeast pheromone processing enzyme Ste13p is regulated by a phosphorylation site in its cytosolic domain. Mol Biol Cell 16:1456–1468
62. Katzmann DJ, Babst M, Emr SD (2001) Ubiquitin-dependent sorting into the multivesicular body pathway requires the function of a conserved endosomal protein sorting complex, ESCRT-I. *Cell* 106:145–155
63. Kim JJ, Lipatova Z, Segev N (2016) Regulation of Golgi cisternal progression by Ypt/Rab GTPases. Dev Cell 36:440–452
64. Kornfeld S (1986) Trafficking of lysosomal enzymes in normal and disease states. J Clin Invest 77:1–6
65. Krahn AH, A Pantazopoulou, and BS Glick (2024) COPI-dependent intra-Golgi recycling at an intermediate stage of cisternal maturation.. *Submitted*
66. Ladinsky MS, Mastronarde DN, McIntosh JR, Howell KE, Staehelin LA (1999) Golgi structure in three dimensions: functional insights from the normal rat kidney cell. J Cell Biol 144:1135–1149
67. Laidlaw KME, Calder G, MacDonald C (2022) Recycling of cell surface membrane proteins from yeast endosomes is regulated by ubiquitinated Ist1. J Cell Biol 221:e202109137
68. Langmeyer L, Borchers AC, Herrmann E, Füllbrunn N, Han Y, Perz A, Auffarth K, Kümmel D, Ungermann C (2020) A conserved and regulated mechanism drives endosomal Rab transition. elife 11:e56090
69. Lewis MJ, Nichols BJ, Prescianotto-Baschong C, Riezman H, Pelham HRB (2000) Specific retrieval of the exocytic SNARE Snc1p from early yeast endosomes. Mol Biol Cell 11:23–38
70. Lin CH, MacGurn JA, Chu T, Stefan CJ, Emr SD (2008) Arrestin-related ubiquitin-ligase adaptors regulate endocytosis and protein turnover at the cell surface. Cell 135:714–725
71. Liu K, Surendhran K, Nothwehr SF, Graham TR (2008) P4-ATPase requirement for AP-1/clathrin function in protein transport from the *trans*-Golgi network and early endosomes. Mol Biol Cell 19:3526–3535
72. Losev E, Reinke CA, Jellen J, Strongin DE, Bevis BJ, Glick BS (2006) Golgi maturation visualized in living yeast. Nature 441:1002–1006
73. Ma M, Burd CG (2020) Retrograde trafficking and plasma membrane recycling pathways of the budding yeast *Saccharomyces cerevisiae*. Traffic 21:45–59
74. MacDonald C, Piper RC (2016) Cell surface recycling in yeast: mechanisms and machineries. Biochem Soc Trans 44:474–478
75. Matsuura-Tokita K, Takeuchi M, Ichihara A, Mikuriya K, Nakano A (2006) Live imaging of yeast Golgi cisternal maturation. Nature 441:1007–1010

76. McNatt MW, McKittrick I, West M, Odorizzi G (2007) Direct binding to Rsp5 mediates ubiquitin-independent sorting of Sna3 via the multivesicular body pathway. Mol Biol Cell 18:697–706
77. Menant A, Barbey R, Thomas D (2006) Substrate-mediated remodeling of methionine transport by multiple ubiquitin-dependent mechanisms in yeast cells. EMBO J 25:4436–4447
78. Mizushima N, Noda T, Yoshimori T, Tanaka Y, Ishii T, George MD, Klionsky DJ, Ohsumi M, Ohsumi Y (1998) A protein conjugation system essential for autophagy. Nature 395:395–398
79. Mogelsvang S, Gomez-Ospina N, Soderholm J, Glick BS, Staehelin LA (2003) Tomographic evidence for continuous turnover of Golgi cisternae in *Pichia pastoris*. Mol Biol Cell 14:2277–2291
80. Mowbrey K, Dacks JB (2009) Evolution and diversity of the Golgi body. FEBS Lett 583:3738–3745
81. Myers MD, Payne GS (2013) Clathrin, adaptors and disease: insights from the yeast *Saccharomyces cerevisiae*. Front Biosci 18:862–891
82. Nagano M, Toshima JY, Siekhaus DE, Toshima J (2019) Rab5-mediated endosome formation is regulated at the *trans*-Golgi network. Commun Biol 2:419
83. Nickerson DP, Russell MR, Lo SY, Chapin HC, Milnes JM, Merz AJ (2012) Termination of isoform-selective Vps21/Rab5 signaling at endolysosomal organelles by Msb3/Gyp3. Traffic 13:1411–1428
84. Nothwehr SF, Hindes AE (1997) The yeast VPS5/GRD2 gene encodes a sorting nexin-1-like protein required for localizing membrane proteins to the late Golgi. J Cell Sci 110:1063–1072
85. Nothwehr SF, Roberts CJ, Stevens TH (1993) Membrane protein retention in the yeast Golgi apparatus: dipeptidyl aminopeptidase a is retained by a cytoplasmic signal containing aromatic residues. J Cell Biol 121:1197–1209
86. Odorizzi G, Babst M, Emr SD (1998a) Fab1p PtdIns(3)P 5-kinase function essential for protein sorting in the multivesicular body. Cell 95:847–858
87. Odorizzi G, Cowles CR, Emr SD (1998b) The AP-3 complex: a coat of many colors. Trends Cell Biol 8:282–288
88. Paczkowski JE, Richardson BC, Fromme JC (2015) Cargo adaptors: structures illuminate mechanisms regulating vesicle biogenesis. Trends Cell Biol 25:408–416
89. Pantazopoulou A, Glick BS (2019) A kinetic view of membrane traffic pathways can transcend the classical view of Golgi compartments. Front Cell Dev Biol 7:153
90. Papanikou E, Day KJ, Austin J 2nd, Glick BS (2015) COPI selectively drives maturation of the early Golgi. elife 4:e13232
91. Papanikou E, Glick BS (2009) The yeast Golgi apparatus: insights and mysteries. FEBS Lett 583:3746–3751
92. Pelham HRB (2002) Insights from yeast endosomes. Curr Opin Cell Biol 14:454–462
93. Petersen L, Bachmann R, Meinerz S, Tanz A, Fischer von Mollard G (2023) Distinct functional domains of the epsin-related Ent5p, a cargo adaptor for the SNARE Tlg2p in transport between endosomes and Golgi. Traffic 24:475–488
94. Powers J, Barlowe C (1998) Transport of Axl2p depends on Erv14p, an ER-vesicle protein related to the *drosophila cornichon* gene product. J Cell Biol 142:1209–1222
95. Prescianotto-Baschong C, Riezman H (1998) Morphology of the yeast endocytic pathway. Mol Biol Cell 9:173–189
96. Prescianotto-Baschong C, Riezman H (2002) Ordering of compartments in the yeast endocytic pathway. Traffic 3:37–49
97. Preuss D, Mulholland J, Franzusoff A, Segev N, Botstein D (1992) Characterization of the *saccharomyces* Golgi complex through the cell cycle by immunoelectron microscopy. Mol Biol Cell 3:789–803
98. Puertollano R, Aguilar RC, Gorshkova I, Crouch RJ, Bonifacino JS (2001) Sorting of mannose 6-phosphate receptors mediated by the GGAs. Science 292:1712–1716
99. Rink J, Ghigo E, Kalaidzidis Y, Zerial M (2005) Rab conversion as a mechanism of progression from early to late endosomes. Cell 122:735–749

100. Ritz AM, Trautwein M, Grassinger F, Spang A (2014) The prion-like domain in the exomer-dependent cargo Pin2 serves as a *trans*-Golgi retention motif. Cell Rep 7:249–260
101. Robinson MS (2004) Adaptable adaptors for coated vesicles. Trends Cell Biol 14:167–174
102. Robinson MS (2015) Forty years of clathrin-coated vesicles. Traffic 16:1210–1238
103. Robinson MS, Antrobus R, Sanger A, Davies AK, Gershlick DC (2024) The role of the AP-1 adaptor complex in outgoing and incoming membrane traffic. J Cell Biol 223:e202310071
104. Rothman JH, Stevens TH (1986) Protein sorting in yeast: mutants defective in vacuole biogenesis mislocalize vacuolar proteins into the late secretory pathway. Cell 47:1041–1051
105. Saffi GT, Botelho RJ (2019) Lysosome fission: planning for an exit. Trends Cell Biol 29:635–646
106. Salminen A, Novick PJ (1987) A ras-like protein is required for a post-Golgi event in yeast secretion. Cell 49:527–538
107. Sato K, Sato M, Nakano A (2001) Rer1p, a retrieval receptor for endoplasmic reticulum membrane proteins, is dynamically localized to the Golgi apparatus by coatomer. J Cell Biol 152:935–944
108. Schu PV, Takegawa K, Fry MJ, Stack JH, Waterfield MD, Emr SD (1983) Phosphatidylinositol 3-kinase encoded by yeast *VPS34* gene essential for protein sorting. Science 260:88–91
109. Scott CC, Vacca F, Gruenberg J (2014) Endosome maturation, transport and functions. Semin Cell Dev Biol 31:2–10
110. Segev N (1991) Mediation of the attachment or fusion step in vesicular transport by the GTP-binding Ypt1 protein. Science 252:1553–1556
111. Semenza JC, Hardwick KG, Dean N, Pelham HR (1990) *ERD2*, a yeast gene required for the receptor-mediated retrieval of luminal ER proteins from the secretory pathway. Cell 61:1349–1357
112. Singer-Krüger B, Stenmark H, Düsterhöft A, Philippsen P, Yoo JS, Gallwitz D, Zerial M (1994) Role of three rab5-like GTPases, Ypt51p, Ypt52p, and Ypt53p, in the endocytic and vacuolar protein sorting pathways of yeast. J Cell Biol 125:283–298
113. Spang A (2015) The road not taken: less traveled roads from the TGN to the plasma membrane. Membranes (Basel) 5:84–98
114. Starr TL, Pagant S, Wang CW, Schekman R (2012) Sorting signals that mediate traffic of chitin synthase III between the TGN/endosomes and to the plasma membrane in yeast. PLoS One 7:e46386
115. Tan PK, Howard JP, Payne GS (1996) The sequence NPFXD defines a new class of endocytosis signal in *Saccharomyces cerevisiae*. J Cell Biol 135:1789–1800
116. Tojima T, Suda Y, Ishii M, Kurokawa K, Nakano A (2019) Spatiotemporal dissection of the *trans*-Golgi network in budding yeast. J Cell Sci 132:jcs231159
117. Tojima T, Suda Y, Jin N, Kurokawa K, Nakano A (2024) Spatiotemporal dissection of the Golgi apparatus and the ER-Golgi intermediate compartment in budding yeast. elife 13:e92900
118. Toshima JY, Nishinoaki S, Sato Y, Yamamoto W, Furukawa D, Siekhaus DE, Sawaguchi A, Toshima J (2014) Bifurcation of the endocytic pathway into Rab5-dependent and -independent transport to the vacuole. Nat Commun 5:3498
119. Toshima JY, Toshima J, Kaksonen M, Martin AC, King DS, Drubin DG (2006) Spatial dynamics of receptor-mediated endocytic trafficking in budding yeast revealed by using fluorescent α-factor derivatives. Proc Natl Acad Sci USA 103:5793–5798
120. Toshima JY, Tsukahara A, Nagano M, Tojima T, Siekhaus DE, Nakano A, Toshima J (2023) The yeast endocytic early/sorting compartment exists as an independent sub-compartment within the *trans*-Golgi network. elife 12:e84850
121. Valdivia RH, Baggott D, Chuang JS, Schekman R (2002) The yeast clathrin adaptor protein complex 1 is required for the efficient retention of a subset of late Golgi membrane proteins. Dev Cell 2:283–294
122. Vida TA, Emr SD (1995) A new vital stain for visualizing vacuolar membrane dynamics and endocytosis in yeast. J Cell Biol 128:779–792

123. Vida TA, Huyer G, Emr SD (1993) Yeast vacuolar proenzymes are sorted in the late Golgi complex and transported to the vacuole via a prevacuolar endosome-like compartment. J Cell Biol 121:1245–1256
124. Viotti C, Bubeck J, Stierhof YD, Krebs M, Langhans M, van den Berg W, van Dongen W, Richter S, Geldner N, Takano J, Jürgens G, de Vries SC, Robinson DG, Schumacher K (2010) Endocytic and secretory traffic in *Arabidopsis* merge in the trans-Golgi network/early endosome, an independent and highly dynamic organelle. Plant Cell 22:1344–1357
125. Voos W, Stevens TH (1998) Retrieval of resident late-Golgi membrane proteins from the prevacuolar compartment of *Saccharomyces cerevisiae* is dependent on the function of Grd19p. J Cell Biol 140:577–590
126. Vowels JJ, Payne GS (1998) A dileucine-like sorting signal directs transport into an AP-3-dependent, clathrin-independent pathway to the yeast vacuole. EMBO J 17:2482–2493
127. Weiskoff AM, Fromme JC (2014) Distinct N-terminal regions of the exomer secretory vesicle cargo Chs3 regulate its trafficking itinerary. Front Cell Dev Biol 2:47
128. Wiederkehr A, Avaro S, Prescianotto-Baschong C, Haguenauer-Tsapis R, Riezman H (2000) The F-box protein Rcy1p is involved in endocytic membrane traffic and recycling out of an early endosome in *Saccharomyces cerevisiae*. J Cell Biol 149:397–410
129. Xu P, Hankins HM, MacDonald C, Erlinger SJ, Frazier MN, Diab NS, Piper RC, Jackson LP, MacGurn JA, Graham TR (2017) COPI mediates recycling of an exocytic SNARE by recognition of a ubiquitin sorting signal. elife 6:e28342
130. Yeung BG, Phan HL, Payne GS (1999) Adaptor complex-independent clathrin function in yeast. Mol Biol Cell 10:3643–3659
131. Zerial M, McBride H (2001) Rab proteins as membrane organizers. Nat Rev Mol Cell Biol 2:107–117
132. Zhdankina O, Strand NL, Redmond JM, Boman AL (2001) Yeast GGA proteins interact with GTP-bound Arf and facilitate transport through the Golgi. Yeast 18:1–18

Chapter 14
Endocytosis and *trans*-Golgi Network in Yeast

Junko Toshima and Jiro Toshima

Abstract Endocytosis is the process by which cells internalize extracellular materials, cell surface proteins, and lipids via vesicles formed by membrane invagination. Endocytosed cargos are first delivered to sorting compartments and then recycled back to the plasma membrane or brought to late endosomal compartments *en route* to the lysosome/vacuole for degradation. The budding *yeast Saccharomyces cerevisiae* has been an important model organism for studying the molecular mechanisms involved in the multi-step process of the endocytic pathway, including clathrin-coated vesicle formation, vesicle internalization, cargo sorting, recycling, and degradation. In the yeast endocytic pathway, the mechanisms of formation and internalization of clathrin-coated vesicles are well conserved in relation to those in mammalian cells. Yeast also has mechanisms similar to those in mammalian cells for the regulation of events in the endo-lysosomal pathway, such as endosome maturation, multi-vesicular body formation, and retrograde transport. In contrast, the structure and properties of the endocytic sorting compartment in yeast differ significantly from those in mammalian cells. Recent studies have revealed that a subregion within the *trans*-Golgi network (TGN) acts as a sorting compartment in yeast. In this chapter, we summarize recent findings regarding the molecular mechanisms of the endocytic pathway, particularly the involvement of the TGN.

Keywords Endocytosis · Endosome · Clathrin · *trans*-Golgi network · Recycling · Actin · Retromer · SNARE · Rab

J. Toshima (✉)
School of Health Science, Tokyo University of Technology, Ota-ku, Tokyo, Japan
e-mail: toshimajk@stf.teu.ac.jp

J. Toshima (✉)
Department of Biological Science and Technology, Tokyo University of Science, Katsushika-ku, Tokyo, Japan
e-mail: jtosiscb@rs.tus.ac.jp

A. Nakano, J. Saraste (eds.), *The Golgi Network, Volume I*, Subcellular Biochemistry 110, https://doi.org/10.1007/978-3-032-06936-8_14

Introduction

Endocytosis is a fundamental cellular function involved in various physiological processes, such as nutrient uptake, receptor internalization, synaptic vesicle recycling, and defense against pathogens. Previous studies of the budding yeast *Saccharomyces cerevisiae* have contributed significantly to the clarification of the mechanisms involved in the endocytic pathway. While many of the mechanisms are highly conserved in yeast and mammalian cells, some mechanisms identified in the former are absent in the latter, or some are present only in mammalian cells. On the basis of findings in mammalian cells, the endocytic pathway has long been understood as a pathway in which endocytic cargos are sequentially transported from endocytic vesicles to early/sorting endosomes, and then to late/multivesicular body (MVB) endosomes and finally to lysosomes. However, recent studies of plants and yeast have revealed that the TGN-mediated route is the main pathway, unlike that in mammalian cells.

The process of clathrin-mediated endocytic vesicle internalization has been shown to be regulated by more than 60 different endocytic proteins in yeast, most of which have similar functions in mammalian cells [67, 152]. By contrast, the mechanisms of the intracellular vesicle transport are considerably more complex in mammalian cells and involve more proteins. For instance, mammalian cells have over 68 Rab GTPases, whereas yeast has only 11 Rab GTPases [14, 134]. Consistent with this, endosomal systems differ significantly between yeast and mammalian cells: mammalian cells have several different types of endosomes, including the early/sorting endosome, the late endosome and the recycling endosome, whereas yeast seems to lack sorting endosomes and instead employs the *trans*-Golgi network (TGN) to fulfill their roles [35, 151].

It has long been debated whether budding yeast has endosomal systems similar to those of mammalian or plant cells, or whether it has its own systems. In budding yeast, the Golgi cisternae lack the stacked structure observed in mammalian and plant cells, and show *cis*-to-*trans* maturation, proceeding thereafter to the TGN [77, 83, 144]. The Golgi/TGN maturation process can be classified into three successive stages: the Golgi stage, the early TGN stage, and the late TGN stage. The early/sorting compartment has been shown to exist in the early TGN [144, 151]. Yeast also has early- and late-stage endosomes, and the Rab5 protein is widely localized to these compartments. Thus, yeast, like plants, has the early/sorting compartments in the TGN, but also has early- to late-stage endosomes, as do animal cells.

This chapter summarizes recent findings regarding the formation and internalization processes of clathrin vesicles, transport to the sorting compartment, endosome formation, and retrograde transport from endosomes in the yeast endocytic pathway.

Formation of Clathrin-Coated Vesicle in Budding Yeast

Clathrin-mediated endocytosis is a major endocytic pathway in yeast, and its molecular mechanisms are well conserved between yeast and mammalian cells. Endocytic proteins are largely classified into three subgroups: coat proteins, actin and its regulatory proteins, and scission-related proteins [67]. First, the mechanism of formation and internalization of clathrin-coated vesicles (CCVs), the first step in the endocytic pathway, will be briefly described.

Coat Proteins

In yeast, coat proteins are divided into four subgroups: early proteins and early/mid/late coat proteins (Fig. 14.1) [78]. Ede1p (Eps15 homolog) and Syp1p (FCHo1/2 homolog) are included in early proteins and are involved in the early stages of clathrin-coated pit formation [135]. Syp1p is an F-BAR domain protein that has the ability to induce membrane curvature at the first initiation step of clathrin-mediated endocytosis [2, 55]. Ede1p comprises a multi-modular domain structure and functions as a scaffold for local accumulation of cargo proteins and other endocytic proteins.

Clathrin and adaptor protein (AP)-2 complex are classified as early coat proteins (Fig. 14.1) [78, 88]. The AP-2 complex localizes at the plasma membrane by interacting with PtdIns(4,5)P_2 and acts as an adaptor for internalization of endocytic cargo proteins [16, 20]. YAP1801p/1802p (CALM/AP180 homologs) also serve as early coat proteins and provide a scaffold for recruitment of clathrin to the plasma membrane [82].

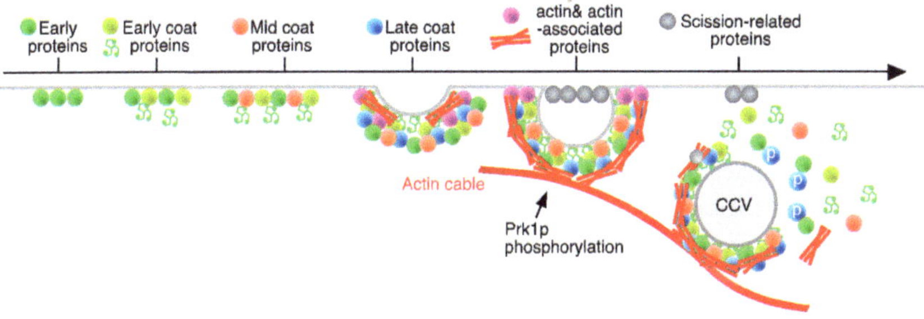

Fig. 14.1 Formation and transport of clathrin-coated vesicles in budding yeast. Early coat proteins are recruited during the immobile phase, which is followed by the ordered assembly of the mid/late coat protein. Actin-associated proteins are also recruited around this time, and they cause actin polymerization to invaginate the plasma membrane. Once the vesicle forms, the scission-related proteins are recruited to the neck of the vesicle. After release from the plasma membrane, the nascent vesicle is immediately uncoated by Prk1/Ark1 phosphorylation. The vesicle then moves rapidly toward the endocytic sorting compartment along actin cables

Sla2p (HIP1R homolog) and Ent1p/2p (Epsin homologs) are classified as mid-coat proteins (Fig. 14.1). Sla2p act as a linker between the plasma membrane and the actin network [137] and can generate F-actin-dependent forces to facilitate membrane invagination at the clathrin-coated pits (CCPs) [7, 46]. Ent1p/2p makes a complex with Sla2p and transmits the force from actin polymerization to the plasma membrane, facilitating membrane invagination [166]. PtdIns(4)P generated by plasma membrane Stt4p PI4-kinase is critical for recruitment of Ent1p/2p to the clathrin-coated pit [164].

Late coat proteins recruit actin regulators at the CCP, leading to the formation of the F-actin network. Pan1p (Eps15/Intersectin homolog) is a key late coat protein linking the clathrin-coated vesicle with the F-actin network, thereby regulating the process of endocytosis [42, 161]. Pan1p makes a complex with several coat proteins and actin regulators, such as Sla1p, End3p, Sla2p, and Arp2/3 complex [39, 142, 146, 161]. Pan1p also directly binds to F-actin with high affinity, and the interaction is negatively regulated by Ark1/Prk1 kinases [147]. Anchoring Pan1p to the peroxisome ectopically triggers most of the events, occurring during the late phase of CCV formation, at the peroxisome [42]. These suggest that Pan1p is a key regulator for initiating, processing, and completing the late phase of endocytosis.

Actin and Its Regulatory Proteins

Actin polymerization is the major force-generating step for membrane reorganization in the process of clathrin-mediated endocytosis (Fig. 14.1) [46, 68]. Yeast has three different F-actin structures: actin patch, actin cable, and the contractile ring [1]. Among them, actin patches play important roles in force generation to promote membrane invagination during CCV formation, while actin cables are required for endocytic vesicle trafficking processes, serving a railway-like function [63, 150]. Actin patch is a branched actin network assembled by the Arp2/3 complex and its regulators known as actin nucleation-promoting factors (NPFs), and actin cable is generated by Formins, Bni1p and Bnr1p, activated by Rho family GTPases [46]. In yeast four NPFs, Las17p (WASP), Myo3p/5p, Pan1p, and Abp1p, have been identified. Among these NPFs, Las17p and Myo3p/5p, which form a complex with yeast WIP (Vrp1p), have been shown to have high activity and play an important role in the initiation of actin assembly at sites of endocytosis [138]. In contrast, Pan1p has a lower NPF activity, and a mutation in the Arp2/3 complex binding region causes only a minor defect in actin polymerization at CCP [138, 147]. The dephosphorylation-mimicking mutant of Pan1p binds to F-actin with high affinity, and its expression causes aberrant actin aggregation, including endocytic vesicles, suggesting that Pan1p is important for maintaining stable interaction between CCV and actin network [147, 148].

Scission-Related Proteins

At the final step of CCV formation, proteins regulating vesicle scission are recruited to the clathrin-coated pits and pinch off the invaginated plasma membrane into the cytoplasm as a CCV (Fig. 14.1) [46, 88]. Dynamin and N-BAR domain proteins, such as Rvs161p/167p (Amphiphysin homologs), play essential roles in this process [15]. In addition, dephosphorylation of PtdIns(4,5)P_2 by the PI phosphatase Sjl2p (Synaptojanin homolog) is necessary for efficient scission of the invaginated membrane by Rvs161p/167p [70].

After the CCV is internalized into the cytosol, the coat proteins dissociate from the vesicle. Dissociation of the coatomer complex from the CCV is regulated by the Ark1p/Prk1p kinases [121, 147]. These kinases are recruited to endocytic sites just after initiation of actin assembly, depending on Abp1p, and phosphorylate several coat proteins, including Sla1p, Ent1/2p, Yap1801/2p, and Pan1p [131]. Phosphorylation of these proteins leads to dissociation of the coat proteins and actin network from the CCV (Fig. 14.1) [147, 148]. Subsequently, these proteins are immediately dephosphorylated by Scd5, a protein phosphatase [19], and recycled for new rounds of CCV formation.

Transport of Endocytic Vesicles to the Endocytic Sorting Compartment

Transport of Endocytic Vesicles to the Sorting Compartment along Actin Cables

In yeast, transport of endocytic vesicles and secretory vesicles is mediated by actin cables that arise from the bud cortex and neck and extend into the mother cell (Pruyne et al., 1998). However, the mechanisms regulating these forms of transport seem to be quite different. Transport of secretory vesicles is dependent on the yeast type V myosins, Myo2/4p, which move toward the actin cable plus ends situated at the bud tip [116, 117], whereas endocytic vesicles move myosin-independently toward the actin cable minus ends at the cell interior (Fig. 14.2) [63, 150].

Yeast Rab11 homologs Ypt31p/32p are required for secretory vesicle formation at the TGN [66], and recruit Myo2p to the vesicles [75]. Ypt31p/32p also recruit Sec2p, a guanine nucleotide exchange factor (GEF) for Sec4p GTPase, thereby further recruiting Sec4p (Rab8/10 homolog) (Fig. 14.2) [101, 158]. In addition, Sec2p binds Sec15p, a subunit of the vesicle tethering complex exocyst [87], and concomitantly Ypt31p/32p are released from the vesicle [89].

On the other hand, the transport of endocytic vesicles occurs in the same direction as actin cable flow, and their migration velocity is almost the same as that of the latter (~400 nm/sec) [63]. In the dephosphorylation-mimicking mutant of Pan1p, actin cables are stably bound to endocytic vesicles, suggesting that Pan1p may act

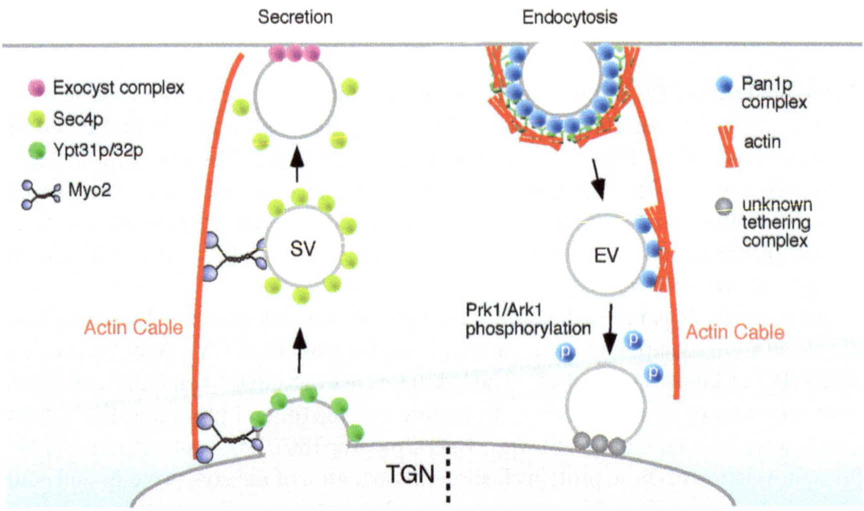

Fig. 14.2 Transport of endocytic and secretory vesicles along actin cables. Pan1p complex on an endocytic vesicle binds to actin to fix the vesicle to the actin cable. After being pinched off from the plasma membrane, the endocytic vesicle moves into the cytosol as a result of actin cable flow. Pan1p phosphorylation by Prk1/Ark1 kinases causes dissociation of the endocytic vesicle from the actin cable, and the vesicle then fuses with the early endosome. A type V Myosin (Myo2p) and specific Rab proteins, Ypt31p/32p and Sec4p, are required for transport of exocytic secretory vesicles. Ypt31p/32p are required for the production of secretory vesicles from the TGN. GTP-bound Ypt31p/32p and Sec4p bind directly to Myo2. Exocyst subunits are also recruited to the vesicle, mediating its fusion with the plasma membrane

as an anchor between endocytic vesicles and actin cables (Fig. 14.2) [148, 166]. How endocytic vesicles moving along actin cables target sorting compartments within the TGN has not been clarified, but since both the secretory and endocytic pathways involve the transport of vesicles between the plasma membrane and the TGN by actin cables, it is reasonable to assume that TGN also interacts with the latter.

Endocytic Sorting Compartment in Budding Yeast

After leaving the plasma membrane, endocytic vesicles that have dissociated the clathrin coat complex are transported to the sorting compartment, where cargos are recycled back to the plasma membrane or brought to late endosomal compartments *en route* to the lysosome/vacuole for degradation (Fig. 14.3) [31, 155]. The structure and properties of the endocytic sorting compartment differ considerably among organisms and can be largely divided into two compartments: early endosomes or the TGN. In mammalian cells, an independent early endosome functions mainly as the sorting compartment for trafficking processes involved in degradation and

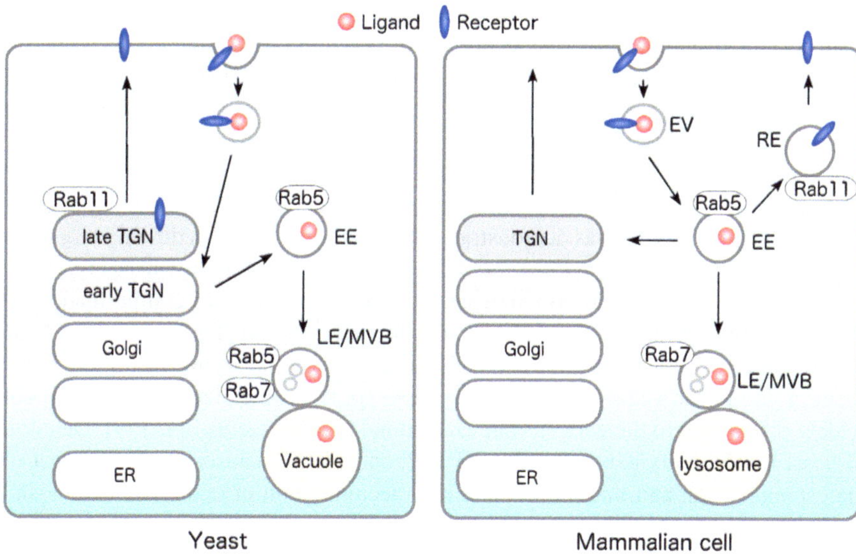

Fig. 14.3 Endocytic pathway in yeast and mammalian cells. In mammalian cells, endocytic vesicles internalized from the plasma membrane are directed to the early/sorting endosome (EE). From there, cargo is transported to the plasma membrane via the recycling endosome (RE), or to the lysosome via the late/MVB endosome. In contrast, in yeast, the endocytosed vesicle directly targets a sub-region within the early TGN. Subsequently, cargo is transported to the vacuole via the late/MVB endosome or recycled back to the plasma membrane

recycling (Fig. 14.3) [49, 50, 61, 153]. The small GTPase, Rab5, a key regulator of early endocytosis, is involved in fusion between early endosomes, late/MVB endosome biogenesis, and endosomal motility [11, 47, 62, 86, 168]. In mammalian cells, three isoforms of Rab5—Rab5a, Rab5b, and Rab5c—are localized at the plasma membrane and early endosome [21]. In contrast, plant cells, such as Arabidopsis (*Arabidopsis thaliana*) or tobacco BY-2 cells, lack early endosomes and instead employ the TGN as an endocytic sorting compartment (Fig. 14.3) [64, 73, 157]. The plant Rab5 homolog (RabF1(Ara7) or Ara6) resides in the TGN and MVB [36, 73], and the TGN—and not the Rab5-positive endosome—is considered as the direct target for endocytic vesicles.

There has been some debate as to whether yeast has a mammalian-type or a plant-type endocytic sorting compartment. Previously, it was thought that budding yeast possesses an endo-lysosomal system that includes early endosomes and late endosomes, similar to those in mammalian cells. This was based on observations that transport to the vacuole passed through at least two biochemically separable endosomal compartments [128]. Additionally, electron microscopy demonstrated that endocytic cargo after internalization was first transported to the tubular endosome-like structure that contains the Q-SNARE Tlg1p (Syntaxin homolog) [107]. However, since Tlg1p and its partner Q-SNARE Tlg2p are localized to the TGN and putative early endosomes [60, 74], it remained unclear whether the TGN or the early endosomes were the direct target of the endocytic vesicle.

In yeast, Golgi cisternae lack the stacked structure observed in mammalian and plant cells, and show *cis*-to-*trans* maturation, thereafter proceeding to the TGN [77, 83, 144, 145]. The Golgi–TGN transition is divided into at least three successive stages: the "Golgi stage," where glycosylation occurs; the "early TGN stage," which receives retrograde traffic; and the "late TGN stage," where transport carriers are produced [144, 145]. Endocytic Q-SNAREs Tlg1p and Tlg2p localize at the early TGN stage, and a recent study using super-resolution confocal live imaging microscopy (SCLIM) [72, 143] has demonstrated that endocytic vesicles directly target the Tlg1p/2p-residing region within the TGN (Tlg2p sub-compartment) (Fig. 14.4) [151]. Endocytic cargos are first delivered to this intra-TGN sub-compartment and then transported to the Rab5-residing endosomes (Fig. 14.4). This suggests that the Tlg2p sub-compartment within the TGN might be the endocytic sorting compartment in yeast. The yeast Rab5 homolog, Vps21p, is not localized to the TGN, but widely distributed to the early-to-late endosomal compartments [44, 149]. Deletion of yeast Rab5 family genes (*VPS21*, YPT52, and *YPT53*) causes a severe defect of endosomal fusion and maturation, and also accumulation of endocytic cargos and vacuolar proteins at early endosomes. This indicates that yeast Rab5 proteins are required for vesicle transport from the TGN to the vacuole, as well as endosomal transport in the endocytic pathway [149]. Thus, yeast appears to have an endosomal structure similar to that of mammalian cells, but uses a specific region of the TGN as the endocytic sorting compartment (Fig. 14.3).

Transport from the TGN to the Vacuole and Mechanism of Endosome Formation

Transport Pathways from TGN to Vacuole (VPS and AP3 Pathways)

Genetic screening in yeast has revealed more than 70 proteins that are necessary for the transport and sorting of newly synthesized proteins to the vacuole [9]. These proteins control two major trafficking pathways from the TGN to the vacuole: the vacuolar protein sorting (VPS) and the adaptor protein (AP)-3-mediated sorting pathways [12, 26, 90]. Many of these proteins are also conserved in higher eukaryotes as VPS proteins. Proteins trafficked via the VPS pathway transit through endosomes *en route* to the vacuole, whereas the AP-3 pathway mediates transport to the vacuole independently of endosomes. Endocytic cargo transport by the VPS pathway requires yeast Rab5 Vps21 function, but transport from the TGN to lysosomes/vacuoles is not completely blocked in cells lacking all Rab5 family members [149]. In cells lacking Rab5, inhibition of the AP-3 pathway almost completely suppresses transport of the endocytosis pathway, suggesting that the endocytic traffic branches into two routes to reach the vacuole: a Rab5-dependent VPS pathway and a Rab5-independent AP-3-mediated pathway [149].

Transport from the TGN to Endosomes in the Endocytic Pathway

It was also believed that in yeast the transport route from the TGN to endosomes converged with the endocytic pathway at late/MVB endosomes [34, 105, 111, 156]. However, based on the recent finding that endocytic vesicles directly target the endocytic sorting compartment within the TGN (Fig. 14.4) [151], it is suggested that the endocytic and biosynthetic pathways share the route from the TGN to the vacuole after they converge at the sorting compartment [35, 151]. Although the TGN serves as an endocytic sorting compartment in both yeast and plants, the mechanism of endocytic cargo transport from the TGN to the vacuole seems to differ between them. Whereas in yeast cargo from the TGN moves to yeast Rab5 (Vps21p)-residing endosomes (Fig. 14.3) [151], in *Arabidopsis* the ESCRT-mediated formation of the MVB begins at the TGN without the mediation

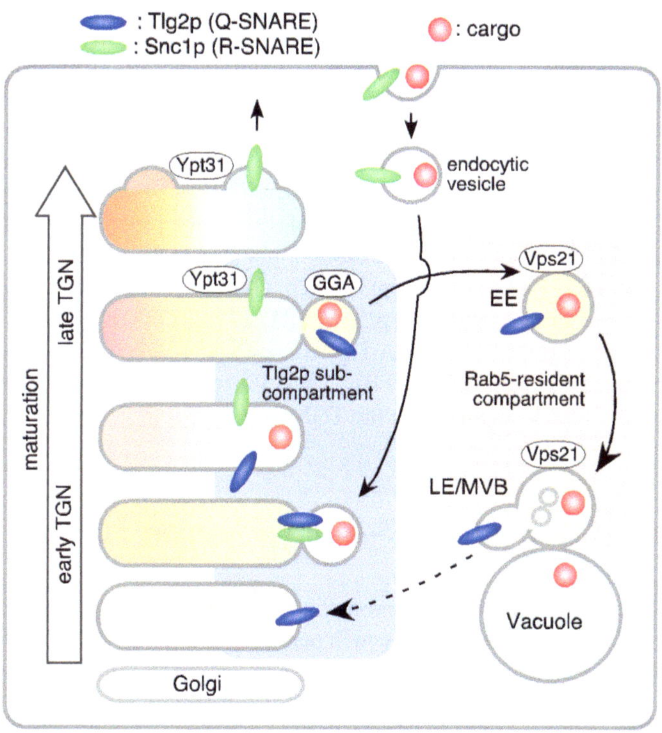

Fig. 14.4 Endocytic sorting and endosomal compartments in yeast. The Golgi/TGN maturation process can be classified into three successive stages: the Golgi stage, the early TGN stage, and the late TGN stage. The Tlg2p sub-compartment exists as a sub-region within the early TGN stage, and endocytosed cargo is first transported to this compartment. The endocytic R-SNARE Snc1p is directed to the Q-SNARE complex, including Tlg2p, transiently localized to the subdomain of the early TGN, and subsequently recycled back to the plasma membrane. In contrast, Tlg2p is transported to the late/MVB endosome via the Vps21p-residing compartment [151]

of endosomes, whereafter the MVB fuses directly with the vacuole [114]. Thus, it seems that *Arabidopsis* lacks early endosomes and that endocytic cargos are directly transported from the TGN to MVB [64, 157].

Transport from the TGN to the endosome is mediated by several different types of CCVs. In yeast, three classes of clathrin adaptors, the AP-1 complex, Gga1p/2p (GGA homologs), and Ent3p/5p (Epsin homologs), have been identified [18, 28, 29, 33, 40]. GGAs and AP-1 form two distinct CCVs, the GGA-enriched vesicle (GGA vesicle), including Ent3p and a minor population of Ent5p, and the AP-1-enriched vesicle (AP-1 vesicle), which includes most of Ent5p and resides at the late TGN (Fig. 14.5) [33, 144, 145]. GGA vesicles are formed approximately slightly earlier than the AP-1 vesicles, and GGA vesicles mediate the transport of cargos from the TGN to the vacuole, whereas AP-1 vesicles mediate intra-Golgi recycling [17, 18]. Several upstream factors that recruit these clathrin adaptors and regulate the progressive formation of GGA and AP-1 vesicles have been clarified. Among them, phosphatidylinositol-4-phosphate (PtdIns(4)P), which is generated by Golgi-resident PI 4-kinase Pik1p, is reported to play a central role in this process (Fig. 14.5) [33, 159, 160]. Pik1p is initially recruited to the Golgi through interaction with Arf1 GTPase and the yeast Frequenin homolog Frq1p [33, 54, 58]. PtdIns(4)P increases GGA localization at the TGN, and GGA in turn further recruits Pik1p [32, 33, 58, 169]. These positive feedback mechanisms facilitate the formation of GGA vesicles at an earlier TGN stage relative to the formation of AP-1 vesicles. Simultaneously, they induce localization of AP-1 and Ent5p at the TGN, leading to sequential formation of AP-1 vesicles at the later stage of the TGN [33]. The yeast Rab11s, Ypt31p/32p, are other important regulators implicated in the formation of transport carriers at the TGN [4, 85, 101]. Ypt31p/32p, Gga1p/2p, and Pik1p have overlapping roles in the recruitment of Ent3p/5p and AP-1 [94]. Ypt31p also activates Arf1p through its GEF, and the activated Arf1p recruits Ypt31p/32p GEF to promote further activation of Ypt31p/32p [85, 110]. Thus, Pik1p and Ypt31p/32p appear to be key regulators of CCV formation at the TGN (Fig. 14.5).

Activation of Vps21p by Rab5-GEF Vps9p at the TGN

Studies using yeast in the 1990s demonstrated that inhibition of the secretory pathway or treatment with Brefeldin A, which perturbs intra-Golgi trafficking [37, 113], causes defects in endocytic transport and accumulation of cargo at the early-stage endocytic compartment [57]. However, it was unclear how inhibition of the secretory pathway affects endocytic transport toward the vacuole. Recent studies have demonstrated that post-Golgi transport mediated by clathrin adaptors plays an important role in the activation of yeast Rab5 Vps21p, and that disturbance of post-Golgi transport by Brefeldin A inhibits the activation of Vps21p [95, 149].

The Vps9 domain-containing proteins, Vps9p and Muk1p, have been identified as specific GEFs for yeast Rab5s [52, 103], and Vps9p functions as a primary Rab5 GEF required for Vps21p-mediated endosome formation [13, 103]. Vps9p can be

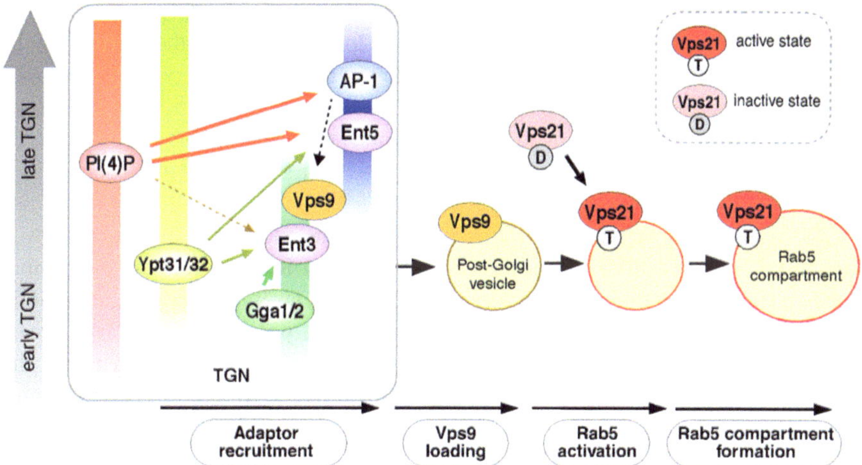

Fig. 14.5 The functional relationship between clathrin adaptors and their upstream regulators in the process of yeast Rab5 activation. The process of yeast Rab5 Vps21p activation via GGA and AP-1-enriched regions. Temporal dynamics of PtdIns(4)P, Ypt31p/32p, and the clathrin adaptors are drawn as thick colored lines. GGA- or AP-1-enriched regions are formed sequentially at the late TGN. PtdIns(4)P produced by Pik1p is required for the recruitment of AP-1 and Ent5p, and Ypt31p/32p recruits Ent3p and Ent5p to the TGN. Gga1p/2p also recruits Ent3p to the GGA-enriched region. Efficient recruitment of the clathrin adaptors by these upstream factors facilitates the loading of Vps9p onto the transport vesicles, whereafter Vps9p recruits and activates Vps21p at the vesicles [94]

recruited to endosomes through the interaction of its CUE domain with ubiquitinated endocytic cargo proteins, thereby promoting the endosomal localization of Vps21p [126]. However, the finding that non-ubiquitinated cargo also requires Vps21p for its transport from the TGN to the vacuole in the VPS pathway suggests that a ubiquitin- and endocytic pathway-independent mechanism exists to recruit Vps21p to the endosome. Interestingly, it has been demonstrated that transport of endocytic vesicles is not essential, but that vesicle transport from the TGN is crucial for Vps21p-mediated endosome formation in yeast [95]. Taken together with the observation that endocytic vesicles directly target a specific region in the TGN [151], it is reasonable to consider that endosome formation is dependent on post-Golgi trafficking.

Ent3p/5p is known to be required for Vps21p activity, and deletion of Ent3p/5p decreases it to about one-third of the level in wild-type cells [95]. In contrast, deletion of GGAs or AP-1 complex, or even both, has a negligible effect on Vps21p activity, whereas deletion of the AP-1 complex in cells lacking Ent3p/5p suppresses Vps21p activity almost completely [94, 95]. Consistent with these observations, it has been demonstrated that Vps9p is first recruited to the TGN dependently on the Arf1 GTPase, and then transported to the endosome to activate Vps21p through Ent3p/5p-mediated vesicle transport [95]. Thus, these adaptors appear to play a distinct role in the process of Vps21p activation: Ent3p/5p and AP-1 are required for loading Vps9p on the vesicles that transport Vps9p to the Vps21p compartment, while GGAs, together with Pik1p and Ypt31p/32p, have a role in the recruitment of Ent3p/5p and AP-1 to the TGN (Fig. 14.5) [94, 95]. In mammalian cells, Rabaptin-5, which is a functional modulator of mammalian Rab5-GEF Rabex-5, has been shown to associate with TGN-residing clathrin adaptors, similarly to yeast Vps9p [84, 97], suggesting that mammalian Rab5 might be activated by the Rabex-5–Rabaptin-5 complex in post-Golgi vesicle transport.

The transition process from the early to the late endosome is regulated by a sequential shift of activity from the early endosomal Rab5 to the late endosomal Rab7, a process termed Rab conversion, which is conserved in many eukaryotic cells [106, 112]. Since the sequential activation from yeast Rab5 Vps21p to yeast Rab7 Ypt7 via Rab7-GEF, Mon1p/Ccz1p, has also been reported in yeast [98], the overall mechanisms regulating the transition from early to late/MVB endosomes appear to exist in yeast.

Localization of Endocytic SNAREs at the TGN

SNARE proteins are key factors that determine organelle identity and operate at the center of fusion reactions [115, 132]. SNARE proteins comprise four classes based on the structures of their SNARE motifs: Qa-, Qb-, Qc-, and R-SNARE [65, 162]. The yeast R-SNAREs Snc1p/2p (VAMP2 homologs), originally identified as proteins residing in secretory vesicles, forms complexes with Qa-SNAREs Sso1p/2p (Syntaxin1 homolog) and Qbc-SNARE Sec9p (or Spo20p) (SNAP25 homolog) to drive fusion of secretory vesicles with the plasma membrane [45, 108]. In addition, Snc1p/2p function as R-SNAREs in endocytic pathways by interacting with the Q-SNARE Tlg1p/Tlg2p/Vti1p complex residing on the TGN (Fig. 14.4) [51], thereby cycling between the TGN and the plasma membrane [6, 74].

As described above, the Q-SNARE Tlg1p/Tlg2p/Vti1p complex functions as an endocytic SNARE complex with R-SNARE Snc1p residing on the endocytic vesicle [51]. The mammalian SNARE protein Syntaxin16 is a functional homolog of the yeast SNARE Tlg2p, and its expression fully complements the mutant phenotypes of yeast *tlg2Δ* mutant [127, 136, 141]. Both Syntaxin16 and Tlg2p bind to the SM protein Vps45 via their N-terminal region, facilitating conversion from a closed to an open conformation [38, 41]. Similar to the yeast endocytic Q-SNARE complex (Tlg1p/Tlg2p/Vti1p), Syntaxin16 localizes to the TGN and forms a complex with Syntaxin6 (Qc-SNARE), which is a homolog of yeast Tlg1p, and mammalian Vti1p (Qb-SNARE). This suggests that yeast and mammalian cells have similar Q-SNARE complexes at the TGN. However, this mammalian Q-SNARE complex does not form a complex with R-SNARE VAMP2 (a homolog of yeast Snc1p). Thus, it remains unclear whether a direct transport pathway from plasma membrane to the TGN exists in mammalian cells.

Retrograde Transport from Endosomes to the TGN or the Plasma Membrane

A portion of the proteins transported to endosomes is usually recycled to the Golgi/TGN by a specific retrieval process [8, 118]. These proteins include receptors and enzymes that cycle between the Golgi/TGN and endosomes, and proteins that are endocytosed, such as the endocytic/exocytic R-SNARE Snc1p and the chitin synthase Chs3p [26, 59, 74, 154]. Previous studies have suggested the existence of at least two distinct retrieval pathways from the endosomes. One is the transport pathway from early endosome to Golgi/TGN, which is the pathway normally followed by Snc1p or Chs3p [30, 104]. However, recent studies have demonstrated that a specific region in the yeast TGN serves as an early endosome, suggesting that this recycling pathway may be employed for direct transport from the TGN to the plasma membrane without the mediation of early endosomes [151].

The other retrieval pathway is the pathway from the late/MVB endosomal to Golgi/TGN transport, which is used by the carboxypeptidase Y receptor Vps10p and the TGN Q-SNAREs, Tlg1p and Tlg2p [69, 105]. This pathway is known to be mediated by the retromer and Snx4p coat complexes. These complexes are recruited to the endosomal surface, where they recognize and sort cargos and deform the membrane into recycling tubules that bud from the endosome and target the Golgi/TGN [81]. A recent study has shown that the SNX-BAR protein Mvp1p, which is required for efficient retrieval of Kex2p and Vps10p, mediates an endosomal recycling pathway that is mechanistically distinct from the retromer and Snx4p pathways [140]. Among these proteins, Snc1p mis-localizes to the vacuolar membrane in cells lacking Snx4p [80], suggesting that a portion of the Snc1p is transported from the TGN to the late/MVB endosomes.

Endocytic Recycling Pathway Through the TGN

In mammalian cells, cargos returning to the plasma membrane can be recycled directly from early endosomes (fast recycling) or indirectly from endosomal recycling compartments, such as the recycling endosome and TGN (slow recycling) (Fig. 14.3) [124, 133]. The slow recycling pathway is mediated by Rab11, which is localized to recycling endosomes [48, 109, 153]. Yeast does not possess independent recycling endosomes but has an endocytic sorting compartment in the TGN, where yeast Rab11 homologs Ypt31p/32p reside, suggesting that this compartment within the TGN functions as a recycling compartment (Fig. 14.4) [151]. Ypt31p/32p are known to be recruited to the compartment during the late TGN stage [151].

Rcy1p is another important regulator implicated in the endocytic recycling pathway in yeast [43]. Rcy1p is an F-box protein that localizes at the late TGN and forms a complex with the GTP-bound form of Ypt31p/32p [23]. Cells lacking the *RCY1* gene show accumulation of endocytic cargos in large compartments, which contain both the endocytic/exocytic R-SNARE Snc1p and Q-SNARE Tlg1p [6, 81]. Although deletion of the *RCY1* gene causes a severe defect in the endocytic recycling pathway, its function is still unclear. Rcy1p has been shown to have genetic and/or physical interactions with several proteins implicated in the recycling pathway, such as Ypt31p/32p, Skp1p (a subunit of SCF ubiquitin ligase complex), and the Drs2p-Cdc50p PS flippase [23, 43, 53].

Endosome-to-Golgi/TGN Retrograde Trafficking Pathways

Endosome-to-Golgi/TGN retrograde trafficking pathways are mediated by three distinct SNX-BAR proteins, the retromer, Snx4p, and Mvp1p [140]. The retromer complex consists of five proteins; Vps5p and Vps17p form a dimer and associate with a complex of Vps26p, Vps29p, and Vps35p [120]. All of these proteins are required for recycling of Vps10p and Tlg1p/2p from late/MVB endosomes, but not for the recycling of Snc1p [74]. The retromer retrieves a variety of Golgi/TGN proteins, and the Vps35p/Vps26p/Vps29p complex recognizes and mediates packaging of the cargos into transport carriers [100]. Consensus sorting signals recognized by this complex consist of short motifs that contain bulky aromatic residues, including YSSL, WKY, and FxFxD. Vps5p and Vps17p are sorting nexin family proteins possessing a Phox-homology (PX) domain, which specifically binds to PtdIns(3)P [25]. The Vps5p-Vps17p dimer binds to endosomes through PX domain-PtdIns(3)P interaction and deforms the membrane into recycling tubules [10, 167].

The Snx4p dimer functions in a retrograde trafficking pathway that is parallel to but independent of the retromer pathway [56]. Three SNX-BAR proteins (Snx4p, Snx41p, Atg20p/Snx42p) form two distinct complexes (Snx4p-Snx41p and Snx4p-Atg20p/Snx42p) that mediate the endosome-to-Golgi/TGN trafficking of cargos such as Snc1p and arginine permease Can1p [99, 119]. Snx4p family proteins are also involved in recycling of the ABC transporter Ste6p and other cell surface proteins to the plasma membrane [71, 125]. The sorting signals recognized by Snx4p-Snx41p and Snx4p-Atg20p/Snx42p have not been identified, although several cargo proteins, including Snc1p, seem to interact with Snx4p through the sorting signals recognized by the retromer complex [56, 139]. Another SNX-BAR protein, Mvp1p, has intrinsic membrane tubulation and vesiculation activity, similar to its human homolog, SNX8, and promotes fission of Vps5p-Vps17p-coated transport carriers from the endosome [140].

Arl1p (Arf-like protein 1) is a highly conserved small GTPase that functions, together with its effectors, as an important regulator of endosome-to-Golgi/TGN retrograde transport [91]. In yeast, recruitment of Arl1p to the *trans*-Golgi requires a second Arf-like GTPase, Arl3p [3, 122]. Arl3p lacks a myristoylation site but is acetylated at the N terminus by the N-terminal acetyltransferase complex [123]. This modification allows Arl3p recruitment to the *trans*-Golgi through interaction with Sys1p [3, 122]. Arl3p and other factors activate the Arl1p-GEF Syt1p by facilitating its phosphorylation, thus promoting GDP-GTP exchange and Arl1 activation

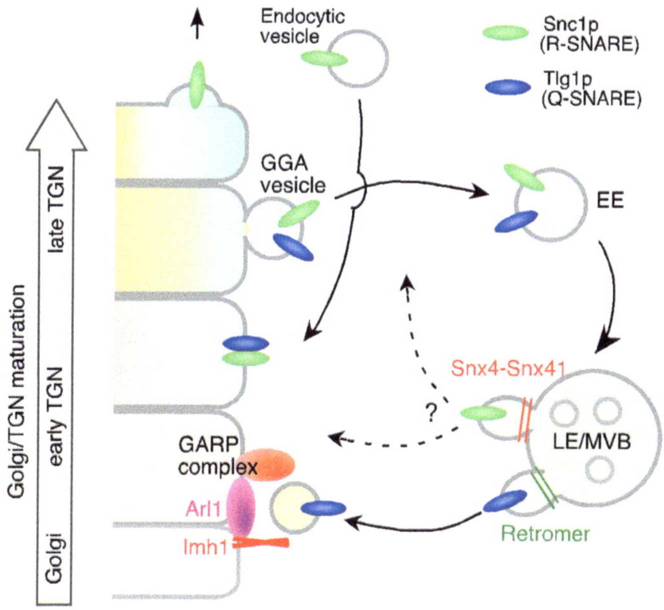

Fig. 14.6 The retrograde pathway from the TGN to the late/MVB endosome. Endocytic R-SNARE Snc1p on an endocytic vesicle binds to the Q-SNARE complex Tlg1p/Tlg2p at the early TGN. Tlg1p/2p are then transported to the endosome via GGA vesicles. After being transported to the endosomes, Tlg1p/2p are recycled back to Golgi/TGN by the retrograde transport mediated by the retromer complex. Retromer complex deforms the endosomal membrane to form cargo-containing recycling tubules/vesicles. The GARP complex is recruited to the TGN through the Arl3-Arl1-Imh1 cascade and interacts with the recycling vesicles derived from endosomes, thereby promoting fusion of the vesicles to the TGN. The Snx4p-Snx41p complex mediates retromer-independent cargo transport such as Snc1p from the late/MVB endosome

[22, 76]. GTP-bound Arl1p further recruits the long coiled-coil protein Imh1p (Golgin homolog) (Fig. 14.6) [92, 93, 163].

Another key regulator involved in vesicle tethering at the *trans*-Golgi is yeast Rab6 Ypt6p. Ypt6p is also critical for the retrograde transport of vesicles from endosomes to the *trans*-Golgi [79]. Ypt6p is activated by its GEF (Ric1p/Rgp1p complex) [129], and promotes endosome-derived vesicle fusion via the recruitment of the GARP (Golgi-associated retrograde protein) tethering complex consisting of four subunits (Vps51p/52p/53p/54p) [27, 102, 122, 130]. Vps52p mediates the interaction between Ypt6p and the GARP complex, and deletion of Ypt6p or GARP inhibits recycling of the endocytic/exocytic R-SNARE Snc1p and the Q-SNAREs Tlg1p/2p from endosomes to the Golgi/TGN [130].

Genetic interactions between components of the Arl1p and Ypt6p pathways suggest that they are parallel pathways with overlapping functions in retrograde transport from endosomes to the Golgi/TGN [5, 102, 165]. Arl1p was also shown to interact with the GARP complex via Vps53p [102, 130]; however, the lack of Arl1p, unlike the lack of Ypt6p, does not affect the localization of the GARP complex at the *trans*-Golgi or the recycling of Snc1p [102]. Thus, the recruitment of the GARP complex to the TGN seems to be mainly mediated by Ypt6p [24]. Overexpression of Arl1p and Imh1p restores endosome-to-TGN trafficking in cells lacking Ypt6p (Chen MBC 2018), suggesting some level of cross-talk between the two pathways.

Conclusion and Future Perspectives

The TGN is a crucial sorting station in the biosynthetic pathways, including the secretory and VPS pathways, and its importance in the yeast endocytosis pathway has been further demonstrated by recent studies. Endocytic cargo is transported directly to the TGN and sorted into degradation or recycling pathways. However, how endocytic vesicles are transported to the TGN and the sorting mechanism of the cargo within the TGN are still largely unknown. On the other hand, the retrograde transport pathways from endosomes to the TGN are well conserved among organisms, and the molecular mechanisms that regulate them are similar. Thus, although the structure of endosomes seems to be different among different organisms, common mechanisms regulating transport from the TGN to the endosomes should also exist. Recent studies suggest that the TGN is not a homogeneous organelle and can be divided into distinct sub-regions that produce specific carriers for efficient cargo sorting. How these specialized sub-regions are formed during Golgi/TGN maturation and how cargo is properly sorted into different carriers are important questions for the future.

Acknowledgments The authors acknowledge the funding support of JSPS KAKENHI GRANT #21 K06087 and JSPS KAKENHI GRANT #22 K061330 to J.T.

Compliance with Ethical Standards This article does not contain any studies with human or animal subjects performed by the authors.

Conflicts of Interest The authors declare that they have no conflicts of interest with the contents of this article.

References

1. Amberg DC (1998) Three-dimensional imaging of the yeast actin cytoskeleton through the budding cell cycle. Mol Biol Cell 9:3259–3262
2. Apel AR, Hoban K, Chuartzman S, Tonikian R, Sidhu S, Schuldiner M, Wendland B, Prosser D (2017) Syp1 regulates the clathrin-mediated and clathrin-independent endocytosis of multiple cargo proteins through a novel sorting motif. Mol Biol Cell 28:2434–2448
3. Behnia R, Panic B, Whyte JR, Munro S (2004) Targeting of the Arf-like GTPase Arl3p to the Golgi requires N-terminal acetylation and the membrane protein Sys1p. Nat Cell Biol 6:405–413

4. Benli M, Doring F, Robinson DG, Yang X, Gallwitz D (1996) Two GTPase isoforms, Ypt31p and Ypt32p, are essential for Golgi function in yeast. EMBO J 15:6460–6475
5. Bensen ES, Yeung BG, Payne GS (2001) Ric1p and the Ypt6p GTPase function in a common pathway required for localization of trans-Golgi network membrane proteins. Mol Biol Cell 12:13–26
6. Best JT, Xu P, McGuire JG, Leahy SN, Graham TR (2020) Yeast synaptobrevin, Snc1, engages distinct routes of postendocytic recycling mediated by a sorting nexin, Rcy1-COPI, and retromer. Mol Biol Cell 31:944–962
7. Boettner DR, Chi RJ, Lemmon SK (2011) Lessons from yeast for clathrin-mediated endocytosis. Nat Cell Biol 14:2–10
8. Bonifacino JS, Rojas R (2006) Retrograde transport from endosomes to the trans-Golgi network. Nat Rev Mol Cell Biol 7:568–579
9. Bowers K, Stevens TH (2005) Protein transport from the late Golgi to the vacuole in the yeast Saccharomyces cerevisiae. Biochim Biophys Acta 1744:438–454
10. Bravo J, Karathanassis D, Pacold CM, Pacold ME, Ellson CD, Anderson KE, Butler PJ, Lavenir I, Perisic O, Hawkins PT, Stephens L, Williams RL (2001) The crystal structure of the PX domain from p40(phox) bound to phosphatidylinositol 3-phosphate. Mol Cell 8:829–839
11. Bucci C, Parton RG, Mather IH, Stunnenberg H, Simons K, Hoflack B, Zerial M (1992) The small GTPase rab5 functions as a regulatory factor in the early endocytic pathway. Cell 70:715–728
12. Burd CG, Babst M, Emr SD (1998) Novel pathways, membrane coats and PI kinase regulation in yeast lysosomal trafficking. Semin Cell Dev Biol 9:527–533
13. Burd CG, Mustol PA, Schu PV, Emr SD (1996) A yeast protein related to a mammalian Ras-binding protein, Vps9p, is required for localization of vacuolar proteins. Mol Cell Biol 16:2369–2377
14. Burri L, Lithgow T (2004) A complete set of SNAREs in yeast. Traffic 5:45–52
15. Campelo F, Malhotra V (2012) Membrane fission: the biogenesis of transport carriers. Annu Rev Biochem 81:407–427
16. Carroll SY, Stimpson HE, Weinberg J, Toret CP, Sun Y, Drubin DG (2012) Analysis of yeast endocytic site formation and maturation through a regulatory transition point. Mol Biol Cell 23:657–668
17. Casler JC, Glick BS (2020) A microscopy-based kinetic analysis of yeast vacuolar protein sorting. elife 9:e56844
18. Casler JC, Johnson N, Krahn AH, Pantazopoulou A, Day KJ, Glick BS (2021) Clathrin adaptors mediate two sequential pathways of intra-Golgi recycling. J Cell Biol 221:e202103199
19. Chang JS, Henry K, Geli MI, Lemmon SK (2006) Cortical recruitment and nuclear-cytoplasmic shuttling of Scd5p, a protein phosphatase-1-targeting protein involved in actin organization and endocytosis. Mol Biol Cell 17:251–262
20. Chapa-y-Lazo B, Allwood EG, Smaczynska-de R II, Snape ML, Ayscough KR (2014) Yeast endocytic adaptor AP-2 binds the stress sensor Mid2 and functions in polarized cell responses. Traffic 15:546–557
21. Chavrier P, Parton RG, Hauri HP, Simons K, Zerial M (1990) Localization of low molecular weight GTP binding proteins to exocytic and endocytic compartments. Cell 62:317–329
22. Chen KY, Tsai PC, Hsu JW, Hsu HC, Fang CY, Chang LC, Tsai YT, Yu CJ, Lee FJ (2010) Syt1p promotes activation of Arl1p at the late Golgi to recruit Imh1p. J Cell Sci 123:3478–3489
23. Chen SH, Chen S, Tokarev AA, Liu F, Jedd G, Segev N (2005) Ypt31/32 GTPases and their novel F-box effector protein Rcy1 regulate protein recycling. Mol Biol Cell 16:178–192
24. Chen YT, Wang IH, Wang YH, Chiu WY, Hu JH, Chen WH, Lee FJS (2019) Action of Arl1 GTPase and golgin Imh1 in Ypt6-independent retrograde transport from endosomes to the trans-Golgi network. Mol Biol Cell 30:1008–1019
25. Chi RJ, Harrison MS, Burd CG (2015) Biogenesis of endosome-derived transport carriers. Cell Mol Life Sci 72:3441–3455

26. Conibear E, Stevens TH (1998) Multiple sorting pathways between the late Golgi and the vacuole in yeast. Biochim Biophys Acta 1404:211–230
27. Conibear E, Stevens TH (2000) Vps52p, Vps53p, and Vps54p form a novel multisubunit complex required for protein sorting at the yeast late Golgi. Mol Biol Cell 11:305–323
28. Copic A, Starr TL, Schekman R (2007) Ent3p and Ent5p exhibit cargo-specific functions in trafficking proteins between the trans-Golgi network and the endosomes in yeast. Mol Biol Cell 18:1803815
29. Costaguta G, Duncan MC, Fernandez GE, Huang GH, Payne GS (2006) Distinct roles for TGN/endosome epsin-like adaptors Ent3p and Ent5p. Mol Biol Cell 17:3907–3920
30. Cullen PJ, Korswagen HC (2011) Sorting nexins provide diversity for retromer-dependent trafficking events. Nat Cell Biol 14:29–37
31. Cullen PJ, Steinberg F (2018) To degrade or not to degrade: mechanisms and significance of endocytic recycling. Nat Rev Mol Cell Biol 19:679–696
32. Daboussi L, Costaguta G, Ghukasyan R, Payne GS (2017) Conserved role for Gga proteins in phosphatidylinositol 4-kinase localization to the trans-Golgi network. Proc Natl Acad Sci USA 114:3433–3438
33. Daboussi L, Costaguta G, Payne GS (2012) Phosphoinositide-mediated clathrin adaptor progression at the trans-Golgi network. Nat Cell Biol 14:239–248
34. Davis NG, Horecka JL, Sprague GF Jr (1993) Cis- and trans-acting functions required for endocytosis of the yeast pheromone receptors. J Cell Biol 122:53–65
35. Day KJ, Casler JC, Glick BS (2018) Budding yeast has a minimal endomembrane system. Dev Cell 44:56–72.e54
36. Dettmer J, Hong-Hermesdorf A, Stierhof YD, Schumacher K (2006) Vacuolar H+-ATPase activity is required for endocytic and secretory trafficking in Arabidopsis. Plant Cell 18:715–730
37. Dinter A, Berger EG (1998) Golgi-disturbing agents. Histochem Cell Biol 109:571–590
38. Dulubova I, Yamaguchi T, Gao Y, Min SW, Huryeva I, Südhof TC, Rizo J (2002) How Tlg2p/syntaxin 16 'snares' Vps45. EMBO J 21:3620–3631
39. Duncan MC, Cope MJ, Goode BL, Wendland B, Drubin DG (2001) Yeast Eps15-like endocytic protein, Pan1p, activates the Arp2/3 complex. Nat Cell Biol 3:687–690
40. Duncan MC, Costaguta G, Payne GS (2003) Yeast epsin-related proteins required for Golgi-endosome traffic define a gamma-adaptin ear-binding motif. Nat Cell Biol 5:77–81
41. Eisemann TJ, Allen F, Lau K, Shimamura GR, Jeffrey PD, Hughson FM (2020) The Sec1/Munc18 protein Vps45 holds the Qa-SNARE Tlg2 in an open conformation. elife 9
42. Enshoji M, Miyano Y, Yoshida N, Nagano M, Watanabe M, Kunihiro M, Siekhaus DE, Toshima JY, Toshima J (2022) Eps15/Pan1p is a master regulator of the late stages of the endocytic pathway. J Cell Biol 221
43. Galan JM, Wiederkehr A, Seol JH, Haguenauer-Tsapis R, Deshaies RJ, Riezman H, Peter M (2001) Skp1p and the F-box protein Rcy1p form a non-SCF complex involved in recycling of the SNARE Snc1p in yeast. Mol Cell Biol 21:3105–3117
44. Gerrard SR, Bryant NJ, Stevens TH (2000) VPS21 controls entry of endocytosed and biosynthetic proteins into the yeast prevacuolar compartment. Mol Biol Cell 11:613–626
45. Gerst JE, Rodgers L, Riggs M, Wigler M (1992) Snc1, a yeast homolog of the synaptic vesicle-associated membrane-protein Synaptobrevin gene family - genetic interactions with the Ras and cap genes. Proc Natl Acad Sci USA 89:4338–4342
46. Goode BL, Eskin JA, Wendland B (2015) Actin and endocytosis in budding yeast. Genetics 199:315–358
47. Gorvel JP, Chavrier P, Zerial M, Gruenberg J (1991) rab5 controls early endosome fusion in vitro. Cell 64:915–925
48. Grant BD, Donaldson JG (2009) Pathways and mechanisms of endocytic recycling. Nat Rev Mol Cell Biol 10:597–608
49. Gruenberg J (2001) The endocytic pathway: a mosaic of domains. Nat Rev Mol Cell Biol 2:721–730

50. Gruenberg J, Maxfield FR (1995) Membrane transport in the endocytic pathway. Curr Opin Cell Biol 7:552–563
51. Gurunathan S, Chapman-Shimshoni D, Trajkovic S, Gerst JE (2000) Yeast exocytic v-SNAREs confer endocytosis. Mol Biol Cell 11:3629–3643
52. Hama H, Tall GG, Horazdovsky BF (1999) Vps9p is a guanine nucleotide exchange factor involved in vesicle-mediated vacuolar protein transport. J Biol Chem 274:15284–15291
53. Hanamatsu H, Fujimura-Kamada K, Yamamoto T, Furuta N, Tanaka K (2014) Interaction of the phospholipid flippase Drs2p with the F-box protein Rcy1p plays an important role in early endosome to trans-Golgi network vesicle transport in yeast. J Biochem 155:51–62
54. Hendricks KB, Wang BQ, Schnieders EA, Thorner J (1999) Yeast homologue of neuronal frequenin is a regulator of phosphatidylinositol-4-OH kinase. Nat Cell Biol 1:234–241
55. Henne WM, Boucrot E, Meinecke M, Evergren E, Vallis Y, Mittal R, McMahon HT (2010) FCHo proteins are nucleators of clathrin-mediated endocytosis. Science 328:1281–1284
56. Hettema EH, Lewis MJ, Black MW, Pelham HR (2003) Retromer and the sorting nexins Snx4/41/42 mediate distinct retrieval pathways from yeast endosomes. EMBO J 22:548–557
57. Hicke L, Zanolari B, Pypaert M, Rohrer J, Riezman H (1997) Transport through the yeast endocytic pathway occurs through morphologically distinct compartments and requires an active secretory pathway and Sec18p/N-ethylmaleimide-sensitive fusion protein. Mol Biol Cell 8:13–31
58. Highland CM, Fromme JC (2021) Arf1 directly recruits the Pik1-Frq1 PI4K complex to regulate the final stages of Golgi maturation. Mol Biol Cell 32:1064–1080
59. Holthuis JC, Nichols BJ, Pelham HR (1998a) The syntaxin Tlg1p mediates trafficking of chitin synthase III to polarized growth sites in yeast. Mol Biol Cell 9:3383–3397
60. Holthuis JCM, Nichols BJ, Dhruvakumar S, Pelham HRB (1998b) Two syntaxin homologues in the TGN/endosomal system of yeast. EMBO J 17:113–126
61. Hopkins CR, Miller K, Beardmore JM (1985) Receptor-mediated endocytosis of transferrin and epidermal growth factor receptors: a comparison of constitutive and ligand-induced uptake. J Cell Sci Suppl 3:173–186
62. Horiuchi H, Giner A, Hoflack B, Zerial M (1995) A GDP/GTP exchange-stimulatory activity for the Rab5-RabGDI complex on clathrin-coated vesicles from bovine brain. J Biol Chem 270:11257–11262
63. Huckaba TM, Gay AC, Pantalena LF, Yang HC, Pon LA (2004) Live cell imaging of the assembly, disassembly, and actin cable-dependent movement of endosomes and actin patches in the budding yeast, Saccharomyces cerevisiae. J Cell Biol 167:519–530
64. Ito E, Uemura T, Ueda T, Nakano A (2016) Distribution of RAB5-positive multivesicular endosomes and the trans-Golgi network in root meristematic cells of Arabidopsis thaliana. Plant Biotechnol (Tokyo) 33:281–286
65. Jahn R, Scheller RH (2006) SNAREs–engines for membrane fusion. Nat Rev Mol Cell Biol 7:631–643
66. Jedd G, Mulholland J, Segev N (1997) Two new Ypt GTPases are required for exit from the yeast trans-Golgi compartment. J Cell Biol 137:563–580
67. Kaksonen M, Roux A (2018) Mechanisms of clathrin-mediated endocytosis. Nat Rev Mol Cell Biol 19:313–326
68. Kaksonen M, Toret CP, Drubin DG (2006) Harnessing actin dynamics for clathrin-mediated endocytosis. Nat Rev Mol Cell Biol 7:404–414
69. Kama R, Robinson M, Gerst JE (2007) Btn2, a Hook1 ortholog and potential batten disease-related protein, mediates late endosome-Golgi protein sorting in yeast. Mol Cell Biol 27:605–621
70. Kishimoto T, Sun Y, Buser C, Liu J, Michelot A, Drubin DG (2011) Determinants of endocytic membrane geometry, stability, and scission. Proc Natl Acad Sci USA 108:E979–E988
71. Krsmanovic T, Pawelec A, Sydor T, Kölling R (2005) Control of Ste6 recycling by ubiquitination in the early endocytic pathway in yeast. Mol Biol Cell 16:2809–2821

72. Kurokawa K, Nakano A (2020) Live-cell imaging by super-resolution confocal live imaging microscopy (SCLIM): simultaneous three-color and four-dimensional live cell imaging with high space and time resolution. Bio-Protocol 10:e3732
73. Lam SK, Siu CL, Hillmer S, Jang S, An G, Robinson DG, Jiang L (2007) Rice SCAMP1 defines clathrin-coated, trans-golgi-located tubular-vesicular structures as an early endosome in tobacco BY-2 cells. Plant Cell 19:296–319
74. Lewis MJ, Nichols BJ, Prescianotto-Baschong C, Riezman H, Pelham HR (2000) Specific retrieval of the exocytic SNARE Snc1p from early yeast endosomes. Mol Biol Cell 11:23–38
75. Lipatova Z, Tokarev AA, Jin Y, Mulholland J, Weisman LS, Segev N (2008) Direct interaction between a myosin V motor and the Rab GTPases Ypt31/32 is required for polarized secretion. Mol Biol Cell 19:4177–4187
76. Liu YW, Huang CF, Huang KB, Lee FJ (2005) Role for Gcs1p in regulation of Arl1p at trans-Golgi compartments. Mol Biol Cell 16:4024–4033
77. Losev E, Reinke CA, Jellen J, Strongin DE, Bevis BJ, Glick BS (2006) Golgi maturation visualized in living yeast. Nature 441:1002–1006
78. Lu R, Drubin DG, Sun Y (2016) Clathrin-mediated endocytosis in budding yeast at a glance. J Cell Sci 129:1531–1536
79. Luo Z, Gallwitz D (2003) Biochemical and genetic evidence for the involvement of yeast Ypt6-GTPase in protein retrieval to different Golgi compartments. J Biol Chem 278:791–799
80. Ma M, Burd CG, Chi RJ (2017) Distinct complexes of yeast Snx4 family SNX-BARs mediate retrograde trafficking of Snc1 and Atg27. Traffic 18:134–144
81. Ma MX, Burd CG (2019) Retrograde trafficking and quality control of yeast synaptobrevin, Snc1, are conferred by its transmembrane domain. Mol Biol Cell 30:1729–1742
82. Maldonado-Baez L, Dores MR, Perkins EM, Drivas TG, Hicke L, Wendland B (2008) Interaction between Epsin/Yap180 adaptors and the scaffolds Ede1/Pan1 is required for endocytosis. Mol Biol Cell 19:2936–2948
83. Matsuura-Tokita K, Takeuchi M, Ichihara A, Mikuriya K, Nakano A (2006) Live imaging of yeast Golgi cisternal maturation. Nature 441:1007–1010
84. Mattera R, Arighi CN, Lodge R, Zerial M, Bonifacino JS (2003) Divalent interaction of the GGAs with the Rabaptin-5-Rabex-5 complex. EMBO J 22:78–88
85. McDonold CM, Fromme JC (2014) Four GTPases differentially regulate the Sec7 Arf-GEF to direct traffic at the trans-golgi network. Dev Cell 30:759–767
86. McLauchlan H, Newell J, Morrice N, Osborne A, West M, Smythe E (1998) A novel role for Rab5-GDI in ligand sequestration into clathrin-coated pits. Curr Biol 8:34–45
87. Medkova M, France YE, Coleman J, Novick P (2006) The Rab exchange factor Sec2p reversibly associates with the exocyst. Mol Biol Cell 17:2757–2769
88. Merrifield CJ, Kaksonen M (2014) Endocytic accessory factors and regulation of clathrin-mediated endocytosis. Cold Spring Harb Perspect Biol 6:a016733
89. Mizuno-Yamasaki E, Medkova M, Coleman J, Novick P (2010) Phosphatidylinositol 4-phosphate controls both membrane recruitment and a regulatory switch of the Rab GEF Sec2p. Dev Cell 18:828–840
90. Mullins C, Bonifacino JS (2001) The molecular machinery for lysosome biogenesis. Bioessays 23:333–343
91. Munro S (2005) The Arf-like GTPase Arl1 and its role in membrane traffic. Biochem Soc Trans 33:601–605
92. Munro S (2011) The golgin coiled-coil proteins of the Golgi apparatus. Cold Spring Harb Perspect Biol 3
93. Munro S, Nichols BJ (1999) The GRIP domain – a novel Golgi-targeting domain found in several coiled-coil proteins. Curr Biol 9:377–380
94. Nagano M, Aoshima K, Shimamura H, Siekhaus DE, Toshima JY, Toshima J (2023) Distinct role of TGN-resident clathrin adaptors for Vps21p activation in the TGN-endosome trafficking pathway. J Cell Sci 136

95. Nagano M, Toshima JY, Siekhaus DE, Toshima J (2019) Rab5-mediated endosome formation is regulated at the trans-Golgi network. Commun Biol 2:419
96. Nakano A (2022) The Golgi apparatus and its next-door neighbors. Front Cell Dev Biol 10:884360
97. Nogi T, Shiba Y, Kawasaki M, Shiba T, Matsugaki N, Igarashi N, Suzuki M, Kato R, Takatsu H, Nakayama K, Wakatsuki S (2002) Structural basis for the accessory protein recruitment by the gamma-adaptin ear domain. Nat Struct Biol 9:527–531
98. Nordmann M, Cabrera M, Perz A, Brocker C, Ostrowicz C, Engelbrecht-Vandre S, Ungermann C (2010) The Mon1-Ccz1 complex is the GEF of the late endosomal Rab7 homolog Ypt7. Curr Biol 20:1654–1659
99. Nothwehr SF, Bruinsma P, Strawn LA (1999) Distinct domains within Vps35p mediate the retrieval of two different cargo proteins from the yeast prevacuolar/endosomal compartment. Mol Biol Cell 10:875–890
100. Nothwehr SF, Ha SA, Bruinsma P (2000) Sorting of yeast membrane proteins into an endosome-to-Golgi pathway involves direct interaction of their cytosolic domains with Vps35p. J Cell Biol 151:297–310
101. Ortiz D, Medkova M, Walch-Solimena C, Novick P (2002) Ypt32 recruits the Sec4p guanine nucleotide exchange factor, Sec2p, to secretory vesicles; evidence for a Rab cascade in yeast. J Cell Biol 157:1005–1015
102. Panic B, Whyte JR, Munro S (2003) The ARF-like GTPases Arl1p and Arl3p act in a pathway that interacts with vesicle-tethering factors at the Golgi apparatus. Curr Biol 13:405–410
103. Paulsel AL, Merz AJ, Nickerson DP (2013) Vps9 family protein Muk1 is the second Rab5 guanosine nucleotide exchange factor in budding yeast. J Biol Chem 288:18162–18171
104. Pelham HR (2002) Insights from yeast endosomes. Curr Opin Cell Biol 14:454–462
105. Piper RC, Cooper AA, Yang H, Stevens TH (1995) VPS27 controls vacuolar and endocytic traffic through a prevacuolar compartment in Saccharomyces cerevisiae. J Cell Biol 131:603–617
106. Podinovskaia M, Spang A (2018) The endosomal network: mediators and regulators of endosome maturation. Prog Mol Subcell Biol 57:1–38
107. Prescianotto-Baschong C, Riezman H (1998) Morphology of the yeast endocytic pathway. Mol Biol Cell 9:173–189
108. Protopopov V, Govindan B, Novick P, Gerst JE (1993) Homologs of the Synaptobrevin vamp family of synaptic vesicle proteins function on the late secretory pathway in saccharomyces-cerevisiae. Cell 74:855–861
109. Ren M, Xu G, Zeng J, De Lemos-Chiarandini C, Adesnik M, Sabatini DD (1998) Hydrolysis of GTP on rab11 is required for the direct delivery of transferrin from the pericentriolar recycling compartment to the cell surface but not from sorting endosomes. Proc Natl Acad Sci USA 95:6187–6192
110. Richardson BC, McDonold CM, Fromme JC (2012) The Sec7 Arf-GEF is recruited to the trans-Golgi network by positive feedback. Dev Cell 22:799–810
111. Rieder SE, Banta LM, Kohrer K, McCaffery JM, Emr SD (1996) Multilamellar endosome-like compartment accumulates in the yeast vps28 vacuolar protein sorting mutant. Mol Biol Cell 7:985–999
112. Rink J, Ghigo E, Kalaidzidis Y, Zerial M (2005) Rab conversion as a mechanism of progression from early to late endosomes. Cell 122:735–749
113. Sata M, Donaldson JG, Moss J, Vaughan M (1998) Brefeldin A-inhibited guanine nucleotide-exchange activity of Sec7 domain from yeast Sec7 with yeast and mammalian ADP ribosylation factors. Proc Natl Acad Sci USA 95:4204–4208
114. Scheuring D, Viotti C, Krüger F, Künzl F, Sturm S, Bubeck J, Hillmer S, Frigerio L, Robinson DG, Pimpl P, Schumacher K (2011) Multivesicular bodies mature from the trans-Golgi network/early endosome in Arabidopsis. Plant Cell 23:3463–3481

115. Schiavo G, Benfenati F, Poulain B, Rossetto O, Polverino de Laureto P, DasGupta BR, Montecucco C (1992) Tetanus and botulinum-B neurotoxins block neurotransmitter release by proteolytic cleavage of synaptobrevin. Nature 359:832–835
116. Schott D, Ho J, Pruyne D, Bretscher A (1999) The COOH-terminal domain of Myo2p, a yeast myosin V, has a direct role in secretory vesicle targeting. J Cell Biol 147:791–808
117. Schott DH, Collins RN, Bretscher A (2002) Secretory vesicle transport velocity in living cells depends on the myosin-V lever arm length. J Cell Biol 156:35–39
118. Seaman MN (2012) The retromer complex – endosomal protein recycling and beyond. J Cell Sci 125:4693–4702
119. Seaman MN, Marcusson EG, Cereghino JL, Emr SD (1997) Endosome to Golgi retrieval of the vacuolar protein sorting receptor, Vps10p, requires the function of the VPS29, VPS30, and VPS35 gene products. J Cell Biol 137:79–92
120. Seaman MN, McCaffery JM, Emr SD (1998) A membrane coat complex essential for endosome-to-Golgi retrograde transport in yeast. J Cell Biol 142:665–681
121. Sekiya-Kawasaki M, Groen AC, Cope MJ, Kaksonen M, Watson HA, Zhang C, Shokat KM, Wendland B, McDonald KL, McCaffery JM, Drubin DG (2003) Dynamic phosphoregulation of the cortical actin cytoskeleton and endocytic machinery revealed by real-time chemical genetic analysis. J Cell Biol 162:765–772
122. Setty SR, Shin ME, Yoshino A, Marks MS, Burd CG (2003) Golgi recruitment of GRIP domain proteins by Arf-like GTPase 1 is regulated by Arf-like GTPase 3. Curr Biol 13:401–404
123. Setty SR, Strochlic TI, Tong AH, Boone C, Burd CG (2004) Golgi targeting of ARF-like GTPase Arl3p requires its Nalpha-acetylation and the integral membrane protein Sys1p. Nat Cell Biol 6:414–419
124. Sheff DR, Daro EA, Hull M, Mellman I (1999) The receptor recycling pathway contains two distinct populations of early endosomes with different sorting functions. J Cell Biol 145:123–139
125. Shi Y, Stefan CJ, Rue SM, Teis D, Emr SD (2011) Two novel WD40 domain-containing proteins, Ere1 and Ere2, function in the retromer-mediated endosomal recycling pathway. Mol Biol Cell 22:4093–4107
126. Shideler T, Nickerson DP, Merz AJ, Odorizzi G (2015) Ubiquitin binding by the CUE domain promotes endosomal localization of the Rab5 GEF Vps9. Mol Biol Cell 26:1345–1356
127. Simonsen A, Bremnes B, Rønning E, Aasland R, Stenmark H (1998) Syntaxin-16, a putative Golgi t-SNARE. Eur J Cell Biol 75:223–231
128. Singer-Krüger B, Frank R, Crausaz F, Riezman H (1993) Partial purification and characterization of early and late endosomes from yeast. Identification of four novel proteins. J Biol Chem 268:14376–14386
129. Siniossoglou S, Peak-Chew SY, Pelham HR (2000) Ric1p and Rgp1p form a complex that catalyses nucleotide exchange on Ypt6p. EMBO J 19:4885–4894
130. Siniossoglou S, Pelham HR (2001) An effector of Ypt6p binds the SNARE Tlg1p and mediates selective fusion of vesicles with late Golgi membranes. EMBO J 20:5991–5998
131. Smythe E, Ayscough KR (2003) The Ark1/Prk1 family of protein kinases. Regulators of endocytosis and the actin skeleton. EMBO Rep 4:246–251
132. Söllner T, Whiteheart SW, Brunner M, Erdjument-Bromage H, Geromanos S, Tempst P, Rothman JE (1993) SNAP receptors implicated in vesicle targeting and fusion. Nature 362:318–324
133. Sönnichsen B, De Renzis S, Nielsen E, Rietdorf J, Zerial M (2000) Distinct membrane domains on endosomes in the recycling pathway visualized by multicolor imaging of Rab4, Rab5, and Rab11. J Cell Biol 149:901–914
134. Stenmark H (2009) Rab GTPases as coordinators of vesicle traffic. Nat Rev Mol Cell Biol 10:513–525
135. Stimpson HE, Toret CP, Cheng AT, Pauly BS, Drubin DG (2009) Early-arriving Syp1p and Ede1p function in endocytic site placement and formation in budding yeast. Mol Biol Cell 20:4640–4651

136. Struthers MS, Shanks SG, MacDonald C, Carpp LN, Drozdowska AM, Kioumourtzoglou D, Furgason ML, Munson M, Bryant NJ (2009) Functional homology of mammalian syntaxin 16 and yeast Tlg2p reveals a conserved regulatory mechanism. J Cell Sci 122:2292–2299
137. Sun Y, Kaksonen M, Madden DT, Schekman R, Drubin DG (2005) Interaction of Sla2p's ANTH domain with PtdIns(4,5)P2 is important for actin-dependent endocytic internalization. Mol Biol Cell 16:717–730
138. Sun Y, Martin AC, Drubin DG (2006) Endocytic internalization in budding yeast requires coordinated actin nucleation and myosin motor activity. Dev Cell 11:33–46
139. Suzuki SW, Emr SD (2018) Membrane protein recycling from the vacuole/lysosome membrane. J Cell Biol 217:1623–1632
140. Suzuki SW, Oishi A, Nikulin N, Jorgensen JR, Baile MG, Emr SD (2021) A PX-BAR protein Mvp1/SNX8 and a dynamin-like GTPase Vps1 drive endosomal recycling. elife 10
141. Tang BL, Low DY, Lee SS, Tan AE, Hong W (1998) Molecular cloning and localization of human syntaxin 16, a member of the syntaxin family of SNARE proteins. Biochem Biophys Res Commun 242:673–679
142. Tang HY, Xu J, Cai M (2000) Pan1p, End3p, and S1a1p, three yeast proteins required for normal cortical actin cytoskeleton organization, associate with each other and play essential roles in cell wall morphogenesis. Mol Cell Biol 20:12–25
143. Tojima T, Miyashiro D, Kosugi Y, Nakano A (2022) Super-resolution live imaging ofcargo traffic through the Golgi apparatus in mammalian cells. Methods Mol, Biol. in press
144. Tojima T, Suda Y, Ishii M, Kurokawa K, Nakano A (2019) Spatiotemporal dissection of the trans-Golgi network in budding yeast. J Cell Sci 132:jcs231159
145. Tojima T, Suda Y, Jin N, Kurokawa K, Nakano A (2024) Spatiotemporal dissection of the Golgi apparatus and the ER-Golgi intermediate compartment in budding yeast. elife 13
146. Toshima J, Toshima JY, Duncan MC, Cope MJ, Sun Y, Martin AC, Anderson S, Yates JR 3rd, Mizuno K, Drubin DG (2007) Negative regulation of yeast Eps15-like Arp2/3 complex activator, Pan1p, by the Hip1R-related protein, Sla2p, during endocytosis. Mol Biol Cell 18:658–668
147. Toshima J, Toshima JY, Martin AC, Drubin DG (2005) Phosphoregulation of Arp2/3-dependent actin assembly during receptor-mediated endocytosis. Nat Cell Biol 7:246–254
148. Toshima JY, Furuya E, Nagano M, Kanno C, Sakamoto Y, Ebihara M, Siekhaus DE, Toshima J (2016) Yeast Eps15-like endocytic protein Pan1p regulates the interaction between endocytic vesicles, endosomes and the actin cytoskeleton. elife 5
149. Toshima JY, Nishinoaki S, Sato Y, Yamamoto W, Furukawa D, Siekhaus DE, Sawaguchi A, Toshima J (2014) Bifurcation of the endocytic pathway into Rab5-dependent and -independent transport to the vacuole. Nat Commun 5:3498
150. Toshima JY, Toshima J, Kaksonen M, Martin AC, King DS, Drubin DG (2006) Spatial dynamics of receptor-mediated endocytic trafficking in budding yeast revealed by using fluorescent alpha-factor derivatives. Proc Natl Acad Sci USA 103:5793–5798
151. Toshima JY, Tsukahara A, Nagano M, Tojima T, Siekhaus DE, Nakano A, Toshima J (2023) The yeast endocytic early/sorting compartment exists as an independent sub-compartment within the trans-Golgi network. elife 12
152. Traub LM (2011) Regarding the amazing choreography of clathrin coats. PLoS Biol 9:e1001037
153. Ullrich O, Reinsch S, Urbé S, Zerial M, Parton RG (1996) Rab11 regulates recycling through the pericentriolar recycling endosome. J Cell Biol 135:913–924
154. Valdivia RH, Baggott D, Chuang JS, Schekman RW (2002) The yeast clathrin adaptor protein complex 1 is required for the efficient retention of a subset of late Golgi membrane proteins. Dev Cell 2:283–294
155. Valencia JP, Goodman K, Otegui MS (2016) Endocytosis and endosomal trafficking in plants. Annu Rev Plant Biol 67(67):309–335
156. Vida TA, Huyer G, Emr SD (1993) Yeast vacuolar proenzymes are sorted in the late Golgi complex and transported to the vacuole via a prevacuolar endosome-like compartment. J Cell Biol 121:1245–1256

157. Viotti C, Bubeck J, Stierhof YD, Krebs M, Langhans M, van den Berg W, van Dongen W, Richter S, Geldner N, Takano J, Jürgens G, de Vries SC, Robinson DG, Schumacher K (2010) Endocytic and secretory traffic in Arabidopsis merge in the trans-Golgi network/early endosome, an independent and highly dynamic organelle. Plant Cell 22:1344–1357
158. Walch-Solimena C, Collins RN, Novick PJ (1997) Sec2p mediates nucleotide exchange on Sec4p and is involved in polarized delivery of post-Golgi vesicles. J Cell Biol 137:1495–1509
159. Walch-Solimena C, Novick P (1999) The yeast phosphatidylinositol-4-OH kinase pik1 regulates secretion at the Golgi. Nat Cell Biol 1:523–525
160. Wang YJ, Wang J, Sun HQ, Martinez M, Sun YX, Macia E, Kirchhausen T, Albanesi JP, Roth MG, Yin HL (2003) Phosphatidylinositol 4 phosphate regulates targeting of clathrin adaptor AP-1 complexes to the Golgi. Cell 114:299–310
161. Wendland B, Emr SD (1998) Pan1p, yeast eps15, functions as a multivalent adaptor that coordinates protein-protein interactions essential for endocytosis. J Cell Biol 141:71–84
162. Wickner W, Schekman R (2008) Membrane fusion. Nat Struct Mol Biol 15:658–664
163. Witkos TM, Lowe M (2015) The Golgin family of coiled-coil tethering proteins. Front Cell Dev Biol 3:86
164. Yamamoto W, Wada S, Nagano M, Aoshima K, Siekhaus DE, Toshima JY, Toshima J (2018) Distinct roles for plasma membrane PtdIns(4)P and PtdIns(4,5)P(2) during receptor-mediated endocytosis in yeast. J Cell Sci 131
165. Yang S, Rosenwald AG (2016) Autophagy in Saccharomyces cerevisiae requires the monomeric GTP-binding proteins, Arl1 and Ypt6. Autophagy 12:1721–1737
166. Yoshida N, Ogura I, Nagano M, Ando T, Toshima JY, Toshima J (2021) Cooperative regulation of endocytic vesicle transport by yeast Eps15-like protein Pan1p and epsins. J Biol Chem 297:101254
167. Yu JW, Lemmon MA (2001) All phox homology (PX) domains from Saccharomyces cerevisiae specifically recognize phosphatidylinositol 3-phosphate. J Biol Chem 276:44179–44184
168. Zerial M, McBride H (2001) Rab proteins as membrane organizers. Nat Rev Mol Cell Biol 2:107–117
169. Zhdankina O, Strand NL, Redmond JM, Boman AL (2001) Yeast GGA proteins interact with GTP-bound Arf and facilitate transport through the Golgi. Yeast 18:1–18

MIX
Papier aus verantwortungsvollen Quellen
Paper from responsible sources
FSC® C105338

If you have any concerns about our products,
you can contact us on
ProductSafety@springernature.com

In case Publisher is established outside the EU,
the EU authorized representative is:
Springer Nature Customer Service Center GmbH
Europaplatz 3, 69115 Heidelberg, Germany

Printed by Libri Plureos GmbH
in Hamburg, Germany